Graduate Texts in Contemporary Physics

Series Editors:

R. Stephen Berry
Joseph L. Birman
Jeffrey W. Lynn
Mark P. Silverman
H. Eugene Stanley
Mikhail Voloshin

Springer
New York
Berlin
Heidelberg
Barcelona
Hong Kong
London
Milan
Paris
Singapore
Tokyo

Graduate Texts in Contemporary Physics

S.T. Ali, J.P. Antoine, and J.P. Gazeau: **Coherent States, Wavelets and Their Generalizations**

A. Auerbach: **Interacting Electrons and Quantum Magnetism**

B. Felsager: **Geometry, Particles, and Fields**

P. Di Francesco, P. Mathieu, and D. Sénéchal: **Conformal Field Theories**

J.H. Hinken: **Superconductor Electronics: Fundamentals and Microwave Applications**

J. Hladik: **Spinors in Physics**

Yu.M. Ivanchenko and A.A. Lisyansky: **Physics of Critical Fluctuations**

M. Kaku: **Introduction to Superstrings and M-Theory, 2nd Edition**

M. Kaku: **Strings, Conformal Fields, and M-Theory, 2nd Edition**

H.V. Klapdor (ed.): **Neutrinos**

J.W. Lynn (ed.): **High-Temperature Superconductivity**

H.J. Metcalf and P. van der Straten: **Laser Cooling and Trapping**

R.N. Mohapatra: **Unification and Supersymmetry: The Frontiers of Quark-Lepton Physics, 2nd Edition**

H. Oberhummer: **Nuclei in the Cosmos**

G.D. Phillies: **Elementary Lectures in Statistical Mechanics**

R.E. Prange and S.M. Girvin (eds.): **The Quantum Hall Effect**

B.M. Smirnov: **Clusters and Small Particles: In Gases and Plasmas**

M. Stone: **The Physics of Quantum Fields**

F.T. Vasko and A.V. Kuznetsov: **Electronic States and Optical Transitions in Semiconductor Heterostructures**

(continued following index)

Syed Twareque Ali
Jean-Pierre Antoine
Jean-Pierre Gazeau

Coherent States, Wavelets and Their Generalizations

With 14 Figures

Springer

Syed Twareque Ali
Department of Mathematics and Statistics
Concordia University
Loyola Campus
7141 Sherbrooke Street West
Montréal, Quebec H4B 1R6
Canada

Jean-Pierre Antoine
Institut de Physique Théorique
Université Catholique de Louvain
Chemin du Cyclotron 2
B-1348 Louvain-la-Neuve
Belgium

Jean-Pierre Gazeau
Laboratoire Physique Théoretique and Matière Condensée
2, Place Jussieu
Université de Paris 7 – Denis Diderot
Paris F-75251
France

Series Editors

R. Stephen Berry
Department of Chemistry
University of Chicago
Chicago, IL 60637
USA

Joseph L. Birman
Department of Physics
City College of CUNY
New York, NY 10031
USA

Jeffrey W. Lynn
Department of Physics
University of Maryland
College Park, MD 20742
USA

Mark P. Silverman
Department of Physics
Trinity College
Hartford, CT 06106
USA

H. Eugene Stanley
Center for Polymer Studies
Physics Department
Boston University
Boston, MA 02215
USA

Mikhail Voloshin
Theoretical Physics Institute
Tate Laboratory of Physics
The University of Minnesota
Minneapolis, MN 55455
USA

Library of Congress Cataloging-in-Publication Data
Ali, S. Twareque (Syed Twareque)
    Coherent states, wavelets and their generalizations / S. Twareque Ali,
Jean-Pierre Antoine, Jean-Pierre Gazeau.
        p.   cm. — (Graduate texts in contemporary physics)
    Includes bibliographical references and index.
    ISBN 0-387-98908-0 (hc : alk. paper)
    1. Coherent states. 2. Wavelets (Mathematics) I. Antoine, Jean-
Pierre. II. Gazeau, Jean-Pierre. III. Title. IV. Series.
QC6.4.C65A55   2000
530.12´4—dc21   99-39821

Printed on acid-free paper.

© 2000 Springer-Verlag New York, Inc.
All rights reserved. This work may not be translated or copied in whole or in part without the written permission of the publisher (Springer-Verlag New York, Inc., 175 Fifth Avenue, New York, NY 10010, USA), except for brief excerpts in connection with reviews or scholarly analysis. Use in connection with any form of information storage and retrieval, electronic adaptation, computer software, or by similar or dissimilar methodology now known or hereafter developed is forbidden.
The use of general descriptive names, trade names, trademarks, etc., in this publication, even if the former are not especially identified, is not to be taken as a sign that such names, as understood by the Trade Marks and Merchandise Marks Act, may accordingly be used freely by anyone.

Production managed by Allan Abrams; manufacturing supervised by Jeffrey Taub.
Photocomposed copy provided from the authors' TeX files.
Printed and bound by Edwards Brothers, Inc., Ann Arbor, MI.
Printed in the United States of America.

9 8 7 6 5 4 3 2 1

ISBN 0-387-98908-0 Springer-Verlag New York Berlin Heidelberg   SPIN 10659217

Preface

Nitya kaaler utshab taba
Bishyer-i-dipaalika
Aami shudhu tar-i-mateer pradeep
Jaalao tahaar shikhaa [1]

— **Tagore**

Should authors feel compelled to justify the writing of yet another book? In an overpopulated world, should parents feel compelled to justify bringing forth yet another child? Perhaps not! But an act of creation is also an act of love, and a love story can always be happily shared.

In writing this book, it has been our feeling that, in all of the wealth of material on coherent states and wavelets, there exists a lack of a discernable, unifying mathematical perspective. The use of wavelets in research and technology has witnessed explosive growth in recent years, while the use of coherent states in numerous areas of theoretical and experimental physics has been an established trend for decades. Yet it is not at all uncommon to find practitioners in either one of the two disciplines who are hardly aware of one discipline's links to the other. Currently, many books are on the market that treat the subject of wavelets from a wide range of perspectives and with windows on one or several areas of a large spectrum

---

[1] Thine is an eternal celebration ... A cosmic Festival of Lights! ... Therein I am a mere flicker of a wicker lamp ... O kindle its flame (my Master!).

of possible applications. On the theory of coherent states, likewise, several excellent monographs, edited collections of papers, synthetic reviews, or specialized articles exist. The emphasis in most of these works is usually on physical applications. In the more mathematical texts, the focus is usually on specific properties — arising either from group theory, holomorphic function theory, and, more recently, differential geometry. The point of view put forward in this book is that both the theory of wavelets and the theory of coherent states can be subsumed into certain broad functional analytic structures, namely, positive operator-valued measures on a Hilbert space and reproducing kernel Hilbert spaces. The specific context in which these structures arise, to generate particular families of coherent states or wavelet transforms, could of course be very diverse, but typically they emanate either from the property of square integrability of certain unitary group representations on Hilbert spaces or from holomorphic structures associated to certain differential manifolds. In talking about square integrable representations, a broad generalization of the concept has been introduced here, moving from the well-known notion of square integrability with respect to the whole group to one based on some of its homogeneous spaces. This generalization, while often implicit in the past in physical discussions of coherent states (notably, in the works of Klauder, Barut and Girardello, or even in the case of the time-honored canonical coherent states, as discovered by Schrödinger), was not readily recognized in the mathematical literature until the work of Gilmore and Perelomov. In this book, this generalization is taken even further, with the result that the classes of coherent states that can be constructed and usefully employed extends to a vast array of physically pertinent groups. Similarly, it is generally known and recognized that wavelets are coherent states arising from the affine group of the real line. Using coherent states of other groups to generate higher dimensional wavelets, or alternative wavelet-like transforms, however, is not a common preoccupation among practitioners of the trade (an exception being the recent book by Torrésani). About a third of the present book is devoted to looking at wavelets from precisely this point of view, thereby displaying the richness of possibilities that exist in this domain. Considerable attention has also been paid in the book to the discretization problem, in particular, with the discussion of τ-wavelets. The interplay between discrete and continuous wavelets is a rich aspect of the theory, which does not seem to have been exploited sufficiently in the past.

In presenting this unifying backdrop, for the understanding of a wide sweep of mathematical and physical structures, it is our hope that the relationship between the two disciplines — wavelets and coherent states — will have been made more transparent, thereby aiding the process of cross-fertilization as well. For graduate students or research workers approaching the disciplines for the first time, such an overall perspective should also make the subject matter easier to assimilate, with the book acting as a dovetailed introduction to both subjects — unfortunately, a frustrating incoherence blurs the existing literature on coherence! Besides being a primer for instruction, the book, of course, is also meant to be a source material for a wide range of very recent results, both in the theory of wavelets and of coherent states. The emphasis is decidedly on the mathematical aspect of the theory, although enough physical examples have been introduced, from time to time, to illustrate the material. While the book is aimed mainly at graduate students and entering research workers in physics and mathematical physics, it is nevertheless hoped that professional physicists and mathematicians will also find it interesting reading, being an area of mathematical physics in which the intermingling of theory and practice is most thorough. Prerequisite for an understanding of most of the material in the book is a familiarity with standard Hilbert space operator techniques and group representation theory, such as every physicist would acknowlegde from the days of graduate quantum mechanics and angular momentum. For the more specialized topics, an attempt has been made to make the treatment self-contained, and, indeed, a large part of the book is devoted to the development of the mathematical formalism.

If a book such as this can make any claims to originality at all, it can mainly be in the manner of its presentation. Beyond that, we believe that a body of material is also presented here (for example, in the use of POV measures or in dealing with the discretization problem) that has not appeared in book form before. An attempt has been made throughout to cite as many references to the original literature as were known — omissions should therefore be attributed to our collective ignorance and we would like to extend our unconditional apologies for any resulting oversight.

This book has grown out of many years of shared research interest and, indeed, camaraderie, between us. Almost all of the material presented here has been touched on in courses, lectures, and seminars given to students and among colleagues at various institutions in Europe, America, Asia, and Africa – notably, in graduate courses and research workshops, given at different times by all of us, in Louvain-la-Neuve, Montréal, Paris, Porto-Novo, Białystok, Dhaka, Fukuoka, Havana, and Prague. One is tempted to say that the geographical diversity here rivals the mathematical menagerie!

To all of our colleagues and students who have participated in these discussions, we would like to extend our heartfelt thanks. In particular, a few colleagues graciously volunteered to critically read parts of the manuscript and to offer numerous suggestions for improvement and clarity. Among them, one ought to especially mention J. Hilgert, G. G. Emch, S. De Bièvre, and J. Renaud. In addition, the figures would not exist without the programming skills of A. Coron, L. Jacques, and P. Vandergheynst (Louvain-la-Neuve), and we thank them all for their gracious help. During the writing of the book, we made numerous reciprocal visits to one another's institutions. To Concordia University, Montréal, the Université Catholique de Louvain, Louvain-la-Neuve, and the Université Paris 7 – Denis Diderot, Paris, we would like to express our appreciation for hospitality and collegiality. The editor from Springer-Verlag, Thomas von Foerster, deserves a special vote of thanks for his cooperation and for the exemplary patience he displayed, even as the event horizon for the completion of the manuscript kept receeding further and further! It goes without saying, however, that all responsibility for errors, imperfections, and residual or outright mistakes is shared jointly by all of us.

Syed Twareque Ali
Jean-Pierre Antoine
Jean-Pierre Gazeau

Montréal, Canada
Louvain-la-Neuve, Belgium
Paris, France

March 1999

Contents

Preface		v
1 Introduction		**1**
2 Canonical Coherent States		**13**
2.1	Minimal uncertainty states	13
2.2	The group-theoretical backdrop	17
2.3	Some functional analytic properties	20
2.4	A complex analytic viewpoint (♣)	24
2.5	Some geometrical considerations	26
2.6	Outlook	27
2.7	Two illustrative examples	27
	2.7.1 A quantization problem (♣)	27
	2.7.2 An application to atomic physics (♣)	29
3 Positive Operator-Valued Measures and Frames		**33**
3.1	Definitions and main properties	34
3.2	The case of a tight frame	40
3.3	Example: A commutative POV measure	41
3.4	Discrete frames	42
4 Some Group Theory		**47**
4.1	Homogeneous spaces, quasi-invariant, and invariant measures	47

	4.2	Induced representations and systems of covariance	54
		4.2.1 Vector coherent states	59
		4.2.2 Discrete series representations of $SU(1,1)$	61
		4.2.3 The regular representations of a group	67
	4.3	An extended Schur's lemma	68
	4.4	Harmonic analysis on locally compact abelian groups . .	70
		4.4.1 Basic notions	70
		4.4.2 Lattices in LCA groups	72
		4.4.3 Sampling in LCA groups	73
	4.5	Lie groups and Lie algebras: A reminder	75
		4.5.1 Lie algebras	75
		4.5.2 Lie groups	77
		4.5.3 Extensions of Lie algebras and Lie groups	82
		4.5.4 Contraction of Lie algebras and Lie groups	85

5 Hilbert Spaces with Reproducing Kernels and Coherent States — 89

5.1	A motivating example	89
5.2	Measurable fields and direct integrals	92
5.3	Reproducing kernel Hilbert spaces	94
	5.3.1 Positive-definite kernels and evaluation maps . . .	94
	5.3.2 Coherent states and POV functions	99
	5.3.3 Some isomorphisms, bases, and ν-selections . . .	101
	5.3.4 A reconstruction problem: Example of a holomorphic map (♣)	104
5.4	Some properties of reproducing kernel Hilbert spaces . .	106

6 Square Integrable and Holomorphic Kernels — 109

6.1	Square integrable kernels	109
6.2	Holomorphic kernels	112
6.3	Coherent states: The holomorphic case	116
	6.3.1 An example of a holomorphic frame (♣)	119
	6.3.2 A nonholomorphic excursion (♣)	122

7 Covariant Coherent States — 127

7.1	Covariant coherent states	128
	7.1.1 A general definition	128
	7.1.2 The Gilmore–Perelomov CS and vector CS	129
	7.1.3 A geometrical setting	133
7.2	Example: The classical theory of coherent states	136
	7.2.1 CS of compact semisimple Lie groups	136
	7.2.2 CS of noncompact semisimple Lie groups	139
	7.2.3 CS of non-semisimple Lie groups	141
7.3	Square integrable covariant CS: The general case	141
	7.3.1 Some further generalizations	144

8	**Coherent States from Square Integrable Representations**		**147**
	8.1	Square integrable group representations	148
	8.2	Orthogonality relations	155
	8.3	The Wigner map	160
	8.4	Modular structures and statistical mechanics	164
9	**Some Examples and Generalizations**		**171**
	9.1	A class of semidirect product groups	171
		9.1.1 Three concrete examples (♣)	176
		9.1.2 A broader setting	179
	9.2	A generalization: α- and V-admissibility	181
		9.2.1 Example of the Galilei group (♣)	186
		9.2.2 CS of the isochronous Galilei group (♣)	190
		9.2.3 Atomic coherent states	196
10	**CS of General Semidirect Product Groups**		**199**
	10.1	Squeezed states (♣)	200
	10.2	Geometry of semidirect product groups	204
		10.2.1 A special class of orbits	204
		10.2.2 The coadjoint orbit structure of Γ	206
		10.2.3 Measures on Γ	209
		10.2.4 Induced representations of semidirect products . .	213
	10.3	CS of semidirect products	214
		10.3.1 Admissible affine sections	220
11	**CS of the Relativity Groups**		**225**
	11.1	The Poincaré groups $\mathcal{P}_+^\uparrow(1,3)$ and $\mathcal{P}_+^\uparrow(1,1)$	225
		11.1.1 The Poincaré group in 1+3 dimensions, $\mathcal{P}_+^\uparrow(1,3)$ (♣)	225
		11.1.2 The Poincaré group in 1+1 dimensions, $\mathcal{P}_+^\uparrow(1,1)$ (♣)	235
		11.1.3 Poincaré CS: The massless case	240
	11.2	The Galilei groups $\tilde{\mathcal{G}}(1,1)$ and $\tilde{\mathcal{G}} \equiv \tilde{\mathcal{G}}(1,3)$ (♣)	242
	11.3	The anti-de Sitter group $SO_o(1,2)$ and its contraction(s) (♣)	245
12	**Wavelets**		**257**
	12.1	A word of motivation	257
	12.2	Derivation and properties of the 1-D continuous wavelet transform (♣)	261
	12.3	A mathematical aside: Extension to distributions	266
	12.4	Interpretation of the continuous wavelet transform	272
		12.4.1 The CWT as phase space representation	272
		12.4.2 Localization properties and physical interpretation of the CWT	273

	12.5	Discretization of the continuous WT: Discrete frames	274
	12.6	Ridges and skeletons	276
	12.7	Applications	278

13 Discrete Wavelet Transforms — 283
13.1 The discrete time or dyadic WT 283
    13.1.1 Multiresolution analysis and orthonormal wavelet bases 284
    13.1.2 Connection with filters and the subband coding scheme 285
    13.1.3 Generalizations 287
    13.1.4 Applications 288
13.2 Towards a fast CWT: Continuous wavelet packets 289
13.3 Wavelets on the finite field $\mathbb{Z}_p$ (♣) 290
13.4 Algebraic wavelets 292
    13.4.1 τ-wavelets of Haar on the line (♣) 292
    13.4.2 Pisot wavelets, etc. (♣) 303

14 Multidimensional Wavelets — 307
14.1 Going to higher dimensions 307
14.2 Mathematical analysis (♣) 308
14.3 The 2-D case 316
    14.3.1 Minimality properties 316
    14.3.2 Interpretation, visualization problems, and calibration 318
    14.3.3 Practical applications of the CWT in two dimensions 322
    14.3.4 The discrete WT in two dimensions 327
    14.3.5 Continuous wavelet packets in two dimensions . . 330

15 Wavelets Related to Other Groups — 331
15.1 Wavelets on the sphere and similar manifolds 331
    15.1.1 The two-sphere (♣) 331
    15.1.2 Generalization to other manifolds 336
15.2 The affine Weyl–Heisenberg group (♣) 338
15.3 The affine or similitude groups of space–time 342
    15.3.1 Kinematical wavelets (♣) 342
    15.3.2 The affine Galilei group (♣) 344
    15.3.3 The (restricted) Schrödinger group (♣) 347
    15.3.4 The affine Poincaré group (♣) 349

16 The Discretization Problem: Frames, Sampling, and All That — 353
16.1 The Weyl–Heisenberg group or canonical CS 354
16.2 Wavelet frames 357

16.3		Frames for affine semidirect products	359
	16.3.1	The affine Weyl–Heisenberg group	359
	16.3.2	The affine Poincaré groups $^{(\clubsuit)}$	360
	16.3.3	Discrete frames for general semidirect products	362
16.4		Groups without dilations: The Poincaré groups $^{(\clubsuit)}$	364
16.5		A group-theoretical approach to discrete wavelet transforms	369
	16.5.1	Generalities on sampling	369
	16.5.2	Wavelets on $\mathbb{Z}_p$ revisited $^{(\clubsuit)}$	370
	16.5.3	Wavelets on a discrete abelian group	372
16.6		Conclusion	387

Conclusion and Outlook — **389**

References — **393**

Index — **415**

1
Introduction

The notion of coherent states (CS[1]) is rooted in quantum physics and its relationship to classical physics. The term 'coherent' itself originates in the current language of quantum optics (for instance, coherent radiation, sources emitting coherently, etc.). It was introduced in the 1960s by Glauber [145], [2] one of the founding fathers of the theory of CS, together with Klauder [189] and Sudarshan [Kl1, 272], in the context of a quantum optical description of coherent light beams emitted by lasers. Since then, coherent states have pervaded nearly all branches of quantum physics — quantum optics, of course, but also nuclear, atomic, and solid-state physics, quantum electrodynamics (the infrared problem), quantization and dequantization problems and path integrals, just to mention a few. It has been said, even convincingly [191], that "coherent states are the natural language of quantum theory!"

As early as 1926, at the very beginning of quantum mechanics, Schrödinger [266] was interested in studying quantum states, which restore the classical behavior of the position operator of a quantum system (in the Heisenberg picture):

$$Q(t) = e^{\frac{i}{\hbar}Ht} Q e^{-\frac{i}{\hbar}Ht}. \tag{1.1}$$

---

[1] The acronym CS will be used throughout this book to mean "coherent state" or "coherent states," depending on the context.

[2] For convenience, we have split the bibliography into two separate lists, books and theses (denoted by letters) and articles (denoted by numbers).

In this relation, $H = P^2/2m + V(Q)$ is the quantum Hamiltonian of the system. Schrödinger understood classical behavior to mean that the average or expected value $\bar{q}(t)$ of the position operator $Q(t)$, in the desired state, would obey the classical equation of motion:

$$m\ddot{\bar{q}}(t) + \overline{\frac{\partial V}{\partial q}} = 0. \tag{1.2}$$

The first example of CS discovered by Schrödinger were of course those pertaining to the harmonic oscillator, $V(q) = \frac{1}{2}m^2\omega^2 q^2$, known universally to physicists, and the prototype of every integrable model. These states $|z\rangle$, parametrized by the complex number z, are defined in such a way that one recovers the familiar sinusoidal solution

$$\langle z \mid Q(t) \mid z \rangle = 2Q_o|z|\cos(\omega t - \varphi), \tag{1.3}$$

where $z = |z|e^{i\varphi}$ and $Q_o = (\hbar/2m\omega)^{1/2}$ is a fundamental quantum length built from the universal constant $\hbar$ and the constants m and ω characterizing the quantum harmonic oscillator under consideration.

In this way, CS mediate a "smooth" transition from classical mechanics to quantum mechanics. One should not be misled, however: CS are rigorously quantum states (witness the constant $\hbar$ appearing in the definition of Q_o), yet they allow for a classical "reading" in a host of quantum situations. This unique qualification results from a set of properties satisfied by these Schrödinger–Klauder–Glauber CS, also called *canonical CS*. The most important among them, and indeed the guiding spirit in the organization of the present volume, are as follows:

P1. The states $|z\rangle$ saturate the Heisenberg inequality:

$$\langle \Delta Q \rangle_z \langle \Delta P \rangle_z = \frac{1}{2}\hbar, \tag{1.4}$$

where $\langle \Delta Q \rangle_z := [\langle z|Q^2|z\rangle - \langle z|Q|z\rangle^2]^{1/2}$.

P2. The states $|z\rangle$ are eigenvectors of the annihilation operator, with eigenvalue z:

$$a|z\rangle = z|z\rangle, \quad z \in \mathbb{C}, \tag{1.5}$$

where $a = (2m\hbar\omega)^{-1/2}(m\omega Q + iP)$.

P3. The states $|z\rangle$ are obtained from the ground state $|0\rangle$ of the harmonic oscillator by a unitary action of the Weyl–Heisenberg group. This is a key Lie group in quantum mechanics, whose Lie algebra is generated by $\{Q, P, I\}$, with $[Q, P] = i\hbar I$ (which implies $[a, a^\dagger] = I$):

$$|z\rangle = e^{(za^\dagger - \bar{z}a)}|0\rangle. \tag{1.6}$$

P4. The CS $\{|z\rangle\}$ constitute an overcomplete family of vectors in the Hilbert space of the states of the harmonic oscillator. This property is encoded in the following resolution of the identity:

$$I = \frac{1}{\pi} \int_{\mathbb{C}} d\operatorname{Re} z \, d\operatorname{Im} z \, |z\rangle\langle z|. \tag{1.7}$$

These four properties are, to various extents, the basis of the many generalizations of the canonical notion of CS, illustrated by the family $\{|z\rangle\}$.

It is important to note here that the properties P1–P4, satisfied by the canonical CS, set them apart radically from ordinary wave packets, familiar from elementary quantum mechanics. Indeed, these latter states are often characterized by good localization at a given instant, either in space *or* in momentum (but not in both), but generically they spread in the course of their time evolution. Moreover, one could not build a continuous family of such wave packets and expect them to generate a resolution of the identity. In this sense, the canonical CS constitute a unique family of wave packets that maintain the same support, both in space *and* in momentum, as exemplified by P1 (coherent states remain coherent!), and also resolve the identity in the sense of P4.

We emphasize, however, that for the various generalizations of CS presented in this book (or elsewhere) there is *a priori* no reason to expect any localization property, either in space or in some other observable, and, even if they do have such a property, they have no reason to preserve it in time. For instance, localization requires the existence of a well-defined position operator, and this is not always available (e.g., in a relativistic quantum mechanics). For a pedagogical review of some of these questions and references to original articles, we refer the reader to [76].

Let us now give some examples of such generalizations, with no pretention of being exhaustive. (Most of these examples are borrowed from the well-known reprint volume edited by Klauder and Skagerstam [Kl2], in which original references may also be found.)

Concerning Property P1, one might mention the approach of Nieto and Simmons [232, 233], who use it as a defining criterion for more general CS adapted to a local potential that has at least one region of confinement.

Property P2 was the takeoff point of Barut and Girardello [65] for constructing CS associated with representations of the group $SL(2,\mathbb{R}) \simeq SU(1,1) \simeq SO_o(1,2)$, which belong to the so-called discrete series. These are representations on Hilbert spaces that can be constructed with the help of a raising operator $a^\dagger$ and its adjoint a, the commutator of the two being equal to the generator X_o of the compact subgroup $SO(2) \simeq U(1)$:

$$[a, a^\dagger] = X_o. \tag{1.8}$$

Thus, the Barut–Girardello CS are again states $|z\rangle$, which satisfy the equation

$$a|z\rangle = z|z\rangle, \; z \in \mathbb{C}. \tag{1.9}$$

Property P3, which has a clear group-theoretical flavor, is at the core of the approach developed independently by Gilmore [144] and Perelomov [246] (and, in spirit, already anticipated by Klauder [189]). The basic ingredients here are a group G, a subgroup H of G, a unitary representation U of G in a Hilbert space $\mathfrak{H}$, and a "probe" $\eta \in \mathfrak{H}$, chosen because of some particular "good" properties that it possesses. A family of CS $\{\eta_x\}$, indexed by the points x of the coset space $X = G/H$, is constructed as follows:

$$\eta_x = U(\sigma(x))\eta, \quad x \in X, \tag{1.10}$$

where $\sigma : G/H \to G$ is a suitable 'section' or lifting $\sigma : X \to G$ – that is, a map associating to each $x \in X$ a certain element $\sigma(x) \in G$, such that the coset $\sigma(x)H$ is precisely x (for Lie groups, this is obviously a fiber bundle construction). In a picturesque way, one may say that the probe η is 'transported' by the group G modulo H.

Finally, Property P4 is in fact, both historically and conceptually, the one that survives (or should survive, presumably after some technical modifications) once one has been forced (however reluctantly!) to abandon some of the other properties — perhaps even all of them. As far as physical applications are concerned, this property has gradually emerged as the one most fundamental for the analysis, or decomposition, of states in the Hilbert space of the problem, or of operators acting on this space. Thus, it will come as no surprise that Property P4 is the *raison d'être* of the present volume. We shall present a comprehensive survey of this point of view in the following pages. Very schematically, given a measure space (X, ν) and a Hilbert space $\mathfrak{H}$, a family of CS $\{\eta_x \mid x \in X\}$ must satisfy an operator identity of the form

$$\int_X |\eta_x\rangle\langle\eta_x| \, d\nu(x) = A, \tag{1.11}$$

where A is a positive, invertible operator on $\mathfrak{H}$. Note that (1.11) must be understood mathematically with a certain bit of caution (for instance, it must be interpreted in a weak sense, that is, in terms of expectation values in states, possibly even resorting to a triplet of spaces $\mathfrak{H}_1 \subset \mathfrak{H} \subset \mathfrak{H}'_1$, in the sense of distribution theory). In the ultimate analysis, what is desired is to make the family $\{\eta_x\}$ operational through the identity (1.11). This means being able to use it as a "frame" through which one reads the information contained in an arbitrary state in $\mathfrak{H}$, in an operator on $\mathfrak{H}$, or in a setup involving both operators and states, such as an evolution equation on $\mathfrak{H}$. At this point, it is legitimate to borrow from the language of quantum physics and say that (1.11) realizes a quantization of the "classical" space (X, ν) and the measurable functions on it. One could, however, get rid of this quantum mechanical dressing as well, which goes back to Schrödinger's original work, and instead switch squarely to the terminology of signal processing. In that setting, $\mathfrak{H}$ is a Hilbert space of finite energy signals and (X, ν) is a space of parameters, suitably chosen for emphasizing certain aspects of the signal that may interest us in particular situations. Every signal contains "noise," but the nature and the amount of noise is different for different signals. In this context, choosing $(X, \nu, \{\eta_x\})$ amounts to selecting a part of the signal that we wish to isolate and interpret, while eliminating a noise that has (once and for all) been deemed useless. Here too, we have in effect chosen a frame. In our opinion, this formal analogy between the "quantization" of (X, ν), namely,

$$x \mapsto |\eta_x\rangle\langle\eta_x|, \tag{1.12}$$

on the one hand, and signal processing, on the other, merits a deeper investigation. It is another goal of the present volume to induce the reader to undertake such a reflection.

A perfect illustration of the deep analogy between quantum mechanics and signal processing is the theory of *wavelets*. This is a relatively recent (about a decade old) approach in signal and image analysis that has undergone an explosive development, with applications to a large number of topics in engineering, applied mathematics, and physics. The crucial point is that wavelets yield a time-frequency or, more properly, a time-scale representation of the signal. The built-in scaling operation makes it a very efficient tool for analyzing singularities in a signal, a function, an image, and so on; that is, the portion of the signal that contains the most significant information.

Now, not surprisingly, wavelets *are* CS, namely, those associated with the affine group of the appropriate dimension, consisting of translations, dilations, and rotations (in dimensions higher than one). In fact, it was Aslaksen and Klauder [56] who first studied the one-dimensional affine group, for the purely quantum mechanical purpose of generalizing Property P1 above to dilations-translations [190], and it was yet another mathematical physicist, Alex Grossmann [156, 159, 160], who discovered the crucial link between the representations of the affine group and an intriguing technique in signal analysis developed by the geophysicist Jean Morlet. Once again, we see here CS bridging the gap between classical physics (this time microseismology in oil prospecting!) and quantum mechanical techniques. This explains why wavelets naturally find their place in a book on coherent states: They constitute one of the most promising developments in the venerable theory of coherent states.

Let us come back for a while to the original motivation for introducing CS, namely, to study quantum reality through a framework formally similar to classical reality. As a matter of fact, a formalism based on Hilbert space or distribution theory (such as a "complete set of commuting observables") is often heavy, cumbersome, replete with countless indices, and perhaps even turns out to be intractable. On the contrary, the use of CS restores to quantum states the appearance of classical points [see (1.11)], and to observables the practicality of a functional formulation in terms of *symbols*. The latter come in two varieties [69, 145], namely, to each observable O, one associates the covariant symbol $\tilde{O}$ and the contravariant symbol $\overset{\circ}{O}$:

$$\tilde{O}(x',x) = \frac{\langle \eta_{x'}|O|\eta_x\rangle}{\langle \eta_{x'}|\eta_x\rangle} \quad \text{and} \quad O = \int_X \overset{\circ}{O}(x)\,|\eta_x\rangle\langle\eta_x|. \tag{1.13}$$

It was precisely this symbolic formulation that enabled Glauber and others [145, 168, 169, 206] to treat a quantized boson or fermion field like a classical field, particularly, for computing correlation functions or other quantities of statistical physics, such as partition functions and derived quantities. Indeed, these quantities are often given as the trace of a product of operators (e.g., the density operator of the system), in which the use of the resolution of the identity (1.11) (with $\langle \eta_x|\eta_x\rangle = 1$) is particularly efficient. For instance:

$$\begin{aligned}
\operatorname{Tr}\left[O_1 O_2 \ldots O_N\right] &= \int_X d\nu(x)\, \langle \eta_x | A^{-1/2} O_1 O_2 \ldots O_N A^{-1/2}|\eta_x\rangle \\
&= \prod_{i=1}^{N} \int_X d\nu(x_i)\, \langle \eta_{x_i}|A^{-1/2} O_i A^{-1/2}|\eta_{x_{i+1}}\rangle \\
&= \prod_{i=1}^{N} \int_X d\nu(x_i)\, \widetilde{(A^{-1/2} O_i A^{-1/2})}(x_i,x_{i+1}),
\end{aligned}$$

where $x_{N+1} \equiv x_1 \equiv x$. Such expressions are convenient for deriving estimates when considering various limits, such as a thermodynamical limit or a path integral. In particular, one can follow the dynamical evolution of a system in a "classical" way, elegantly going back to the study of classical "trajectories" in the space X. It is a striking fact that the latter, which is often a genuine phase space or a space of classical states, is being used as a parameter space for quantum events. We encounter here, in a purely operational fashion, a debate whose formulation goes back to the beginnings of quantum theory, namely, the preeminence given by the physicist — or the signal engineer — to aspects of reality that appear most immediate on the phenomenological level and are then called "classical" or "quantal," depending on one's practice or epistemological preference.

So much for the physical background of the book. Indeed, the reader will readily sense a strong mathematical bias throughout the work. The reason for this is twofold. On the one hand, there are several review papers or books on the market that cover the physical developments of the theory of CS. Besides the Klauder–Skagerstam reprint volume already quoted [Kl2], we also ought to mention the review article by Zhang, Feng, and Gilmore [289] and the proceedings of the 1993 Oak Ridge Conference, entitled *Coherent States: Past, Present and Future* [Fen], celebrating the 50th anniversary of the Glauber–Klauder CS. On the other hand, the mathematical developments around the concept of CS have been equally spectacular, if less well-known among physicists. In fact, the only comprehensive book on the topic is the monograph of Perelomov [Per], going back to 1986. As we shall see, far-reaching generalizations of the "standard" CS have appeared in the meantime, which substantially change the perspective. A recent survey of these developments is given in our review article [19].

Coming back to the title of the Oak Ridge volume, it is not our intention (nor is it within our competence) to write here a detailed history of the past. Concerning the present, a good testimony to the immediate pertinence of our topic is the status of CS for the hydrogen atom. As early as his 1926 work, Schrödinger [266] had tried to construct CS appropriate to the hydrogen atom problem, but without success. His search was motivated by the quasi-classical character of the canonical CS, which made them very desirable for studying the quantization of a classical dynamical system, a point that is still pertinent today. What Schrödinger sought, but could not obtain, was a set of well-localized wave packets, concentrated around the classical Kepler orbits, and thus permitting an easy transition from the classical to the quantum description of the hydrogen atom. Several systems of CS have been proposed since Schrödinger's original work, derived mainly from the groups $SU(1,1)$ and $SO(4,2)$ (the latter is the full dynamical group of the H-atom; the former is the subgroup describing its radial motion only), but none of them is fully convincing (see [194] for a list of references).

In this context, we may note that the two most celebrated models of classical *and* quantum physics, namely the r^2 harmonic oscillator potential (stability around an equilibrium position) and the $1/r$ Kepler potential (Gauss's theorem) are at the same time mathematically very close (they may even be transformed into each other — see below), and yet profoundly different from the point of view of physical interpretation. The confining property of the former potential leads to a purely discrete, semibounded spectrum, whereas the latter simultaneously exhibits (at least, for the radial part) a bounded discrete spectrum and a positive continuum. It is a highly nontrivial task to construct CS for the H-atom from a superposition of its (generalized) eigenstates, so as to satisfy Properties P1 and P4.

Of course, an alternative approach is possible, exploiting the mathematical equivalence between the harmonic oscillator and the H-atom, regularized at the origin and restricted to negative energies. Indeed, the so-called OC-transformation maps one problem into the other, interchanging energy and coupling constant [Kem, 131]. Thus there appears a natural basis in the state space of the H-atom, the *Sturmian basis* $\{\psi_{i,\lambda}(\lambda\mathbf{x})\}$, where $i \equiv (n, l, m)$ is the standard triple of quantum numbers [130, 258]. This basis is orthonormal with respect to the measure $r^{-1}\, d\mathbf{x}$ and the basis vectors depend on the dilation parameter λ as follows:

$$\psi_{i,\lambda}(\lambda\mathbf{x}) = [U(\lambda)h_i](\mathbf{x}) = \lambda^{3/2}\, h_i(\lambda\mathbf{x}).$$

Choosing $\lambda = 1/n$ for $i = (n, l, m)$ yields precisely the *incomplete* set of H-atom bound states. The dynamical symmetries $SO(4,1)$ and $SO(4,2)$ of the Keplerian potential are in fact the ones that describe best the Sturmian system, in the sense that they reflect the symmetries of the harmonic oscillator. The function h_i is the eigenfunction of the operator $r(1 - \Delta)$ corresponding to the eigenvalue n. Thus it is natural to construct "Sturmian" CS related to $SO(4,1)$ or $SO(4,2)$ (see for instance [290]), for this construction is fully in the spirit of the canonical CS, or at least of the Barut–Girardello CS [65]. Within the same circle of ideas, we may also mention the CS built in [239], again based on $SO(4,2)$ and close to the spirit of the present book. In fact, we feel that the Sturmian approach has not been totally exploited in the construction of CS. For instance, Sturmian functions could be used as probes for building orbits coherent under the action of a relativity group, such as the Galilei or the affine Galilei group. [CS for these groups will be discussed at length in Chapter 11, Section 11.2, and Chapter 15, Section 15.3.2, respectively.]

However, no matter how elegant they are, all these CS give only a very imperfect answer to the physical problems raised by the recent experiments in atomic physics. Indeed, the Kepler wave packets, that Schrödinger looked for, have by now been seen *experimentally*, and they exhibit the fascinating phenomenon of *revival*: A state, initially well localized, spreads all over its Kepler orbit, but, after a large number of revolutions, reconcentrates itself where it began, passing in between through various multilocalized stages (see [228, 287] and the popular account in [229]). Moreover, the story underwent a new development when Klauder introduced a new set of CS for the H-atom [194], abandoning the group-theoretical underpinning, but imposing *time stability* (a CS must remain a CS throughout its time evolution), as well as *continuity* in the labels, that is, continuity of the map $x \mapsto \eta_x$ (this condition was in fact always assumed, at least implicitly). While these new states do not yet fully answer the original question (they do not follow the classical Kepler orbits with little or no dispersion [68]), they open the door to a whole host of new CS, adapted to sytems with either a discrete or a continuous spectrum, or both, and are specified by insisting on classical action–angle variables [142]. So the present stage is still very lively!

As for the future, we will not take the risk of making definite predictions. Only one thing is certain: CS will continue to occupy a central position in quantum theories, and new developments may be expected both in the mathematical aspects and for specific physical applications.

A word now about the organization of the book. First, let us emphasize that we will skip a number of topics that the reader might expect, because of their current fashionable character. To mention a few, we will *not* discuss superCS (that is, CS associated with supergroups or supersymmetries), q-deformed CS, nor the use of CS in quantization processes or for defining path integrals. Some of these topics have been described, for instance, in our review paper [19] or in the Klauder–Skagerstam volume [Kl2]

Let us now go into the details. Quite naturally, we start, in Chapter 2, with a comprehensive survey of the canonical CS, discussing in detail properties P1–P4 above. In particular, P3 means that the canonical coherent states can be obtained by acting on the oscillator ground state with the operators of a unitary representation U of the Weyl–Heisenberg group G_{WH}. It turns out that the representation U has the property of *square integrability* — the matrix elements of $U(g)$ are square integrable functions of g [the integration being performed with respect to the left (or right) invariant Haar measure on G_{WH}]. Furthermore, the physical states (or rays) associated with the CS are not indexed by elements of G_{WH} itself, but by points in the coset space $X = G_{WH}/Z$, where Z is the center of G_{WH} and isomorphic to the unit circle. All of these aspects will lead us to far-reaching generalizations. In addition, the canonical CS have remarkable functional analytic and meromorphic properties. For example, using them, one can derive the Fock–Bargmann representation of quantum mechanics on Hilbert spaces of analytic functions.

The stage being thus set, we begin our study by four mostly mathematical chapters (Chapters 3–6). As we said above, the resolution of the operator (1.11) is the central object of the theory. Mathematically, it is equivalent to the existence of a *reproducing kernel* $K(x,y) = \langle \eta_x | A^{-1} \eta_y \rangle$ (for the moment, we assume A^{-1} to be bounded), which means that the Hilbert space $\mathfrak{H}$ is realized as a space of functions $f(x) \equiv \langle \eta_x | f \rangle$ and one has

$$f(x) = \int_X K(x,y)\, f(y)\, d\nu(y), \ \forall f \in \mathfrak{H}. \tag{1.14}$$

Thus, the basic framework of our theory is a *reproducing kernel Hilbert space*. Hilbert spaces of this type have many attractive properties, and reproducing kernels are well-studied objects in mathematics [Mes, 55], which may be defined on quite general measure spaces. Besides, they arise naturally in large classes of complex domains [70], leading to preferred families of vectors in the Hilbert space that generate resolutions of the identity.

As a first step (Chapter 3), we review a number of mathematical topics needed in the sequel, centered on the notion of positive operator-valued (POV) measures. Then, in Chapter 4, we give a quick reminder of a number of concepts and results from group theory, such as (quasi-)invariant measures on homogeneous spaces, induced group representations, localization properties, harmonic analysis on abelian groups, Lie groups and Lie algebras, their extensions and contractions, and so on. In Chapter 5, we investigate systematically reproducing kernel Hilbert spaces. Particular attention is given, in Chapter 6, to the two classes of square integrable and holomorphic reproducing kernels.

Thus, equipped with the necessary mathematical machinery, we are in a position to state our definition of a coherent state (Chapter 7). It is general enough to cover such widely different situations as Gilmore–Perelomov CS, abstract CS, or discrete frames in Hilbert space. In the same chapter, which occupies a pivotal position in the book, we also describe briefly what may be called the classical theory of coherent states, largely due to Gilmore and Perelomov. It basically deals with the CS associated with representations of simple groups, as examplified by $SU(2)$ and $SU(1,1)$ in the compact and noncompact cases, respectively. Our treatment is rather short here, except for a typical physical application in each case, and we refer to the monograph of Perelomov [246] for further information.

From here on, we concentrate on CS associated with group representations. Although the theory described so far is more general, this is the class that is readily applicable in specific physical situations. Chapter 8 covers in detail the class of CS most studied in the literature, namely, the CS associated with square integrable group representations. Given a locally compact group or a Lie group G, and a square integrable representation U of G (if one such exists) in some Hilbert space $\mathfrak{H}$, fix a vector $\eta \in \mathfrak{H}$ (possibly satisfying an additional *admissibility condition*). Then, a CS system, associated to U, is defined to be the following set of vectors in $\mathfrak{H}$:

$$S = \{\eta_g = U(g)\eta \,|\, g \in G\}. \qquad (1.15)$$

In other words, coherent states are the elements of the *orbit* of η under the (square integrable) representation $U(G)$. They possess a number of remarkable properties, in particular, they satisfy very useful orthogonality relations, allowing one to derive a generalized Wigner map [briefly, this amounts to lifting the analysis from the Hilbert space $\mathfrak{H}$ to the Hilbert space $\mathcal{B}_2(\mathfrak{H})$ of all Hilbert–Schmidt operators on $\mathfrak{H}$]. While this is the basic construction, the approach of Gilmore [144] and Perelomov [246] introduces an additional degree of freedom in that it indexes the CS by points, not of G itself, but of $X_\eta = G/H_\eta$, where H_η is the subgroup of elements of G leaving η invariant up to a phase. The resulting CS enjoy all of the desirable properties of the canonical CS. Actually, a slight generalization of the notion of admissibility permits one to cover more general situations, such as the Galilei group or the so-called vector CS.

The following chapters are basically devoted to applications of the preceding theory and its generalizations to various groups of physical interest. In Chapters 9 and 10, we turn to a class of great physical interest, namely Lie groups of the semidirect product type, $G = V \rtimes S$, where V is a vector space (typically, $\mathbb{R}^n$) and S is a group of automorphisms of V. It is possible to give a unified treatment of all such groups, since their representations may be constructed by Mackey's method of induced representations [Ma1, Ma2]. In addition, one can decide when such a representation is square integrable, thanks to work by Kleppner–Lipsman [23, 197], Bernier–Taylor [71], Führ [127], and Aniello et al. [23]. The physical interest of this class is that it covers the so-called relativity groups, that is, the kinematical groups associated to the Galilean, Newtonian, and Einsteinian relativities [58], which are fully treated in Chapter 11. The problem is that these groups have *no* square integrable representations in the usual sense; hence the preceding theory is not applicable as it stands. There exist, however, families of vectors in certain representation spaces of these groups, which strongly resemble coherent states, and appear in an essential way in localization and quantization problems, but are not obtainable by the Gilmore–Perelomov method [7, 21, 250, 251]. A possible way to incorporate all of these families of states within a unified framework of *generalized coherent states* was suggested in [4] and later developed in a series of papers [9] [16]. The point of departure here is the observation that, while, in the Gilmore–Perelomov framework, CS are indexed by points of $X_\eta = G/H_\eta$, in many instances (particularly for the relativity groups), more general classes of CS may be constructed using a suitable homogeneous space $X = G/H$, where H is a closed subgroup of G that does not necessarily coincide with H_η. Then, one needs to find a *section* $\sigma : X \to G$, and one ends with the general definition of CS given in Chapter 7; that is, the vectors $\eta_{\sigma(x)} = U(\sigma(x))\eta, x \in X$. As a prototype of the general theory, we treat in detail the Poincaré group, probably the most important group of quantum physics. Comparable information about the other relativity groups may be found in the literature, in particular, our earlier papers [9]–[16]. We also discuss the relationship between the various cases, the precise link being the notion of group contraction and its extension to group representations.

Chapters 12–15 present a survey of wavelets, first in one dimension, then in higher dimensions (mostly two, for the purposes of image analysis). As said before, wavelets are the CS associated with the affine groups: the affine group or the $ax + b$ group of the line and the similitude group of $\mathbb{R}^n$ for $n > 1$, consisting of translations, dilations, and rotations. Note that, in both cases, the group is a semidirect product:

$$G_{\text{aff}}^{(+)} = \mathbb{R} \rtimes \mathbb{R}_*^{(+)}, \quad SIM(n) = \mathbb{R}^n \rtimes (\mathbb{R}_*^+ \times SO(n)),$$

where $\mathbb{R}^n$ corresponds to translations, $\mathbb{R}_*^{(+)}$ corresponds to dilations ($a \neq 0$ and $a > 0$, respectively), and $SO(n)$ corresponds to rotations. While stressing the group-theoretical origin of the (continuous) wavelet transform (WT), we also describe in Chapter 12 various concrete aspects, such as the explicit construction of wavelets, the explanation of the efficiency of the WT in signal and image analysis, a brief survey of physical and technological applications, and so on. In addition, we discuss a number of mathematical properties, such as the minimal uncertainty property [i.e., the generalization of P1 above], or the extension of the WT to distributions (largely following [173]). We also give, in Chapter 13, a brief account of the so-called discrete WT, based on the notion of multiresolution analysis, which is based on a totally different philosophy and more widely known in applications. A rather new development, which we also touch on in this chapter, is the construction of wavelets based on irrational numbers, such as the golden mean τ or the so-called Pisot numbers. Although only limited wavelet families of this kind have been constructed so far (the analogue of Haar wavelets), this approach offers promise for the analysis of fractals or various quasi-periodic patterns.

After surveying, in Chapter 14, the case of multidimensional wavelets, we turn in Chapter 15 to generalized wavelets, namely, CS associated with other affine groups, obtained by adjoining dilations to the corresponding kinematical groups (Galilei, Poincaré) or to the Weyl–Heisenberg group.

Finally, Chapter 16 touches on the crucial aspect of the discretization of the wavelet or the CS transform. The application of any kind of CS to a concrete problem requires numerical computations, and this means that all formulas have to be discretized: Signals and images are discretely sampled, integrals are replaced by sums, and the WT is thus also a digital signal. This transition raises deep mathematical questions, some of which we discuss as well as briefly describing the results obtained so far in the literature. We conclude with an account of the sampling problem, which then leads to a generalization of wavelet analysis to abstract abelian groups. In this way, one is approaching the goal of giving a general, group-related formulation of the discrete wavelet transform and the multiresolution formalism. This aspect touches on work in progress, so that no definitive conclusions may yet be drawn.

A final word of warning before proceeding. This book is about mathematical physics; hence, its audience consists of both mathematicians and physicists. Of course, it is impossible to please everyone at the same time. So, we have on some occasions dwelled on mathematics longer than expected, following the internal dynamics of the topic at hand. Conversely, we have, at times, indulged in a thorough discussion of certain physical examples or applications. We apologize to both communities for these various wanderings, our sole excuse being our desire to illustrate the general theory and make the text livelier.

This in fact characterizes the spirit of the book. We have conceived it more as a research monograph, designed for self-study, rather than a textbook suitable for the classroom. We have therefore *not* included any standard end-of-chapter exercises. Instead, one will find dispersed throughout the book a large number of worked out examples, which could be thought of as being research level problems. For the convenience of the reader, we have marked them systematically with a clubsuit symbol, ♣, hoping that this somewhat unorthodox method of identification will make easier the study of sometimes difficult material.

2
Canonical Coherent States

This chapter is devoted to a fairly detailed examination of the quintessential example of coherent states — the *canonical coherent states*. It is fair to say that the entire subject of coherent states developed by analogy from this example. As mentioned in Chapter 1, this set of states, or rays in the Hilbert space of a quantum mechanical system, was originally discovered by Schrödinger transition from quantum to classical mechanics. They are endowed with a remarkable array of interesting properties, some of which we shall survey in this chapter. Apart from initiating the discussion, this will also help us in motivating the various mathematical directions in which one can try to generalize the notion of a CS.

Throughout the book, all Hilbert spaces will be taken to be separable and defined over the complexes. Thus, if $\mathfrak{H}$ is a Hilbert space, its dimension, denoted $\dim \mathfrak{H}$, is (countably) infinite. Also, for any two vectors ϕ, ψ in $\mathfrak{H}$, their scalar product $\langle \phi | \psi \rangle$ will be taken to be antilinear in the first variable ϕ and linear in the second variable ψ (the standard physicist's convention). Unless otherwise stated, we shall use the natural system of units, in which $c = \hbar = 1$.

2.1 Minimal uncertainty states

Recall that the quantum kinematics of a free n-particle system is based on the existence of an irreducible representation of the *canonical commutation* relations (CCR),

$$[Q_i, P_j] = iI\delta_{ij}, \quad i,j = 1,2,\ldots,n, \tag{2.1}$$

on a Hilbert space $\mathfrak{H}$. (Here I denotes the identity operator on $\mathfrak{H}$.) If n is finite, then, according to the well-known uniqueness theorem of von Neumann [vNe], up to unitary equivalence, there is only one irreducible representation of (2.1) by self-adjoint operators, on a (separable, complex) Hilbert space (see Section 2.2 for a more precise statement). Furthermore, the CCR imply that, for any state vector ψ in $\mathfrak{H}$ (note, $\|\psi\| = 1$), the Heisenberg *uncertainty relations* hold:

$$\langle \Delta Q_i \rangle_\psi \langle \Delta P_i \rangle_\psi \geq \frac{1}{2}, \quad i = 1,2,\ldots,n, \tag{2.2}$$

where, for an arbitrary operator A on $\mathfrak{H}$,

$$\langle \Delta A \rangle_\psi = [\langle \psi | A^2 \psi \rangle - |\langle \psi A \psi \rangle|^2]^{\frac{1}{2}} \tag{2.3}$$

is its standard deviation in the state ψ (to be precise, the vector ψ must belong to the domain of the operator A^2). As already pointed out by Schrödinger (see also [vNe, Bie]), the Hilbert space contains an entire family of states, $\eta^\mathbf{s}$, labeled by a vector parameter $\mathbf{s} = (s_1, s_2, \ldots, s_n) \in \mathbb{R}^n$, each one of which saturates the uncertainty relations (2.2):

$$\langle \Delta Q_i \rangle_{\eta^\mathbf{s}} \langle \Delta P_i \rangle_{\eta^\mathbf{s}} = \frac{1}{2}, \quad i = 1,2,\ldots,n. \tag{2.4}$$

We call these vectors *minimal uncertainty states* (MUSTs). In the configuration space, or Schrödinger representation of the CCR, in which

$$\mathfrak{H} = L^2(\mathbb{R}^n, d\mathbf{x}), \quad \mathbf{x} = (x_1, x_2, \ldots, x_n),$$
$$(Q_i \psi)(\mathbf{x}) = x_i \psi(\mathbf{x}), \quad (P_i \psi)(\mathbf{x}) = -i\frac{\partial}{\partial x_i}\psi(\mathbf{x}), \tag{2.5}$$

the MUSTs, $\eta^\mathbf{s}$, are just the Gaussian wave packets

$$\eta^\mathbf{s}(\mathbf{x}) = \prod_{i=1}^n (\pi s_i^2)^{-\frac{1}{4}} \exp[-\frac{x_i^2}{2s_i^2}]. \tag{2.6}$$

Not surprisingly, quantum systems in these states display behavior very close to classical systems. More generally, a larger family of states exists, namely, *gaussons* or *Gaussian pure states* [268], which exhibits the minimal uncertainty property. These latter states $\eta_{\mathbf{q},\mathbf{p}}^{U,V}$ are parametrized by two vectors, $\mathbf{q} = (q_1, q_2, \ldots, q_n)$, $\mathbf{p} = (p_1, p_2, \ldots, p_n) \in \mathbb{R}^n$ and two real $n \times n$ matrices U and V, of which U is positive definite. In the Schrödinger representation,

$$\eta_{\mathbf{q},\mathbf{p}}^{U,V}(\mathbf{x}) = \pi^{-\frac{n}{4}}[\det U]^{\frac{1}{4}} \exp\left[i(\mathbf{x} - \frac{\mathbf{q}}{2}) \cdot \mathbf{p}\right]$$
$$\times \exp\left[-\frac{1}{2}(\mathbf{x} - \mathbf{q}) \cdot (U + iV)(\mathbf{x} - \mathbf{q})\right]. \tag{2.7}$$

2.1. Minimal uncertainty states

In the optical literature, states of the type $\eta_{\mathbf{q},\mathbf{p}}^{U,V}$, for which U is a diagonal matrix but *not* the identity matrix, are called *squeezed states* [75, 282]. Note that, when $\mathbf{q} = \mathbf{p} = 0$ and U is diagonal, with eigenvalues $1/s_i^2$, $i = 1, 2, \ldots, n$, the gaussons (2.7) are exactly the MUSTs (2.6). Moreover, if T denotes the orthogonal matrix that diagonalizes U, i.e., $TUT^{-1} = D$, where D is the matrix of eigenvalues of U, then, defining the vectors $\mathbf{x}' = T\mathbf{x}$, $\mathbf{q}' = T\mathbf{q}$, $\mathbf{p}' = T\mathbf{p}$, and the matrix $V' = TVT^{-1}$, we may rewrite (2.7) as

$$\eta_{\mathbf{q}',\mathbf{p}'}^{D,V'}(\mathbf{x}') = \pi^{-\frac{n}{4}} [\det D]^{\frac{1}{4}} \exp\left[i\left(\mathbf{x}' - \frac{\mathbf{q}'}{2}\right) \cdot \mathbf{p}'\right]$$
$$\times \exp\left[-\frac{1}{2}(\mathbf{x}' - \mathbf{q}') \cdot (D + iV')(\mathbf{x}' - \mathbf{q}')\right]. \quad (2.8)$$

It is clear from this relation that, if Q_i', P_i', $i = 1, 2, \ldots, n$, are the components of the rotated vector operators, $\mathbf{Q}' = T^{-1}\mathbf{Q}$, $\mathbf{P}' = T^{-1}\mathbf{P}$, where $\mathbf{Q} = (Q_1, Q_2, \ldots, Q_n)$, $\mathbf{P} = (P_1, P_2, \ldots, P_n)$ are the vector operators of position and momentum, respectively [see (2.5)], then

$$\langle \Delta Q_i' \rangle_{\eta_{\mathbf{q},\mathbf{p}}^{U,V}} \langle \Delta P_i' \rangle_{\eta_{\mathbf{q},\mathbf{p}}^{U,V}} = \frac{1}{2}, \quad i = 1, 2, \ldots, n. \quad (2.9)$$

To examine some properties of the MUSTs (2.6), take $n = 1$, for simplicity (of notation), and define the *creation* and *annihilation operators*,

$$a^\dagger = \frac{1}{\sqrt{2}}(s^{-1}Q - isP), \qquad a = \frac{1}{\sqrt{2}}(s^{-1}Q + isP),$$
$$[a, a^\dagger] = 1. \quad (2.10)$$

Using these operators and the MUST η^s, for a fixed $s \in \mathbb{R}$, we can generate a very interesting class of other MUSTs [which of course is a subclass of (2.7) and was already noticed by von Neumann [vNe]]. To do so, define the complex variable

$$z = x + iy = \frac{1}{\sqrt{2}}(s^{-1}q - isp), \qquad (q, p) \in \mathbb{R}^2 \quad (2.11)$$

and write

$$\eta^s = |0\rangle. \quad (2.12)$$

(Note that $a|0\rangle = 0$.) Also, let $\{|n\rangle\}_{n=0}^\infty$ be the normalized eigenstates of the *number operator* $N = a^\dagger a$:

$$N|n\rangle = n|n\rangle, \quad |n\rangle = \frac{(a^\dagger)^n}{\sqrt{n!}}|0\rangle, \quad \langle m|n\rangle = \delta_{mn}. \quad (2.13)$$

Then, the set of states in $\mathfrak{H}$,

$$|z\rangle = \exp\left[-\frac{|z|^2}{2} + za^\dagger\right]|0\rangle$$

$$= \exp\left[-\frac{|z|^2}{2}\right] \sum_{n=0}^{\infty} \frac{z^n}{\sqrt{n!}} |n\rangle, \qquad (2.14)$$

for all $z \in \mathbb{C}$, have the eigenvalue property

$$a|z\rangle = z|z\rangle. \qquad (2.15)$$

It is straightforward to verify that each one of these states $|z\rangle$ is again a MUST satisfying (2.4).

Suppose now that we have a quantized electromagnetic field (in a box), and let $a_k^\dagger, a_k$, $k = 0, \pm 1, \pm 2, \ldots$, be the creation and annihilation operators for the various Fourier modes k. Then, in the states

$$|\{z_k\}\rangle = \bigotimes_k |z_k\rangle,$$

the electromagnetic field behaves "classically." More precisely [145], the correlation functions for the field factorize in these states. Thus, let $x = (\mathbf{x}, t)$ be a space-time point, and let $\mathbf{E}^+(x)$ be the positive frequency part of the quantized electric field (note: $\mathbf{E}^-(x) = \mathbf{E}^+(x)^*$ is the negative frequency part of the field). Then,

$$\mathbf{E}^+(x)|\{z_k\}\rangle = \underline{\mathcal{E}}(x)|\{z_k\}\rangle,$$

where $\underline{\mathcal{E}}$ is a three-vector-valued function of x, giving the observed field strength at the point x. Let ρ be the density matrix,

$$\rho = |\{z_k\}\rangle\langle\{z_k\}|,$$

and $G^{(n)}_{\mu_1,\mu_2,\ldots,\mu_{2n}}$ be the *correlation functions*,

$$G^{(n)}_{\mu_1,\mu_2,\ldots,\mu_{2n}}(x_1, x_2, \ldots, x_{2n}) =$$
$$= \text{Tr}[\rho E^-_{\mu_1}(x_1) \ldots E^-_{\mu_n}(x_n) E^+_{\mu_{n+1}}(x_{n+1}) \ldots E^+_{\mu_{2n}}(x_{2n})], \qquad (2.16)$$

where $E^\pm_{\mu_k}$ denotes the μ_kth component of $\mathbf{E}^\pm$. It is then easily verified that

$$G^{(n)}_{\mu_1,\mu_2,\ldots,\mu_{2n}}(x_1, x_2, \ldots, x_{2n}) = \prod_{k=1}^{n} \overline{\mathcal{E}}_{\mu_k}(x_k) \prod_{\ell=n+1}^{2n} \mathcal{E}_{\mu_\ell}(x_\ell). \qquad (2.17)$$

It is because of this factorizability property that the states $|\{z_k\}\rangle$ or the MUSTs $|z\rangle$ were called *coherent states*. In the current mathematical literature, however (though not always in the optical literature), the term coherent state is used to designate an entire array of other mathematically related states, which do not necessarily display either the factorizability property (2.17) or the minimal uncertainty property (2.4). We shall reserve the term *canonical coherent states* for the MUSTs (2.14).

In order to bring out some additional properties of the canonical CS $|z\rangle$, let us use (2.11) to write

$$|\overline{z}\rangle = \eta^s_{\sigma(q,p)}, \qquad (2.18)$$

where $\bar{z}$ is just the complex conjugate of z and z and q, p are related by (2.11). The significance of the σ in this notation will become clear in a while. A short computation shows that

$$\begin{aligned}\langle \eta^s_{\sigma(q,p)}|Q|\eta^s_{\sigma(q,p)}\rangle &= q, \\ \langle \eta^s_{\sigma(q,p)}|P|\eta^s_{\sigma(q,p)}\rangle &= p.\end{aligned} \quad (2.19)$$

In other words, the MUST $\eta^s_{\sigma(q,p)}$ is a translated Gaussian wave packet, centered at the point q in position and p in momentum space. Explicitly, as a vector in $L^2(\mathbb{R}, dx)$,

$$\eta^s_{\sigma(q,p)}(x) = (\pi s^2)^{-\frac{1}{4}} \exp\left[-i(\frac{q}{2} - x)p\right] \exp\left[-\frac{(x-q)^2}{2s^2}\right], \quad (2.20)$$

which should be compared to (2.7).

[In (2.18), we have departed from the physicist's convention and used $\bar{z}$ instead of z to denote the CS. This is because we shall later want to represent them as holomorphic, rather than antiholomorphic, functions of z — see (2.60) below — and our Hilbert space scalar product is linear in the second variable.]

2.2 The group-theoretical backdrop

A group-theoretical property of $|z\rangle$ emerges if we use the Baker–Campbell–Hausdorff identity,

$$e^{A+B} = e^{-\frac{1}{2}[A,B]} e^A e^B, \quad (2.21)$$

for two operators A, B, the commutator $[A, B]$ of which commutes with both A and B, and the fact that $a^n|0\rangle = 0$, $n \geq 1$, to write $|z\rangle$ in (2.14) as

$$|z\rangle = \exp[-\frac{1}{2}|z|^2] \, e^{za^\dagger} e^{\bar{z}a}|0\rangle = e^{za^\dagger - \bar{z}a}|0\rangle. \quad (2.22)$$

In terms of (q, p), this is

$$\eta^s_{\sigma(q,p)} = e^{i(pQ - qP)} \eta^s \equiv U(q,p)\eta^s, \quad (2.23)$$

where $\forall (q,p) \in \mathbb{R}^2$, the operators $U(q,p) = e^{i(pQ-qP)}$ are, of course, unitary. Moreover, we have the integral relation,

$$\frac{1}{2\pi} \int_{\mathbb{R}^2} |\eta^s_{\sigma(q,p)}\rangle \langle \eta^s_{\sigma(q,p)}| \, dqdp = I. \quad (2.24)$$

The convergence of the above integral is in the *weak sense* (see Section 3.1), i.e., for any two vectors ϕ, ψ in the Hilbert space $\mathfrak{H}$,

$$\frac{1}{2\pi} \int_{\mathbb{R}^2} \langle \phi|\eta^s_{\sigma(q,p)}\rangle \langle \eta^s_{\sigma(q,p)}|\psi\rangle \, dqdp = \langle \phi|\psi\rangle. \quad (2.25)$$

To check the validity of this relation, we use (2.20) to obtain

$$\langle\phi|\eta^s_{\sigma(q,p)}\rangle = (\pi s^2)^{-\frac{1}{4}} e^{-\frac{ipq}{2}} \int_{\mathbb{R}} \overline{\phi(x)} \exp[ipx] \exp[-\frac{(x-q)^2}{2s^2}] \, dx. \quad (2.26)$$

Hence, the left-hand side of (2.25) becomes

$$\frac{1}{2\pi\sqrt{\pi}s} \int_{\mathbb{R}^2} dq\, dp \int_{\mathbb{R}^2} dx\, dx'\, \overline{\phi(x)} \exp[ip(x-x')]$$
$$\times \exp[-\frac{(x-q)^2}{2s^2} - \frac{(x'-q)^2}{2s^2}]\psi(x'). \quad (2.27)$$

Using the representation

$$\frac{1}{2\pi} \int_{\mathbb{R}} e^{ip(x-x')} \, dp = \delta(x-x'), \quad (2.28)$$

for the δ-distribution, and performing the integration over x', the above integral becomes

$$\frac{1}{\sqrt{\pi}s} \int_{\mathbb{R}^2} \overline{\phi(x)} \exp[-\frac{(x-q)^2}{s^2}]\psi(x) \, dq\, dx$$
$$= \frac{1}{\sqrt{\pi}} \int_{\mathbb{R}} \exp[-q^2]\, dq \int_{\mathbb{R}} \overline{\phi(x)}\psi(x)\, dx$$
$$= \langle\phi|\psi\rangle. \quad (2.29)$$

The relation (2.24) is called the *resolution of the identity* generated by the canonical CS.

The operators $U(q,p)$ in (2.23) arise from a unitary, irreducible representation (UIR) of the *Weyl–Heisenberg group*, G_{WH}, which is a central extension of the group of translations of the two-dimensional Euclidean plane. The UIR in question is the unitary representation of G_{WH} that integrates the CCR (2.1). An arbitrary element g of G_{WH} is of the form

$$g = (\theta, q, p), \quad \theta \in \mathbb{R}, \quad (q,p) \in \mathbb{R}^2,$$

with multiplication law,

$$g_1 g_2 = (\theta_1 + \theta_2 + \xi((q_1,p_1);(q_2,p_2)), q_1+q_2, p_1+p_2), \quad (2.30)$$

where ξ is the multiplier function

$$\xi((q_1,p_1);(q_2,p_2)) = \frac{1}{2}(p_1 q_2 - p_2 q_1). \quad (2.31)$$

Any infinite-dimensional UIR, U^λ, of G_{WH} is characterized by a real number $\lambda \neq 0$ (in addition, there are also degenerate, 1-D UIRs corresponding to $\lambda = 0$, but they are irrelevant here [Per]) and may be realized on the same Hilbert space $\mathfrak{H}$ as the one carrying an irreducible representation of the CCR:

$$U^\lambda(\theta,q,p) = e^{i\lambda\theta} U^\lambda(q,p) := e^{i\lambda(\theta-\frac{pq}{2})} e^{i\lambda pQ} e^{-i\lambda qP}. \quad (2.32)$$

2.2. The group-theoretical backdrop

If $\mathfrak{H} = L^2(\mathbb{R}, dx)$, these operators are defined by the action

$$(U^\lambda(\theta, q, p)\phi)(x) = e^{i\lambda\theta} e^{i\lambda p(x-\frac{q}{2})} \phi(x-q), \qquad \phi \in L^2(\mathbb{R}, dx). \qquad (2.33)$$

Thus, the three operators, I, Q, P, appear now as the infinitesimal generators of this representation and are realized as:

$$(Q\phi)(x) = x\phi(x), \quad (P\phi)(x) = -\frac{i}{\lambda}\frac{\partial\phi}{\partial x}(x), \quad [Q, P] = \frac{i}{\lambda}I. \qquad (2.34)$$

For our purposes, we take for λ the specific value, $\lambda = \frac{1}{\hbar} = 1$, and simply write U for the corresponding representation.

Denoting the phase subgroup of G_{WH} [the subgroup of elements $g = (\theta, 0, 0)$, $\theta \in \mathbb{R}$], by Θ, it is easily seen that the left coset space G_{WH}/Θ can be identified with $\mathbb{R}^2$ and a general element in it parametrized by (q, p). In terms of this parametrization, G_{WH}/Θ carries the *invariant* measure

$$d\nu(q, p) = \frac{dq\,dp}{2\pi}. \qquad (2.35)$$

The function

$$\sigma : G_{WH}/\Theta \to G_{WH}, \quad \sigma(q, p) = (0, q, p) \qquad (2.36)$$

then defines a *section* in the group G_{WH}, now viewed as a *fiber bundle*, over the base space G_{WH}/Θ, having fibers isomorphic to Θ. Thus, the family of canonical CS is the set,

$$\mathfrak{S}_\sigma = \{\eta^s_{\sigma(q,p)} = U(\sigma(q, p))\eta^s \mid (q, p) \in G_{WH}/\Theta\}, \qquad (2.37)$$

and the operator integral in (2.24) becomes

$$\int_{G_{WH}/\Theta} |\eta^s_{\sigma(q,p)}\rangle\langle\eta^s_{\sigma(q,p)}|\, d\nu(q, p) = I. \qquad (2.38)$$

In other words, the CS $\eta^s_{\sigma(q,p)}$ are labeled by the points (q, p) in the *homogeneous space* G_{WH}/Θ of the Weyl–Heisenberg group, and they are obtained by the action of the unitary operators $U(\sigma(q, p))$, of a UIR of G_{WH}, on a fixed vector $\eta^s \in \mathfrak{H}$. The resolution of the identity equation (2.38) is then a statement of the *square integrability* of the UIR, U, with respect to the homogeneous space G_{WH}/Θ. This way of looking at coherent states turns out to be extremely fruitful. Indeed, one could ask if it might not be possible to use this idea to generalize the notion of a CS and to build families of such states, using UIRs of groups other than the Weyl–Heisenberg group, making sure in the process that basic relations of the type (2.36), (2.37), and (2.38) are still fulfilled. We shall see that this is indeed possible and that such an approach yields a powerful generalization of the notion of a coherent state.

Two remarks are in order before proceeding. First, and not surprisingly, the same canonical CS may be obtained from the oscillator group $H(4)$, which is the group with the Lie algebra generated by $\{a, a^\dagger, N = a^\dagger a, I\}$.

2.3 Some functional analytic properties

The resolution of the identity, given by the operator integral in (2.38), leads to some interesting functional analytic properties of the CS, $\eta^s_{\sigma(q,p)}$. These properties can be studied in their abstract forms and used to obtain a generalization of the notion of a CS, independently of any group-theoretical implications.

Let $\tilde{\mathfrak{H}} = L^2(G_{WH}/\Theta, d\nu)$ be the Hilbert space of all complex-valued functions on G_{WH}/Θ that are square integrable with respect to $d\nu$. Then, (2.38) implies that functions $\Phi : G_{WH}/\Theta \to \mathbb{C}$ of the type

$$\Phi(q,p) = \langle \eta^s_{\sigma(q,p)} | \phi \rangle, \tag{2.39}$$

for $\phi \in \mathfrak{H}$, define elements in $\tilde{\mathfrak{H}}$, and, moreover, writing $W : \mathfrak{H} \to \tilde{\mathfrak{H}}$ for the linear map that associates, via (2.39), an element ϕ in $\mathfrak{H}$ with an element Φ in $\tilde{\mathfrak{H}}$ (i.e., $W\phi = \Phi$), we see that W is an *isometry* or a norm-preserving linear map:

$$\|W\phi\|^2 = \|\Phi\|^2 = \int_{G_{WH}/\Theta} |\Phi(q,p)|^2 \, d\nu(q,p) = \|\phi\|^2. \tag{2.40}$$

The range of this isometry, which we denote by $\mathfrak{H}_K$,

$$\mathfrak{H}_K = W\mathfrak{H} \subset \tilde{\mathfrak{H}}, \tag{2.41}$$

is a closed subspace of $\tilde{\mathfrak{H}}$, and, furthermore, it is a *reproducing kernel Hilbert space*. To understand the meaning of this, consider the function $K(q,p;\, q',p')$ defined on $G_{WH}/\Theta \times G_{WH}/\Theta$,

$$\begin{aligned}
K(q,p;\, q',p') &= \langle \eta^s_{\sigma(q,p)} | \eta^s_{\sigma(q',p')} \rangle \\
&= \exp[-\frac{i}{2}(pq' - p'q)] \, \exp[-\frac{s^2}{4}(p-p')^2] \, \exp[-\frac{1}{4s^2}(q-q')^2] \\
&= \exp[z\bar{z}' - \frac{1}{2}|z|^2 - \frac{1}{2}|z'|^2] \\
&= \langle \bar{z} | \bar{z}' \rangle = K(z, \bar{z}'),
\end{aligned} \tag{2.42}$$

the third and fourth equalities following from (2.18) and (2.20). The function K is a *reproducing kernel*, a name that reflects the reproducing

2.3. Some functional analytic properties

property satisfied by any vector $\Phi \in \mathfrak{H}_K$:

$$\Phi(q,p) = \int_{G_{WH}/\Theta} K(q,p; q',p')\, \Phi(q',p')\, d\nu(q',p'). \tag{2.43}$$

The function K enjoys the following properties.

1. Hermiticity

$$K(q,p; q',p') = \overline{K(q',p'; q,p)}. \tag{2.44}$$

2. Positivity

$$K(q,p; q,p) > 0. \tag{2.45}$$

3. Idempotence

$$\int_{G_{WH}/\Theta} K(q,p; q'',p'') K(q'',p''; q',p')\, d\nu(q'',p'') = K(q,p; q',p'). \tag{2.46}$$

The above relations hold for all $(q,p), (q',p') \in G_{WH}/\Theta$. Condition (2.46) is a consequence of (2.38) and also called the *square integrability property* of K. All three relations are the transcription of the fact that the orthogonal projection operator $\mathbb{P}_K$ of $\mathfrak{H}$ onto $\mathfrak{H}_K$ is an integral operator, with kernel $K(q,p; q',p')$.

It is easy to see that the kernel K actually determines the Hilbert space $\mathfrak{H}_K$. Comparing (2.39) and (2.42), we see that

$$(W \eta^s_{\sigma(q',p')})(q,p) = K(q,p; q',p'); \tag{2.47}$$

in other words, for fixed (q',p'), the function $(q,p) \mapsto K(q,p; q',p')$ is simply the image in $\mathfrak{H}_K$ of the CS $\eta^s_{\sigma(q',p')}$ under the isometry (2.39). Additionally, if Φ is an element of the Hilbert space $\mathfrak{H}_K$, it is necessarily of the form (2.39). Hence, multiplying both sides of that equation by $\eta^s_{\sigma(q,p)}$ and integrating, we get, upon using (2.38),

$$\phi = \int_{G_{WH}/\Theta} \Phi(q,p) \eta^s_{\sigma(q,p)}\, d\nu(q,p). \tag{2.48}$$

This shows that the set of vectors $\eta^s_{\sigma(q,p)}$, $(q,p) \in G_{WH}/\Theta$, is *overcomplete* in $\mathfrak{H}$, and, hence, since W is an isometry, the set of vectors,

$$\xi_{\sigma(q,p)} = W \eta^s_{\sigma(q,p)}, \qquad \xi_{\sigma(q,p)}(q',p') = K(q',p'; q,p), \tag{2.49}$$

[for all $(q,p) \in G_{WH}/\Theta$] is overcomplete in $\mathfrak{H}_K$. [Note that the vectors $\xi_{\sigma(q,p)}$ are the same CS as the $\eta^s_{\sigma(q,p)}$, but now written as vectors in the Hilbert space of functions $\mathfrak{H}_K$.] The term overcompleteness is to be understood in the following way: Since $\mathfrak{H}_K$ is a separable Hilbert space, it is always possible to choose a countable basis $\{\eta_i\}_{i=1}^\infty$ in it and to express any vector $\phi \in \mathfrak{H}$ as a linear combination of the vectors in this basis. By contrast, the family of CS, $\mathfrak{S}_\sigma$ in (2.37) is labeled by a pair of continuous parameters (q,p), and (2.48), or equivalently (2.38), is the statement of the fact that any vector ϕ can be expressed in terms of the vectors in this family. Clearly, it should be possible to choose a countable set of vectors $\{\eta^s_{\sigma(q_i,p_i)}\}_{i=1}^\infty$ from $\mathfrak{S}_\sigma$ and still obtain a basis for $\mathfrak{H}$. This is in fact possible, and many different discretizations exist. The most familiar situation is that in which the set of points $\{q_i,p_i\}$ is a lattice, such that the area of the unit cell is smaller than a critical value (to be sure, the resulting set of CS is then *overcomplete*). The determination of adequate subsets $\{q_i,p_i\}$ leads to very interesting mathematical problems, for instance, in number theory and in the theory of analytic functions. These considerations are part of the general problem of CS discretization, which we shall tackle in Chapter 16.

Equation (2.39) also implies a boundedness property for the functions Φ in the reproducing kernel Hilbert space $\mathfrak{H}_K$. Indeed, using the unitarity of $U(\sigma(q,p))$,

$$|\Phi(q,p)| \leq \|\eta\|\,\|\phi\|, \quad \forall (q,p) \in G_{WH}/\Theta, \tag{2.50}$$

implying that the vectors in $\mathfrak{H}_K$ are all bounded functions. More importantly, this also shows the continuity of the linear map

$$E_K(q,p) : \mathfrak{H}_K \to \mathbb{C}, \qquad E_K(q,p)\Phi = \Phi(q,p), \tag{2.51}$$

which simply evaluates each function $\Phi \in \mathfrak{H}_K$ at the point (q,p) and is therefore called an *evaluation map*. As we shall see later, this can in fact be taken to be the defining property of a reproducing kernel Hilbert space and used to arrive at coherent states via a relation of the type (2.47).

The CS $\eta^s_{\sigma(q,p)}$, along with the resolution of the identity relation (2.38) can be used to obtain a useful family of *localization operators* on the *phase space* $\Gamma = G_{WH}/\Theta$. Indeed, relations such as (2.19) tend to indicate that the CS $\eta^s_{\sigma(q,p)}$ do in some sense describe the localization properties of the quantum system in the phase space Γ. To pursue this point a little further, denote by Δ an arbitrary Borel set in Γ, considered as a measure space, and let $\mathcal{B}(\Gamma)$ denote the σ-algebra of all Borel sets of Γ. Define the positive, bounded operator

$$a(\Delta) = \int_\Delta |\eta^s_{\sigma(q,p)}\rangle\langle\eta^s_{\sigma(q,p)}|\,d\nu(q,p). \tag{2.52}$$

This family of operators, as Δ runs through $\mathcal{B}(\Gamma)$, enjoys certain measure theoretical properties, as follows.

1. If J is a countable index set and Δ_i, $i \in J$, are mutually disjoint elements of $\mathcal{B}(\Gamma)$, i.e., $\Delta_i \cap \Delta_j = \emptyset$, for $i \neq j$ ($\emptyset$ denoting the empty set), then

$$a(\cup_{i \in J}\Delta_i) = \sum_{i \in J} a(\Delta_i), \tag{2.53}$$

the sum being understood to converge weakly.

2. Normalization:

$$a(\Gamma) = I, \quad \text{also} \quad a(\emptyset) = 0. \tag{2.54}$$

Such a family of operators $a(\Delta)$ is said to constitute a *normalized, positive operator-valued (POV) measure* on $\mathfrak{H}$. Using the isometry W in (2.39) and the CS $\xi_{\sigma(q,p)}$ in (2.49), we obtain the normalized POV measure $a_K(\Delta)$ on $\mathfrak{H}_K$:

$$\begin{aligned}a_K(\Delta) &= \int_\Delta |\xi_{\sigma(q,p)}\rangle\langle\xi_{\sigma(q,p)}|\, d\nu(q,p) \\ &= W a(\Delta) W^*.\end{aligned} \tag{2.55}$$

Note that

$$\begin{aligned}a_K(\Gamma) &= \int_{G_{WH}/\Theta} |\xi_{\sigma(q,p)}\rangle\langle\xi_{\sigma(q,p)}|\, d\nu(q,p) \\ &= \mathbb{P}_K,\end{aligned} \tag{2.56}$$

where $\mathbb{P}_K$ is the projection operator, $\mathfrak{H}_K = \mathbb{P}_K \widetilde{\mathfrak{H}}$, which projects onto the reproducing kernel subspace $\mathfrak{H}_K$ of $\widetilde{\mathfrak{H}}$.

If $\Psi \in \mathfrak{H}_K$ is an arbitrary state vector, and $\Psi = W\psi$, $\psi \in \mathfrak{H}$, then by (2.55) and (2.39),

$$\langle \Psi | a_K(\Delta) \Psi \rangle = \langle \psi | a(\Delta) \psi \rangle = \int_\Delta |\Psi(q,p)|^2\, d\nu(q,p). \tag{2.57}$$

This means that, if $\Psi(q,p)$ is considered as the phase space wave function of the system, then $a_K(\Delta)$ is the *operator of localization* in the region Δ of phase space. Of course, in order to interpret $|\Psi(q,p)|^2$ as a phase space probability density, an appropriate concept of joint measurement of position and momentum has to be developed. This can in fact be done (see, for example, [Bus, Pru, 4]). Here, let us just indicate, without proof, an interesting fact that reinforces the interpretation of the $a_K(\Delta)$ as localization operators. On $\mathfrak{H}_K$, define the two unbounded operators Q_K and P_K,

$$\begin{aligned}\langle \Psi | Q_K \Phi \rangle &= \int_{\mathbb{R}^2} \overline{\Psi(q,p)}\, q \Phi(q,p)\, d\nu(q,p), \\ \langle \Psi | P_K \Phi \rangle &= \int_{\mathbb{R}^2} \overline{\Psi(q,p)}\, p \Phi(q,p)\, d\nu(q,p),\end{aligned} \tag{2.58}$$

on vectors Ψ, Φ chosen from appropriate dense sets in $\mathfrak{H}_K$. Then, it can be shown (see, e.g., [4]) that

$$[Q_K, P_K] = iI_K, \quad I_K = \text{identity operator on}\ \mathfrak{H}_K. \tag{2.59}$$

Thus, multiplication by q and p, respectively, yield the position and momentum operators on $\mathfrak{H}_K$.

Mathematically, the virtue of the above functional analytic description of the coherent states $\eta^s_{\sigma(q,p)}$ is that it points to another possibility of generalization: We would like to associate CS to arbitrary reproducing kernel

2.4 A complex analytic viewpoint (♣)

To bring out some complex analytic properties of the canonical CS, let $\phi \in \mathfrak{H}$ be an arbitrary vector. Computing its scalar product with the CS $|\bar{z}\rangle$ using (2.14), we get

$$\langle \bar{z}|\phi\rangle = \exp[-\frac{|z|^2}{2}] \sum_{n=0}^{\infty} \frac{\langle n|\phi\rangle}{\sqrt{n!}} z^n$$

$$= \exp[-\frac{|z|^2}{2}] f(z). \qquad (2.60)$$

Here f is an analytic function of the complex variable z. In terms of $z, \bar{z}$, we may write the measure (2.35) as

$$d\nu(q,p) = \frac{dz \wedge d\bar{z}}{2\pi i}, \qquad (2.61)$$

and let us define the new measure

$$d\mu(z,\bar{z}) = \exp[-|z|^2] \frac{dz \wedge d\bar{z}}{2\pi i}. \qquad (2.62)$$

In this notation, "∧" denotes the *exterior product* of the two differentials dz and $d\bar{z}$ (considered as *one-forms* on the *complex manifold* $\mathbb{C}$). In measure theoretic terms, the quantity $idz \wedge d\bar{z}/2$ simply represents the Lebesgue measure $dxdy$, $z = x + iy$, on $\mathbb{C}$. Comparing (2.60) and (2.61) with (2.39) and (2.40), we see that $\mathfrak{H}_K$ can be identified with the Hilbert space of all analytic functions in z that are square integrable with respect to $d\mu$. Let $\mathfrak{H}_{hol}$ denote this Hilbert space. Then, the linear map

$$W_{hol} : \mathfrak{H} \to \mathfrak{H}_{hol}, \qquad (W_{hol}\phi)(z) = \exp[\frac{|z|^2}{2}] \langle\bar{z}|\phi\rangle, \qquad (2.63)$$

is an isometry. Using (2.18) and (2.20), we can compute the vectors

$$f_{\sigma(q,p)} = W_{hol}\, \eta^s_{\sigma(q,p)} = W_{hol}\,|\bar{z}\rangle = \exp[-\frac{|z|^2}{2}] \zeta_{\bar{z}}, \qquad (2.64)$$

which are the images of the $\eta^s_{\sigma(q,p)}$ in $\mathfrak{H}_{hol}$. The vectors $\zeta_{\bar{z}} \in \mathfrak{H}_{hol}$ represent the analytic functions [see (2.42)]:

$$\zeta_{\bar{z}}(z') = e^{z'\bar{z}} = \exp[\frac{1}{2}(|z|^2 + |z'|^2)] K(z',\bar{z}). \qquad (2.65)$$

From this, it is clear that the function $K_{hol} : \mathbb{C} \times \mathbb{C} \to \mathbb{C}$,

$$K_{hol}(z',\bar{z}) = \langle \zeta_{\bar{z}'}|\zeta_{\bar{z}}\rangle_{\mathfrak{H}_{hol}} = e^{z'\bar{z}}, \qquad (2.66)$$

is a reproducing kernel for $\mathfrak{H}_{hol}$. (Here, $\langle\cdot|\cdot\rangle_{\mathfrak{H}_{hol}}$ denotes the scalar product in $\mathfrak{H}_{hol}$.) Indeed, for any $f \in \mathfrak{H}_{hol}$ and $z \in \mathbb{C}$,

$$\int_{\mathbb{C}} K_{hol}(z,\bar{z}')f(z')\,d\mu(z',\bar{z}') - f(z) = \langle \zeta_{\bar{z}}|f\rangle_{\mathfrak{H}_{hol}}. \tag{2.67}$$

Note that, in view of (2.42),

$$K_{hol}(z',\bar{z}) = \exp\left[\frac{1}{2}(|z|^2 + |z'|^2)\right] K(z',\bar{z}). \tag{2.68}$$

Furthermore, the vectors $\zeta_{\bar{z}}$ satisfy the resolution of the identity relation on $\mathfrak{H}_{hol}$:

$$\int_{\mathbb{C}} |\zeta_{\bar{z}}\rangle\langle\zeta_{\bar{z}}|\,d\mu(z,\bar{z}) = I_{\mathfrak{H}_{hol}}. \tag{2.69}$$

The MUST $\eta^s = |0\rangle$ is represented as the constant vector

$$W_{hol}\,\eta^s = u_0 = K_{hol}(\cdot,0), \qquad u_0(z) = 1, \quad \forall z \in \mathbb{C}, \tag{2.70}$$

in $\mathfrak{H}_{hol}$.

Since $K_{hol}(z,\bar{z}) = \|\zeta_{\bar{z}}\|^2$, (2.67) implies that

$$|f(z)| \le [K_{hol}(z,\bar{z})]^{\frac{1}{2}}\|f\|. \tag{2.71}$$

Hence, the evaluation map

$$E_{hol}(z) : \mathfrak{H}_{hol} \to \mathbb{C}, \qquad E_{hol}(z)f = f(z), \tag{2.72}$$

is continuous. Indeed, the CS $\zeta_{\bar{z}}$ could have been obtained by using this fact alone, i.e., by defining it to be the vector that, for arbitrary $f \in \mathfrak{H}_{hol}$, gives $f(z) = \langle \zeta_{\bar{z}}|f\rangle$. Such a construction would be independent of any group-theoretical considerations, and it is intrinsic to complex manifolds admitting *Kähler structures* (see Section 2.5 below).

The representation $\exp[-|z|^2/2]\,\zeta_{\bar{z}}$ of the CS on the space of holomorphic functions $\mathfrak{H}_{hol}$ is known among physicists as the *Fock–Bargmann representation*, and the Hilbert space $\mathfrak{H}_{hol}$ is known as the *Bargmann space of entire analytic functions*. The operators a, $a^\dagger$, in this representation, are given by

$$(af)(z) = \frac{\partial f}{\partial z}(z), \qquad (a^\dagger f)(z) = zf(z), \qquad f \in \mathfrak{H}_{hol}. \tag{2.73}$$

The basis vectors $|n\rangle \in \mathfrak{H}$, [see (2.13)] are mapped by W_{hol} to the vectors

$$W_{hol}\,|n\rangle = u_n, \qquad u_n(z) = \frac{z^n}{\sqrt{n!}}, \tag{2.74}$$

in $\mathfrak{H}_{hol}$. Equation (2.66) then implies:

$$K_{hol}(z',\bar{z}) = \sum_{n=0}^{\infty} u_n(z')\overline{u_n(z)}. \tag{2.75}$$

2.5 Some geometrical considerations

As already pointed out, the existence of the CS $\zeta_{\bar{z}}$ can be traced back to certain intrinsic geometrical properties of $\mathbb{C}$, considered as a 1-D, complex *Kähler manifold*. While we do not intend, at this point, to discuss this notion in any depth, it is still possible to get a general idea of what is involved. To begin with, $\mathbb{C}$ may be thought of as being either a 1-D complex manifold or a 2-D real manifold $\mathbb{R}^2$, equipped with a complex structure. In the first case, one works with the *holomorphic coordinate* z (or the *antiholomorphic coordinate* $\bar{z}$). In the second case, one uses the real coordinates q, p. Considered as a real manifold, $\mathbb{R}^2$ is *symplectic*; i.e., it comes equipped with a closed, nondegenerate *two-form* [compare with (2.35) and (2.61)]

$$\Omega = dq \wedge dp = \frac{1}{i} dz \wedge d\bar{z}. \tag{2.76}$$

While considered as a complex manifold, $\mathbb{C}$ admits the *Kähler potential* function:

$$\Phi(z', \bar{z}) = z'\bar{z}, \tag{2.77}$$

from which the two-form emerges upon differentiation:

$$\Omega = \frac{1}{i} \frac{\partial^2 \Phi(z, \bar{z})}{\partial z \partial \bar{z}} dz \wedge d\bar{z}. \tag{2.78}$$

Similarly, the Kähler potential also determines the reproducing kernel:

$$K_{hol}(z', \bar{z}) = \exp[\Phi(z', \bar{z})], \tag{2.79}$$

while the measure $d\mu$ [see (2.62)], defining the Hilbert space $\mathfrak{H}_{hol}$ of holomorphic functions, is given in terms of it by

$$d\mu(z, \bar{z}) = \exp[-\Phi(z, \bar{z})] \frac{dz \wedge d\bar{z}}{2\pi i}. \tag{2.80}$$

Continuing, if we define the complex *one-form*

$$\Theta = -i\partial_{\bar{z}}\Phi(z, \bar{z}) = -iz d\bar{z}, \tag{2.81}$$

we get

$$\Omega = \partial_z \Theta, \tag{2.82}$$

where $\partial_z, \partial_{\bar{z}}$ denote (exterior) differentiation with respect to z and $\bar{z}$, respectively. It appears, therefore, that it is the Kähler structure of $\mathbb{C}$ (or the fact that it comes equipped with the Kähler potential Φ) that leads to the existence of the Hilbert space $\mathfrak{H}_{hol}$ of holomorphic functions and, consequently, the CS $\zeta_{\bar{z}}$ [the appearance of these latter being a consequence of the continuity of the evaluation map (2.72)]. Once again, this situation is generic to all Kähler manifolds.

Let $\mathbb{P}(z)$ be the one-dimensional projection operator onto the vector subspace of $\mathfrak{H}_{hol}$ generated by the vector $\zeta_{\bar{z}}$, and denote this subspace by $\mathfrak{H}_{hol}(z)$. The collection of all of these one-dimensional subspaces, as z ranges over $\mathbb{C}$, defines a (holomorphic) *line bundle* over the manifold $\mathbb{C}$ — a structure which is intimately related to the existence of a *geometric prequantization* of $\mathbb{C}$. We hasten to add, however, that, while a complex Kähler structure is in some sense ideally suited to the existence of a geometric prequantization, a family of CS may define a geometric prequantization even in the absence of such a structure.

2.6 Outlook

We have quickly gleaned through a number of illustrative properties of the canonical coherent states. As mentioned earlier, each one of these properties can be taken as the starting point for a generalization of the notion of a CS. From a purely physical point of view, for example, it is useful to look for generalizations that preserve the minimal uncertainty property. In doing so, it is useful to exploit some of the group-theoretical properties as well. Mathematical generalizations could be based on group-theoretical, analytic, or related geometrical properties. We shall attempt to describe a bit of all of these various possibilities, and, along the way, we shall be naturally led to some powerful applications of the theory of CS to wavelets and signal analysis. Before digging further into the mathematics, however, we shall exhibit two examples of physical applications of the canonical CS.

2.7 Two illustrative examples

To end this chapter, we leave the reader with two examples illustrating the use of the canonical coherent states in quantization theory and in atomic physics, respectively.

2.7.1 A quantization problem (♣)

In classical mechanics, observables are real-valued functions on phase space, and they form an algebra with respect to a product defined by the Poisson bracket. The observables of quantum mechanics are self-adjoint operators on a Hilbert space, forming an algebra with respect to the commutator bracket, divided by $i\hbar$ (note that $\hbar$ is restored in this section). [Strictly speaking, this statement is only valid on a common invariant domain in the Hilbert space, because the observables are in general represented by unbounded operators; for the moment, we shall forget this mathematical complication, which is the origin of alternative formulations of quantum mechanics, based on rigged Hilbert spaces (see [25] for a recent review).] A quantization of a classical system is a linear map $f \mapsto O_f$ of the classical observables f to self-adjoint operators O_f in such a way that the Poisson bracket $\{f, g\}$ of two classical observables is mapped to $(i\hbar)^{-1}[O_f, O_g]$. One also tries to ensure, in the process, that some particular subalgebra of the quantized observables, chosen for physical reasons, be irreducibly realized on the Hilbert space. Using the canonical CS, we now construct such a map in a simple physical situation.

Consider a classical particle of mass m, having a single degree of freedom, moving on the configuration space $\mathbb{R}$ and having the phase space $\mathbb{R}^2$. On the Hilbert space $\mathfrak{H} = L^2(\mathbb{R}, dx)$, take the set of CS

$$\eta_{\sigma(q,p)} = \exp[\frac{i}{\hbar}(x - \frac{q}{2})p]\eta(x - q), \qquad (2.83)$$

obtained by setting $s = 1$ in (2.20) and corresponding to a function f of the variables (q, p), define the formal operator:

$$O_f = \int_{\mathbb{R}^2} f(q,p)|\eta_{\sigma(q,p)}\rangle\langle\eta_{\sigma(q,p)}|\, d\nu(q,p), \qquad d\nu(q,p) = \frac{dqdp}{2\pi\hbar} \qquad (2.84)$$

(this operator is often denoted $\widehat{f}$, but this notation is somewhat confusing). In general, the operator O_f defined in this way will be unbounded and technical questions involving domains have to be addressed, as noted above. Assuming that O_f can be defined on a dense set, however, its action on a vector ϕ, taken from this set, is given by the integral operator relation:

$$(O_f\phi)(x) = \frac{1}{h}\int_{\mathbb{R}^2} dqdp\, f(q,p)\, [\int_{\mathbb{R}} dx'\, e^{-\frac{i}{\hbar}(x'-x)p}\eta(x'-q)\phi(x')]\, \eta(x-q). \qquad (2.85)$$

From this, it follows that, if $f(q,p) = f(q)$ is a function of q alone, then O_f is the operator of multiplication by the function $|\eta|^2 * f$ (the asterisk denotes a convolution):

$$|\eta|^2 * f(x) = \int_{\mathbb{R}} |\eta(x-q)|^2 f(q)\, dq. \qquad (2.86)$$

Similarly, if $f(q,p) = f(p)$ is a function of p alone, then O_f is the (in general, pseudo-) differential operator

$$O_f = f(-i\hbar\frac{\partial}{\partial x}) \qquad (2.87)$$

(formally, if $f(q)$ is written as a power series in q, then O_f is obtained by replacing q by $-i\hbar\partial/\partial x$). In particular, taking $f(q) = q$ and $f(p) = p$, we get

$$(O_q\phi)(x) = x\phi(x), \qquad (O_p\phi)(x) = -i\hbar\frac{\partial\phi}{\partial x}(x), \qquad (2.88)$$

while, if $f = H$, the harmonic oscillator Hamiltonian,

$$H = \frac{p^2}{2m} + \frac{m^2\omega^2}{2}q^2, \qquad (2.89)$$

then
$$O_H = -\frac{\hbar^2}{2m}\frac{d^2}{dx^2} + \frac{m^2\omega^2}{2}x^2 + C, \qquad (2.90)$$

where C is the constant
$$C = \frac{\sqrt{\pi}}{4}\hbar^{\frac{3}{2}}m^2\omega^2, \qquad (2.91)$$

which simply changes the ground state energy.

We see in this example that this method of quantization yields the expected result, in that the Poisson bracket $\{q,p\}$ is properly mapped to the commutator bracket $(i\hbar)^{-1}[O_q, O_p]$, and the algebra generated by O_q, O_p and I is irreducibly represented on $\mathfrak{H}$. Thus, formally at least, the use of the canonical CS leads to the same quantization result as ordinarily obtained by making the substitutions, $q \to$ "multiplication by x" and $p \to -i\hbar \partial/\partial x$. Details on this quantization technique may be found in [10],[69],[235], and [250].

2.7.2 An application to atomic physics (♣)

As another illustration of the efficiency of CS methods, we consider the example of a system of N two-level atoms interacting with a radiation field. In the Dicke model [Scu, 109], the atoms have fixed positions in a 1-D box of length L, and they are supposed to be far enough from each other so that their mutual interaction is negligible. Nevertheless, the assembly has to be thought of as a single quantum entity, because of its collective inteaction with the radiation field. In [168], Hepp and Lieb have obtained some exact results for the thermodynamic properties of this system in the limit $N \to +\infty$, $L \to +\infty$, $N/L = $ const. In particular, they have shown the existence of a second-order phase transition, radiance $\to$ superradiance, at a certain critical temperature T_c, when the atom-field coupling is strong enough. This coupling is supposed to hold in the so-called dipole approximation. On the other hand, the box length L is supposed to be sufficiently small, compared to the radiation wavelength, so that all of the atoms experience the same field. In the *rotating-wave approximation* [Scu], the Hamiltonian of the system reads (in units $\hbar = c = 1$ and standard notation, H instead of O_H)

$$H = H_o + H_I, \qquad (2.92)$$

where H_o and H_I are the free and interaction Hamiltonians, respectively:

$$\begin{aligned} H_o &= H_{rad} + H_{at} \\ &= \sum_k \nu_k a_k^\dagger a_k + \frac{1}{2}\omega \sum_{j=1}^N \sigma_j^z, \end{aligned} \qquad (2.93)$$

$$H_I = \frac{1}{2\sqrt{L}}\left[(\sum_k \lambda'_k a_k)(\sum_{j=1}^N \sigma_j^+) + (\sum_k \lambda'_k a_k^\dagger)(\sum_{j=1}^N \sigma_j^-)\right]. \quad (2.94)$$

Here, $a_k^\dagger$ and a_k are the creation and annihilation operators, respectively, for the kth radiation mode with frequency ν_k, ω is the two-level energy shift, and λ' is the atom-field coupling strength. Two-level atoms are described by Pauli matrices

$$\sigma_j^\pm = \sigma_j^x \pm i\sigma_j^y, \quad (2.95)$$

$$\sigma_j^x = \begin{pmatrix} 0 & 1 \\ 1 & 0 \end{pmatrix}_j, \quad \sigma_j^y = \begin{pmatrix} 0 & -i \\ i & 0 \end{pmatrix}_j, \quad \sigma_j^z = \begin{pmatrix} 1 & 0 \\ 0 & -1 \end{pmatrix}_j.$$

The Dicke model deals with a single mode of frequency ν. The corresponding Hamiltonian reads

$$H = a^\dagger a + \sum_{j=1}^N \left[\frac{1}{2}\epsilon\sigma_j^z + \frac{\lambda}{2\sqrt{N}}(a\sigma_j^+ + a^\dagger\sigma_j^-)\right], \quad (2.96)$$

where we have used the modified parameters

$$\epsilon = \omega/\nu, \quad \lambda = \lambda'\sqrt{\rho}/\nu, \quad \rho = N/L, \quad (2.97)$$

better suited to thermodynamic considerations.

The thermodynamic properties of the system are encoded in the partition function

$$Z(N,T) = \operatorname{Tr} e^{-\beta H}, \quad \beta = \frac{1}{k_B T} \quad (2.98)$$

(k_B is the Boltzmann constant and T is the absolute temperature). It is precisely at this point, in the explicit computation of the partition function, that the canonical CS of the single-mode field fully play their simplifying role [283]. As usual, we denote them by $|z\rangle$, $a|z\rangle = z|z\rangle$, $z \in \mathbb{C}$. Since, for a trace class operator A, we have

$$\operatorname{Tr} A = \sum_{n\geq 0} \langle n|A|n\rangle = \int \frac{d^2z}{\pi} \langle z|A|z\rangle, \quad (2.99)$$

with $d^2z = d\operatorname{Re} z\, d\operatorname{Im} z$, the expression (2.98) for the partition function becomes

$$Z(N,T) = \sum_{s_1=\pm 1} \cdots \sum_{s_N=\pm 1} \int \frac{d^2z}{\pi} \langle s_1\ldots s_N|\langle z|e^{-\beta H}|z\rangle|s_1\ldots s_N\rangle. \quad (2.100)$$

The sums in (2.100) are taken over all possible atomic sites (for simplicity, we are avoiding a tensor-product notation). Thus, it is necessary to estimate

2.7. Two illustrative examples

the partial matrix element

$$\langle z | e^{-\beta H} | z \rangle = \sum_{r=0}^{\infty} \frac{(-\beta)^r}{r!} \langle z | H^r | z \rangle. \tag{2.101}$$

After rescaling the mode operators as

$$b = \frac{a}{\sqrt{N}}, \quad b^\dagger = \frac{a^\dagger}{\sqrt{N}},$$

so that

$$bb^\dagger = b^\dagger b + \frac{1}{N}, \tag{2.102}$$

we rearrange terms in the expansion of H^r, using normal ordering, to show that

$$\langle z | H^r | z \rangle = \bar{z} z + \sum_{j=1}^{N} h_j^r + O(\frac{1}{N}), \tag{2.103}$$

where the h_j's are the individual atomic Hamiltonians

$$h_j = \frac{1}{2} \epsilon \sigma_j^z + \frac{\lambda}{2\sqrt{N}} (z \sigma_j^+ + \bar{z} \sigma_j^-).$$

Thus, we get the estimate

$$\langle z | e^{-\beta H} | z \rangle = e^{-\beta \bar{z} z} \, e^{-\beta \sum_{j=1}^{N} h_j} + O(\frac{1}{N}), \tag{2.104}$$

and, hence, for the partition function (2.100),

$$Z(N,T) = \int_C \frac{d^2 z}{\pi} e^{-\beta |z|^2} \sum_{s_1 = \pm 1} \cdots \sum_{s_N = \pm 1} \left(\prod_{j=1}^{N} \langle s_j | e^{-\beta h_j} | s_j \rangle \right) + O(\frac{1}{N})$$

$$= \int_C \frac{d^2 z}{\pi} e^{-\beta |z|^2} (\operatorname{Tr} e^{-\beta h})^N + O(\frac{1}{N}) \tag{2.105}$$

Since the generic atomic Hamiltonian,

$$h = \frac{1}{2} \epsilon \sigma^z + \frac{\lambda}{2\sqrt{N}} (z \sigma^+ + \bar{z} \sigma^-),$$

has the eigenvalues

$$\pm \frac{1}{2} \epsilon (1 + 4\lambda^2 |z|^2 / \epsilon^2 N)^{1/2},$$

we get, upon performing the angular integration and making use of standard asymptotic methods,

$$Z(N,T) = \int_C \frac{d^2 z}{\pi} e^{-\beta |z|^2} \left(2 \cosh \frac{1}{2} \beta \epsilon (1 + 4\lambda^2 \frac{|z|^2}{\epsilon^2 N})^{1/2} \right)^N + O(\frac{1}{N})$$

$$\approx \text{const.} \sqrt{N} \max_{0 \le \frac{|z|^2}{N} \le \infty} \exp N\varphi(\frac{|z|^2}{N}), \quad \text{for large } N, \quad (2.106)$$

where

$$\varphi(y) = -\beta y + \ln\left(2\cosh\frac{1}{2}\beta\epsilon(1 + 4\frac{\lambda^2}{\epsilon^2}y)^{1/2}\right).$$

What remains to be done now is to obtain the value of y, which maximizes $\varphi(y)$, according to whether the coupling λ is strong ($\lambda > \sqrt{\epsilon}$) or weak ($\lambda < \sqrt{\epsilon}$). In the former case, one obtains a critical temperature T_c given by

$$\epsilon/\lambda^2 = \tanh\frac{\epsilon}{2k_B T_c}, \quad (2.107)$$

for which the system jumps from a "normal" state, at $T > T_c$, to a superradiative state at $T < T_c$, whereas there is no such phase transition for weak coupling. As with every phase transition, a physical quantity presents a discontinuity at $T > T_c$, in the present case, the specific heat.

This is better understood through the following expressions, obtained using similar coherent state methods, for the average number of photons per atom:

$$\langle\left(\frac{a^\dagger a}{N}\right)^r\rangle = \delta_{r0}, \quad (2.108)$$

for $\lambda^2 > \epsilon$ and all β, or for $\lambda^2 > 0$, $\beta < \beta_c$, while

$$\langle\left(\frac{a^\dagger a}{N}\right)^r\rangle = (\lambda^2\sigma^2 - \epsilon^2/4\lambda)^r, \quad (2.109)$$

for $\lambda^2 > \epsilon$ and $\beta > \beta_c$, where σ is such that

$$2\sigma = \tanh\beta\lambda^2\sigma \ne 0.$$

These expressions clarify the terminology. In the "normal" radiant state, in which (2.108) holds, the number of photons emitted goes to zero as $N \to \infty$. This is not so in the superradiant regime (2.109), in which an infinite number of photons is emitted, as a consequence of the coherence of the maser light, a truly collective effect.

Note that a generalization to multimode fields is straightforward, the same method based on the use of the canonical CS being again useful for carrying out explicit computations.

3
Positive Operator-Valued Measures and Frames

This chapter, and the three succeeding it, constitute a mathematical interlude, preparing the ground for the formal definition of a coherent state in Chapter 7 and the subsequent development of the general theory. As should be clear already, from a look at the last chapter, in order to define CS mathematically and obtain a synthetic overview of the different contexts in which they appear, it is necessary to understand a bit about positive operator-valued (POV) measures on Hilbert spaces and their close connection with certain types of group representations. In Chapter 2, we have also encountered examples of reproducing kernels and reproducing kernel Hilbert spaces, which in turn are intimately connected with the notion of POV measures and, hence, coherent states. In this chapter, we gather together the relevant mathematical concepts and results about POV measures. In the next chapter, we will do the same for the theory of groups and group representations. Chapters 5 and 6 will then be devoted to a study of reproducing kernel Hilbert spaces. The treatment is necessarily condensed, but we give ample reference to more exhaustive literature. Although the mathematically initiated reader may wish to skip these four mathematical chapters, the discussion of many of the topics here is sufficiently different from their treatment in standard texts to warrant at least a cursory glance at it.

As in the previous chapter, let $\mathfrak{H}$ be a separable, complex Hilbert space. The set of all bounded, linear operators on $\mathfrak{H}$ will be denoted by $\mathcal{L}(\mathfrak{H})$, and its subset of positive elements will be denoted by $\mathcal{L}(\mathfrak{H})^+$. Note that a bounded operator A is an element of $\mathcal{L}(\mathfrak{H})^+$ if and only if A is self-adjoint and $\langle \phi | A \phi \rangle \geq 0$, for all vectors ϕ in $\mathfrak{H}$. In particular, $A = P$ is an (orthogonal) projection operator if $P = P^* = P^2$. We shall not have much occasion to deal with unbounded or positive unbounded operators. Throughout, and as much as possible, we have attempted to illustrate new concepts with simple examples.

3.1 Definitions and main properties

The following is a brief recapitulation of some aspects of the theory of POV measures. We only mention those concepts that will be needed in the sequel. A more detailed treatment may be found in, for example, [Be1]. Let X be a metrizable, locally compact space. (All group spaces and parameter spaces for defining CS will be of this type. Metrizability is a technical assumption, which entails coinciding of Baire and Borel subsets of X.) The *Borel sets* $\mathcal{B}(X)$ of X consist of elements of the σ-algebra formed by its open sets. A *positive Borel measure* ν on $\mathcal{B}(X)$ is a map $\nu : \mathcal{B}(X) \to \overline{\mathbb{R}^+} = \mathbb{R}^+ \cup \{\infty\}$ satisfying the following properties.

1. For the empty set $\emptyset$,
$$\nu(\emptyset) = 0. \tag{3.1}$$

2. If J is a countable index set and Δ_i, $i \in J$, are mutually disjoint elements of $\mathcal{B}(X)$ (i.e., $\Delta_i \cap \Delta_j = \emptyset$, for $i \neq j$), then
$$\nu(\cup_{i \in J} \Delta_i) = \sum_{i \in J} \nu(\Delta_i). \tag{3.2}$$

If, in addition, ν satisfies
$$\nu(\Delta) = \sup_{C \subseteq \Delta,\, C\,=\,compact} \nu(C), \tag{3.3}$$

then it is called a *regular* Borel measure. Unless the contrary is stated, all Borel measures will be assumed to be regular. The measure ν is said to be *bounded* or *finite* if $\nu(X) < \infty$.

While referring to a measure ν, we shall use either of the two notations, ν or $d\nu$. The measure of a set Δ, however, will always be written as $\nu(\Delta)$. The most common example of a Borel measure is the Lebesgue measure, $d\nu = dx$, on the real line $\mathbb{R}$. The Borel sets in this case are the elements of the σ-algebra formed by open intervals $\Delta = (a,b)$ in $\mathbb{R}$, with $\nu(\Delta) = b - a$. Furthermore, if $\rho(x)$, $x \in \mathbb{R}$, is a positive density function, then $d\mu(x) \equiv \mu(dx) = \rho(x)\,dx$ defines another Borel measure on $\mathbb{R}$. Of course, the Lebesgue measure is not finite, but, if the density ρ has a finite integral over $\mathbb{R}$, then $d\mu$ is a finite measure.

Definition 3.1.1 (POV measure): *Let $\mathfrak{H}$ be a Hilbert space and X be a locally compact space. A positive operator-valued (POV) measure is a map $a : \mathcal{B}(X) \to \mathcal{L}(\mathfrak{H})^+$, satisfying:*

1. $a(\emptyset) = 0$.

2. For a countable index set J and mutually disjoint elements Δ_i, $(i \in J)$, in $\mathcal{B}(X)$,
$$a(\cup_{i \in J} \Delta_i) = \sum_{i \in J} a(\Delta_i). \tag{3.4}$$

The above sum is understood to converge weakly, which means that, if ϕ and ψ are arbitrary elements of $\mathfrak{H}$, then the complex sum $\sum_{i \in J} \langle \phi | a(\Delta_i) \psi \rangle$ converges. This is the same as saying that the POV measure a is equivalent to the entire collection of positive Borel measures, $\mu_\phi(\Delta) = \langle \phi | a(\Delta) \phi \rangle$, for all $\phi \in \mathfrak{H}$.

The POV measure a is said to be *regular* if, for each $\phi \in \mathfrak{H}$, the measure μ_ϕ is regular. It is said to be *bounded* if
$$a(X) = A \in \mathcal{L}(\mathfrak{H})^+. \tag{3.5}$$

This just means that each one of the positive Borel measures μ_ϕ above is bounded. In particular, the POV measure is said to be *normalized* if
$$a(X) = I \; (= \text{identity operator on } \mathfrak{H}). \tag{3.6}$$

Let $F : X \to \mathcal{L}(\mathfrak{H})^+$ be a weakly measurable function; i.e., for each $x \in X$, $F(x)$ is a bounded, positive operator, and, for arbitrary $\phi, \psi \in \mathfrak{H}$, the complex function $x \mapsto \langle \phi | F(x) \psi \rangle$ is measurable in $\mathcal{B}(X)$ [that is, the inverse image of any Borel subset of $\mathbb{C}$ belongs to $\mathcal{B}(X)$]. Using F and the regular Borel measure ν, we may define the regular POV measure a, such that for all $\Delta \in \mathcal{B}(X)$,
$$a(\Delta) = \int_\Delta F(x) \, d\nu(x). \tag{3.7}$$

The operator function F is then called a *density* for the POV measure a. We shall assume in all such cases that the support of the measure ν is all of X. (This is not a severe restriction on the class of measures used, since one can always delete the complement of the support of ν in X and work with a smaller space X', which is just the support of ν). Furthermore, we shall mainly be concerned with bounded POV measures with densities, such that the operator
$$A = \int_X F(x) \, d\nu(x) \tag{3.8}$$
has an inverse that is densely defined on $\mathfrak{H}$. (This technical requirement will ensure, in situations that we shall encounter, that using F it is possible to define a family of vectors that span $\mathfrak{H}$.) The operator A is sometimes referred to as the *resolution operator*. We will not, in general, assume A^{-1} to be bounded. The following special case, however, will be of particular interest to us. Assume that, for all $x \in X$, the operators $F(x)$ have the same finite rank n. Then, for each x, there exists a set of linearly independent

vectors, η_x^i, $i = 1, 2, \ldots, n$, in $\mathfrak{H}$ for which the map $x \mapsto \eta_x^i$ is measurable and

$$F(x) = \sum_{i=1}^{n} |\eta_x^i\rangle\langle\eta_x^i|, \qquad (3.9)$$

so that (3.8) can be rewritten as

$$A = \sum_{i=1}^{n} \int_X |\eta_x^i\rangle\langle\eta_x^i|\, d\nu(x). \qquad (3.10)$$

Definition 3.1.2 (Frame): *A set of vectors η_x^i, $i = 1, 2, \ldots, n$, $x \in X$, constitutes a rank-n frame, denoted $\mathcal{F}\{\eta_x^i, A, n\}$, if, for each $x \in X$, the vectors η_x^i, $i = 1, 2, \ldots, n$ are linearly independent, and if they satisfy (3.10), with both A and A^{-1} being bounded positive operators on $\mathfrak{H}$. If $A = \lambda I$, for some $\lambda > 0$, the frame is said to be* tight.

The frame condition (3.10) may alternatively be written in the following more familiar form:

$$\mathsf{m}(A)\|f\|^2 \leq \sum_{i=1}^{n} \int_X |\langle\eta_x^i|f\rangle|^2\, d\nu(x) \leq \mathsf{M}(A)\|f\|^2, \; \forall f \in \mathfrak{H}, \qquad (3.11)$$

where $\mathsf{m}(A)$ and $\mathsf{M}(A)$ denote, respectively, the infimum and the supremum of the spectrum of A. These numbers, which satisfy $\mathsf{m}(A) > 0$, $\mathsf{M}(A) < \infty$, are usually called the *frame bounds*. [Note: It is customary, in the literature, to denote frame bounds by the letters A and B, but this would conflict with our use of the letter A for the resolution operator (3.8).] In case the frame is tight, $\mathsf{m}(A) = \mathsf{M}(A)$. When $n = 1$, we occasionally write, for simplicity, $\mathcal{F}\{\eta_x, A\} \equiv \mathcal{F}\{\eta_x, A, 1\}$.

All this takes a more familiar shape if the space X is discrete, with ν a counting measure. Indeed the relation (3.10) then reads

$$A = \sum_{i=1}^{n} \sum_{x \in X} |\eta_x^i\rangle\langle\eta_x^i|, \qquad (3.12)$$

and the frame is discrete — the type usually encountered in the literature. We shall take up the notion of a discrete frame in detail in Section 3.4.

A POV measure a is said to be *commutative* if $[a(\Delta), a(\Delta')] = 0$, for all $\Delta, \Delta' \in \mathcal{B}(X)$, i.e., if all operators $a(\Delta)$, $\Delta \in \mathcal{B}(X)$, mutually commute. A POV measure $a = P$ is called a *projection-valued* (PV) measure if $P(\Delta)$ is a projection operator for each $\Delta \in \mathcal{B}(X)$. A PV measure is necessarily commutative.

Let a be a POV measure and μ be a positive Borel measure on X. We say that a is *smooth* with respect to μ if they have the same sets of measure zero; i.e., for any $\Delta \in \mathcal{B}(X)$, $a(\Delta) = 0$ implies $\mu(\Delta) = 0$ and vice versa. Given a bounded POV measure a, it is always possible to find a Borel measure μ with respect to which it is smooth. Indeed, let $\{\phi_n\}_{i=1}^{N}$ be an orthonormal basis for $\mathfrak{H}$. Then,

$$\mu(\Delta) = \sum_{n=1}^{N} \frac{1}{2^n} \langle \phi_n | a(\Delta) \phi_n \rangle \tag{3.13}$$

defines a positive, bounded Borel measure $[\mu(X) \leq \|a(X)\|]$. Clearly, $a(\Delta) = 0$ implies $\mu(\Delta) = 0$. Next, since each term in the sum on the right-hand side of (3.13) is positive, $\mu(\Delta) = 0$ implies $\langle \phi_n | a(\Delta) \phi_n \rangle = 0$ for all n. On the other hand, since $a(\Delta)$ is a positive operator, there exists $b \in \mathcal{L}(\mathfrak{H})$ for which $b^2 = a(\Delta)$. Thus, $\|b\phi_n\|^2 = 0$, for all n, implying that $b = 0$, whence $a(\Delta) = 0$, proving that a is smooth with respect to μ. Note that, if a is a normalized POV measure, that is, $a(X) = I$, then $\mu(X) = 1$ and μ becomes a probability measure.

Examples of POV measures

We have already seen examples of POV measures in Chapter 2. In fact (2.52) defines a POV measure, with density $F(q,p) = |\eta^s_{\sigma(q,p)}\rangle\langle\eta^s_{\sigma(q,p)}|$, on the Borel sets of the phase space Γ. This is an example of a noncommutative POV measure, and, indeed, we shall mostly be concerned with POV measures of this particular type in this book. Examples of commutative POV measures are also easily obtained. Let $\mathfrak{H} = L^2(\mathbb{R}, dx)$, and denote by χ_Δ the characteristic function of the set $\Delta \in \mathcal{B}(\mathbb{R})$:

$$\chi_\Delta(x) = \begin{cases} 1, & \text{if } x \in \Delta, \\ 0, & \text{otherwise}. \end{cases}$$

Then, the operators $P(\Delta)$, such that

$$(P(\Delta)\phi)(x) = \chi_\Delta(x)\phi(x), \qquad \phi \in L^2(\mathbb{R}, dx), \tag{3.14}$$

define a normalized PV measure (the above relation is assumed to hold for almost all $x \in \mathbb{R}$, with respect to the Lebesgue measure). More generally, let $f : \mathbb{R} \to \mathbb{R}^+$ be an integrable function satisfying

$$\int_\mathbb{R} f(x)\, dx = 1.$$

For arbitrary $\Delta \in \mathcal{B}(\mathbb{R})$, define the operator $a(\Delta)$ on $L^2(\mathbb{R}, dx)$ by

$$(a(\Delta)\phi)(x) = (\chi_\Delta * f)(x)\phi(x), \qquad \phi \in L^2(\mathbb{R}, dx). \tag{3.15}$$

Here, as in (2.86), $*$ denotes convolution:

$$(\chi_\Delta * f)(x) = \int_\mathbb{R} \chi_\Delta(x-y) f(y)\, dy = -\int_\Delta f(x-y)\, dy.$$

Then, it is easy to verify that a is a normalized, commutative POV measure.

For general normalized, commutative POV measures, there is a representation theorem [2, 3] that says that every such measure can in a sense be expressed in the above manner. To motivate this result, we start with an example. Let us rewrite (3.15) somewhat differently: For each $y \in \mathbb{R}$, define the PV measure P_y by

$$(P_y(\Delta)\phi)(x) = \chi_\Delta(x-y)\phi(x),$$

and let ν be the Borel measure

$$d\nu(x) = f(x)\, dx, \qquad \nu(\mathbb{R}) = 1.$$

Then, it is not hard to see that (3.15) may be reexpressed as

$$(a(\Delta)\phi)(x) = (\chi_\Delta * \nu)\phi(x),$$

and, as a weak integral, $a(\Delta)$ has the representation

$$a(\Delta) = \int_{\mathbb{R}} P_y(\Delta)\, d\nu(y). \tag{3.16}$$

This shows that the commutative POV measure a can be expressed as a *probability average* (since ν is a probability measure) over PV measures. It is this sort of a representation, as a probability average, which can be generalized to arbitrary regular, normalized, commutative POV measures. We briefly describe below the general situation.

As before, let $\mathfrak{H}$ be an abstract Hilbert space, X be a metrizable, locally compact space and a be a normalized, regular, commutative POV measure on $\mathcal{B}(X)$. Take the set of positive operators $\{a(\Delta) \mid \Delta \in \mathcal{B}(X)\}$ and form all possible finite products and complex linear combinations of these. Denote the resulting set by $\mathcal{S}$ and its commutant by $\mathcal{S}'$. This latter is the set of all operators in $\mathcal{L}(\mathfrak{H})$ that commute with every element of $\mathcal{S}$. The commutant of $\mathcal{S}'$ (i.e., the double commutant $\mathcal{S}''$ of $\mathcal{S}$), which we denote by $\mathfrak{A}(a)$, is then a von Neumann algebra and, hence, a weakly closed algebra of operators in $\mathcal{L}(\mathfrak{H})$. (Recall that a von Neumann algebra $\mathfrak{A}$ is a *-invariant set of bounded operators on $\mathfrak{H}$ that is an algebra under the operator product and equal to its double commutant, $\mathfrak{A} = \mathfrak{A}''$. Such an algebra is necessarily weakly closed [Dix1, Ta2]. We will have more to say about this in Chapter 8, Section 8.4.)

Consider now the set $\widehat{\mathfrak{S}}$ of all PV measures on $\mathcal{B}(X)$, the ranges of which lie in $\mathfrak{A}(a)$. Thus, the PV measure P is an element of $\widehat{\mathfrak{S}}$ if and only if $P(\Delta) \in \mathfrak{A}(a)$ for all $\Delta \in \mathcal{B}(X)$. There is a natural structure of a measure space on $\widehat{\mathfrak{S}}$ [3], so that one can define ordinary Borel measures on it. Let $\overline{\nu}$ be such a positive measure on $\widehat{\mathfrak{S}}$, with $\overline{\nu}(\widehat{\mathfrak{S}}) = 1$ (i.e., $\overline{\nu}$ is a probability measure on $\widehat{\mathfrak{S}}$). Then the operators $\overline{a}(\Delta), \Delta \in \mathcal{B}(X)$, defined through the integral representation

$$\overline{a}(\Delta) = \int_{\widehat{\mathfrak{S}}} P(\Delta)\, d\overline{\nu}(P), \quad \text{or, symbolically,} \quad \overline{a} = \int_{\widehat{\mathfrak{S}}} P\, d\overline{\nu}(P),$$

generate a normalized POV measure. Once again, the convergence of the above integral is in the weak topology of $\mathcal{L}(\mathfrak{H})$, meaning that, for arbitrary

$\phi, \psi \in \mathfrak{H}$,

$$\langle \phi | \bar{a}(\Delta) \psi \rangle = \int_{\widehat{\mathfrak{S}}} \langle \phi | P(\Delta) \psi \rangle \, d\bar{\nu}(P).$$

The interesting fact about the above representation of the POV measure $\bar{a}$, as a probability average over the PV measures P, is that $\bar{a}$ and $\bar{\nu}$ determine each other uniquely. Thus, there is a unique probability measure $\bar{\nu} = \nu$ for which $\bar{a} = a$. This result, which we state precisely below, is an example of an *extreme point representation* in the sense of Choquet [Phe].

Theorem 3.1.3: *If a is a normalized, regular, commutative POV measure on $\mathcal{B}(X)$, there is a unique probability Borel measure ν on $\widehat{\mathfrak{S}}$ such that*

$$a(\Delta) = \int_{\widehat{\mathfrak{S}}} P(\Delta) \, d\nu(P), \tag{3.17}$$

the integral converging weakly. The measure ν is unaltered if $\mathfrak{A}(a)$ is replaced by any other von Neumann algebra that contains $\mathfrak{A}(a)$.

For arbitrary POV measures, commutative or not, there is a general theorem, due to Naĭmark [227], which states that every such measure can be embedded in a PV measure on an enlarged Hilbert space. Applications of this theorem will show up in various contexts in later chapters, warranting a full statement of it here.

Theorem 3.1.4 (Naĭmark's extension theorem): *Let a be an arbitrary, normalized POV measure on $\mathcal{B}(X)$. Then there are a Hilbert space $\widetilde{\mathfrak{H}}$, an isometric embedding $W : \mathfrak{H} \to \widetilde{\mathfrak{H}}$ and a PV measure $\widetilde{P}$ on $\mathcal{B}(X)$, with values in $\mathcal{L}(\widetilde{\mathfrak{H}})$, such that*

$$W a(\Delta) W^{-1} = \mathbb{P} \widetilde{P}(\Delta) \mathbb{P}, \tag{3.18}$$

for all $\Delta \in \mathcal{B}(X)$, where $\mathbb{P}$ is the projection operator

$$\mathbb{P} \widetilde{\mathfrak{H}} = W \mathfrak{H}. \tag{3.19}$$

The extension $\{\widetilde{P}, \widetilde{\mathfrak{H}}\}$ of $\{a, \mathfrak{H}\}$ can be chosen to be minimal in the sense that the set of vectors

$$\mathcal{S} = \{\widetilde{P}(\Delta)\Phi \mid \Delta \in \mathcal{B}(X), \Phi \in W\mathfrak{H}\} \tag{3.20}$$

is dense in $\widetilde{\mathfrak{H}}$. This minimal extension is unique up to unitary equivalence.

(Note that, in (3.18), the inverse map W^{-1} is defined on the range of W in $\widetilde{\mathfrak{H}}$.)

For illustrative purposes, we work out the Naĭmark extensions for the cases in which a defines a frame, with $A = I$ [see (3.9)–(3.10)], and a is commutative, so that the extreme point representation of Theorem 3.1.3 is valid.

3.2 The case of a tight frame

For the tight frame $\mathcal{F}\{\eta_x^i, I, n\}$, consider the Hilbert space $\widetilde{\mathfrak{H}} = \mathbb{C}^n \otimes L^2(X, d\nu)$. This space consists of all ν-measurable functions $\Phi : X \to \mathbb{C}^n$, such that

$$\|\Phi\|^2 = \sum_{i=1}^n \int_X |\Phi^i(x)|^2 \, d\nu(x) < \infty,$$

$\Phi^i(x)$ denoting the ith component of $\Phi(x)$ in $\mathbb{C}^n$. Let $W : \mathfrak{H} \to \widetilde{\mathfrak{H}}$ be the linear map

$$(W\phi)^i(x) = \Phi^i(x) = \langle \eta_x^i | \phi \rangle, \quad i = 1, 2, \ldots, n, \quad x \in X. \tag{3.21}$$

Then, W is an isometry. Indeed,

$$\sum_{i=1}^n \int_X |\Phi^i(x)|^2 \, d\nu(x) = \sum_{i=1}^n \int_X \langle \phi | \eta_x^i \rangle_{\mathfrak{H}} \langle \eta_x^i | \phi \rangle_{\mathfrak{H}} \, d\nu(x),$$

which, in view of the weak convergence of the integral (3.10), implies that

$$\|\Phi\|_{\widetilde{\mathfrak{H}}}^2 = \|\phi\|_{\mathfrak{H}}^2. \tag{3.22}$$

(The notation $\| \cdots \|_{\mathfrak{K}}$ means that the norm is taken in the Hilbert space $\mathfrak{K}$. Similarly, $\langle \cdot | \cdot \rangle_{\mathfrak{K}}$ denotes the scalar product in $\mathfrak{K}$.) If Φ is an arbitrary vector in the range, $W\mathfrak{H}$, of W in $\widetilde{\mathfrak{H}}$, it is easily verified that the inverse map W^{-1} acts on it as

$$W^{-1}\Phi = \sum_{i=1}^n \int_X \Phi^i(x) \eta_x^i \, d\nu(x) \in \mathfrak{H}, \tag{3.23}$$

giving a *reconstruction formula* for $\phi = W^{-1}\Phi$, when the coefficients $\Phi^i(x) = \langle \eta_x^i | \phi \rangle$, $i = 1, 2, \ldots, n$, $x \in X$ are known. On $\widetilde{\mathfrak{H}}$, define the projection operators $\widetilde{P}(\Delta), \Delta \in \mathcal{B}(X)$, by

$$(\widetilde{P}(\Delta)\Phi)(x) = \chi_\Delta \Phi(x), \quad \Phi \in \widetilde{\mathfrak{H}}, \tag{3.24}$$

where χ_Δ is the characteristic function of the set Δ. Clearly $\widetilde{P}$ is a PV measure on $\mathcal{B}(X)$ with values in $\mathcal{L}(\widetilde{\mathfrak{H}})$. We then have the following result.

Theorem 3.2.1: *The PV measure $\widetilde{P}$ extends the POV measure,*

$$a(\Delta) = \sum_{i=1}^n \int_\Delta |\eta_x^i\rangle \langle \eta_x^i| \, d\nu(x), \quad \Delta \in \mathcal{B}(X), \tag{3.25}$$

minimally in the sense of Naĭmark.

Proof. Let $\Phi, \Phi' \in \widetilde{\mathfrak{H}}$ be arbitrary. Then there are vectors $\phi, \phi' \in \mathfrak{H}$ such that

$$\phi = W^{-1}\mathbb{P}\Phi, \quad \phi' = W^{-1}\mathbb{P}\Phi'.$$

Hence, using (3.24) and the definition of the scalar product on $\widetilde{\mathfrak{H}}$,

$$\begin{aligned}
\langle \Phi | \mathbb{P}\widetilde{P}(\Lambda)\mathbb{P}\Phi' \rangle_{\widetilde{\mathfrak{H}}} &= \sum_{i=1}^{n} \int_{X} \chi_{\Delta}(x) \overline{(W\phi)^{i}(x)} (W\phi')^{i}(x)\, d\nu(x) \\
&= \sum_{i=1}^{n} \int_{\Delta} \langle \phi | \eta_{x}^{i} \rangle_{\mathfrak{H}} \langle \eta_{x}^{i} | \phi' \rangle_{\mathfrak{H}}, \quad \text{by (3.21)} \\
&= \langle \phi | a(\Delta) \phi' \rangle_{\mathfrak{H}}, \quad \text{by (3.7) and (3.9)} \\
&= \langle W^{-1}\mathbb{P}\Phi | a(\Delta) W^{-1}\mathbb{P}\Phi' \rangle_{\mathfrak{H}}.
\end{aligned}$$

Since W is an isometry from $\mathfrak{H}$ to $W\mathfrak{H} \subset \widetilde{\mathfrak{H}}$, and Φ, Φ' are arbitrary, the last equality is immediately seen to imply (3.18). Thus, $\{\widetilde{P}, \widetilde{\mathfrak{H}}\}$ is an extension of $\{a, \mathfrak{H}\}$. To see that this extension is minimal, it is enough to prove that, if $\widehat{\Phi} \in \widetilde{\mathfrak{H}}$ is a vector for which $\langle \widehat{\Phi} | \widetilde{P}(\Delta)\Phi \rangle_{\widetilde{\mathfrak{H}}} = 0$, for all $\Delta \in \mathcal{B}(X)$ and all $\Phi \in W\mathfrak{H}$ [see (3.20)], then $\widehat{\Phi} = 0$. Now,

$$\langle \widehat{\Phi} | \widetilde{P}(\Delta)\Phi \rangle_{\widetilde{\mathfrak{H}}} = \sum_{i=1}^{n} \int_{\Delta} \overline{\widehat{\Phi}^{i}(x)} \Phi^{i}(x)\, d\nu(x),$$

and, hence, the vanishing of the left-hand side of this equation, for all $\Delta \in \mathcal{B}(X)$, implies that

$$\sum_{i=1}^{n} \overline{\widehat{\Phi}^{i}(x)} \Phi^{i}(x) = \sum_{i=1}^{n} \overline{\widehat{\Phi}^{i}(x)} \langle \eta_{x}^{i} | \phi \rangle = 0,$$

for almost all $x \in X$ (with respect to the measure ν), where $\phi = W^{-1}\Phi$. Thus, since $\phi \in \mathfrak{H}$ is arbitrary, this means that

$$\sum_{i=1}^{n} \widehat{\Phi}^{i}(x) \eta_{x}^{i} = 0,$$

for almost all $x \in X$. The linear independence of the η_{x}^{i}, $i = 1, 2, \ldots, n$, then shows that $\widehat{\Phi}^{i}(x) = 0$, for all i and almost all $x \in X$. Hence, $\Phi = 0$. □

3.3 Example: A commutative POV measure

To illustrate the same construction for the case in which a is commutative and has the representation (3.17), let $\widetilde{\mathfrak{H}} = \mathfrak{H} \otimes L^{2}(\widehat{\mathfrak{S}}, d\nu)$. This is the Hilbert space of all functions $\Phi : \widehat{\mathfrak{S}} \to \mathfrak{H}$ satisfying

$$\|\Phi\|_{\widetilde{\mathfrak{H}}}^{2} = \int_{\widehat{\mathfrak{S}}} \|\Phi(P)\|_{\mathfrak{H}}^{2}\, d\nu(P) < \infty,$$

and let $\mathbb{I} \in L^{2}(\widehat{\mathfrak{S}}, d\nu)$ be the function such that $\mathbb{I}(P) = 1$ for all $P \in \widehat{\mathfrak{S}}$. The linear map $W : \mathfrak{H} \to \widetilde{\mathfrak{H}}$ with

$$W\phi = \phi \otimes \mathbb{I}, \tag{3.26}$$

is an isometry, since
$$\nu(\widehat{\mathfrak{S}}) = \int_{\widehat{\mathfrak{S}}} \mathbb{I}(P)\, d\nu(P) = 1.$$

The projection operator $\mathbb{P}$ in (3.19) is now computed to be
$$\mathbb{P}\Phi = \left[\int_{\widehat{\mathfrak{S}}} \Phi(P)\, d\nu(P)\right] \otimes \mathbb{I}. \tag{3.27}$$

We define a PV measure $\widetilde{P}$ on $\mathcal{B}(X)$, with values in $\mathcal{L}(\widetilde{\mathfrak{H}})$, as
$$(\widetilde{P}(\Delta)\Phi)(P) = P(\Delta)\Phi(P), \quad \Phi \in \widetilde{\mathfrak{H}}, \quad \Delta \in \mathcal{B}(X), \tag{3.28}$$
the above relation being assumed to hold for almost all $P \in \widehat{\mathfrak{S}}$ (with respect to the measure ν). Once again, one can see that $\widetilde{P}$ extends a minimally, in the sense of Naĭmark. Indeed, to prove that $\widetilde{P}$ satisfies (3.18), note that for all $\phi \in \mathfrak{H}$ and $\Delta \in (B)\widehat{\mathfrak{S}}$,

$$\begin{aligned}
(\mathbb{P}\widetilde{P}(\Delta)\phi \otimes \mathbb{I})(P) &= \left[\int_{\widehat{\mathfrak{S}}} (\widetilde{P}(\Delta)\phi \otimes \mathbb{I})(P)\, d\nu(P)\right] \otimes \mathbb{I}), \quad \text{by (3.27)} \\
&= \left[\int_{\widehat{\mathfrak{S}}} P(\Delta)\phi\mathbb{I}(P)\, d\nu(P)\right] \otimes \mathbb{I}), \quad \text{by (3.28)} \\
&= (a(\Delta)\phi) \otimes \mathbb{I}, \quad \text{by (3.17)}
\end{aligned}$$

(as a weak integral). From this (3.18) is seen to follow. The proof of minimality uses a technical result on von Neumann algebras [2], and we omit it.

3.4 Discrete frames

As mentioned in Section 3.1, an important case for applications is that of a *discrete frame*, which corresponds to a discrete space X, equipped with a counting measure ν. The theory of discrete frames was originally developed in the 1950s in the context of nonharmonic Fourier series [113], and it regained prominence recently in the context of wavelet theory [Dau, 97, 167] (see Chapters 12 and 16). In view of their practical importance, it is fitting to discuss them here at some length.

A countable family of vectors $\{\psi_j\}$ in a Hilbert space $\mathfrak{H}$ is called a *(discrete) frame* if there are two positive constants m, M, with $0 < \mathsf{m} < \mathsf{M} < \infty$, such that
$$\mathsf{m}\,\|\phi\|^2 \leq \sum_{j=1}^{\infty} |\langle \psi_j | \phi \rangle|^2 \leq \mathsf{M}\,\|\phi\|^2, \quad \forall \phi \in \mathfrak{H}. \tag{3.29}$$

The two constants m, M are the *frame bounds*. If $\mathsf{m} = \mathsf{M} > 1$, the frame is said to be *tight*. Of course, if $\mathsf{m} = \mathsf{M} = 1$, and $\|\psi_j\| = 1, \forall j$, the set $\{\psi_j\}$ is simply an orthonormal basis.

3.4. Discrete frames

It should be clear that the definition of (discrete) frame just given coincides with that given in (3.12) above, namely,

$$A = \sum_j |\psi_j\rangle\langle\psi_j|, \tag{3.30}$$

that is,

$$\langle\phi|A\phi\rangle = \sum_j |\langle\psi_j|\phi\rangle|^2. \tag{3.31}$$

Thus, the frame condition (3.11), which says that A and A^{-1} are both bounded, coincides with (3.29), with frame bounds $\mathsf{m} = m(A)$, $\mathsf{M} = M(A)$, denoting, respectively, the infimum and the supremum of the spectrum of A. In the notation introduced above, the frame would be denoted as $\mathcal{F}\{\psi_j, A\}$.

The properties of a frame are best discussed in terms of the frame operator $F : \mathfrak{H} \to \ell^2$, defined by

$$F : \phi \mapsto \{\langle\psi_j|\phi\rangle\}.$$

Then one has $A = F^*F$ and the inequalities (3.29) may be written as

$$\mathsf{m}I \leq F^*F \leq \mathsf{M}I, \tag{3.32}$$

where I is the identity operator. This in turn implies that F^*F is invertible and

$$\mathsf{M}^{-1}I \leq (F^*F)^{-1} \leq \mathsf{m}^{-1}I. \tag{3.33}$$

Define now, for each $n \in \mathbb{N}$,

$$\widetilde{\psi}_j = A^{-1}\psi_j = (F^*F)^{-1}\psi_j, \tag{3.34}$$

so that $\psi_j = A\widetilde{\psi}_j$. Then the following theorem is true:

Theorem 3.4.1: *The vectors $\{\widetilde{\psi}_j\}$ constitute a frame, with frame bounds M^{-1}, m^{-1} and frame operator $\widetilde{F} = F(F^*F)^{-1}$. In addition, the expansion*

$$\phi(x) = \sum_{j=1}^{\infty} \langle\psi_j|\phi\rangle \, \widetilde{\psi}_j(x) \tag{3.35}$$

converges strongly in $\mathfrak{H}$, i.e., $\widetilde{F}^*F = I$.

Proof. That $\{\widetilde{\psi}_j\}$ is the frame described in the statement results from the equalities

$$\sum_j |\langle\widetilde{\psi}_j|\phi\rangle|^2 = \sum_j |\langle(F^*F)^{-1}\psi_j|\phi\rangle|^2$$
$$= \sum_j |\langle\psi_j|(F^*F)^{-1}\phi\rangle|^2$$
$$= \|F(F^*F)^{-1}\phi\|^2 = \langle\phi|(F^*F)^{-1}\phi\rangle$$

and the inequalities (3.29). Furthermore,
$$\widetilde{F}^*F = (F^*F)^{-1}F^*F = I,$$
that is, (3.35) is an identity. □

In other words, the duality between the two frames may be written as $\widetilde{F}^*F = F^*\widetilde{F} = I$ or, explicitly,
$$\sum_j |\psi_j\rangle\langle\widetilde{\psi}_j| = \sum_j |\widetilde{\psi}_j\rangle\langle\psi_j| = I. \tag{3.36}$$

The frame $\mathcal{F}\{\widetilde{\psi}_j, A^{-1}\}$ is called the *dual* or *reciprocal frame* of $\mathcal{F}\{\psi_j, A\}$. This notion is crucial for applications. In the case of wavelet expansions, it is the basis of the so-called biorthogonal scheme [Dau], briefly discussed in Section 13.1.3.

The important point here is that, for all practical purposes, a good frame is almost as good as an orthonormal basis. By 'good frame,' we mean that the expansion (3.35) converges sufficiently fast. How would one estimate the speed of this convergence? By (3.35), we need to compute $\widetilde{\psi}_j = (F^*F)^{-1}\psi_j$. If M and m are close to each other, $A = F^*F$ is close to $\frac{1}{2}(M+m)I$; hence, A^{-1} is close to $\frac{2}{M+m}I$, and, thus, $\widetilde{\psi}_j$ is close to $\frac{2}{M+m}\psi_j$. Hence, we may write
$$\phi = \frac{2}{M+m}\sum_j \langle\psi_j|\phi\rangle\psi_j + R\phi, \tag{3.37}$$
where
$$R = I - \frac{2}{M+m}A. \tag{3.38}$$
Therefore,
$$\begin{aligned}A^{-1} &= \frac{2}{M+m}(I-R)^{-1} \\ &= \frac{2}{M+m}\sum_{k=0}^{\infty} R^k.\end{aligned} \tag{3.39}$$

The series converges in norm, since, by (3.38),
$$-\frac{M-m}{M+m}I \leq R \leq \frac{M-m}{M+m}I, \tag{3.40}$$
which implies
$$\|R\| \leq \frac{M-m}{M+m} = \frac{M/m - 1}{M/m + 1}.$$

As a consequence, the expansion (3.35) converges as a power series in $|M/m - 1|$. Thus, the frame is good if $|M/m - 1| \ll 1$, in particular, if it is tight. To the first order, the expansion (3.35) becomes

$$\phi = \frac{2}{M+m} \sum_j \langle \psi_j | \phi \rangle \psi_j. \tag{3.41}$$

The quantity

$$w(\mathcal{F}) = \frac{M-m}{M+m} \tag{3.42}$$

is called the *width* or the *snugness* of the frame $\mathcal{F}$. It measures the lack of tightness, since $w(\mathcal{F}) = 0$ iff the frame $\mathcal{F}$ is tight. Notice that a frame and its dual have the same width. More details on frames may be found in [Dau],[98], and [167].

An interesting question of *discretization* now arises, namely, what is the connection between continuous frames and discrete ones? As we shall see in Chapter 12, Section 12.5, the practical implementation of the wavelet transform (a special case of coherent states) in signal processing requires the selection of a discrete set of points in the transform space. Indeed, all formulas must generally be evaluated numerically, and a computer is an intrinsically discrete object (even finite!). But this operation must be performed in such a way that no information is lost. This requirement then immediately leads to the determination of a discrete frame. The same situation actually prevails for any CS transform, of the type we shall develop at length in this book. Putting the question in general terms, it now reads: Given a continuous frame $\{\psi_x, A_c, x \in X\}$, based on a general space X, can one find a discrete set of points $\{x_j \in X, j \in J\}$ such that $\{\psi_{x_j}, A_d, j \in J\}$ is a discrete frame, possibly with a different width? Moreover, *a priori* the two operators A_c, A_d need not coincide.

Besides its obvious mathematical interest, a positive answer to this question is crucial for practical applications of CS. In view of its importance, we shall devote Chapter 16 to the discretization problem, listing on the way the explicit results that are known in particular examples.

4
Some Group Theory

In this chapter, we introduce a few concepts from the theory of groups, Lie algebras, transformation spaces, and group representations, presenting them in a form and notation adapted to the aims of this book. (A good source for more detailed information is, for example, [Bar].)

4.1 Homogeneous spaces, quasi-invariant, and invariant measures

Let G be a locally compact (metrizable) group — in fact, most of the time, we shall take it to be a Lie group (additional information, specific to Lie groups and Lie algebras, is given in Section 4.5). Suppose that X is a *transformation space* for G, also called a *G-space*. This means that there is defined on X an action (we shall only consider a left action) of the group: $G \times X \to X$, to be written $(g, x) \mapsto gx$ and assumed to be continuous in the topologies of G and X. If G is a Lie group, X will be taken to be a manifold and the above action assumed *smooth*, i.e., infinitely differentiable. We shall mostly be concerned with the case in which the action of G on X is *transitive*, which means that given $x, y \in X$, it is always possible to find $g \in G$, which solves the equation $y = gx$. The space X will then be called a *transitive G-space* or a *homogeneous space*. The *stability subgroup* of a point $x \in X$ is the set

$$H_x = \{g \in G | gx = x\}. \tag{4.1}$$

This is a closed subgroup of G. As an example of a G-space, let H be a closed subgroup of G, and consider the left coset space G/H. Elements of G/H are of the form gH, for $g \in G$, and G itself has the transitive left action $gH \mapsto g'gH$ (for arbitrary $g' \in G$) on G/H. If G is a Lie group and H is a Lie subgroup of G, then G/H has a natural structure of a manifold and the above action is smooth. In fact, such a transformation space is the generic example of a homogeneous space. Indeed, let X be any homogeneous space of G, and let $x_0 \in X$ be fixed. Denote by H the stability subgroup of x_0. Since G acts transitively on X, for any $x \in X$, we can find $g \in G$, for which $x = gx_0$. The identification $x \mapsto gH$ is then a homeomorphism (diffeomorphism, if G is a Lie group) between X and G/H. Let X be a transformation space for G and $x \in G$. The *orbit* of x under G is the set

$$Gx = \{y = gx | g \in G\} \subset X. \tag{4.2}$$

If X is a homogeneous space, it corresponds to a single orbit under G.

The group G carries a left and a right invariant Haar measure. We shall denote the left Haar measure by μ and use it systematically. The right Haar measure, when used, will be denoted by μ_r. If $\mu = \mu_r$, the group is called *unimodular*. In general, μ and μ_r are different but equivalent measures; i.e., they have the same null sets. Thus, there exists a measurable function $\boldsymbol{\Delta} : G \to \mathbb{R}^+$, such that

$$d\mu(g) = \boldsymbol{\Delta}(g) d\mu_r(g). \tag{4.3}$$

This function, called the *modular function* of the group, is an $\mathbb{R}^+$-valued character satisfying, for μ-almost all $g, g_1, g_2 \in G$,

$$\left. \begin{array}{rl} \boldsymbol{\Delta}(g) &> 0, \\ \boldsymbol{\Delta}(e) &= 1, \quad e = \text{identity element of } G, \\ \boldsymbol{\Delta}(g_1 g_2) &= \boldsymbol{\Delta}(g_1)\boldsymbol{\Delta}(g_2). \end{array} \right\} \tag{4.4}$$

Furthermore, for μ-almost all $g, g' \in G$, the following relations hold:

$$\left. \begin{array}{rl} d\mu_r(g) &= \boldsymbol{\Delta}(g^{-1})\, d\mu(g) = d\mu(g^{-1}), \\ d\mu(gg') &= \boldsymbol{\Delta}(g')\, d\mu(g). \end{array} \right\} \tag{4.5}$$

While the group itself always carries a left (and a right) invariant Haar measure, the homogeneous space X need not carry any measure invariant under the action $x \mapsto gx$. *Quasi-invariant* measures, however, always exist on X. The measure ν on X is said to be quasi-invariant if ν and ν_g are equivalent measures, for all $g \in G$, where ν_g is defined to be the measure obtained by the natural action of g on ν:

$$\nu_g(\Delta) = \nu(g\Delta), \quad \Delta \in \mathcal{B}(X). \tag{4.6}$$

The Radon–Nikodym derivative of ν_g with respect to ν,

$$\lambda(g, x) = \frac{d\nu_g(x)}{d\nu(x)}, \tag{4.7}$$

4.1. Homogeneous spaces, quasi-invariant, and invariant measures

is then a *cocycle*, $\lambda : G \times X \to \mathbb{R}^+$, with the properties

$$\left.\begin{array}{rcl}\lambda(g_1 g_2, x) & = & \lambda(g_1, x)\lambda(g_2, g_1^{-1}x), \\ \lambda(e, x) & = & 1,\end{array}\right\} \quad (4.8)$$

the above equations holding for μ-almost all $g_1, g_2 \in G$, and ν-almost all $x \in X$. (Note that all the measures ν_g, $g \in G$, belong to the same measure class; i.e., they all have the same measure-zero sets.) Given a cocycle such as above, however, it is always possible to find a strict cocycle [i.e., one for which the conditions (4.8) hold for all $g_1, g_2 \in G$, and all $x \in X$], which is equal to the given cocycle almost everywhere. We shall always assume that such a strict cocycle has been chosen.

Let H be a closed subgroup of G, and write $X = G/H$ and $\pi : G \to X$, the *canonical surjection* (or projection map), $\pi(g) = gH$. A *section* on X is a map $\sigma : X \to G$, satisfying $\pi(\sigma(x)) = x$, for all $x \in X$. While it is always possible to find sections that are Borel maps, this is not the case if one insists on smooth sections (in the case in which G is Lie group). On the other hand, *local smooth sections* always exist. This means that, given $x \in X$, it is always possible to find an open set $U \subset X$, with $x \in U$, and a smooth section $\sigma_U : U \to G$. (It ought to be noted, however, that U can be quite arbitrary.) More than that, if X is a connected manifold, one may assume the open set $U \subset X$ to be *dense* in X. (Note that $X \setminus U$ is then a set of ν-measure zero.)

A simple example (♣)

Let us illustrate, with a simple example, the various measures appearing on a group and on homogeneous spaces. Consider the group G_{ut} of all 2×2 real, upper triangular matrices with nonzero determinant:

$$g = \begin{pmatrix} x & y \\ 0 & z \end{pmatrix}, \quad x, y, z \in \mathbb{R}, \quad xz \neq 0. \quad (4.9)$$

The left and right invariant Haar measures for G_{ut} can be easily computed as

$$d\mu(g) = \frac{dx\, dy\, dz}{|x^2 z|}, \quad d\mu_r(g) = \frac{dx\, dy\, dz}{|xz^2|}, \quad (4.10)$$

from which we find that

$$\Delta(g) = \left|\frac{z}{x}\right|. \quad (4.11)$$

Now, let H_1 be the subgroup of G_{ut}, which consists of all elements of the type:

$$h = \begin{pmatrix} x & 0 \\ 0 & z \end{pmatrix}. \quad (4.12)$$

50   4. Some Group Theory

Since an arbitrary $g \in G_{ut}$ can be written as

$$g = \begin{pmatrix} x & y \\ 0 & z \end{pmatrix} = \begin{pmatrix} 1 & \frac{y}{z} \\ 0 & 1 \end{pmatrix} \begin{pmatrix} x & 0 \\ 0 & z \end{pmatrix},$$

we see that the coset space G_{ut}/H_1 can be parametrized by a single variable $u \, (= \frac{y}{z}) \in \mathbb{R}$. Alternatively, an element in G_{ut}/H_1 can be represented by a matrix

$$\mathfrak{u} = \begin{pmatrix} 1 & u \\ 0 & 1 \end{pmatrix} \in G_{ut}. \tag{4.13}$$

To obtain the transformation properties of $\mathfrak{u}$ under the action of $g \in G_{ut}$, note that

$$g\mathfrak{u} = \begin{pmatrix} 1 & \frac{xu+y}{z} \\ 0 & 1 \end{pmatrix} \begin{pmatrix} x & 0 \\ 0 & z \end{pmatrix}.$$

This gives the transformation rule,

$$u \mapsto gu = \frac{xu+y}{z}.$$

Thus, the Lebesgue measure du is *quasi-invariant* on $G_{ut}/H_1 \simeq \mathbb{R}$, with

$$d(g^{-1}u) = \lambda(g, u) \, du = \frac{z}{x} \, du. \tag{4.14}$$

It is also easy to see that no *invariant* measure exists on G_{ut}/H_1. Consider, however, the subgroup H_2 of G_{ut}, which consists of all matrices of the type

$$h = \begin{pmatrix} 1 & y \\ 0 & 1 \end{pmatrix}, \qquad y \in \mathbb{R}. \tag{4.15}$$

Then proceeding as before, it can be seen that the coset space G_{ut}/H_2 is identifiable with all matrices of the type

$$\mathfrak{v} = \begin{pmatrix} x & 0 \\ 0 & y \end{pmatrix} \in G_{ut}.$$

Since this is topologically isomorphic to $\mathbb{R}^2$, with the origin taken out, we find that G_{ut}/H_2 carries the *invariant* measure,

$$d\nu(x,y) = \frac{dx \, dy}{|xy|}. \tag{4.16}$$

Finally, note that, when we chose to represent an element of G_{ut}/H_1 by the matrix $\mathfrak{u}$ in (4.13), we actually made the association $\mathfrak{u}H_1 \mapsto \mathfrak{u}$; in other words, we chose a section

$$\sigma_0 : G_{ut}/H_1 \simeq \mathbb{R} \to G_{ut}, \qquad \sigma_0(u) = \begin{pmatrix} 1 & u \\ 0 & 1 \end{pmatrix}. \tag{4.17}$$

An entire class of other sections σ may then be obtained from σ_0 by writing

$$\sigma(u) = \sigma_0(u) \begin{pmatrix} f_1(u) & 0 \\ 0 & f_2(u) \end{pmatrix} = \begin{pmatrix} f_1(u) & u \\ 0 & f_2(u) \end{pmatrix}, \tag{4.18}$$

4.1. Homogeneous spaces, quasi-invariant, and invariant measures 51

where f_1 and f_2 are real-valued, nowhere vanishing Borel functions.

Returning to the general discussion, it is possible, using a global Borel section σ and a quasi-invariant measure ν on the homogeneous space $X = G/H$, to construct another quasi-invariant measure ν_σ, which in a sense is the standard quasi-invariant measure for the chosen section. Indeed, using a strict cocycle λ, let us write

$$d\nu_\sigma(x) = \lambda(\sigma(x), x) \, d\nu(x). \qquad (4.19)$$

The measure ν_σ inherits interesting properties from ν_g. Indeed, since σ is a Borel section, λ is a Borel function, and thus ν_σ is a Borel measure. Also, let $x_0 \in X$ be the point that is stable under H, so that

$$x = \sigma(x)x_0. \qquad (4.20)$$

Furthermore, if $f : X \to \mathbb{C}$ is an arbitrary Borel function, then the defining relation (4.6) implies, for all $g \in G$,

$$\int_X f(x) \, d\nu_g(x) = \int_X f(gx) \, d\nu(x) = \int_X f(x)\lambda(g, x) \, d\nu(x)$$
$$= \int_X f(x) \, d\nu(g^{-1}x). \qquad (4.21)$$

Alternatively,

$$d\nu_g(x) = \lambda(g, x) \, d\nu(x) = d\nu(g^{-1}x). \qquad (4.22)$$

From these relations, we may now derive the following properties of ν_σ:

1. ν_σ is a quasi-invariant measure. Indeed, this follows immediately from (4.6) and the fact that $\lambda(g, x) > 0$, for all g, x. Also, from (4.6) and (4.22),

$$d\nu_\sigma(g^{-1}x) = \lambda(\sigma(g^{-1}x), x)\lambda(g, x) \, d\nu(x). \qquad (4.23)$$

2. If X admits an (left) invariant measure m, then ν_σ is a scalar multiple of m. Indeed, in this case one can find a function f, which can be taken to be positive everywhere, such that the quasi-invariant measure ν satisfies the relation $d\nu(x) = f(x) \, dm(x)$. Thus,

$$d\nu_g(x) = d\nu(g^{-1}x) = f(g^{-1}x) \, dm(x),$$

implying that

$$\lambda(g, x) = \frac{d\nu_g(x)}{d\nu(x)} = \frac{f(g^{-1}x)}{f(x)}.$$

Setting $g = \sigma(x)$, and using (4.20),

$$\lambda(\sigma(x), x) = \frac{f(x_0)}{f(x)},$$

so that
$$d\nu_\sigma(x) = f(x_0)\, dm(x). \tag{4.24}$$

3. ν_σ is independent of the quasi-invariant measure ν used to define it. To see this, let $\tilde{\nu}$ be another quasi-invariant measure and set
$$d\tilde{\nu}_\sigma = \tilde{\lambda}(\sigma(x), x)\, d\tilde{\nu}(x), \qquad \tilde{\lambda}(\sigma(x), x) = \frac{d\tilde{\nu}_g(x)}{d\tilde{\nu}(x)}.$$

Since both ν and $\tilde{\nu}$ are quasi-invariant measures, there is a Borel function $f : X \to \mathbb{R}^+$, $f(x) > 0$, for all $x \in X$, such that
$$d\tilde{\nu}(x) = f(x)\, d\nu(x),$$
and, hence,
$$d\tilde{\nu}_g(x) = d\tilde{\nu}(g^{-1}x) = f(g^{-1}x)\, d\nu(g^{-1}x).$$

Thus,
$$\tilde{\lambda}(g, x) = \frac{d\tilde{\nu}(g^{-1}x)}{d\tilde{\nu}(x)} = \frac{f(g^{-1}x)}{f(x)} \lambda(g, x).$$

(This relation has a natural meaning in the language of cohomology [Asc, Hil]; namely, λ and $\tilde{\lambda}$ are related by a *coboundary* and, therefore, are *cohomologically equivalent*.) This implies that
$$\tilde{\lambda}(\sigma(x), x)\, d\tilde{\nu}(x) = f(x_0)\lambda(\sigma(x), x)\, d\nu(x);$$
that is,
$$\tilde{\nu}_\sigma = f(x_0)\nu_\sigma,$$
and $f(x_0)$ is a constant.

4. The measure $[\nu_\sigma]_\sigma$, formed by applying the prescription (4.19) to ν_σ is a constant multiple of ν_σ. Since ν_σ is a quasi-invariant measure, this assertion is a consequence of the previous result. To prove it directly, let us write
$$\frac{d\nu_\sigma(g^{-1}x)}{d\nu_\sigma(x)} = \lambda_\sigma(g, x).$$

Then,
$$d[\nu_\sigma]_\sigma(x) = \lambda_\sigma(\sigma(x), x)\, d\nu_\sigma(x).$$

But,
$$\lambda_\sigma(g, x) = \frac{d\nu_\sigma(g^{-1}x)}{d\nu_\sigma(x)} = \frac{\lambda(\sigma(g^{-1}x), g^{-1}x)}{\lambda(\sigma(x), x)} \cdot \frac{d\nu(g^{-1}x)}{d\nu(x)},$$
implying that
$$d[\nu_\sigma]_\sigma(x) = \frac{\lambda(\sigma(\sigma(x)^{-1}x), \sigma(x)^{-1}x)}{\lambda(\sigma(x), x)} \lambda(\sigma(x), x)\, d\nu_\sigma(x)$$

4.1. Homogeneous spaces, quasi-invariant, and invariant measures   53

$$= \lambda(\sigma(x_0), x_0) \, d\nu_\sigma(x), \qquad \text{by (4.20)}.$$

Since $\sigma(x_0) \in H$ is fixed, the result follows.

The above results show that, once the section σ is fixed, the "natural" quasi-invariant measure to use on X is ν_σ. In fact, this is the measure we shall use in Chapter 7, Section 7.1, to give a general definition of coherent states.

An example using the affine group (♣)

As an example of an explicit construction of the measure ν_σ, consider the connected affine group G_+, also called the $ax+b$ group, consisting of transformations of $\mathbb{R}$ of the type $x \mapsto ax + b$, $x \in \mathbb{R}$, where $a > 0$, $b \in \mathbb{R}$. Writing

$$g = (b, a) \in G_+, \tag{4.25}$$

one has the multiplication law

$$g_1 g_2 = (b_1 + a_1 b_2, a_1 a_2). \tag{4.26}$$

Consider the subgroup H of G_+,

$$H = \{g \in G_+ \mid g = (0, a), \ a \in \mathbb{R}^+\}. \tag{4.27}$$

Then, $G_+/H \simeq \mathbb{R}$, since, for $(b, a) \in G_+$,

$$(b, a) = (b, 1)(0, a), \qquad b \in \mathbb{R}. \tag{4.28}$$

Also, since, for $x \in \mathbb{R}$,

$$(b, a)(x, 1) = (ax + b, 1)(0, a), \tag{4.29}$$

the action of G_+ on the coset space G_+/H, parametrized globally by points in $\mathbb{R}$, can be written as

$$gx = ax + b, \qquad g = (b, a) \in G_+. \tag{4.30}$$

Thus, $\mathbb{R}$ is a homogeneous space for G_+; however, there is no invariant measure on $\mathbb{R}$ under the action (4.30). On the other hand, for any Borel function $\rho : \mathbb{R} \to \mathbb{R}^+$, $\rho(x) \neq 0$ (for all x), the measure

$$d\nu(x) = \rho(x) \, dx \tag{4.31}$$

is quasi-invariant. Indeed, for $g = (a, b) \in G_+$,

$$d\nu_g(x) = d\nu(g^{-1}x) = \rho\left(\frac{x}{a} - \frac{b}{a}\right) \frac{1}{a} \, dx, \tag{4.32}$$

so that

$$\lambda(g, x) = \frac{d\nu_g(x)}{d\nu(x)} = \frac{\rho\left(\frac{x}{a} - \frac{b}{a}\right)}{a\rho(x)}. \tag{4.33}$$

54   4. Some Group Theory

A (global) section $\sigma_0 : \mathbb{R} \to G_+$ is now defined by

$$\sigma_0(x) = (x, 1), \tag{4.34}$$

and, in view of (4.28), any other section $\sigma : \mathbb{R} \to G_+$ can be expressed in terms of σ_0 as

$$\sigma(x) = \sigma_0(x)(0, f(x)) = (x, f(x)), \tag{4.35}$$

where $f : \mathbb{R} \to \mathbb{R}^+$, $f(x) \neq 0$ (for all x). Using such a general section, we compute the measure ν_σ to be [see (4.33)]

$$d\nu_\sigma(x) = \lambda(\sigma(x), x) \, d\nu(x) = \frac{\rho(0)}{f(x)} \, dx. \tag{4.36}$$

Note that this also explicitly demonstrates the independence of ν_σ of the quasi-invariant measure ν in (4.31) with which we started. The measure ν_σ does depend, however, on σ, through f. In fact, choosing $f(x) = \rho(0)/\rho(x)$, we may actually make $\nu_\sigma = \nu$.

4.2 Induced representations and systems of covariance

The inducing construction of Mackey [210] is a method for obtaining unitary representations of locally compact groups starting from known representations of subgroups. We present here a very brief account of the construction of such a representation, mainly to set up the notation and terminology that we shall be using, and later in Chapter 10, Section 10.2.4, we shall give a more detailed account of the inducing construction for semi-direct product type of groups. (We assume that the reader has at least a basic idea of the representation theory of locally compact groups and in particular the theory of induced representations. Detailed and highly readable accounts of the latter may be found in, for example, [Kir],[Lip], or [Var].)

Suppose that the group G has a representation by unitary operators $U(g)$, $g \in G$, on a Hilbert space $\mathfrak{H}$. We shall consider strongly continuous, unitary representations only. Strong continuity means that, for all $\phi \in \mathfrak{H}$,

$$\|(U(g) - I)\phi\| \to 0 \text{ as } g \to e.$$

The method of induced representations constructs such strongly continuous representations of G, starting from strongly continuous unitary representations of subgroups of G. The following is an outline of the procedure, adapted to our special needs.

Let H be a closed subgroup of G and $X = G/H$, let ν be a quasi-invariant measure on X, and $\lambda(g, \cdot)$ be the Radon–Nikodym derivative of the transformed measure ν_g, $g \in G$ with respect to ν, as defined in (4.7).

4.2. Induced representations and systems of covariance

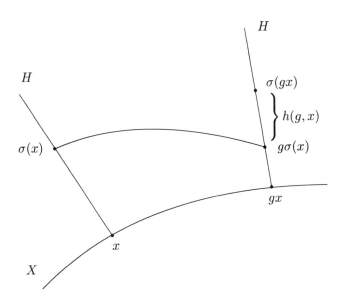

Figure 4.1. Origin of the cocycle $h(g,x)$.

Fix a Borel section $\sigma : X \to G$. For $g \in G$ and $x \in X$, we may write

$$g\sigma(x) = \sigma(gx)h(g,x), \quad \text{where} \quad h(g,x) = \sigma(gx)^{-1}g\sigma(x) \in H. \quad (4.37)$$

Here, $h : G \times X \to H$ is again a cocycle, with $h'(g,x) = [h(g^{-1},x)]^{-1}$ satisfying conditions similar to (4.8):

$$\left.\begin{array}{rcl} h'(g_1 g_2, x) &=& h'(g_1,x)h'(g_2, g_1^{-1}x), \\ h'(e,x) &=& e, \end{array}\right\} \quad (4.38)$$

for all $g_1, g_2 \in G$ and all $x \in X$ [see comment following (4.8) and Figure 4.1]. Suppose that H has a strongly continuous unitary representation $h \mapsto V(h)$, $h \in H$, on a Hilbert space $\mathfrak{K}$. Denoting by $\mathcal{U}(\mathfrak{K})$ the group of all unitary operators on $\mathfrak{K}$, define the map $B : G \times X \to \mathcal{U}(\mathfrak{K})$ by

$$B(g,x) = [\lambda(g,x)]^{\frac{1}{2}} V(h(g^{-1},x))^{-1}. \quad (4.39)$$

It is not hard to see that B also satisfies cocycle conditions for all $g_1, g_2 \in G$ and $x \in X$ (again, after adjustment on null sets, if necessary):

$$\left.\begin{array}{rcl} B(g_1 g_2, x) &=& B(g_1,x)B(g_2, g_1^{-1}x), \\ B(e,x) &=& I_{\mathfrak{K}} \; (= \text{identity operator on } \mathfrak{K}). \end{array}\right\} \quad (4.40)$$

Consider next the Hilbert space $\widetilde{\mathfrak{H}} = \mathfrak{K} \otimes L^2(X, d\nu)$, of functions $\Phi : X \to \mathfrak{K}$, which are square integrable in the norm

$$\|\Phi\|_{\widetilde{\mathfrak{H}}}^2 = \int_X \|\Phi(x)\|_{\mathfrak{K}}^2 \, d\nu(x).$$

It is straightforward to verify that the operators $\widetilde{U}(g)$, $g \in G$, defined by

$$(\widetilde{U}(g)\Phi)(x) = B(g, x)\Phi(g^{-1}x), \qquad \Phi \in \widetilde{\mathfrak{H}}, \tag{4.41}$$

are unitary on $\widetilde{\mathfrak{H}}$. Moreover, they define a strongly continuous representation of G. The representation $g \mapsto \widetilde{U}(g)$ so constructed is called the representation of G *induced* from the representation V of the subgroup H. In general, $\widetilde{U}$ is not irreducible, even when V is irreducible. The specific choice of the Borel section σ and the quasi-invariant measure ν is inconsequential, since a different choice leads to a unitarily equivalent representation.

There is an invariant formulation of the inducing construction, which avoids the use of the section σ and, in fact, is the form in which the construction appears most often in the mathematical literature. We briefly mention this construction, since we shall be using it also in the sequel. Let us denote by $[g]$, the element $x = gH$ in the coset space $X = G/H$, and consider the Hilbert space $L^2_{\mathfrak{K}}(X, d\nu)$, of all measurable functions $\mathbf{f} : G \to \mathfrak{K}$, which satisfy the condition,

$$\mathbf{f}(gh) = V(h)^{-1}\mathbf{f}(g), \qquad g \in G, \quad h \in H, \tag{4.42}$$

and are square integrable in the sense that

$$\int_X \|\mathbf{f}(g)\|_{\mathfrak{K}}^2 \, d\nu([g]) < \infty. \tag{4.43}$$

It is easily verified that, in view of (4.42), the integrand in (4.43) is indeed a function over X, so that the integral is well defined, as is also the corresponding scalar product. Now define a linear map, $\mathbf{W} : \widetilde{\mathfrak{H}} = \mathfrak{K} \otimes L^2(X, d\nu) \to L^2_{\mathfrak{K}}(X, d\nu)$, by

$$(\mathbf{W}\Phi)(g) = \mathbf{f}(g) = V(h)^{-1}\Phi(x), \qquad x = [g] \in X, \tag{4.44}$$

where $h \in H$ is determined from $g = \sigma([g])h$. (Note that, for given $g \in G$ and a fixed section σ, this decomposition is unique.) It is straightforward to verify that this map is onto and, in fact, unitary. The image of the induced representation $\widetilde{U}$, on $L^2_{\mathfrak{K}}(X, d\nu)$ under this unitary map, then has a simple covariant form. Indeed, writing $^V U(g) = \mathbf{W}\widetilde{U}(g)\mathbf{W}^{-1}$, a straightforward computation, using (4.37), (4.39), and (4.41), yields

$$(^V U(g)\, \mathbf{f})(g') = \lambda(g, x)^{\frac{1}{2}} \, \mathbf{f}(g^{-1}g'), \tag{4.45}$$

λ being the cocycle defined in (4.7).

Every induced representation has a canonically associated PV measure, which in a sense determines the representation. To see this, consider the following PV measure $\tilde{P}$, defined on the Borel sets of the homogeneous space X:

$$(\tilde{P}(\Delta)\Phi)(x) = \chi_\Delta(x)\Phi(x). \tag{4.46}$$

The operators $\tilde{P}(\Delta)$ transform covariantly under the action of the $\tilde{U}(g)$:

$$\tilde{U}(g)\tilde{P}(\Delta)\tilde{U}(g)^* = \tilde{P}(g\Delta), \tag{4.47}$$

where $g\Delta$ is the transform of the set Δ under the natural action of G on X. A pair $\{\tilde{U},\tilde{P}\}$ satisfying (4.46) is called a *system of imprimitivity* based on X. Thus, associated to an induced representation, there is always a canonical system of imprimitivity, and, as asserted in Theorem 4.2.2, this system of imprimitivity actually determines the representation. Before coming to that, however, it will be useful to also consider the following, more general concept:

Definition 4.2.1 (System of covariance): *A system of covariance based on X is a pair $\{U,a\}$, consisting of a unitary representation U of a locally compact group G, and a normalized POV measure a, defined on the Borel sets Δ of a G-space X, such that the relation*

$$U(g)a(\Delta)U(g)^* = a(g\Delta) \tag{4.48}$$

holds for all $g \in G$ and $\Delta \in \mathcal{B}(X)$. In particular, if $a = P$ is a PV measure, the system of covariance is called a system of imprimitivity. The system $\{U,a\}$ is called a transitive system of covariance, if the action of G on X is transitive.

As mentioned earlier, a system of imprimitivity is determinative of an induced representation. The precise sense in which this is to be understood is brought out by the next theorem, due to Mackey [210].

Theorem 4.2.2 (Mackey's imprimitivity theorem): *Let $\{U,P\}$ be a transitive system of imprimitivity, based on the homogeneous space X of the locally compact group G. Then, there exist a closed subgroup H of G, a Hilbert space $\mathfrak{K}$, and a continuous unitary representation V of H on $\mathfrak{K}$, such that the given system is unitarily equivalent to the canonical system of imprimitivity $\{\tilde{U},\tilde{P}\}$, arising from the representation $\tilde{U}$ of G induced from V.*

Explicitly, the above theorem states that, if $\mathfrak{H}$ is the Hilbert space of the representation U, and $\tilde{\mathfrak{H}}$ that of the induced representation $\tilde{U}$, then there exists a unitary map $W : \mathfrak{H} \to \tilde{\mathfrak{H}}$, such that

$$\left.\begin{array}{rl} WU(g)W^{-1} &= \tilde{U}(g), \quad g \in G, \\ WP(\Delta)W^{-1} &= \tilde{P}(\Delta), \quad \Delta \in \mathcal{B}(X). \end{array}\right\} \tag{4.49}$$

While a similar result does not hold for POV measures, as a consequence of Naĭmark's extension theorem (Theorem 3.1.4), it is possible to show that every transitive system of covariance can always be embedded into a system of imprimitivity. Indeed, we have the following result [87, 230, 267]:

Theorem 4.2.3: *If $\{U, a\}$ is a transitive system of covariance, then U is a subrepresentation of an induced representation.*

Proof. Let $\{U, a\}$ act on the Hilbert space $\mathfrak{H}$. Extend a to $\widetilde{P}$ on the enlarged Hilbert space $\widetilde{\mathfrak{H}}$ using Theorem 3.1.4, and let W and $\mathbb{P}$ be as in Naĭmark's theorem [see (3.18) and (3.19)]. To extend $U(g)$ to a unitary operator $\widetilde{U}(g)$ on $\widetilde{\mathfrak{H}}$, we proceed as follows: Since the set of vectors

$$\mathcal{S} = \{\widetilde{P}(\Delta)\Phi \mid \Delta \in \mathcal{B}(X), \Phi \in W\mathfrak{H}\}$$

[see (3.20)] is dense $\widetilde{\mathfrak{H}}$, it is enough to describe the action of $\widetilde{U}(g)$ on this set. Define

$$\widetilde{U}(g)\widetilde{P}(\Delta)\Phi = \widetilde{P}(g\Delta)WU(g)W^{-1}\Phi, \quad \Delta \in \mathcal{B}(X), \quad \Phi \in W\mathfrak{H}. \quad (4.50)$$

Then,

$$\begin{aligned}
\|\widetilde{U}(g)\widetilde{P}(\Delta)\Phi\|_{\widetilde{\mathfrak{H}}}^2 &= \langle \widetilde{P}(g\Delta)WU(g)W^{-1}\Phi | \widetilde{P}(g\Delta)WU(g)W^{-1}\Phi\rangle_{\widetilde{\mathfrak{H}}} \\
&= \langle WU(g)W^{-1}\Phi | \widetilde{P}(g\Delta)WU(g)W^{-1}\Phi\rangle_{\widetilde{\mathfrak{H}}}, \\
&\quad \text{since } \widetilde{P}(g\Delta) \text{ is a projection operator} \\
&= \langle \mathbb{P}WU(g)W^{-1}\Phi | \widetilde{P}(g\Delta)WU(g)W^{-1}\Phi\rangle_{\widetilde{\mathfrak{H}}} \\
&= \langle WU(g)W^{-1}\Phi | Wa(g\Delta)U(g)W^{-1}\Phi\rangle_{\widetilde{\mathfrak{H}}}, \quad \text{by (3.18)} \\
&= \langle \Phi | Wa(\Delta)W^{-1}\Phi\rangle_{\widetilde{\mathfrak{H}}}, \quad \text{by (4.48)} \\
&= \langle \Phi | \widetilde{P}(\Delta)\Phi\rangle_{\widetilde{\mathfrak{H}}} = \|\widetilde{P}(\Delta)\Phi\|^2.
\end{aligned}$$

Thus, $\widetilde{U}(g)$ is unitary.

Next, for $\Delta, \Delta' \in \mathcal{B}(X)$,

$$\begin{aligned}
\widetilde{U}(g)\widetilde{P}(\Delta)\widetilde{U}(g)^*\widetilde{P}(\Delta')\Phi &= \widetilde{U}(g)\widetilde{P}(\Delta)\widetilde{P}(g^{-1}\Delta')WU(g)^*W^{-1}\Phi, \quad \text{by (4.50)} \\
&= \widetilde{U}(g)\widetilde{P}(\Delta \cap g^{-1}\Delta')WU(g)^*W^{-1}\Phi \\
&= \widetilde{P}(g(\Delta \cap g^{-1}\Delta'))WU(g)W^{-1}WU(g)^*W^{-1}\Phi \\
&= \widetilde{P}(g\Delta \cap \Delta')\Phi = \widetilde{P}(g\Delta)\widetilde{P}(\Delta')\Phi.
\end{aligned}$$

Thus, the imprimitivity condition (4.47) is satisfied on the dense set $\mathcal{S}$, and, hence, it is satisfied on all of $\widetilde{\mathfrak{H}}$. In other words, the pair $\{\widetilde{U}, \widetilde{P}\}$ is a system of imprimitivity that minimally extends the system of covariance $\{U, a\}$. □

To illustrate the usefulness of this result, and to convince the reader of its pertinence to the subject matter of this book, we work out a few examples in some detail.

4.2.1 Vector coherent states

Vector coherent states will be considered in their proper context in Chapter 7, Section 7.1.2. Here, we obtain a special class of such vectors as an illustration of the relationship between systems of covariance and induced representations. Let $g \mapsto U(g)$ be a continuous unitary representation of the (locally compact group) G on the Hilbert space $\mathfrak{H}$. Let $\mathfrak{K} \subset \mathfrak{H}$ be a subspace of dimension $n < \infty$. Let $H \subset G$ consist of all elements $h \in G$ for which $U(h)\mathfrak{K} \subset \mathfrak{K}$ (i.e., $\mathfrak{K}$ is stable under H). If $\mathbb{P}_\mathfrak{K}$ is the projection operator $\mathbb{P}_\mathfrak{K}\mathfrak{H} = \mathfrak{K}$, then

$$[U(h), \mathbb{P}_\mathfrak{K}] = 0, \qquad h \in H. \tag{4.51}$$

It is not hard to see that H is a closed subgroup of G. Let $F \in \mathcal{L}(\mathfrak{K})$ be a strictly positive operator, satisfying

$$[U(h), F] = 0, \qquad h \in H. \tag{4.52}$$

Note that, if the restriction of the representation $h \mapsto U(h)$ of H to $\mathfrak{K}$ is irreducible, then $F = \lambda \mathbb{P}_\mathfrak{K}$, for some $\lambda > 0$. In general, one can find an orthonormal basis $u^i \in \mathfrak{K}$, $i = 1, 2, \ldots, n$, and numbers $\lambda_i > 0$, such that

$$F = \sum_{i=1}^n \lambda_i |u^i\rangle\langle u^i| = \sum_{i=1}^n |\eta^i\rangle\langle \eta^i|, \qquad \eta^i = \lambda_i^{\frac{1}{2}} u^i, \tag{4.53}$$

and

$$U(h)u^i = \sum_{j=1}^n V(h)_{ji} u^j, \tag{4.54}$$

where the $V(h)_{ji}$ are the matrix elements of the $n \times n$ unitary matrices $V(h)$, $h \in H$, realizing a unitary representation of H on $\mathbb{C}^n$. Let $X = G/H$, and suppose that it carries the (left) *invariant* measure ν. Let $\sigma : X \to G$ be a global Borel section and $h : G \times X \to H$ be a cocycle defined as in (4.37). Define the set of vectors in $\mathfrak{H}$:

$$\mathfrak{S} = \{\eta^i_{\sigma(x)} = U(\sigma(x))\eta^i \mid i = 1, 2, \ldots, n; \ x \in X\}. \tag{4.55}$$

Then, by virtue of (4.52) and (4.37), the operator

$$F(x) = \sum_{i=1}^n |\eta^i_{\sigma(x)}\rangle\langle \eta^i_{\sigma(x)}| = U(\sigma(x))FU(\sigma(x))^* \tag{4.56}$$

has the property

$$U(g)F(x)U(g)^* = F(gx), \qquad x \in X, \ g \in G. \tag{4.57}$$

The invariance of F under H, as follows from (4.52), implies that $F(x)$ does not depend on the section σ. Indeed, if $\sigma' : X \to G$ is another section, then there is a Borel function $h : X \to H$ such that $\sigma'(x) = \sigma(x)h(x)$, and,

hence,
$$\sum_{i=1}^{n} |\eta^i_{\sigma'(x)}\rangle\langle\eta^i_{\sigma'(x)}| = U(\sigma'(x))FU(\sigma'(x))^* = F(x).$$

Suppose now that the vectors $\mathfrak{S}$ form a tight frame $\mathcal{F}\{\eta^i_{\sigma(x)}, I, n\}$, so that

$$\sum_{i=1}^{n} \int_X |\eta^i_{\sigma(x)}\rangle\langle\eta^i_{\sigma(x)}| \, d\nu(x) = I. \tag{4.58}$$

[If the representation $g \mapsto U(g)$ is irreducible, then just the convergence of the integral on the left-hand side of the above equation to an operator A would imply $A = \lambda I$, $\lambda > 0$. This is because (4.57) and the invariance of the measure ν together imply $U(g)AU(g)^* = A$; and, hence, A commutes with every $U(g)$, $g \in \mathfrak{H}$. Thus, for an irreducible representation, the frame is necessarily tight.] The associated normalized POV measure

$$a(\Delta) = \sum_{i=1}^{n} \int_\Delta |\eta^i_{\sigma(x)}\rangle\langle\eta^i_{\sigma(x)}| \, d\nu(x) \tag{4.59}$$

is independent of σ, and, furthermore, for any $g \in G$,

$$\begin{aligned} U(g)a(\Delta)U(g)^* &= \int_\Delta F(gx) \, d\nu(x), \quad \text{by (4.57)} \\ &= \int_{g\Delta} F(x) \, d\nu(x), \quad \text{by the invariance of } \nu \\ &= a(g\Delta). \end{aligned}$$

Thus, $\{U, a\}$ is a transititive system of covariance. By Theorem 4.2.3, U must be a subrepresentation of an induced representation. To work this out directly, we extend a to the PV measure $\widehat{P}$, on the Hilbert space $\widetilde{\mathfrak{H}} = \mathbb{C}^n \otimes L^2(X, d\nu)$, following the construction of Section 3.2 [see (3.24)]. Thus, $W : \mathfrak{H} \to \widetilde{\mathfrak{H}}$,

$$(W\phi)^i(x) = \langle\eta^i_{\sigma(x)}|\phi\rangle = \Phi^i(x), \quad i = 1, 2, \ldots, n,$$

is a unitary embedding of $\mathfrak{H}$ onto a subspace $\mathfrak{H}_\eta = W\mathfrak{H}$ of $\widetilde{\mathfrak{H}}$. Writing

$$U_\eta(g) = WU(g)W^{-1}, \tag{4.60}$$

we find, for any $\Phi \in \mathfrak{H}_\eta$,

$$\begin{aligned} (U_\eta(g)\Phi)^i(x) &= \langle\eta^i_{\sigma(x)}|U(g)W^{-1}\Phi\rangle_{\mathfrak{H}} \\ &= \langle U(h(g^{-1},x))\eta^i_{\sigma(g^{-1}x)}|W^{-1}\Phi\rangle_{\mathfrak{H}}, \quad \text{by (4.37)} \\ &= \sum_{j=1}^{n}[V(h(g^{-1},x))^*]_{ij}\Phi^j(g^{-1}x), \tag{4.61} \end{aligned}$$

by (4.54) and the unitarity of $V(h)$, $h \in H$. Defining $\widetilde{U}(g)$ on all of $\mathfrak{H}$ by
$$(\widetilde{U}(g)\Phi)(x) = V(h(g^{-1}, x))^*\Phi(g^{-1}x), \qquad (4.62)$$
and comparing with (4.41), we see that $\widetilde{U}$ is unitary, and is in fact the representation of G that is induced from the representation V of H on $\mathbb{C}^n$. Furthermore, $\{\widetilde{U}, \widetilde{P}\}$, with $\widetilde{P}$ as in (3.24), is a system of imprimitivity that minimally extends $\{U, a\}$.

The vectors $\mathfrak{S}$ in (4.55) are called *vector coherent states* [262]. The above analysis shows that such states, when they form a tight frame, arise from representations U that are subrepresentations of induced representations.

4.2.2 Discrete series representations of $SU(1,1)$

Representations of the group $SU(1,1)$, which belong to the discrete series, provide illustrations of both the inducing construction and systems of covariance. The group $SU(1,1)$ consists of all 2×2 complex matrices g of the type

$$g = \begin{pmatrix} \alpha & \beta \\ \overline{\beta} & \overline{\alpha} \end{pmatrix}, \quad \alpha, \beta \in \mathbb{C}, \quad \det g = |\alpha|^2 - |\beta|^2 = 1. \qquad (4.63)$$

The inverse element is

$$g^{-1} = \begin{pmatrix} \overline{\alpha} & -\beta \\ -\overline{\beta} & \alpha \end{pmatrix}, \qquad (4.64)$$

while the elements $k \in SU(1,1)$ of the type

$$k = \begin{pmatrix} e^{i\frac{\phi}{2}} & 0 \\ 0 & e^{-i\frac{\phi}{2}} \end{pmatrix}, \quad 0 \leq \phi < 4\pi, \qquad (4.65)$$

constitute the *maximal compact subgroup* K of $SU(1,1)$. The *Cartan decomposition* of an arbitrary group element, $g = pk$, $k \in K$, is then obtained (see Section 4.5.2) as

$$g = \begin{pmatrix} \alpha & \beta \\ \overline{\beta} & \overline{\alpha} \end{pmatrix} = |\alpha| \begin{pmatrix} 1 & z \\ \overline{z} & 1 \end{pmatrix} \begin{pmatrix} \frac{\alpha}{|\alpha|} & 0 \\ 0 & \frac{\overline{\alpha}}{|\alpha|} \end{pmatrix},$$
$$z = \beta \overline{\alpha}^{-1}, \quad |\alpha| = (1 - |z|^2)^{-\frac{1}{2}}, \qquad (4.66)$$

from which it is seen that the coset space $SU(1,1)/K$ is homeomorphic to the open unit disc $\mathcal{D} = \{z \in \mathbb{C} \,|\, |z| < 1\}$ of the complex plane. Furthermore, since

$$g \begin{pmatrix} 1 & z \\ \overline{z} & 1 \end{pmatrix} = |\alpha'| \begin{pmatrix} 1 & z' \\ \overline{z}' & 1 \end{pmatrix} \begin{pmatrix} \frac{\alpha'}{|\alpha'|} & 0 \\ 0 & \frac{\overline{\alpha}'}{|\alpha'|} \end{pmatrix},$$

where

$$\alpha' = \beta \overline{z} + \alpha, \qquad z' = \frac{\alpha z + \beta}{\overline{\beta} z + \overline{\alpha}},$$

the action of $SU(1,1)$ on $\mathcal{D}$ is captured by the fractional holomorphic transformation

$$z \mapsto z' = gz = \frac{\alpha z + \beta}{\bar{\beta} z + \bar{\alpha}}. \tag{4.67}$$

The invariant measure on $\mathcal{D}$, under this transformation, is now easily computed. Indeed, we have

$$dz' = \frac{dz}{(\bar{\beta} z + \bar{\alpha})^2}, \tag{4.68}$$

and, hence,

$$dz' \wedge d\bar{z}' = \frac{dz \wedge d\bar{z}}{|\bar{\beta} z + \bar{\alpha}|^4}. \tag{4.69}$$

On the other hand, by (4.67),

$$1 - |z'|^2 = \frac{1 - |z|^2}{|\bar{\beta} z + \bar{\alpha}|^2}, \tag{4.70}$$

so that

$$\frac{1}{|\bar{\beta} z + \bar{\alpha}|^4} = \left(\frac{1 - |z'|^2}{1 - |z|^2}\right)^2. \tag{4.71}$$

Using this in (4.69), we get

$$\frac{dz' \wedge d\bar{z}'}{(1 - |z'|^2)^2} = \frac{dz \wedge d\bar{z}}{(1 - |z|^2)^2}. \tag{4.72}$$

Thus, the measure

$$d\nu(z, \bar{z}) = \frac{1}{2\pi i} \frac{dz \wedge d\bar{z}}{(1 - |z|^2)^2} \tag{4.73}$$

on $\mathcal{D}$ is invariant under the action (4.67) of $SU(1,1)$.

Note that, as a complex manifold, $\mathcal{D}$ admits a Kähler structure determined by the potential function

$$\Phi(z', \bar{z}) = -\log(1 - z'\bar{z}), \tag{4.74}$$

in terms of which we obtain the invariant two-form

$$\Omega = \frac{1}{i} \frac{\partial^2 \Phi(z, \bar{z})}{\partial z \partial \bar{z}} dz \wedge d\bar{z} = \frac{1}{i} \frac{dz \wedge d\bar{z}}{(1 - |z|^2)^2}, \tag{4.75}$$

which is closed, i.e., $d\Omega = 0$, and determines the measure $d\nu$.

Next, for each $j = 1, 3/2, 2, 5/2, \ldots$, let us define a *quasi-invariant* measure on $\mathcal{D}$,

$$d\mu_j(z, \bar{z}) = \exp[-2j\,\Phi(z,\bar{z})]\,d\nu(z,\bar{z}) = (1 - |z|^2)^{2j} d\nu(z, \bar{z})$$
$$= (1 - |z|^2)^{2j-2} \frac{dz \wedge d\bar{z}}{2\pi i}, \tag{4.76}$$

4.2. Induced representations and systems of covariance

and consider the Hilbert space

$$\mathfrak{H}^j = L^2(\mathcal{D}, (2j-1)d\mu_j), \tag{4.77}$$

of all complex measurable functions f on $\mathcal{D}$, which are square integrable with respect to $(2j-1)d\mu_j$; i.e.,

$$\|f\|_j^2 = (2j-1)\int_\mathcal{D} |f(z)|^2 \, d\mu_j(z, \bar{z}) < \infty. \tag{4.78}$$

(More properly, we should write $f(z, \bar{z})$ rather than $f(z)$ in the above, but we shall ignore this technicality.) The factor $(2j-1)$ is a normalization constant that ensures that

$$\|u_0\|_j^2 = 1, \quad \text{where} \quad u_0(z) = 1, \; \forall z \in \mathcal{D}.$$

Indeed, using polar coordinates, in which

$$\frac{dz \wedge d\bar{z}}{2\pi i} = \frac{r\,dr\,d\theta}{\pi},$$

we immediately find

$$\|u_0\|_j^2 = \frac{2j-1}{\pi} \int_0^{2\pi} d\theta \int_0^1 (1-r^2)^{2j-2}\,r\,dr = 1.$$

An induced representation $g \mapsto U^j(g)$ of $SU(1,1)$ can now be constructed, for each j, following the procedure outlined in Section 4.2. Let V^j be the (irreducible) representation of the subgroup K, of dimension 1,

$$V^j(k) = \exp[-i(j\phi)], \tag{4.79}$$

where the element $k \in K$ is given by (4.65). We proceed to compute [see (4.39)]

$$B(g, z) = [\lambda(g, z)]^{\frac{1}{2}} V^j(h(g^{-1}, z))^{-1} \tag{4.80}$$

for an arbitrary group element $g \in SU(1,1)$. Take the section $\sigma : \mathcal{D} \to SU(1,1)$,

$$\sigma(z) = \frac{1}{\sqrt{1-|z|^2}} \begin{pmatrix} 1 & z \\ \bar{z} & 1 \end{pmatrix}, \tag{4.81}$$

and note that, since

$$g^{-1}\sigma(z) = \frac{1}{\sqrt{1-|z|^2}} \begin{pmatrix} -\beta\bar{z}+\bar{\alpha} & \bar{\alpha}z-\beta \\ \alpha\bar{z}-\bar{\beta} & -\bar{\beta}z+\alpha \end{pmatrix},$$

the coset decomposition (4.66) implies that

$$h(g^{-1}, z) = \begin{pmatrix} \frac{-\beta\bar{z}+\bar{\alpha}}{|-\beta\bar{z}+\bar{\alpha}|} & 0 \\ 0 & \frac{-\bar{\beta}z+\alpha}{|-\bar{\beta}z+\alpha|} \end{pmatrix}. \tag{4.82}$$

64   4. Some Group Theory

Hence,

$$V^j(h(g^{-1},z)) = \frac{(-\bar{\beta}z+\alpha)^{2j}}{|-\bar{\beta}z+\bar{\alpha}|^{2j}} \Rightarrow V^j(h(g^{-1},z))^{-1} = \frac{|\alpha-\bar{\beta}z|^{2j}}{(\alpha-\bar{\beta}z)^{2j}}. \quad (4.83)$$

Next, for the quasi-invariant measure μ_j, (4.22) and (4.76) together imply that

$$\lambda(g,z) = \frac{d\mu_j(g^{-1}z, \overline{g^{-1}z})}{d\mu_j} = \left(\frac{1-|g^{-1}z|^2}{1-|z|^2}\right)^{2j}$$

$$= \frac{1}{|\alpha-\bar{\beta}z|^{4j}}, \quad (4.84)$$

the last equality being obtained by taking $z' = g^{-1}z$ in (4.71). Combining (4.80) with (4.76) and (4.84) leads to

$$B(g,z) = (\alpha-\bar{\beta}z)^{-2j}. \quad (4.85)$$

Consequently, inserting (4.67), with g replaced by g^{-1}, and (4.85) into (4.41), we obtain the representation U^j of $SU(1,1)$, on the Hilbert space $\mathfrak{H}^j$, which is induced from the representation V^j of the subgroup K:

$$(U^j(g)f)(z) = (\alpha-\bar{\beta}z)^{-2j} f\left(\frac{\bar{\alpha}z-\beta}{\alpha-\bar{\beta}z}\right), \quad f \in \mathfrak{H}^j. \quad (4.86)$$

Being an induced representation, U^j is unitary. It is not irreducible, however, and we shall presently construct an irreducible subrepresentation of it. First, let us note a few general features.

The restriction of U^j to the subgroup K is seen to be

$$(U^j(k)f)(z) = e^{-ij\phi} f(e^{-i\phi}z) = V^j(k)f(e^{-i\phi}z), \quad k \in K. \quad (4.87)$$

In particular, if $f(z) = z^n$, for some positive integer n, then

$$(U^j(k)f)(z) = e^{-(j+n)\phi} f(z), \quad k \in K. \quad (4.88)$$

Similarly, for arbitrary $f \in \mathfrak{H}^j$,

$$(U^j(\sigma(z))f)(z') = \frac{(1-|z|^2)^j}{(1-z'\bar{z})^{2j}} f\left(\frac{z'-z}{1-z'\bar{z}}\right). \quad (4.89)$$

Defining the projection operators $P(\Delta)$ on $\mathfrak{H}^j$,

$$(P(\Delta)f)(z) = \chi_\Delta(z) f(z),$$

it is easily checked that

$$U^j(g)P(\Delta)U^j(g)^* = P(g\Delta), \quad g \in SU(1,1), \quad (4.90)$$

giving a system of imprimitivity.

In order to isolate an irreducible subrepresentation of $\mathfrak{H}^j$, let $\mathfrak{H}^j_{hol}$ denote the set of all elements $f \in \mathfrak{H}^j$, which are *holomorphic* in z. Theorem 6.2.2

4.2. Induced representations and systems of covariance

then allows us to infer that $\mathfrak{H}_{hol}^j$ is a Hilbert subspace of $\mathfrak{H}^j$. Consider the function $K_{hol}^j : \mathbb{C} \times \mathbb{C} \to \mathbb{C}$,

$$K_{hol}^j(z', \bar{z}) = (1 - z'\bar{z})^{-2j} = \exp\left[2j\Phi(z', \bar{z})\right], \tag{4.91}$$

where $\Phi(z', \bar{z})$ is the potential function defined in (4.74). Since $|z'| < 1$, $|z| < 1$, the middle term in this equation can be expanded in an infinite series to obtain

$$K_{hol}^j(z', \bar{z}) = \sum_{n=0}^{\infty} u_n(z') \, \overline{u_n(z)}, \tag{4.92}$$

where the u_n are the monomials

$$u_n(z) = \left[\frac{\Gamma(2j+n)}{\Gamma(n+1)\,\Gamma(2j)}\right]^{\frac{1}{2}} z^n, \qquad n = 0, 1, 2, \ldots. \tag{4.93}$$

Clearly, $u_n \in \mathfrak{H}_{hol}^j$, for all n, and every function $f \in \mathfrak{H}_{hol}^j$ can be written in terms of them. Also, it is not hard to check that

$$(2j - 1) \int_D K_{hol}^j(z', \bar{w}) K_{hol}^j(w, \bar{z}) \, d\mu_j(w, \bar{w}) = K_{hol}^j(z', \bar{z}), \tag{4.94}$$

and we recognize here the condition for K_{hol}^j to be a reproducing kernel (see Section 2.3). The elements $\zeta_{\bar{z}}^j \in \mathfrak{H}_{hol}^j$, generated using K_{hol}^j,

$$\zeta_{\bar{z}}^j(z') = K_{hol}^j(z', \bar{z}) = (1 - z'\bar{z})^{-2j}, \tag{4.95}$$

that is,

$$\zeta_{\bar{z}}^j = \sum_{n=0}^{\infty} \overline{u_n(z)} \, u_n, \tag{4.96}$$

enjoy the properties

$$\left.\begin{array}{rcl} \langle \zeta_{\bar{z}}^j | \zeta_{\bar{z}'}^j \rangle & = & K_{hol}^j(z', \bar{z}) \\ \langle \zeta_{\bar{z}}^j | f \rangle & = & f(z), \quad f \in \mathfrak{H}_{hol}^j \end{array}\right\}, \tag{4.97}$$

as can easily be checked using (4.94). Furthermore, these properties also ensure that the resolution of the identity relation

$$(2j - 1) \int_D |\zeta_{\bar{z}}^j\rangle\langle\zeta_{\bar{z}}^j| \, d\mu_j(z, \bar{z}) = I_{hol} \tag{4.98}$$

is satisfied (where I_{hol} is the identity operator on $\mathfrak{H}_{hol}^j$), and thus the vectors $\zeta_{\bar{z}}^j$, $z \in D$, form an overcomplete set in $\mathfrak{H}_{hol}^j$. Using (4.89), let us define a second set of vectors,

$$\eta_{\sigma(z)} = (2j - 1)^{\frac{1}{2}} U^j(\sigma(z)) u_0 = (2j - 1)^{\frac{1}{2}} (1 - |z|^2)^j \zeta_{\bar{z}}^j, \tag{4.99}$$

and note that they also satisfy a resolution of the identity, but with respect to the invariant measure ν on $\mathcal{D}$ (see (4.73)):

$$\int_{\mathcal{D}} |\eta_{\sigma(z)}\rangle\langle\eta_{\sigma(z)}|\, d\nu(z,\bar{z}) = I_{hol}. \qquad (4.100)$$

The vectors $\eta_{\sigma(z)}$ are generated by acting on the single vector u_0 with the operators $U^j(\sigma(z))$ of the group representation. On the other hand, being essentially the vectors $\zeta_{\bar{z}}^j$ (i.e., up to a scale factor), the $\eta_{\sigma(z)}$ also span $\mathfrak{H}_{hol}^j$. Thus, $\mathfrak{H}_{hol}^j$ is the smallest closed subspace of $\mathfrak{H}^j$, containing u_0 and invariant under U^j. Furthermore, as will be proved in Lemma 4.2.4, every closed subspace of $\mathfrak{H}_{hol}^j$ that is stable under U^j and different from the trivial subspace $\{0\}$ must contain the vector u_0. Thus, the representation U^j restricted to $\mathfrak{H}_{hol}^j$ is irreducible. (Actually, this is a special case of a more general result [198].) Denote this restriction by U_{hol}^j. For $j = 1, 3/2, 2, \ldots$, these are representations in the so-called *discrete series* representations of $SU(1,1)$. The vectors $\eta_{\sigma(z)}$, $z \in \mathcal{D}$, are the coherent states of the discrete series representations of $SU(1,1)$ or CS associated to the unit disc [Per].

For each Borel set Δ of $\mathcal{D}$, now construct the positive operator

$$a(\Delta) = \int_{\Delta} |\eta_{\sigma(z)}\rangle\langle\eta_{\sigma(z)}|\, d\nu(z,\bar{z}) = (2j-1)\int_{\Delta} |\zeta_{\bar{z}}^j\rangle\langle\zeta_{\bar{z}}^j|\, d\mu_j(z,\bar{z}). \qquad (4.101)$$

Since, for arbitrary $g \in SU(1,1)$, $g\sigma(z) = \sigma(gz)h(g,z)$ [see (4.82)], it follows from (4.89) and the invariance of the measure ν that the covariance relation,

$$U_{hol}^j(g)a(\Delta)U_{hol}^j(g)^* = a(g\Delta), \qquad \Delta \in \mathcal{B}(\mathcal{D}), \qquad (4.102)$$

is satisfied. Thus, $\{U_{hol}^j, a\}$ is a transitive system of covariance, and of course we have here a subrepresentation of an induced representation, in conformity with Theorem 4.2.3. Moreover, the system of imprimitivity $\{U^j, P\}$ in (4.90) is its minimal Naĭmark extension, as is evident from the discussion in Sections 3.2 and 4.2.1. Note, finally, that the projection operator $\mathbb{P}_{hol}$ corresponding to the subspace $\mathfrak{H}_{hol}^j \subset \mathfrak{H}^j$ acts as

$$(\mathbb{P}_{hol}f)(z) = (2j-1)\int_{\mathcal{D}} K_{hol}^j(z,\bar{w})f(w)\, d\mu_j(w,\bar{w}) \qquad (4.103)$$

[this is a straightforward consequence of (4.94)], and, hence, $a(\Delta)$ and $P(\Delta)$ are related through

$$\mathbb{P}_{hol}P(\Delta)\mathbb{P}_{hol} = a(\Delta), \qquad \Delta \in \mathcal{B}(\mathcal{D}), \qquad (4.104)$$

as required by (3.18). (Note: in the context of that equation, $\mathfrak{H}_{hol}^j = \mathfrak{H}$, $\mathfrak{H}^j = \widetilde{\mathfrak{H}}$, and W is just the inclusion map $\mathfrak{H}_{hol}^j \hookrightarrow \mathfrak{H}^j$.)

We end this section by proving a technical result used to show that U_{hol}^j is irreducible [Sug].

Lemma 4.2.4: *Let $\mathfrak{H}_0 \subset \mathfrak{H}_{hol}^j$ be a nontrivial, closed subspace that is stable under U_{hol}^j. Then, $u_0 \in \mathfrak{H}_0$.*

Proof. Let $f \in \mathfrak{H}_0$, and write $f(z) = \sum_{n=0}^{\infty} a_n z^n$. By (4.88),

$$(U_{hol}^j(k)f)(z) = e^{-ij\phi} \sum_{n=0}^{\infty} a_n e^{-n\phi} z^n, \quad \forall k \in K.$$

Thus,

$$\frac{1}{2\pi}\int_0^{2\pi} e^{ij\phi}(U^j(k)f)(z)\, d\phi = a_0 = f(0), \quad \forall z \in \mathcal{D}.$$

Since $\mathfrak{H}_0$ is stable under U_{hol}^j, it follows that $U_{hol}^j(k)f \in \mathfrak{H}_0$. Also,

$$\frac{1}{2\pi}\int_0^{2\pi} e^{ij\phi} U_{hol}^j(k)f \, d\phi \in \mathfrak{H}_0,$$

since $\mathfrak{H}_0$ is closed and the integral is the limit of Riemann sums of the type

$$\sum_m e^{ij\phi_m} U_{hol}^j(k_m) f \, \frac{\phi_{m+1} - \phi_m}{2\pi}, \quad \text{where} \quad k_m = \begin{pmatrix} e^{i\frac{\phi_m}{2}} & 0 \\ 0 & e^{-i\frac{\phi_m}{2}} \end{pmatrix},$$

with each $U^j(k_m)f \in \mathfrak{H}_0$. Thus, the function $f(0)u_0$ belongs to $\mathfrak{H}_0$, so that $u_0 \in \mathfrak{H}_0$, provided $f(0) \neq 0$. We still need to show, however, that it is possible to find an $f \in \mathfrak{H}_0$ with $a_0 = f(0) \neq 0$. Since $\mathfrak{H}_0 \neq \{0\}$, one can find an element $\tilde{f} \in \mathfrak{H}_0$ and a point $z_0 \in \mathcal{D}$ such that $\tilde{f}(z_0) \neq 0$. Moreover, the action of $SU(1,1)$ on $\mathcal{D}$ being transitive, there is a $g \in SU(1,1)$ for which $g^{-1}0 = z_0$. Hence, by (4.86),

$$(U_{hol}^j(g)\tilde{f})(0) = \alpha^{-2j}\tilde{f}(z_0) \neq 0,$$

since $\alpha \neq 0$. Because $U^j(g)\tilde{f}$ is again an element in $\mathfrak{H}_0$, the desired result follows upon writing $f = U_{hol}^j(g)\tilde{f}$. □

4.2.3 The regular representations of a group

There are two representations of a locally compact group G, both induced representations, which are of great importance in harmonic analysis and will reappear later in our development of the theory of CS. These are the so-called regular representations of G. Let μ be the left Haar measure on G, and consider the trivial subgroup $H = \{e\}$, consisting of just the identity element. The representation of G induced by the trivial representation of H is carried by the Hilbert space $L^2(G, d\mu)$. Denoting this representation by U_ℓ, we have, for all $f \in L^2(G, d\mu)$,

$$(U_\ell(g)f)(g') = f(g^{-1}g'), \quad g, g' \in G. \qquad (4.105)$$

This representation is called the *left regular representation* of G. Similarly, using the right Haar measure μ_r and the Hilbert space $L^2(G, d\mu_r)$, we can construct another unitary representation U_r as

$$(U_r(g)f)(g') = f(g'g), \qquad g, g' \in G, \tag{4.106}$$

for all $f \in L^2(G, d\mu_r)$. This representation is called the *right regular representation* of G. In general, these representations are reducible. On the other hand, U_ℓ and U_r are unitarily equivalent representations. Indeed, the map

$$V : L^2(G, d\mu) \to L^2(G, d\mu_r), \qquad (Vf)(g) = f(g^{-1}), \qquad g \in G, \tag{4.107}$$

is easily seen [using (4.5)] to be unitary, and

$$VU_\ell(g)V^{-1} = U_r(g), \qquad g \in G. \tag{4.108}$$

The regular representation U_r can also be realized on the Hilbert space $L^2(G, d\mu)$, rather than on $L^2(G, d\mu_r)$, using the fact that μ and μ_r are related by the modular function $\boldsymbol{\Delta}$ through (4.3). Thus, the map

$$W : L^2(G, d\mu_r) \to L^2(G, d\mu), \qquad (Wf)(g) = \boldsymbol{\Delta}(g)^{-\frac{1}{2}} f(g) \tag{4.109}$$

is unitary, and, for all $f \in L^2(G, d\mu)$,

$$(\overline{U}_r(g)f)(g') = \boldsymbol{\Delta}(g)^{\frac{1}{2}} f(g'g), \quad \text{where} \quad \overline{U}_r(g) = WU_r(g)W^{-1}, \quad g \in G. \tag{4.110}$$

From this, we see that the left and right regular representations commute:

$$[U_\ell(g_1), \overline{U}_r(g_2)] = 0 \qquad g_1, g_2 \in G. \tag{4.111}$$

Clearly, the two representations U_ℓ and $\overline{U}_r$ on $L^2(G, d\mu)$ are also unitarily equivalent. More interesting, however, is the map $J : L^2(G, d\mu) \to L^2(G, d\mu)$,

$$\begin{aligned}(Jf)(g) &= \overline{f(g^{-1})}\boldsymbol{\Delta}(g)^{-\frac{1}{2}}, \qquad J^2 = I \\ JU_\ell(g)J &= \overline{U}_r(g), \qquad g \in G,\end{aligned} \tag{4.112}$$

which is an *antiunitary isomorphism*.

4.3 An extended Schur's lemma

In harmonic analysis, the irreducibility of a unitary group representation is usually determined by an application of Schur's lemma (see, for example, [Kir]). We discuss this lemma in some detail here, since we shall require an extended version of it when dealing with square integrable representations in Chapter 8.

Lemma 4.3.1 (Classical Schur's lemma): *Let U be a continuous unitary irreducible representation of G on the Hilbert space $\mathfrak{H}$. If $T \in \mathcal{L}(\mathfrak{H})$, and T commutes with $U(g)$, for all $g \in G$, then $T = \lambda I$, for some $\lambda \in \mathbb{C}$.*

4.3. An extended Schur's lemma

A general proof of this theorem may, for example, be found in [Sug]. Actually, there is a more general version of this lemma, which we now state and prove and then use to prove a further extended version of it. Let U_1 and U_2 be two representations of G on the Hilbert spaces $\mathfrak{H}_1$ and $\mathfrak{H}_2$, respectively. A linear map $T: \mathfrak{H}_1 \to \mathfrak{H}_2$ is said to *intertwine* U_1 and U_2 if

$$TU_1(g) = U_2(g)T, \qquad g \in G. \tag{4.113}$$

Given two Hilbert spaces $\mathfrak{H}_1$ and $\mathfrak{H}_2$, a linear map $T: \mathfrak{H}_1 \to \mathfrak{H}_2$ is said to be a *multiple of an isometry* if there exists $\lambda > 0$ such that

$$\|T\phi\|_{\mathfrak{H}_2}^2 = \lambda \|\phi\|_{\mathfrak{H}_1}^2, \qquad \phi \in \mathfrak{H}_1. \tag{4.114}$$

Lemma 4.3.2 (Generalized Schur's lemma): *Let U_1 be a unitary irreducible representation of G on $\mathfrak{H}_1$ and U_2 be a unitary, but not necessarily irreducible, representation of G on $\mathfrak{H}_2$. Let $T: \mathfrak{H}_1 \to \mathfrak{H}_2$ be a bounded linear map that intertwines U_1 and U_2. Then, T is either null or a multiple of an isometry.*

Proof. Consider the adjoint map $T^*: \mathfrak{H}_2 \to \mathfrak{H}_1$, and take the adjoint of the relation (4.113), getting

$$U_1(g)T^* = T^* U_2(g), \qquad g \in G, \tag{4.115}$$

since $U_j^*(g) = U_j(g^{-1})$. From this and (4.113), we obtain

$$T^* T U_1(g) = T^* U_2(g) T = U_1(g) T^* T, \qquad g \in G. \tag{4.116}$$

Now, U_1 is unitary and irreducible, and $T^*T: \mathfrak{H}_1 \to \mathfrak{H}_1$ commutes with it. Hence, by the classical Schur's lemma, $T^*T = \lambda I$, for some $\lambda \in \mathbb{C}$; that is, T is either null or a multiple of an isometry. $\square$

The next extended version of Schur's lemma, which we now state and prove, has been adapted from [159]. This is the version that we shall eventually require.

Lemma 4.3.3 (Extended Schur's lemma): *Let U_1 be a unitary irreducible representation of G on $\mathfrak{H}_1$ and U_2 be a unitary, but not necessarily irreducible, representation of G on $\mathfrak{H}_2$. Let $T: \mathfrak{H}_1 \to \mathfrak{H}_2$ be a closed linear map, the domain $\mathcal{D}(T)$ of which is dense in $\mathfrak{H}_1$ and stable under U_1 [i.e., $U_1(g)\phi \in \mathcal{D}(T)$, for all $g \in G$ and $\phi \in \mathcal{D}(T)$], and suppose that T intertwines U_1 and U_2. Then, T is either null or a multiple of an isometry.*

Proof. (The idea of this proof is to convert $\mathcal{D}(T)$ into a Hilbert space by equipping it with the *graph norm* of T, and then to use the generalized Schur's lemma above.) On $\mathcal{D}(T)$, define the new scalar product

$$\langle \phi | \psi \rangle_T = \langle \phi | \psi \rangle_{\mathfrak{H}_1} + \langle T\phi | T\psi \rangle_{\mathfrak{H}_2}, \qquad \|\phi\|_T^2 = \|\phi\|_{\mathfrak{H}_1}^2 + \|T\phi\|_{\mathfrak{H}_2}^2, \tag{4.117}$$

for all $\phi, \psi, \in \mathcal{D}(T)$. Then, since T is closed, $\mathcal{D}(T)$ equipped with this scalar product is a Hilbert space, which we denote by $\mathfrak{H}_T$. Consider T as a linear map $T : \mathfrak{H}_T \to \mathfrak{H}_2$. Then,

$$\|T\| = \sup_{\phi \in \mathfrak{H}_T} \frac{\|T\phi\|_{\mathfrak{H}_2}}{\|\phi\|_T} = \sup_{\phi \in \mathfrak{H}_T} \frac{\|T\phi\|_{\mathfrak{H}_2}}{[\|\phi\|_{\mathfrak{H}_1}^2 + \|T\phi\|_{\mathfrak{H}_2}^2]^{\frac{1}{2}}}.$$

Thus, $\|T\| \le 1$, so that $T : \mathfrak{H}_T \to \mathfrak{H}_2$ is a bounded linear map. Next, it is seen that the representation U_1 is unitary on $\mathfrak{H}_T$. Indeed, for all $\phi \in \mathfrak{H}_T$ and $g \in G$,

$$\begin{aligned}\|U_1(g)\phi\|_T^2 &= \|U_1(g)\phi\|_{\mathfrak{H}_1}^2 + \|TU_1(g)\phi\|_{\mathfrak{H}_2}^2 = \|\phi\|_{\mathfrak{H}_1}^2 + \|U_2(g)T\phi\|_{\mathfrak{H}_2}^2 \\ &= \|\phi\|_{\mathfrak{H}_1}^2 + \|T\phi\|_{\mathfrak{H}_2}^2 = \|\phi\|_T^2.\end{aligned}$$

Also, for all $g \in G$, the restriction of $U_1(g)$ to $\mathfrak{H}_T$ is surjective (i.e., onto $\mathfrak{H}_T$), so that, for all $\phi \in \mathfrak{H}_T$, $U_1(g)\phi = \psi$ implies $\psi \in \mathfrak{H}_T$. This follows from the stability of $\mathcal{D}(T)$ under $U_1(g)$ and the fact that $U_1(g^{-1})\psi = U_1(g^{-1})U_1(g)\phi = \phi$. Thus, by Lemma 4.3.2, $T : \mathfrak{H}_T \to \mathfrak{H}_2$ is either a multiple of an isometry, in which case there exists a $\lambda > 0$ such that, for all $\phi \in \mathfrak{H}_T$,

$$\|T\phi\|_{\mathfrak{H}_2}^2 = \lambda \|\phi\|_T^2 = \lambda \|\phi\|_{\mathfrak{H}_1}^2 + \lambda \|T\phi\|_{\mathfrak{H}_2}^2,$$

or else T is null. In any case, $\lambda \ne 1$, and

$$\|T\phi\|_{\mathfrak{H}_2}^2 = \frac{\lambda}{1-\lambda}\|\phi\|_{\mathfrak{H}_1}^2, \qquad \lambda \ge 0.$$

Thus, $T : \mathcal{D}(T) \to \mathfrak{H}_2$ is either null and, hence, null on all of $\mathfrak{H}_1$, or else it is a multiple of an isometry, in which case, by virtue of the closedness of T, it extends to a multiple of the isometry from $\mathfrak{H}_1$ to $\mathfrak{H}_2$. □

As a corollary, if $\mathfrak{H}_1 = \mathfrak{H}_2$ and $U_1 = U_2$, then, as a consequence of the classical Schur's lemma, T is a multiple of the identity.

4.4 Harmonic analysis on locally compact abelian groups

4.4.1 Basic notions

We give in this section a summary of some basic concepts and useful results on locally compact abelian (LCA for short) groups. Further information may be found in [Rud] or any textbook on harmonic analysis (for instance, [Fol]).

Given an LCA group G (whose group operation shall be written additively), we denote by μ_G the Haar measure on G. When G is discrete, μ_G is naturally the counting measure, and discrete versions of relations that we shall define on the whole group G will then hold.

4.4. Harmonic analysis on locally compact abelian groups

Let $\widehat{G}$ be the unitary dual of G, that is, the set of all its unitary characters. We use the notation
$$\omega(x) = \langle \omega, x \rangle, \quad x \in G, \omega \in \widehat{G},$$
to express the duality pairing between G and $\widehat{G}$. We again denote the group law on $\widehat{G}$ additively: For any $\chi, \chi' \in \widehat{G}$, we write $\langle \chi + \chi', x \rangle = \langle \chi, x \rangle \langle \chi', x \rangle$, $x \in G$. On the dual group $\widehat{G}$, we take the Haar measure to be the dual of the given Haar measure on G.

The Fourier transform is defined as usual: Given $f \in L^1(G)$, its Fourier transform is the function $\mathcal{F}f = \widehat{f} \in L^\infty(\widehat{G})$ defined by
$$\widehat{f}(\omega) = \int_G f(x) \overline{\langle \omega, x \rangle} \, d\mu(x). \tag{4.118}$$

It is a standard result (see, e.g., [Rud] for more details) that the Fourier transform so defined extends to an isometry between $L^2(G)$ and $L^2(\widehat{G})$. More precisely, there exists a measure $\mu_{\widehat{G}}$ on $\widehat{G}$ (the dual Haar measure) such that the following Plancherel formula holds: For all $f \in L^2(G)$, one has $\widehat{f} \in L^2(\widehat{G})$ and
$$\int_{\widehat{G}} \left| \widehat{f}(\omega) \right|^2 d\mu_{\widehat{G}}(\omega) = \int_G |f(x)|^2 d\mu(x). \tag{4.119}$$

This Fourier transform has the usual properties with respect to the operations of translations and modulation. Here, the translation of parameter $b \in G$ is the unitary operator T_b defined on $L^2(G)$ by
$$(T_b f)(x) = f(x - b),$$
and the modulation of parameter $b \in G$ is the unitary operator E_b defined on $L^2(\widehat{G})$ by
$$(E_b \widehat{f})(\omega) = \overline{\langle \omega, b \rangle} \, \widehat{f}(\omega).$$

The maps $b \xmapsto{T} T_b$ and $b \xmapsto{E} E_b$ provide two unitarily equivalent representations of G, with respective carrier spaces $L^2(G)$ and $L^2(\widehat{G})$), in fact, the regular representation of G. The associated interwining operator is the Fourier transform. Explicitly, we have
$$\mathcal{F} T_b = E_b \mathcal{F} \quad \text{for all} \quad b \in G.$$

If f and g are two functions over G, their convolution product is defined as
$$(f * g)(x) = \int_G f(x - y) g(y) d\mu(y), \tag{4.120}$$
if the integral on the right-hand side converges. The convolution product is associative, and, since G is abelian, it is also commutative; that is,
$$(f * g) * h = f * (g * h), \quad f * g = g * f.$$

In the Fourier space, the convolution is given by pointwise multiplication:
$$\widehat{f * g} = \hat{f}\hat{g}.$$
The standard example, of course, is $G = SO(2) \simeq S^1 \equiv \mathbb{T}$, the unit circle, with Haar measure $d\theta/2\pi$. The unitary characters of G are the trigonometric functions
$$\langle \chi_m, \theta \rangle = e^{im\theta}, \quad m \in \mathbb{Z},$$
so that $\hat{G} = \mathbb{Z}$ and the harmonic analysis on $SO(2)$ is the theory of Fourier series. Note that part of the theory extends to nonabelian groups. If G is compact, $\hat{G}$ is discrete, and harmonic analysis leads to expansions in appropriate special functions. For instance, in the case of $SO(3)$, one gets the theory of spherical harmonics, which indeed yields Fourier expansions on the two-sphere S^2.

4.4.2 Lattices in LCA groups

Let $\Gamma \subset G$ be any subgroup of G, and let $\Gamma^\perp \subset \hat{G}$ be its annihilator, defined as
$$\Gamma^\perp = \{\omega \in \hat{G} | \langle \omega, x \rangle = 0 \text{ for all } x \in \Gamma\}. \tag{4.121}$$
$\Gamma^\perp$ is a closed subgroup of $\hat{G}$. We always have $\Gamma \subset (\Gamma^\perp)^\perp$, and the reverse inclusion is true if Γ is itself a closed subgroup of G.

There is a remarkable duality between subgroups and quotient groups of a locally compact abelian group, as stated in the following theorem, which extends the Pontrjagin duality theorem (see [Rud, Theorem 2.1.2] or [Fol, Theorem 4.39]).

Theorem 4.4.1 (Pontrjagin duality): *Let Γ be a closed subgroup of G, and define*
$$\Phi : \widehat{G/\Gamma} \to \Gamma^\perp \quad \text{and} \quad \Psi : \hat{G}/\Gamma^\perp \to \hat{\Gamma}$$
by $\Phi(\eta) = \eta \circ q$, $\Psi(\omega \Gamma^\perp) = \omega|_\Gamma$, *where* $q : G \longrightarrow G/\Gamma$ *is the canonical projection. Then, Φ and Ψ are isomorphisms of topological groups.*

Note that the quotient groups G/Γ and $\hat{G}/\Gamma^\perp$ are both locally compact abelian groups. The main idea of Theorem 4.4.1 is that, on the one hand, the characters of G/Γ can be identified with the subset $\Gamma^\perp$ of $\hat{G}$ and, on the other hand, $\hat{\Gamma}$ can be identified with the quotient $\hat{G}/\Gamma^\perp$. Note also that the surjectivity of Ψ yields a sort of Hahn–Banach theorem for locally compact abelian groups, in the sense that every character on Γ extends to a character of G.

Let us now suppose that Γ is a lattice in G; that is, Γ is a discrete subgroup of G and the quotient G/Γ is compact. By a *fundamental domain* of Γ in G, we mean a μ-measurable set $\Omega \subset G$, such that, for each $x \in G$,

$\Omega \cap (x+\Gamma)$ consists of a single point. Given a lattice $\Gamma \subset G$, it can be shown that a fundamental domain Ω of Γ always exists in G. It is standard to take $\Omega = G/\Gamma$, but many other choices are possible. The *lattice size*, which we denote by $s(\Gamma)$, is defined as the measure of a fundamental domain, that is, $s(\Gamma) = \mu(\Omega)$, and it is independent of the particular choice of Ω. The quantity $s(\Gamma)^{-1}$ then serves as a measure of the density of Γ. Taking $\Omega = G/\Gamma$ allows us to take the measure naturally inherited from G as the measure of G/Γ; we normalize it to be a probability measure such that the Weyl formula (see, e.g., [Fol] for details) holds:

$$\int_G f(x)\, d\mu(x) = \int_{G/\Gamma} \sum_{\gamma \in \Gamma} f(x+\gamma)\, d\mu_{G/\Gamma}(x), \quad \forall f \in L^1(G). \tag{4.122}$$

If Γ is a lattice in G, then its annihilator $\Gamma^\perp$ is a lattice in $\widehat{G}$ and $s(\Gamma)s(\Gamma^\perp) = 1$ (see, for instance, [154, Lemma 6.2.3]). If Γ_i ($i = 1, 2$) are two lattices in G, such that $\Gamma_2 \subset \Gamma_1$, then in the dual space we have $\Gamma_1^\perp \subset \Gamma_2^\perp$; furthermore, in all of the practical situations we know, it is always possible to choose a fundamental domain Ω_i of Γ_i, $i = 1, 2$, in such a way that, as subsets of G, they satisfy $\Omega_1 \subset \Omega_2$.

The whole discussion on the existence and choices of fundamental domains may be repeated in the dual space. In particular, we have the following version of the Weyl formula:

$$\int_{\widehat{G}} f(\chi)\, d\mu_{\widehat{G}}(\chi) = \int_{\widehat{G}/\Gamma^\perp} \sum_{\lambda \in \Gamma^\perp} f(\chi+\lambda)\, d\mu_{G/\Gamma^\perp}(\chi), \quad \forall f \in L^1(\widehat{G}). \tag{4.123}$$

The canonical example is $G = \mathbb{R}$. The unitary characters χ_θ of $\mathbb{R}$ are given by the pure oscillations

$$\langle \chi_\theta, t \rangle = e^{i\theta t}.$$

Each character of $\mathbb{R}$ can, therefore, be identified with a point $\theta \in \mathbb{R}$ that is interpreted as a frequency, and $\widehat{\mathbb{R}}$ is thus identified with $\mathbb{R}$. We take the Lebesgue measure dx as the Haar measure on $\mathbb{R}$; if $d\theta$ denotes the Lebesgue measure on $\widehat{\mathbb{R}}$, then the dual measure of dx is $d\theta/2\pi$.

Let $\Gamma = \mathbb{Z}$; then $\Gamma^\perp = 2\pi\mathbb{Z}$. By Theorem 4.4.1, we have $\widehat{\mathbb{Z}} \simeq \widehat{\mathbb{R}}/\mathbb{Z}^\perp$; the dual of $\mathbb{Z}$ can, therefore, be identified with the unit circle $\mathbb{T} = \mathbb{R}/2\pi\mathbb{Z}$. A fundamental domain of $\mathbb{Z}^\perp$ in $\widehat{\mathbb{R}}$ can be chosen to be $\Omega = [-\pi, \pi)$, and then $s(\mathbb{Z}^\perp) = 1 = s(\mathbb{Z})$. The Haar measure on $\mathbb{Z}$ is the counting measure, and its dual measure on $\mathbb{T}$ is then $d\theta/2\pi$.

4.4.3 Sampling in LCA groups

A natural map from functions defined over the whole group G to functions over the lattice Γ is given by the *sampling operator* (perfect sampling in the terminology of [Hol]) Ξ_Γ associated to Γ, defined as

$$\Xi_\Gamma f = f_{|\Gamma},$$

where f is a function with suitable properties (i.e., for which point values make sense).

A function f over G is said to be Γ-periodic if, for all $x \in G$,

$$f(x + \gamma) = f(x), \text{ for all } \gamma \in \Gamma.$$

Such a function can be identified with a function over G/Γ in the usual way. The periodization operator $P : L^1(G) \to L^1(G/\Gamma)$ is defined as

$$(Pf)(x) = \sum_{\gamma \in \Gamma} f(x + \gamma). \tag{4.124}$$

By the Weyl formula (4.122), we have

$$\langle \phi \mid f \rangle_{L^2(G)} = \langle \phi \mid Pf \rangle_{L^2(G/\Gamma)}, \text{ for any } \Gamma\text{-periodic function } \phi. \tag{4.125}$$

In other words, the periodization operator allows us to write the scalar product of a Γ-periodic function ϕ with an arbitrary function f over G as a scalar product over G/Γ. In the particular case in which $G = \mathbb{R}$ and $\Gamma = \mathbb{Z}$, the Fourier transform of a sequence is a $\mathbb{Z}^\perp$-periodic function, i.e., a 2π-periodic function.

We close this section by giving the useful Poisson summation formula, which links sampling to periodization via the Fourier transform. The proof can be found in most treatises on harmonic analysis (see [Fol],[154], and also [Hol]).

Theorem 4.4.2 (Poisson summation formula): *Let $\Gamma \subset G$ be a lattice. Assume that $f \in L^1(G) \cap \mathcal{F}^{-1}L^1(\widehat{G})$. Then the following is true.*

1. *The Γ-periodized $\phi(\dot{x}) = \sum_{\gamma \in \Gamma} f(x + \gamma)$ of f is in $L^1(G/\Gamma)$ (where $\dot{x} = x + \Gamma$) and for $\eta \in \widehat{G/\Gamma} \simeq \Gamma^\perp$,*

$$\widehat{\phi}(\eta) = \frac{1}{s(\Gamma)} \widehat{f}(\eta). \tag{4.126}$$

2. *We have*

$$\sum_{\gamma \in \Gamma} f(x + \gamma) = \frac{1}{s(\Gamma)} \sum_{\eta \in \Gamma^\perp} \widehat{f}(\eta)\langle \eta, x \rangle \quad a.e. \tag{4.127}$$

$$\sum_{\chi \in \Gamma^\perp} \widehat{f}(\omega + \chi) = s(\Gamma) \sum_{x \in \Gamma} f(x)\overline{\langle \omega, x \rangle} \quad a.e. \tag{4.128}$$

3. *If, in addition, $\sum_{\eta \in \Gamma^\perp} |\widehat{f}(\eta)|^2 < \infty$, then $\phi \in L^2(G/\Gamma)$ and the relations (4.127) and (4.128) hold in the L^2-sense.*

For the canonical example, $G = \mathbb{R}$, $\Gamma = \mathbb{Z}$, one recovers, of course, the usual Poisson summation formula familiar in signal processing [Lyn, Pap] and in harmonic analysis [Kah].

4.5 Lie groups and Lie algebras: A reminder

The theory of Lie groups, Lie algebras, and their representations is widely known and many excellent books cover it, for instance [Hel], [Jac], or [Kna]. Nevertheless, it is useful to recall here a few basic facts that will be used later in this book, and at the same time fix our notation.

4.5.1 Lie algebras

Let $\mathfrak{g}$ be a complex Lie algebra, that is, a complex vector space with an antisymmetric bracket $[.,.]$ that satisfies the Jacobi identity

$$[[X,Y],Z] + [[Y,Z],X] + [[Z,X],Y] = 0, \ \forall X, Y, Z \in \mathfrak{g}. \tag{4.129}$$

For $X, Y \in \mathfrak{g}$, the relation $(\mathrm{ad} X)(Y) = [X, Y]$ gives a linear map $\mathrm{ad}: \mathfrak{g} \to \mathrm{End}\,\mathfrak{g}$ (endomorphisms of $\mathfrak{g}$), called the *adjoint representation* of $\mathfrak{g}$. Next, if $\dim \mathfrak{g} < \infty$, it makes sense to define

$$B(X, Y) = \mathrm{Tr}[(\mathrm{ad} X)(\mathrm{ad} Y)], \ X, Y \in \mathfrak{g}. \tag{4.130}$$

B is a symmetric bilinear form on $\mathfrak{g}$, called the *Killing form* of $\mathfrak{g}$. Alternatively, one may choose a basis $\{X_j, j = 1, \ldots, n\}$ in $\mathfrak{g}$, in terms of which the commutation relations read

$$[X_i, X_j] = \sum_{k=1}^{n} c_{ij}^k X_k, \ i,j = 1, \ldots, n, \tag{4.131}$$

where c_{ij}^k are called the structure constants and $n = \dim \mathfrak{g}$. Then it is easy to see that $g_{ij} = \sum_{k,m=1}^{n} c_{ik}^m c_{jm}^k = B(X_i, X_j)$ defines a metric on $\mathfrak{g}$, called the Cartan–Killing metric.

The Lie algebra $\mathfrak{g}$ is said to be *simple* (respectively, *semisimple*) if it does not contain any nontrivial ideal (respectively, abelian ideal). A semisimple Lie algebra may be decomposed into a direct sum of simple ones. Furthermore, $\mathfrak{g}$ is semisimple iff the Killing form is nondegenerate (Cartan's criterion).

Let $\mathfrak{g}$ be semisimple. Choose in $\mathfrak{g}$ a Cartan subalgebra $\mathfrak{h}$, i.e. a maximal nilpotent subalgebra (it is in fact maximal abelian and unique up to conjugation). The dimension ℓ of $\mathfrak{h}$ is called the *rank* of $\mathfrak{g}$. A *root* of $\mathfrak{g}$ with respect to $\mathfrak{h}$ is a linear form on $\mathfrak{h}$, $\alpha \in \mathfrak{h}^*$, for which there exists $X \neq 0$ in $\mathfrak{g}$ such that $(\mathrm{ad} H)X = \alpha(H)X, \ \forall H \in \mathfrak{h}$.

Then, according to the fundamental work of Cartan and Chevalley, one can find a basis $\{H_i, E_\alpha\}$ of $\mathfrak{g}$, with the following properties. $\{H_j, j = 1, \ldots, \ell\}$ is a basis of $\mathfrak{h}$, and each generator E_α is indexed by a nonzero root α, in such a way that the commutation relations (4.131) may be written in the following form:

$$\begin{aligned}[][H_i, H_j] &= 0, \ j = 1, \ldots, \ell, \\ [H_i, E_\alpha] &= \alpha(H_i) E_\alpha, \ i = 1, \ldots, \ell, \end{aligned} \tag{4.132}$$

$$[E_\alpha, E_{-\alpha}] = H_\alpha \equiv \sum_{i=1}^{\ell} \alpha^i H_i \in \mathfrak{h},$$

$$[E_\alpha, E_\beta] = N_{\alpha\beta} E_{\alpha+\beta},$$

where $N_{\alpha\beta} = 0$ if $\alpha + \beta$ is not a root. Let Δ denote the set of roots of $\mathfrak{g}$. Note that the nonzero roots come in pairs $\alpha \in \Delta \Leftrightarrow -\alpha \in \Delta$, and no other nonzero multiple of a root is a root. Accordingly, the set of nonzero roots may be split into a subset Δ_+ of positive roots and the corresponding subset of negative roots $\Delta_- = \{-\alpha, \alpha \in \Delta_+\}$. The set Δ_+ is contained in a simplex (convex pyramid) in $\mathfrak{h}^*$, the edges of which are the so-called simple positive roots, i.e., positive roots that cannot be decomposed as the sum of two other positive roots [Jac, Kna, 26]. Of course, the same holds for Δ_-. The consideration of root systems is the basis of the Cartan classification of simple Lie algebras into four infinite series $A_\ell, B_\ell, C_\ell, D_\ell$ and five exceptional algebras G_2, F_4, E_6, E_7, E_8 [Jac].

In addition to the Lie algebra $\mathfrak{g}$, one may also consider the universal enveloping algebra $\mathfrak{U}(\mathfrak{g})$, which consists of all polynomials in the elements of $\mathfrak{g}$, modulo the commutation relations (4.131). Of special interest are the so-called Casimir elements, which generate the center of $\mathfrak{U}(\mathfrak{g})$, and in particular, the quadratic Casimir element $C_2 = \sum_{i,k=1}^{n} g_{ik} X^i X^k$, where $\{X^i\}$ is the basis of $\mathfrak{g}$ dual to $\{X_i\}$, i,e, $B(X^i, X_j) = \delta^i_j$ and g_{ik} is the Cartan–Killing metric. The element C_2 does not depend on the choice of the basis $\{X_i\}$. In a Cartan–Chevalley basis $\{H_j, E_\alpha\}$, one gets

$$C_2 = \sum_{j=1}^{\ell} (H_j)^2 + \sum_{\alpha \in \Delta} E_\alpha E_{-\alpha}. \qquad (4.133)$$

In the same way, one defines a real Lie algebra as a real vector space with an antisymmetric bracket $[.,.]$ that satisfies the Jacobi identity. The two concepts are closely related. If $\mathfrak{g}$ is a real Lie algebra, one can define its *complexification* $\mathfrak{g}^c$ by complexifying it as a vector space and extending the Lie bracket by linearity. If dim $\mathfrak{g} = n$, the complex dimension of $\mathfrak{g}^c$ is still n, but its real dimension is $2n$. Conversely, if $\mathfrak{g}$ is a complex Lie algebra, and one restricts its parameter space to real numbers, one obtains a real form $\mathfrak{g}_r$, i.e., a real Lie algebra whose complexification is again $\mathfrak{g}$. A given complex Lie algebra has in general several nonisomorphic real forms (also classified by Cartan), among them a unique compact one, characterized by the fact that the Cartan–Killing metric g_{ij} is negative definite. For instance, the complex Lie algebra A_2 yields two real forms, $\mathfrak{su}(3)$, the compact one, and $\mathfrak{su}(2,1)$.

For physical applications, it is not so much the Lie algebras themselves that matter, but their representations in Hilbert spaces. The key ingredient for building the latter, also due to Cartan, is the notion of a *weight vector*. Let T be a representation of the Lie algebra $\mathfrak{g}$ in the Hilbert space $\mathfrak{H}$, that is, a linear map of $\mathfrak{g}$ into the operators on $\mathfrak{H}$, such that

$$T([X,Y]) = [T(X), T(Y)], \; \forall X, Y \in \mathfrak{g}. \tag{4.134}$$

The representation T is called *hermitian* if $T(X^*) = T(X)^*$, for every $X \in \mathfrak{g}$.

By Schur's lemma, it follows that the Casimir elements of $\mathfrak{U}(\mathfrak{g})$ can be simultaneously diagonalized in any irreducible hermitian representation of $\mathfrak{g}$, and in fact their eigenvalues characterize the representation uniquely.

Let $\{H_j, E_\alpha\}$ be a Cartan–Chevalley basis of the complexified Lie algebra $\mathfrak{g}^c$ of $\mathfrak{g}$. Then a *weight* for T is a linear form $\lambda \in \mathfrak{h}^*$, for which a nonzero vector $|\lambda\rangle \in \mathfrak{H}$ exists, such that $[T(H) - \alpha(H)]^n|\lambda\rangle = 0$, for all $H \in \mathfrak{h}$ and some n. Then $|\lambda\rangle$ is called a weight vector if $n = 1$, that is, $T(H)|\lambda\rangle = \alpha(H)|\lambda\rangle$, $\forall H \in \mathfrak{h}$. In particular, $|\lambda\rangle$ is called a *highest* (respectively *lowest*) *weight vector* if

$$T(E_{\alpha_+})|\lambda\rangle = 0, \; \forall \alpha_+ \in \Delta_+, \; (\text{resp. } T(E_{\alpha_-})|\lambda\rangle = 0, \; \forall \alpha_- \in \Delta_-), \tag{4.135}$$

where Δ_+, respectively Δ_-, is the set of the positive, resp. negative, roots. The interest of this notion is the fundamental result of Cartan, which says that the irreducible, finite-dimensional, hermitian representations of simple Lie algebras are in one-to-one correspondence with highest weight vectors (see [Hum], [Jac], and [26] for the geometrical construction of representations in those terms).

We illustrate this on the simplest example, namely $\mathfrak{su}(2)$, with the familiar angular momentum basis $\{J_o, J_+, J_-\}$ and $\Delta_\pm = \{J_\pm\}$. In the unitary spin-j irreducible representation $D^{(j)}$ of $SU(2)$, of dimension $2j + 1$, with standard basis $\{|j, m\rangle \mid m = -j, \ldots, j\}$, the highest (respectively, lowest) weight vector $|j, j\rangle$ (respectively, $|j, -j\rangle$) satisfies the relation $J_+|j, j\rangle = 0$ (respectively, $J_-|j, -j\rangle = 0$) and has an isotropy subalgebra $\mathfrak{u}(1)$ (by this, we mean the subalgebra that annihilates the given vector). Notice that here $C_2 = J_o^2 + J_+J_- + J_-J_+$ is indeed diagonal in $D^{(j)}$, with eigenvalue $j(j+1)$.

4.5.2 Lie groups

A Lie group G may be defined in several ways [Hel, Kna], for instance, as a smooth manifold with a group structure, such that the group operations $(g_1, g_2) \mapsto g_1 g_2$, $g \mapsto g^{-1}$ are C^k, for some $k \geq 2$. This actually implies that all group operations are in fact (real) analytic. A Lie group is said to be *simple*, respectively, *semisimple*, if it does not have any nontrivial invariant subgroup, respectively, abelian invariant subgroup.

Let G be a Lie group. For $g \in G$, consider the map $L_g : G \to G$ with $L_g(g') = gg'$. The derivative of this map at $g' \in G$, denoted $T_{g'}(L_g)$, sets up an isomorphism $T_{g'}(L_g) : T_{g'}G \to T_{gg'}G$ between the tangent spaces at g and gg'. For any vector $X \in T_eG$ (the tangent space at the identity), let us define a vector field $\widehat{X}$ on G by $\widehat{X}_g = T_e(L_g)X$. Such a vector field is said to be left invariant. It can be demonstrated that the usual Lie bracket $[\widehat{X},\widehat{Y}]$ of two left invariant vector fields is again a left invariant vector field. Using this fact, we see that, if $\widehat{Y}_g = T_e(L_g)Y$, $Y \in T_eG$, the relation $[X,Y] = [\widehat{X},\widehat{Y}]$ defines a Lie bracket on the tangent space, T_eG, at the identity of the Lie group G. Thus, equipped with this bracket relation, T_eG becomes a Lie algebra that we denote by $\mathfrak{g}$.

Next, it can be shown that, for any $X \in \mathfrak{g}$, there exists a unique analytic homomorphism, $t \mapsto \theta_x(t)$ of $\mathbb{R}$ into G such that

$$\dot{\theta}_x(0) = \frac{d}{dt}\theta_x(t)|_{t=0} = X,$$

and, for each $X \in \mathfrak{g}$, we then write $\exp X = \theta_x(1)$. The map $X \mapsto \exp X$ is called the *exponential map*, and it defines a homeomorphism between an open neighborhood N_0 of the origin $0 \in \mathfrak{g}$ and an open neighborhood N_e of the identity element e of G. Each $\theta_x(t)$ defines a one-parameter subgroup of G, with infinitesimal generator X,

$$\theta_x(t) = \exp(tX),$$

and every one-parameter subgroup is obtained in this way. If G is a matrix group, the elements X of the Lie algebra are also matrices and the exponential map comes out in terms of matrix exponentials.

Using this tool, the fundamental theorems of Lie may then be sketched as follows.

1. Every Lie group G has a unique Lie algebra $\mathfrak{g}$, obtained as the vector space of infinitesimal generators of all one parameter subgroups, in other words, the tangent space T_eG, at the identity element of G.

2. Given a Lie algebra $\mathfrak{g}$, one may associate to it in a unique way, by the exponential map $X \mapsto \exp X$, a connected and simply connected Lie group $\overline{G}$, with Lie algebra $\mathfrak{g}$ ($\overline{G}$ is called the universal covering of G). Any other connected Lie group G with the same Lie algebra $\mathfrak{g}$ is of the form $G = \overline{G}/D$, where D is an invariant discrete subgroup of $\overline{G}$.

Furthermore, a Lie group G is simple, respectively, semisimple, if and only if its Lie algebra $\mathfrak{g}$ is simple, respectively, semisimple.

Here again, one may start with a *real* Lie group G, build its complexification G^c, and find all real forms of G^c. One, and only one, of them is compact (it corresponds, of course, to the compact real form of the Lie algebra). For instance, the complex Lie group $SL(2,\mathbb{C})$ has two real forms, $SU(2)$ and $SU(1,1)$, the former being compact.

A Lie group has natural actions on its Lie algebra and its dual. These are the *adjoint* and *coadjoint* actions, respectively, and may be understood in terms of the exponential map. For $g \in G$, $g' \mapsto gg'g^{-1}$ defines a differentiable map from G to itself. The derivative of this map at $g' = e$

4.5. Lie groups and Lie algebras: A reminder

is an invertible linear transformation, Ad_g, of $T_e G$ (or, equivalently, of $\mathfrak{g}$) onto itself, giving the adjoint action. Thus, for $t \in (-\varepsilon, \varepsilon)$, for some $\varepsilon > 0$, such that $\exp(tX) \in G$ and $X \in \mathfrak{g}$,

$$Y = \frac{d}{dt}[g \exp(tX) \, g^{-1}]|_{t=0} := \mathrm{Ad}_g(X) \quad (4.136)$$

is a tangent vector in $T_e G = \mathfrak{g}$. If G is a matrix group, the adjoint action is simply

$$\mathrm{Ad}_g(X) = g X g^{-1}. \quad (4.137)$$

Now considering $g \mapsto \mathrm{Ad}_g$ as a function on G with values in $\mathrm{End}\,\mathfrak{g}$, its derivative at the identity, $g = e$, defines a linear map $\mathrm{ad} : \mathfrak{g} \to \mathrm{End}\,\mathfrak{g}$. Thus, if $g = \exp X$, then $\mathrm{Ad}_g = \exp(\mathrm{ad} X)$, and it can be verified that $(\mathrm{ad} X)(Y) = [X, Y]$.

The corresponding coadjoint actions are now obtained by dualization: The coadjoint action $\mathrm{Ad}_g^\#$ of $g \in G$ on the dual $\mathfrak{g}^*$, of the Lie algebra, is given by

$$<\mathrm{Ad}_g^\#(X^*);\, X> = <X^*;\, \mathrm{Ad}_{g^{-1}}(X)>, \qquad X^* \in \mathfrak{g}^*, \quad X \in \mathfrak{g}, \quad (4.138)$$

where $<\cdot\,;\,\cdot> \equiv <\cdot\,;\,\cdot>_{\mathfrak{g}^*,\mathfrak{g}}$ denotes the dual pairing between $\mathfrak{g}^*$ and $\mathfrak{g}$. Once again, the (negative of the) derivative of the map $g \mapsto \mathrm{Ad}_g^\#$ at $g = e$ is a linear transformation $\mathrm{ad}^\# : \mathfrak{g} \to \mathrm{End}\,\mathfrak{g}^*$, such that, for any $X \in \mathfrak{g}$, $(\mathrm{ad}^\#)(X)$ is the map,

$$<(\mathrm{ad}^\#)(X)(X^*);\, Y> = <X^*;\, (\mathrm{ad})(X)(Y)>, \qquad X^* \in \mathfrak{g}^*, \quad Y \in \mathfrak{g}. \quad (4.139)$$

Clearly, $\mathrm{Ad}_g^\# = \exp(-\mathrm{ad}^\# X)$. If we introduce a basis in $\mathfrak{g}$ and represent Ad_g by a matrix in this basis, then, in terms of the dual basis in $\mathfrak{g}^*$, $\mathrm{Ad}_g^\#$ is represented by the transposed inverse of this matrix.

Under the coadjoint action, the vector space $\mathfrak{g}^*$ splits into a union of disjoint *coadjoint orbits*

$$\mathcal{O}_{X^*} = \{\mathrm{Ad}_g^\#(X^*) \mid g \in G\}. \quad (4.140)$$

According to the Kirillov–Souriau–Kostant theory (see [Gui] or [Woo]), each coadjoint orbit carries a natural symplectic structure. This arises as follows: Let Y^* be an arbitrary point on the coadjoint orbit $\mathcal{O}_{X^*}$ and $t \mapsto s(t)$, $t \in (-\varepsilon, \varepsilon)$, be a smooth curve in $\mathcal{O}_{X^*}$ passing through Y^*. We take $s(t)$ to be of the form

$$s(t) = \mathrm{Ad}_{g(t)}^\# Y^*, \qquad g(t) = \exp(tX), \quad X \in \mathfrak{g}.$$

(Note that $s(0) = Y^*$.) Differentiating with respect to t,

$$\dot{s}(0) = (\mathrm{ad}^\#)(X)(Y^*) := v_X,$$

is a tangent vector to the orbit at the point Y^*. Let

$$\mathfrak{g}_{Y^*} = \{X \in \mathfrak{g} \mid \cdot (\mathrm{ad}^\# X)(Y^*) = 0\}, \quad (4.141)$$

be the annihilator of Y^*. Then, denoting elements in $\mathfrak{g}/\mathfrak{g}_{Y^*}$ by $X \oplus \mathfrak{g}_{Y^*}$, $X \in \mathfrak{g}$, it is clear that the map

$$i_{Y^*} : \mathfrak{g}/\mathfrak{g}_{Y^*} \to T_{Y^*}\mathcal{O}_{X^*}, \qquad i_{Y^*}(X \oplus \mathfrak{g}_{Y^*}) = v_X = (\mathrm{ad}^{\#} X)(Y^*),$$

is a vector space isomorphism. This enables us to identify the tangent space $T_{Y^*}\mathcal{O}_{X^*}$ to the orbit $\mathcal{O}_{X^*}$ at the point Y^*, with $\mathfrak{g}/\mathfrak{g}_{Y^*}$, for all $Y^* \in \mathcal{O}_{X^*}$. For any $Y^* \in \mathcal{O}_{X^*}$, consider now the the antisymmetric bilinear functional Ω_{Y^*}, defined on the tangent space $T_{Y^*}\mathcal{O}_{X^*}$,

$$\Omega_{Y^*}(v_X, v_Y) = \langle Y^* \,;\, [X, Y] \rangle. \tag{4.142}$$

Since

$$\langle (\mathrm{ad}^{\#})(X)(Y^*) \,;\, Y \rangle = \langle Y^* \,;\, [X, Y] \rangle,$$

the expression in (4.142) is well defined and nondegenerate on $T_{Y^*}\mathcal{O}_{X^*}$. Thus, we have defined an antisymmetric bilinear form Ω on the coadjoint orbit $\mathcal{O}_{X^*}$, given at any point $Y^* \in \mathcal{O}_{X^*}$ by (4.142). This form is nondegenerate and closed, i.e., its exterior derivative, $d\Omega = 0$, which follows from the Jacobi identity (4.129) satisfied by the elements of the Lie algebra $\mathfrak{g}$. Hence, Ω defines a symplectic structure on the manifold $\mathcal{O}_{X^*}$. In addition, Ω is G-invariant. This implies, in particular, that the orbit is of even dimension and carries a natural G-invariant (Liouville) measure. Therefore, a coadjoint orbit is a natural candidate for realizing the *phase space* of a classical system and, hence, a starting point for a quantization procedure. Explicit examples of coadjoint orbits for semidirect product groups will be worked out in detail in Section 10.2, where they will turn out to be natural parameter spaces for building coherent states.

Semisimple Lie groups have several interesting decompositions. In the sequel, we are going to use three of them, the Cartan, the Iwasawa, and the Gauss decompositions.

1. Cartan decomposition

This is the simplest case. Given a semisimple Lie group G, its real Lie algebra $\mathfrak{g}$ always possesses a *Cartan involution*, that is, an automorphism $\theta : \mathfrak{g} \to \mathfrak{g}$, with square equal to the identity:

$$\theta[X, Y] = [\theta(X), \theta(Y)], \ \forall\, X, Y \in \mathfrak{g}, \quad \theta^2 = I, \tag{4.143}$$

and such that the symmetric bilinear form $B_\theta(X, Y) = -B(X, \theta Y)$ is positive definite, where B is the Cartan–Killing form. Then, the Cartan involution θ yields an eigenspace decomposition

$$\mathfrak{g} = \mathfrak{k} \oplus \mathfrak{p} \tag{4.144}$$

of $\mathfrak{g}$ into $+1$ and -1 eigenspaces. It follows that

$$[\mathfrak{k}, \mathfrak{k}] \subseteq \mathfrak{k}, \quad [\mathfrak{k}, \mathfrak{p}] \subseteq \mathfrak{p}, \quad [\mathfrak{p}, \mathfrak{p}] \subseteq \mathfrak{k}. \tag{4.145}$$

Assume for simplicity that the center of G is finite, and let K denote the analytic subgroup of G with Lie algebra $\mathfrak{k}$. Then [Kna], (i) K is closed and maximal compact; (ii) there exists a Lie group automorphism Θ of G, with differential θ, such that $\Theta^2 = 1$ and the subgroup fixed by Θ is K; and (iii) the mapping $K \times \mathfrak{p} \to G$ given by $(k, X) \mapsto k \exp X$ is a diffeomorphism onto. One may as well write the diffeomorphism as $(k, X) \mapsto \exp X \, k \equiv p \, k$, and in that form we recognize the Cartan decomposition given explicitly in (4.66) for $SU(1,1)$.

2. *Iwasawa decomposition*

Any connected semisimple Lie group G has an Iwasawa decomposition into three closed subgroups, namely $G = KAN$, where K is a maximal compact subgroup, A is abelian, and N is nilpotent, and the last two are simply connected [Kna]. This means that every element $g \in G$ admits a *unique* factorization $g = kan$, $k \in K, a \in A, n \in N$, and the multiplication map $K \times A \times N \to G$ given by $(k, a, n) \mapsto kan$ is a diffeomorphism onto.

Assume that G has a finite center. Let M be the centralizer of A in K, that is, $M = \{k \in K : ka = ak, \forall a \in A\}$ (if the center of G is not finite, the definition of M is slightly more involved). Then, $P = MAN$ is a closed subgroup of G, called the minimal parabolic subgroup. The interest of this subgroup is that the quotient manifold $\mathcal{X} = G/P \sim K/M$ carries the unitary irreducible representations of the principal series of G (which are induced representations), in the sense that these representations are realized in the Hilbert space $L^2(\mathcal{X}, d\nu)$, where ν is the natural G-invariant measure. To give a concrete example, take $G = SO_o(3,1)$, the Lorentz group. Then, the Iwasawa decomposition reads as

$$SO_o(3,1) = SO(3) \cdot A \cdot N, \tag{4.146}$$

where $A \sim SO_o(1,1) \sim \mathbb{R}$ is the subgroup of Lorentz boosts in the z-direction and $N \sim \mathbb{C}$ has dimension two and is abelian. In this case, $M = SO(2)$, the subgroup of rotations around the z-axis, so that $\mathcal{X} = G/P \sim SO(3)/SO(2) \sim S^2$, the two-sphere. This example is the geometrical framework for the construction of wavelets on the sphere S^2, which we shall describe in Chapter 15, Section 15.1.

A closely related decomposition is the so-called KAK decomposition [Kna]. Again, let G be a semisimple Lie group with a finite center, K be a maximal compact subgroup, and $G = KAN$ be the corresponding Iwasawa decomposition. Then, every element in G has a decomposition as $k_1 a k_2$ with $k_1, k_2 \in K$ and $a \in A$. This decomposition is in general *not* unique, but a is unique up to conjugation. A familiar example of a KAK decomposition is the expression of a general rotation $\gamma \in SO(3)$ as the product of three rotations, parametrized by the Euler angles

$$\gamma = m(\psi) \, u(\theta) \, m(\varphi), \tag{4.147}$$

where m and u denote rotations around the z-axis and the y-axis, respectively.

3. Gauss decomposition

Again, let G be a semisimple Lie group and $G^c = \exp \mathfrak{g}^c$ be the corresponding complexified group. If $\mathfrak{b}$ is a subalgebra of $\mathfrak{g}^c$, we call it *maximal* (in the sense of Perelomov [Per]) if $\mathfrak{b} \oplus \bar{\mathfrak{b}} = \mathfrak{g}^c$, where $\bar{\mathfrak{b}}$ is the conjugate of $\mathfrak{b}$ in $\mathfrak{g}^c$. Let $\{H_j, E_\alpha\}$ a Cartan–Chevalley basis of the complexified Lie algebra $\mathfrak{g}^c$.

The complex group G^c possesses remarkable subgroups, namely,

1. H^c, the Cartan subgroup, which is generated by $\{H_j\}$.

2. $B_\pm$, called the Borel subgroups, which are maximal connected, solvable, subgroups, corresponding to the subalgebras $\mathfrak{b}_\pm$, generated by $\{H_j, E_\alpha \,|\, \alpha \in \Delta_\pm\}$; if $\mathfrak{b}$ is maximal, then $\mathfrak{b}_+ = \mathfrak{b}$ and $\mathfrak{b}_- = \bar{\mathfrak{b}}$ generate Borel subgroups.

3. $Z_\pm$, the connected nilpotent subgroups generated by $\{E_{\alpha_\pm} \,|\, \alpha_\pm \in \Delta_\pm\}$.

The interest of these subgroups is that almost all elements of G^c admit a *Gauss decomposition*:

$$g = z_+ h z_- = b_+ z_- = z_+ b_-, \quad z_\pm \in Z_\pm, \quad h \in H^c, \quad b_\pm \in B_\pm. \quad (4.148)$$

It follows that the quotients $X_+ = G^c / B_-$ and $X_- = B_+ \backslash G^c$ are compact complex homogeneous manifolds, on which G^c acts by holomorphic transformations. This is the crucial fact behind the geometry of some CS, of group-theoretical origin, as first pointed out in [238].

4.5.3 Extensions of Lie algebras and Lie groups

Exactly as Mackey induction is a method for building representations of a group from representations of a subgroup, it is useful to have a method for constructing a group G from two smaller ones, one of them at least becoming a closed subgroup of G. Several possibilities are available, of increasing sophistication.

1. Direct product

This is the most trivial solution, which amounts to glue the two groups together without interaction. Given two (topological or Lie) groups G_1, G_2, their direct product $G = G_1 \times G_2$ is simply their Cartesian product, endowed with the group law:

$$(g_1, g_2)(g_1', g_2') = (g_1 g_1', g_2 g_2'), \quad g_1, g_1' \in G_1, \, g_2, g_2' \in G_2. \quad (4.149)$$

With the obvious identifications $g_1 \sim (g_1, e_2)$, $g_2 \sim (e_1, g_2)$, where e_j denotes the neutral element of G_j, $j = 1, 2$, it is clear that both G_1 and G_2

are invariant subgroups of $G_1 \times G_2$. In the case of Lie groups, the notion of direct product corresponds to that of direct sum of the corresponding Lie algebras, $\mathfrak{g} = \mathfrak{g}_1 \oplus \mathfrak{g}_2$, and again both $\mathfrak{g}_1$ and $\mathfrak{g}_2$ are ideals of $\mathfrak{g}$.

2. Semidirect product

A more interesting construction arises when one of the groups, say, G_2, acts on the other one, G_1, by automorphisms. More precisely, one has a homomorphism α from G_2 into the group $\operatorname{Aut} G_1$ of automorphisms of G_1. Although the general definition may be given as in the first case, we will consider in the sequel only the case in which $G_1 \equiv V$ is abelian, in fact, a vector space (hence, group operations are noted additively), and $G_2 \equiv S$ is a subgroup of $\operatorname{Aut} V$. Then, we define the semidirect product $G = V \rtimes S$ as the Cartesian product, endowed with the group law:

$$(v, s)(v', s') = (v + \alpha_s(v'), ss'), \quad v, v' \in V, \ s, s' \in S. \tag{4.150}$$

According to the law (4.150), the neutral element of G is $(0, e_S)$ and the inverse of (v, s) is $(v, s)^{-1} = (-\alpha_s^{-1}(v), s^{-1}) = (-\alpha_{s^{-1}}(v), s^{-1})$. It is easy to check that V is an invariant subgroup of G, while S is not in general. As a matter of fact, S is invariant iff the automorphism α is trivial, i.e., the product is direct. Indeed, one has readily

$$(v, s)(v', e_S)(v, s)^{-1} = (\alpha_s(v'), e_S) \in V$$

and

$$(v, s)(0, s')(v, s)^{-1} = (v, ss')(-\alpha_s^{-1}(v), s^{-1}) = (v - \alpha_{ss's^{-1}}(v), ss's^{-1}).$$

In addition to the Weyl–Heisenberg group $G_{WH} = \mathbb{R} \rtimes \mathbb{R}^2$ that we discussed in Chapter 2, examples of semidirect products of this type are the following groups, which we will study in detail in the sequel:

1. the Euclidean group $E(n) = \mathbb{R}^n \rtimes SO(n)$;

2. the Poincaré group $\mathcal{P}_+^\uparrow(1,3) = \mathbb{R}^n \rtimes SO_o(1,3)$, where the second factor is the Lorentz group;

3. the similitude group $SIM(n) = \mathbb{R}^n \rtimes (\mathbb{R}_*^+ \times SO(n))$, where $\mathbb{R}_*^+$ is the group of dilations, whereas $SO(n)$ denotes the rotations, as in the first example (since these two operations commute, one gets here a direct product).

3. Group extension

An even more general concept is that of extension. We say that G is an *extension* of G_1 by G_2 if $G_1 \simeq G/G_2$. Thus, here G_2 is an invariant subgroup of G, but G_1 need not even be a subgroup. The extension is said to be *central* if G_2 is contained in the center of G.

The most efficient way of studying group extensions is by using the language of exact sequences [Bar, 220]. Let $\{H_k, k \in I \subset \mathbb{N}\}$ be a sequence of groups, and, for each $k \in I$, let $i_k : H_{k-1} \to H_k$ be a homomorphism:

$$\cdots \to H_{k-1} \xrightarrow{i_k} H_k \xrightarrow{i_{k+1}} H_{k+1} \to \cdots$$

Then, the sequence $\{H_k, k \in I\}$ is *exact* if, for each $k \in I$, $\text{Im}\, i_k = \text{Ker}\, i_{k+1} \subset H_k$. Let $f : G \to G'$ be any homomorphism from G into another group G'. Then, an obvious example of an exact sequence is

$$1 \to \text{Ker} f \to G \xrightarrow{f} \text{Im} f \to 1,$$

where 1 denotes the trivial group with one element. Other ones are

$$1 \to Z(G) \to G \to \text{Int}\, G \to 1,$$

where $Z(G)$ and $\text{Int}\, G$ denote, respectively, the center and the group of inner automorphisms (i.e., conjugations) of G, and

$$1 \to \text{Int}\, G \to \text{Aut}\, G \to \text{Out}\, G \to 1,$$

where $\text{Out}\, G \simeq \text{Aut}\, G/\text{Int}\, G$ denotes the group of outer automorphisms of G.

Using this language, we say that G is an extension of G_1 by G_2 if the following sequence is exact:

$$1 \to G_2 \xrightarrow{i} G \xrightarrow{\pi} G_1 \to 1. \qquad (4.151)$$

Now, the exactness of the first triple, $1 \to G_2 \xrightarrow{i} G$, means that i is an injection; that is, G_2 is a subgroup of G. Similarly, the exactness of $G \xrightarrow{\pi} G_1 \to 1$ means that π is a surjection. Then, the exactness of the central triple means precisely that $G_2 \simeq \text{Ker}\, \pi$, i.e., $G_1 = \text{Im}\, \pi \simeq G/G_2$, as announced.

As a consequence of the definition, each element $g \in G$ induces an automorphism of G_2 by conjugation, $h \mapsto ghg^{-1}$, $h \in G_2$, and thus one gets a homomorphism from G into $\text{Aut}\, G_2$. From this, one deduces several additional exact sequences, and one gets:

$$\begin{array}{ccccccccc}
 & & 1 & & & & & & \\
 & & \downarrow & & & & & & \\
 & & Z(G_2) & & & & & & \\
 & & \downarrow & & & & & & \\
1 & \to & G_2 & \to & G & \to & G_1 & \to & 1 \\
 & & \downarrow & & \downarrow & & \downarrow & & \\
1 & \to & \text{Int}\, G_2 & \to & \text{Aut}\, G_2 & \to & \text{Out}\, G_2 & \to & 1. \\
 & & \downarrow & & & & & & \\
 & & 1 & & & & & &
\end{array} \qquad (4.152)$$

This diagram simplifies if G_2 is abelian, which is the only case we shall need. Then, indeed, $Z(G_2) = G_2$, so that $\text{Int}\, G_2 = 1$ and $\text{Aut}\, G_2 = \text{Out}\, G_2$. Thus,

(4.152) reduces to the scheme:

$$1 \to G_2 \xrightarrow{i} G \xrightarrow{\pi} G_1 \to 1. \atop \downarrow \alpha \atop \mathrm{Aut}\, G_2 \qquad (4.153)$$

In these terms, the extension problem may be formulated as follows: Given a group G_1, an abelian group G_2, and an homomorphism $\alpha : G_1 \to \mathrm{Aut}\, G_2$, to find all groups G such that the conjugation by $g \in G$ induces on G_2 the automorphism $\alpha \circ \pi(g)$. Whereas the general extension problem (4.152) may have no solution for a given homomorphism $\alpha : G_1 \to \mathrm{Out}\, G_2$, the restricted problem (4.153) always has one (trivial) solution, namely, the semidirect product $G = G_2 \rtimes G_1$, which corresponds to the double exact sequence

$$1 \to G_2 \xrightarrow{i} G \underset{\sigma}{\overset{\pi}{\rightleftarrows}} G_1 \rightleftarrows 1, \qquad (4.154)$$

where the homomorphism $\sigma : G_1 \to G$ is an injection (so that G_1 is now a subgroup of G), such that $\pi \circ \sigma = 1$, the identity [in that case, one says that the exact sequence (4.154) splits].

The canonical example of central extension is, of course, the Weyl–Heisenberg group $G_{WH} = \mathbb{R} \rtimes \mathbb{R}^2$. Besides that one, the most familiar case is the construction of the full or extended Galilei group $\widetilde{\mathcal{G}}$ ($\equiv G$) as a central extension of the ordinary Galilei group $\mathcal{G}$ ($\equiv G_1$) by $\mathbb{R}$ ($\equiv G_2$). The elements of $\widetilde{\mathcal{G}}$ are pairs $\tilde{g} = (\theta, g)$, with $\theta \in \mathbb{R}$ and $g \in \mathcal{G}$, and the group law reads

$$\tilde{g}\tilde{g}' = (\theta, g)(\theta', g') = (\theta + \theta' + \xi(g, g'), gg'),$$

where $\xi : \mathcal{G} \times \mathcal{G} \to \mathbb{R}$ is again a multiplier, as in (2.31). This example will be treated in great detail in Chapter 9, Section 9.2.1. The problem of central extensions will also reappear in a crucial fashion in the study of CS for the affine Galilei group, which is an extension of a dilation group of dimension 2 by the Galilei group, which will be discussed in Chapter 15, Section 15.3.2.

4.5.4 Contraction of Lie algebras and Lie groups

We conclude this section by recalling some basic facts concerning the process of contraction, both for Lie algebras and Lie groups and their representations. Let $\mathfrak{g}_1 = (V, [.,.]_1)$ and $\mathfrak{g}_2 = (V, [.,.]_2)$ be two Lie algebras on the same vector space V. We say that $\mathfrak{g}_2$ is a *contraction* of $\mathfrak{g}_1$ if there is a one-parameter family of invertible linear mappings $\phi_\epsilon, \epsilon \in (0, 1]$, from V to V such that

$$\lim_{\epsilon \to 0} \phi_\epsilon^{-1} [\phi_\epsilon X, \phi_\epsilon Y]_1 = [X, Y]_2, \ \forall\, X, Y \in V. \qquad (4.155)$$

The limit (4.155) defines a new Lie algebra structure on V, which is not isomorphic to the original one. In the case of an Inönü–Wigner contraction [181], a particular subalgebra of $\mathfrak{g}_1$ is conserved throughout the process. More precisely, suppose $\mathfrak{g}_1$ decomposes into a subalgebra $\mathfrak{s}$ in and a vector subspace $\mathfrak{v}_c$, that is

$$\mathfrak{g}_1 = \mathfrak{s} + \mathfrak{v}_c, \tag{4.156}$$

in such a way that

$$[\mathfrak{s},\mathfrak{s}]_2 \subset \mathfrak{s}, \quad [\mathfrak{v}_c,\mathfrak{v}_c]_2 = 0, \quad [\mathfrak{s},\mathfrak{v}_c] \subset \mathfrak{v}_c. \tag{4.157}$$

Using (4.156), we can decompose any $X \in V$ as

$$X = X_\mathfrak{s} + X_c, \quad X_\mathfrak{s} \in \mathfrak{s}, \quad X_c \in \mathfrak{v}_c,$$

and define the contracting mappings

$$\phi_\epsilon(X) = X_\mathfrak{s} + \epsilon X_c.$$

Then, applying (4.155) does not affect the subalgebra $\mathfrak{s}$. We say in this case that we have a contraction of $\mathfrak{g}_1$ along $\mathfrak{s}$.

The contraction process may be lifted to the corresponding Lie groups [111, 112]. Again, let $\mathfrak{g}_1$ and $\mathfrak{g}_2$ be two Lie algebrasn such that $\mathfrak{g}_2$ is a contraction of $\mathfrak{g}_1$. Let G_1 be the simply connected Lie group with Lie algebra $\mathfrak{g}_1$. Let S be the subgroup of G_1 whose Lie algebra is $\mathfrak{s}$ in the decomposition (4.156). Defining the semidirect product

$$G_2 \equiv V_c \rtimes S, \quad V_c = \exp \mathfrak{v}_c \simeq \mathfrak{v}_c,$$

one easily checks that $\mathfrak{g}_2 = (V, [.,.]_2)$ is the Lie algebra of G_2. Consider now the family of maps $\Pi_\epsilon : G_2 \to G_1$ given by

$$\Pi_\epsilon : (v,s) \mapsto \left(\exp_{G_1} \epsilon v\right) \cdot s. \tag{4.158}$$

They play essentially the same role as the maps ϕ_ϵ of (4.155) at the level of the corresponding groups, namely,

$$\lim_{\epsilon \to 0} \Pi_\epsilon^{-1}\left(\Pi_\epsilon(g) \overset{1}{\circ} \Pi_\epsilon(g')\right) = g \overset{2}{\circ} g', \quad \forall g, g' \in G_2, \tag{4.159}$$

where $\overset{1}{\circ}, \overset{2}{\circ}$ denote the product in G_1, G_2, respectively. Indeed, one checks readily that $T_e\Pi_\epsilon = \phi_\epsilon$, where $T_e\Pi_\epsilon$ is the derivative of Π_ϵ evaluated at the neutral element of G_1. It is easy to see on (4.158) that the subgroup S is preserved during the contraction.

Classical examples of group contraction are that of $SO(3)$ into the 2-D Euclidean group $E(2) = \mathbb{R}^2 \rtimes SO(2)$ and those of the relativity groups

$$\begin{array}{c} SO_o(1,4) \\ \kappa \to 0 \searrow \\ \mathcal{P}_+^\uparrow(1,3) \overset{c\to\infty}{\Longrightarrow} \widetilde{\mathcal{G}}(1,3), \\ \nearrow \\ SO_o(2,3) \end{array} \tag{4.160}$$

where $SO_o(1,4)$ is the de Sitter group, $SO_o(2,3)$ is the anti-de Sitter group, $\mathcal{P}_+^\uparrow(1,3)$ is the Poincaré group and $\widetilde{\mathcal{G}}(1,3)$ is the extended Galilei group, whereas κ is the curvature of the anti-de Sitter universe and, of course, c is the speed of light. We will study these examples in Chapter 11.

The last step is to apply the contraction method to group representations. Whereas contractions of Lie algebras and Lie groups are relatively ancient and well known [181, 264], the extension of the procedure to group representations is rather recent [221]. A rigorous version has been given by Dooley [111, 112], which we follow. The additional difficulty here is that the representation space itself varies during the contraction.

Let G_2 be a contraction of G_1, defined by the contraction map $\Pi_\epsilon : G_2 \to G_1$, and let U be a representation of G_2 in a Hilbert space $\mathfrak{H}$. Suppose that, for each $\epsilon \in (0,1]$, we have a representation $\{\mathfrak{H}_\epsilon, U_\epsilon\}$ of G_1, a dense subspace $\mathcal{D}_\epsilon$ of $\mathfrak{H}$ and a linear injective map $I_\epsilon : \mathfrak{H}_\epsilon \to \mathcal{D}_\epsilon$ (this map is called a *precontraction* in [255]). Then, one says that the representation U of G_2 is a contraction of the family of representations $\{U_\epsilon\}$ of G_1 if one can find a dense subspace $\mathcal{D}$ of $\mathfrak{H}$ such that, for all $\phi \in \mathcal{D}$ and $g \in G_2$, for every ϵ small enough, $\phi \in \mathcal{D}_\epsilon$, $U_\epsilon(\Pi_\epsilon(g)) I_\epsilon^{-1}\phi \in I_\epsilon^{-1}(\mathcal{D}_\epsilon)$ and

$$\lim_{\epsilon \to 0} \|I_\epsilon U_\epsilon(\Pi_\epsilon g) I_\epsilon^{-1}\phi - U(g)\phi\|_{\mathfrak{H}} = 0. \tag{4.161}$$

This definition is not yet fully satisfactory, however, in the sense that the vector ϕ is fixed, independently of ϵ, and this is not always natural. An example will be discussed in Chapter 11, Section 11.3. There, in the contraction of $SU(1,1)$ towards the Poincaré group $\mathcal{P}_+^\uparrow(1,1)$, the states that contract correctly (namely, the CS) do indeed depend on the contraction parameter. Hence, we replace in (4.161) the fixed vector ϕ by $I_\epsilon\phi_\epsilon$ and demand [255]:

$$\lim_{\epsilon \to 0} \|I_\epsilon U_\epsilon(\Pi_\epsilon g) \phi_\epsilon - U(g)I_\epsilon\phi_\epsilon\|_{\mathfrak{H}} = 0. \tag{4.162}$$

Then, one has to find the appropriate domain of contraction, that is, the set of vectors $\phi_\epsilon \in \mathcal{D}_\epsilon$, such that the limit $\lim_{\epsilon \to 0} I_\epsilon\phi_\epsilon$ exists in some suitable sense.

As we will see in Chapter 11, this (extended) notion of contraction of group representations gives a precise mathematical meaning to the relationship between the three relativity groups mentioned above, at the quantum level. As a consequence, it yields also the convergence into one another of the corresponding CS systems.

5
Hilbert Spaces with Reproducing Kernels and Coherent States

This chapter is somewhat technical in nature. On the other hand, the treatment of reproducing kernel Hilbert spaces given below is rather different from that normally found in the literature, and the level of generality adopted assures immediate applicability of the concept to the various geometric and functional analytic contexts in which they are later required. Unfortunately, the level of technicality apparent in our treatment of reproducing kernels is rather high, but it was deemed necessary for the development of the theory. We have tried to lighten the reading by including illustrative examples.

It must have become clear by now that reproducing kernels and Hilbert spaces possessing such kernels are at the heart of the theory of CS, and we have seen several examples of them in Chapter 2. A systematic development of some of the main features of such spaces is therefore in order here.

5.1 A motivating example

In order to motivate the discussion, let us go back to the canonical coherent states,

$$\eta^s_{\sigma(q,p)} = e^{i(pQ-qP)}\eta^s,$$

as defined in (2.23). Each one of these vectors, $\eta^s_{\sigma(q,p)}$, defines a one-dimensional subspace, $\mathfrak{K}_{(q,p)} = \mathbb{C}[\eta^s_{\sigma(q,p)}]$, of the underlying Hilbert space $\mathfrak{H}$ (which is also the space spanned by these vectors). As a subspace

90   5. Hilbert Spaces with Reproducing Kernels and Coherent States

of $\mathfrak{H}$, the space $\mathfrak{K}_{(q,p)}$ is itself a (one-dimensional) Hilbert space, under the same scalar product as on $\mathfrak{H}$. The collection of all of these spaces, $\{\mathfrak{K}_{(q,p)} \mid (q,p) \in \mathbb{R}^2\}$, forms a measurable field of Hilbert spaces, in a sense to be made precise below. (As a matter of fact, in this case, the set $\{\mathfrak{K}_{(q,p)} \mid (q,p) \in \mathbb{R}^2\}$ actually defines a line bundle, with base space $\mathbb{R}^2$, and fiber $\mathfrak{K}_{(q,p)}$, above each point $(q,p) \in \mathbb{R}^2$.) Consider now the operator

$$\widehat{K}(q,p;\, q'p') = |\eta^s_{\sigma(q,p)}\rangle\langle\eta^s_{\sigma(q,p)}|\eta^s_{\sigma(q',p')}\rangle\langle\eta^s_{\sigma(q',p')}| \quad (5.1)$$

on $\mathfrak{H}$, defined for each choice of pairs, (q,p) and (q',p'), of points in $\mathbb{R}^2$. Using the positive operators (in this case, simply one-dimensional projection operators)

$$F(q,p) = |\eta^s_{\sigma(q,p)}\rangle\langle\eta^s_{\sigma(q,p)}|, \quad (q,p) \in \mathbb{R}^2, \quad (5.2)$$

(5.1) may be rewritten in the form,

$$\widehat{K}(q,p;\, q',p') = F(q,p)^{\frac{1}{2}} F(q',p')^{\frac{1}{2}}. \quad (5.3)$$

Since, for any vector $\phi \in \mathfrak{K}_{(q',p')}$, the vector $\widehat{K}(q,p;\, q'p')\phi$ lies in $\mathfrak{K}_{(q,p)}$, we may see $\widehat{K}(q,p;\, q'p')$ as a linear map from $\mathfrak{K}_{(q',p')}$ to $\mathfrak{K}_{(q,p)}$, which satisfies

$$\widehat{K}(q,p;\, q'p')^* = \widehat{K}(q',p';\, q,p),$$

and, for $(q,p) = (q',p')$, is a strictly positive operator on $\mathfrak{K}_{(q',p')}$. Such a mapping is called a positive-definite kernel. Next, for each $(q,p) \in \mathbb{R}^2$, let us define a vector-valued function $\widehat{\xi}_{\sigma(q,p)}$ that associates to each $(q',p') \in \mathbb{R}^2$ the vector,

$$\widehat{\xi}_{\sigma(q,p)}(q',p') = \widehat{K}(q',p';\, q,p)\eta^s_{\sigma(q,p)},$$

in $\mathfrak{K}_{(q',p')}$, and equip the linear span of all of these vector functions $\widehat{\xi}_{\sigma(q,p)}$, with the scalar product:

$$\langle\widehat{\xi}_{\sigma(q,p)}|\widehat{\xi}_{\sigma(q',p')}\rangle_{\widehat{K}} = \langle\eta^s_{(q,p)}|\eta^s_{(q',p')}\rangle_{\mathfrak{H}}. \quad (5.4)$$

The closure of this linear span, in the above scalar product, is then a Hilbert space, which we denote by $\widehat{\mathfrak{H}}_{\widehat{K}}$. Let us emphasize again that this is a space of vector-valued functions. The set of vectors $\widehat{\xi}_{\sigma(q,p)}$, $(q,p) \in \mathbb{R}^2$, is clearly total in $\widehat{\mathfrak{H}}_{\widehat{K}}$, and the association $\eta^s_{(q,p)} \mapsto \widehat{\xi}_{\sigma(q,p)}$ sets up a bijective isometry W between the spaces $\mathfrak{H}$ and $\widehat{\mathfrak{H}}_{\widehat{K}}$. The Hilbert space $\widehat{\mathfrak{H}}_{\widehat{K}}$ is a reproducing kernel Hilbert space with the reproducing kernel given by (5.1). The reproducing property of the kernel is expressed by the fact that if $\widehat{\Phi}$ is any element in $\widehat{\mathfrak{H}}_{\widehat{K}}$ [i.e., $\widehat{\Phi}$ is a vector-valued function with $\widehat{\Phi}(q,p) \in \mathfrak{K}_{(q,p)}$], then, for any $\phi \in \mathfrak{K}_{(q,p)}$,

$$\langle\phi|\widehat{\Phi}(q,p)\rangle_{\mathfrak{K}_{(q,p)}} = \langle\widehat{K}(\,\cdot\,;q,p)\phi|\widehat{\Phi}\rangle_{\widehat{K}}, \quad (5.5)$$

where $\widehat{K}(\,\cdot\,;q,p)\phi$ is the element, $\widehat{\Psi} \in \widehat{\mathfrak{H}}_{\widehat{K}}$, such that $\widehat{\Psi}(q',p') = \widehat{K}(q',p'\,;q,p)\phi$. This also means that the evaluation map $E_{\widehat{K}}(q,p) : \widehat{\mathfrak{H}}_{\widehat{K}} \to$

5.1. A motivating example

$\mathfrak{K}_{(q,p)}$, which associates to each $\widehat{\Phi} \in \widehat{\mathfrak{H}}_{\widehat{K}}$ the vector $E_{\widehat{K}}(q,p)\widehat{\Phi} = \widehat{\Phi}(q,p)$, is continuous as a linear map. Indeed, if $E_{\widehat{K}}(q,p)^*$ denotes the dual of this map, then it is not hard to see that

$$\widehat{K}(q,p;q'p') = E_{\widehat{K}}(q,p)E_{\widehat{K}}(q',p')^* \quad \text{and} \quad F_{\widehat{K}}(q,p) = E_{\widehat{K}}(q,p)^* E_{\widehat{K}}(q,p), \tag{5.6}$$

where $F_{\widehat{K}}(q,p)$ is the image of the operator $F(q,p)$ under the isometry $W : \mathfrak{H} \to \widehat{\mathfrak{H}}_{\widehat{K}}$.

Finally, note that, if we choose in each one of the Hilbert spaces $\mathfrak{K}_{(q,p)}$, the basis vector $\eta^s_{\sigma(q,p)}$ and write the reproducing kernel $\widehat{K}(q,p;q',p')$ in this basis, then we get precisely the reproducing kernel $K(q,p;q',p') = \langle \eta^s_{\sigma(q,p)} | \eta^s_{\sigma(q',p')} \rangle$, encountered in Chapter 2 [see (2.42)].

The reproducing kernel defined above has a square integrability property, arising from the resolution of the identity condition satisfied by $F(q,p)$ on $\mathfrak{H}$ [see (2.38)]:

$$\int_{\mathbf{R}^2} F(q,p) \frac{dq\,dp}{2\pi} = I. \tag{5.7}$$

Indeed, this implies that the reproducing kernel $\widehat{K}(q,p;q'p')$ satisfies

$$\int_{\mathbf{R}^2} \widehat{K}(q,p;q''p'')\widehat{K}(q'',p'';q'p') \frac{dq''\,dp''}{2\pi} = \widehat{K}(q,p;q'p'), \tag{5.8}$$

which then says that the scalar product (5.4) in the reproducing kernel Hilbert space $\widehat{\mathfrak{H}}_{\widehat{K}}$ may be rewritten as an L^2-type scalar product:

$$\langle \widehat{\xi}_{\sigma(q,p)} | \widehat{\xi}_{\sigma(q,p)} \rangle_K = \int_{\mathbf{R}^2} \langle \widehat{\xi}_{\sigma(q,p)}(q',p') | \widehat{\xi}_{\sigma(q,p)}(q',p') \rangle_{\mathfrak{K}_{(q',p')}} \frac{dq'\,dp'}{2\pi}. \tag{5.9}$$

In the rest of this chapter, we shall put all of these observations on a wider footing, in obtaining a comprehensive generalization of the concept of a reproducing kernel Hilbert space. Whenever we have a family of CS, there is an associated reproducing kernel Hilbert space, and the basic ingredients in constructing such a space are either POV measures or measurable families of Hilbert spaces. Before leaving this example, we would like to emphasize that a reproducing kernel, and an associated Hilbert space, can always be defined as above, whenever we are given a positive-definite kernel on a measurable family of Hilbert spaces. The resulting reproducing kernel Hilbert space, while necessarily a space of (vector-valued) functions, does not have to be an L^2-space. The possibility of embedding it into an L^2-type of space requires, in addition, the existence of a resolution of the identity type of relation, such as in (5.7).

5.2 Measurable fields and direct integrals

Let us begin with a locally compact space X equipped with a Borel measure. For technical convenience, we shall assume throughout that the *support* of ν is *all* of X. This means, for instance, that, if $U \subset X$ is any *open* set, and $f : U \to \mathbb{R}^+$ a continuous, positive function, then

$$\int_U f(x)\, d\nu(x) = 0 \quad \Rightarrow \quad f(x) = 0, \quad \forall x \in U.$$

Suppose that to each $x \in X$ we associate a Hilbert space $\mathfrak{K}_x$. Let $\langle \,\cdot\, | \,\cdot\, \rangle_x$ and $\|\ldots\|_x$ denote the scalar product and norm, respectively, in $\mathfrak{K}_x$. (The Cartesian product space $\prod_{x \in X} \mathfrak{K}_x$ has a natural vector space structure that we assume for it. The following discussion on measurable fields and direct integrals of Hilbert spaces has been adapted from [Ta2].)

Definition 5.2.1: *The family $\{\mathfrak{K}_x \,|\, x \in X\}$ is called a measurable field of Hilbert spaces if there exists a subspace $\mathfrak{M}$ of the product space $\prod_{x \in X} \mathfrak{K}_x$ such that*

1. *for each $\Phi \in \mathfrak{M}$, the positive, real-valued function $x \mapsto \|\Phi(x)\|_x$ on X is ν-measurable;*

2. *if, for any $\Phi \in \prod_{x \in X} \mathfrak{K}_x$, the complex-valued functions $x \mapsto \langle \Phi(x) | \Psi(x) \rangle_x$, for all $\Psi \in \mathfrak{M}$, are ν-measurable, then $\Phi \in \mathfrak{M}$; and*

3. *there exists a countable subset $\{\Phi_n\}_{n=1}^{\infty}$ of $\mathfrak{M}$ such that for each $x \in X$ the set of vectors $\{\Phi(x)_n\}_{n=1}^{\infty}$ is total in $\mathfrak{K}_x$.*

Elements in $\mathfrak{M}$ are called *ν-measurable vector fields*, and the set $\{\Phi_n\}_{n=1}^{\infty}$ is called a *fundamental sequence* of ν-measurable vector fields. Condition 3 also implies that each $\mathfrak{K}_x$ is a separable Hilbert space, and, indeed, we shall mostly deal with the case in which they are all finite dimensional. Measurable fields of Hilbert spaces are useful for building *direct integrals* of Hilbert spaces, which are generalizations of direct sums. The next two lemmas give criteria for identifying measurable fields of Hilbert spaces and the existence of a very convenient type of fundamental sequences.

Lemma 5.2.2: *Let $\{\Phi_n\}_{n=1}^{\infty} \subset \prod_{x \in X} \mathfrak{K}_x$ satisfy the following:*

1. *for each m and n, the function $x \mapsto \langle \Phi_m(x) | \Phi_n(x) \rangle_x$ on X is ν-measurable; and*

2. *for each $x \in X$, the sequence of vectors $\{\Phi_n(x)\}_{n=1}^{\infty}$ is total in $\mathfrak{K}_x$.*

Then the set

$$\mathfrak{M} = \{\Psi \in \prod_{x \in X} \mathfrak{K}_x \,|\, x \mapsto \langle \Phi_n(x) | \Psi(x) \rangle_x \text{ is } \nu\text{-measurable for all } n\}$$

5.2. Measurable fields and direct integrals

satisfies conditions 1, 2 and 3 of Definition 5.2.1, and hence $\{\mathfrak{K}_x \,|\, x \in X\}$ is a measurable field of Hilbert spaces and $\mathfrak{M}$ is a ν-measurable field of vectors.

Lemma 5.2.3: Let $\{\mathfrak{K}_x \,|\, x \in X\}$ be a measurable field of Hilbert spaces on $\{X,\nu\}$. Then the function $x \mapsto N(x) = \dim \mathfrak{K}_x$ on X is measurable, and there exists a fundamental sequence $\{\Phi_n\}_{n=1}^{\infty}$ of measurable vector fields such that

1. $\{\Phi_n(x)\}_{n=1}^{N(x)}$ is an orthonormal basis for $\mathfrak{K}_x$, $\forall x \in X$;
2. $\Phi_{N(x)+k}(x) = 0$, $k = 1, 2, \ldots$, if $N(x) < \infty$.

Suppose now that we are given $\{X, \nu\}$, as before, and a measurable field of Hilbert spaces $\{\mathfrak{K}_x \,|\, x \in X\}$ along with the set of ν-measurable vector fields $\mathfrak{M}$. Let $\widetilde{\mathfrak{H}} \subset \mathfrak{M}$ be the collection of all (ν-equivalence classes of) vector fields Φ satisfying

$$\|\Phi\|^2 := \int_X \|\Phi(x)\|_x^2 \, d\nu(x) < \infty, \tag{5.10}$$

and define on it the scalar product

$$\langle \Phi | \Psi \rangle = \int_X \langle \Phi(x) | \Psi(x) \rangle_x \, d\nu(x), \quad \Phi, \Psi \in \widetilde{\mathfrak{H}}. \tag{5.11}$$

(As usual, this scalar product is nondegenerate on equivalence classes.) It can be shown that $\widetilde{\mathfrak{H}}$ is complete in the norm (5.10), and, hence, equipped with the scalar product (5.11), it becomes a Hilbert space. We call $\widetilde{\mathfrak{H}}$ the *direct integral* of the measurable field of Hilbert spaces $\{\mathfrak{K}_x \,|\, x \in X\}$ and write

$$\widetilde{\mathfrak{H}} = \int_X^{\oplus} \mathfrak{K}_x \, d\nu(x). \tag{5.12}$$

If $\mathfrak{K}_x = \mathfrak{K}$, some fixed Hilbert space, for all $x \in X$, the field $\{\mathfrak{K}_x \,|\, x \in X\}$ is called a *constant field of Hilbert spaces* and then it is easily seen that (as a unitary equivalence)

$$\int_X^{\oplus} \mathfrak{K}_x \, d\nu(x) \cong \mathfrak{K} \otimes L^2(X, d\nu). \tag{5.13}$$

Example using a POV function

The following example of the construction of a measurable field of Hilbert spaces and their direct integral will be extremely valuable to us in the sequel. Let $\mathfrak{H}$ be an arbitrary Hilbert space and let $\{X,\nu\}$ be as above. Let $F : X \to \mathcal{L}(\mathfrak{H})^+$ be a weakly measurable, positive operator-valued function on X, i.e., a function of the variable x, such that its values $F(x)$ are positive operators on $\mathfrak{H}$ and, for arbitrary $\phi, \psi \in \mathfrak{H}$, the complex-valued function

$x \mapsto \langle\phi|F(x)\psi\rangle$ is measurable, and assume that $F(x) \neq 0$, $\forall x \in X$. (Note, we are also assuming that the support of ν is all of X.) An example of such a function is the map $(q,p) \mapsto F(q,p) = |\eta^s_{\sigma(q,p)}\rangle\langle\eta^s_{\sigma(q,p)}|$ defined in (5.2). We now construct a measurable field of Hilbert spaces and a direct integral associated in a natural way to the operator-valued function F. Since $F(x)$ is a positive operator, it has a *square root*, denoted $F(x)^{\frac{1}{2}}$, with the property, $F(x)^{\frac{1}{2}} F(x)^{\frac{1}{2}} = F(x)$. For each $x \in X$, consider the range of this square root operator:

$$\operatorname{Ran} F(x)^{\frac{1}{2}} = \{F(x)^{\frac{1}{2}}\phi \mid \phi \in \mathfrak{H}\} \subset \mathfrak{H}. \tag{5.14}$$

In general, this is not a closed subspace of $\mathfrak{H}$; let $\mathfrak{K}_x$ be its closure. Denote by $\|\ldots\|_x$ and $\langle \cdot | \cdot \rangle_x$ the restrictions of the norm and scalar product, respectively, of $\mathfrak{H}$ to $\mathfrak{K}_x$. (Note that, in particular, if $F(x)$ has finite rank, then $\operatorname{Ran} F(x)^{\frac{1}{2}} = \mathfrak{K}_x$.) Let $\{\phi_n\}_{n=1}^{\infty}$ be a basis of $\mathfrak{H}$, and consider the sequence of vectors $\Phi_n \in \prod_{x \in X} \mathfrak{K}_x$, $n = 1, 2, \ldots$, with

$$\Phi_n(x) = F(x)^{\frac{1}{2}}\phi_n, \quad n = 1, 2, \ldots, \quad x \in X. \tag{5.15}$$

The measurability of the function $x \mapsto F(x)$ now implies that $\{\Phi_n\}_{n=1}^{\infty}$ satisfies the conditions of Lemma 5.2.2. Hence, we can construct the set of ν-measurable fields $\mathfrak{M}$ following that lemma and obtain a direct integral Hilbert space, as in (5.12).

5.3 Reproducing kernel Hilbert spaces

Using the foregoing notion of measurable fields of Hilbert spaces, we now proceed to discuss reproducing kernel Hilbert spaces of equivalence classes of measurable fields. The presentation given here borrows heavily from [88], [162], and [Mes], although we introduce a few convenient modifications which bring out the connection between measurable POV functions and reproducing kernel Hilbert spaces (see Theorem 5.3.8).

5.3.1 *Positive-definite kernels and evaluation maps*

The starting point is once more a locally compact space X equipped with a regular Borel measure ν, the support of which is all of X, a measurable field of Hilbert spaces $\{\mathfrak{K}_x \mid x \in X\}$ and an associated set $\mathfrak{M}$ of ν-measurable vector fields. We assume that $\dim \mathfrak{K}_x$ is a *fixed finite number* n for all $x \in X$. (The assumption of finiteness of n is made for technical convenience and can be relaxed if necessary.) For $x, y \in X$, denote by $\mathcal{L}(\mathfrak{K}_x, \mathfrak{K}_y)$ the space of all (bounded) linear maps from $\mathfrak{K}_x$ to $\mathfrak{K}_y$. If $A \in \mathcal{L}(\mathfrak{K}_x, \mathfrak{K}_y)$ and $A^* : \mathfrak{K}_y \to \mathfrak{K}_x$ is the adjoint map,

$$\langle A^* v_y | v_x \rangle_x = \langle v_y | A v_x \rangle_y, \quad v_x \in \mathfrak{K}_x, \ v_y \in \mathfrak{K}_y,$$

then $A^* \in \mathcal{L}(\mathfrak{K}_y, \mathfrak{K}_x)$. [Note: $\mathcal{L}(\mathfrak{K}_x, \mathfrak{K}_x) = \mathcal{L}(\mathfrak{K}_x)$.]

Definition 5.3.1 (Measurable positive-definite kernel): *A measurable positive-definite ν-kernel, on a measurable field of Hilbert spaces $\{\mathfrak{K}_x \,|\, x \in X\}$, is a map that associates with each pair of points $(x,y) \in X \times X$ an element $K(x,y) \in \mathcal{L}(\mathfrak{K}_x, \mathfrak{K}_y)$ and has the following properties:*

1. *if $x_1, x_2, \ldots, x_N$ is any finite set of points in X and $v_i \in \mathfrak{K}_{x_i}$, $i = 1, 2, \ldots, N$, are arbitrary vectors, then*

$$\sum_{i,j=1}^{N} \langle v_i | K(x_i, x_j) v_j \rangle \geq 0; \tag{5.16}$$

2. *for each $x \in X$, the operator $K(x,x) \in \mathcal{L}(\mathfrak{K}_x)$ is strictly positive-definite, i.e.,*

$$K(x,x) v_x = 0 \Leftrightarrow v_x = 0; \tag{5.17}$$

3. *for fixed $y \in X$ and $v_y \in \mathfrak{K}_y$, the function $x \mapsto K(x,y) v_y$, denoted $K(\cdot, y) v_y$, is an element of the set $\mathfrak{M}$ of the ν-measurable vector fields associated with $\{\mathfrak{K}_x \,|\, x \in X\}$.*

Note that conditions 1 and 2 imply that, for all $x, y \in X$,

$$K(x,y) = K(y,x)^*. \tag{5.18}$$

To check the validity of this last relation, let $v_x \in \mathfrak{K}_x$ and $v_y \in \mathfrak{K}_y$ be arbitrary. By (5.16),

$$\langle v_x | K(x,x) v_x \rangle + \langle v_y | K(y,y) v_y \rangle + \langle v_x | K(x,y) v_y \rangle + \langle v_y | K(y,x) v_x \rangle \geq 0.$$

Hence, by (5.17), the quantity

$$\langle v_x | K(x,y) v_y \rangle + \langle v_y | K(y,x) v_x \rangle$$

is real for all $v_x \in \mathfrak{K}_x$ and $v_y \in \mathfrak{K}_y$; i.e.,

$$\langle v_x | K(x,y) v_y \rangle + \overline{\langle v_x | K(x,y)^* v_y \rangle}$$

is real. This can only be true, however, if (5.18) holds.

In the usual definition of positive kernels and the construction of reproducing kernel Hilbert spaces, using such kernels, the strict positivity condition (5.17) is not postulated. We assume it here, however, to eliminate unnecessary technicalities later.

For $\Phi \in \mathfrak{M}$, let $[\Phi]$ denote its ν-equivalence class; i.e., if $\Psi \in \mathfrak{M}$, then $\Psi \in [\Phi]$ if and only if $\Psi(x) = \Phi(x)$ for ν-almost all $x \in X$. In other words, any two elements of $\mathfrak{M}$, differing only on a set of ν-measure zero, belong to the same ν-equivalence class. The measurable, positive-definite ν-kernel K is said to be *admissible* if, for any $x, y \in X$ and $v_x \in \mathfrak{K}_x$, $v_y \in \mathfrak{K}_y$, the equality $[K(\cdot, y) v_y] = [K(\cdot, x) v_x]$ holds if and only if $K(z, y) v_y = K(z, x) v_x$ for *all* $z \in X$. Let $[\mathfrak{M}]$ denote the set of all ν-equivalence classes of $\mathfrak{M}$, and

96   5. Hilbert Spaces with Reproducing Kernels and Coherent States

similarly, if $\mathfrak{F}$ is a subspace of $\mathfrak{M}$, we shall denote by $[\mathfrak{F}]$ the corresponding subspace of $[\mathfrak{M}]$.

Definition 5.3.2: *A ν-selection on a linear subspace $[\mathfrak{F}]$ of $[\mathfrak{M}]$ is a linear map $\sigma : [\mathfrak{F}] \to \mathfrak{F}$, such that, for all $[\Phi] \in [\mathfrak{F}]$, one has $[\sigma([\Phi])] = [\Phi]$.*

To understand the significance of the above definition, we note that, given $[\mathfrak{F}]$, it is always possible to choose an element $\Phi \in \mathfrak{F}$ as representative of an equivalence class $[\Phi] \in [\mathfrak{F}]$. It is not always possible, however, to make the correspondence $[\Phi] \to \Phi$ linear for the entire subspace $[\mathfrak{F}]$. The existence of a ν-selection on a particular subspace $[\mathfrak{F}]$ ensures this linearity. When a ν-selection exists on a subspace $[\mathfrak{F}]$ of $[\mathfrak{M}]$, we shall simply write Φ for $[\Phi]$, with the understanding that $\sigma([\Phi]) = \Phi$.

Theorem 5.3.3: *Let K be an admissible, positive-definite ν-kernel on the measurable field of Hilbert spaces $\{\mathfrak{K}_x \mid x \in X\}$. Then, there exists a unique Hilbert space $\mathfrak{H}_K \subset [\mathfrak{M}]$ of ν-equivalence classes of measurable vector fields satisfying the following conditions:*

1. *the set*
$$\mathcal{S}_K = \{[K(\cdot, x)v_x] \mid x \in X, \quad v_x \in \mathfrak{K}_x\} \tag{5.19}$$
is total in $\mathfrak{H}_K$;

2. *there exists a ν-selection $\sigma : \mathfrak{H}_K \to \mathfrak{M}$ such that the linear map*
$$E_K(x) : \mathfrak{H}_K \to \mathfrak{K}_x, \qquad E_K(x)[\Phi] = \sigma([\Phi])(x), \tag{5.20}$$
is bounded (in the norms of $\mathfrak{H}_K$ and $\mathfrak{K}_x$) for each $x \in X$;

3. *if $E_K(x)^* : \mathfrak{K}_x \to \mathfrak{H}_K$ is the dual map of $E_K(x)$, then*
$$K(x, y) = E_K(x) E_K(y)^*. \tag{5.21}$$

Proof. Consider finite sums of vectors in $\mathcal{S}_K$ of the type
$$[\Phi^N] = \sum_{i=1}^{N} [K(\cdot, x_i) v_i], \quad [\Psi^M] = \sum_{j=1}^{M} [K(\cdot, y_j) w_j],$$
$$x_i, y_j \in X, \quad v_i \in \mathfrak{K}_{x_i}, \quad w_j \in \mathfrak{K}_{y_j}.$$

Using these, we may define a *unique sesquilinear form*,
$$\langle [\Phi^N] \mid [\Psi^M] \rangle_K = \sum_{i=1}^{N} \sum_{j=1}^{M} \langle v_i \mid K(x_i, y_j) w_j \rangle_{x_i}, \tag{5.22}$$

or, equivalently,
$$\langle [K(\cdot, x) v_x] \mid [K(\cdot, y) w_y] \rangle_K = \langle v_x \mid K(x, y) w_y \rangle_x, \tag{5.23}$$

on the linear span of $\mathcal{S}_K$. [This is obviously a *reproducing* condition, analogous to (5.8).] Indeed, the admissibility of K shows first of all that

5.3. Reproducing kernel Hilbert spaces   97

$\langle\,\cdot\,|\,\cdot\,\rangle_K$ is well defined; for, if $[K(\cdot,x')v_{x'}] = [K(\cdot,x)v_x]$ and $[K(\cdot,y')w_{y'}] = [K(\cdot,y)w_y]$, then

$$\begin{aligned}\langle v_{x'}|K(x',y')w_{y'}\rangle_{x'} &= \langle v_{x'}|K(x',y)w_y\rangle_{x'} = \langle K(y,x')v_{x'}|w_y\rangle_y \\ &= \langle K(y,x)v_x|w_y\rangle_y = \langle v_x|K(x,y)w_y\rangle_x,\end{aligned}$$

which is the consistency condition required by (5.23). Furthermore, $\langle\,\cdot\,|\,\cdot\,\rangle_K$ is *nondegenerate*, since, for each $x \in X$,

$$\begin{aligned}\|\Phi^N(x)\|_x^2 &= \sum_{i,j=1}^N \langle K(x,x_i)v_i|K(x,x_j)v_j\rangle_x \\ &= \sum_{i,j=1}^N \langle[K(\cdot,x)K(x,x_i)v_i]\,|\,[K(\cdot,x_j)v_j]\rangle_K \\ &= \langle[K(\cdot,x)\Phi^N(x)]\,|\,[\Phi^N]\rangle_K \le \|[K(\cdot,x)\Phi^N(x)]\|_K\,\|[\Phi^N]\|_K,\end{aligned}$$

in virtue of the Cauchy–Schwarz inequality (with an obvious meaning for $\|\ldots\|_K$). But,

$$\begin{aligned}\|[K(\cdot,x)\Phi^N(x)]\|_K^2 &= \langle[K(\cdot,x)\Phi^N(x)]\,|\,[K(\cdot,x)\Phi^N(x)]\rangle_K \\ &= \langle\Phi^N(x)|K(x,x)\Phi^N(x)\rangle_K \\ &= \|K(x,x)^{\frac{1}{2}}\Phi^N(x)\|_x^2\,.\end{aligned}$$

Thus,

$$\|\Phi^N(x)\|_x \le \|K(x,x)\|^{\frac{1}{2}}\,\|[\Phi^N]\|_K, \qquad (5.24)$$

and, hence, $\|[\Phi^N]\|_K = 0$, so that $\Phi^N(x) = 0$ for all $x \in X$. On the other hand, $\Phi^N(x) = 0, \forall x \in X$ surely implies that $\|[\Phi^N]\|_K = 0$.

Completing the linear span of $\mathcal{S}_K$ in the scalar product (5.22), we obtain a Hilbert space $\mathfrak{H}_K$. To see that $\mathfrak{H}_K \subset [\mathfrak{M}]$, let $[\Phi^N] = \sum_{j\in I_N}[K(\cdot,x_j)v_j]$, $N = 1,2,3,\ldots$, where I_N is a finite index set for each N, be a Cauchy sequence in the linear span of $\mathcal{S}_K$. Then, (5.24) shows that, for each $x \in X$, $\Phi^N(x)$ is a Cauchy sequence in $\mathfrak{K}_x$. Since $\mathfrak{K}_x$ is a Hilbert space, there is, for each $x \in X$, a $\Phi(x) \in \mathfrak{K}_x$ such that $\lim_{N\to\infty}\|\Phi^N(x) - \Phi(x)\|_x = 0$. Moreover, standard measure theoretic arguments show that $\Phi \in \mathfrak{M}$. Similarly, letting $\|[\Phi]\| = \lim_{N\to\infty}\|[\Phi^N]\|$, it is easily seen that $[\Phi] \in \mathfrak{H}_K$ and that $\|[\Phi^N] - [\Phi]\|_K \to 0$.

To prove the existence of a ν-selection on $\mathfrak{H}_K$, define a pointwise map σ on vectors $[\Phi^N] \in \mathfrak{H}_K$ as

$$\sigma([\Phi^N])(x) = \Phi^N(x), \qquad x \in X. \qquad (5.25)$$

Then, by (5.24),

$$\|\sigma([\Phi^N])(x)\|_x \le \|K(x,x)\|^{\frac{1}{2}}\,\|[\Phi^N]\|_K,$$

and therefore σ can be extended to all of $\mathfrak{H}_K$. Thus, the norm estimate

$$\|\sigma([\Phi])(x)\|_x \le \|K(x,x)\|^{\frac{1}{2}}\,\|[\Phi]\|_K, \quad [\Phi] \in \mathfrak{H}_K, \qquad (5.26)$$

is generally valid, and, consequently, σ defines a ν-selection on $\mathfrak{H}_K$. This also proves the boundedness of the evaluation map $E_K(x)$ in (5.20) as well as of the dual map $E_K(x)^*$. Next, for arbitrary $x, y \in X$, $v_x \in \mathfrak{K}_x$ and $w_y \in \mathfrak{K}_y$,

$$\langle v_x | K(x,y) w_y \rangle_x = \langle v_x | E_K(x)[K(\cdot, y) w_y] \rangle_x = \langle E_K(x)^* v_x | [K(\cdot, y) w_y] \rangle_K.$$

From this, using (5.23) and the fact that $\mathcal{S}_K$ is total in $\mathfrak{H}_K$, we get

$$E_K(x)^* v_x = [K(\cdot, x) v_x]. \tag{5.27}$$

Hence,

$$E_K(x) E_K(y)^* w_y = E_K(x)[K(\cdot, y) w_y] = K(x, y) w_y,$$

proving (5.21). □

From now on, we assume that $n < \infty$. It is not hard to see that the following three conditions are equivalent (recall that we are assuming that $\dim \mathfrak{K}_x = n < \infty$, for all $x \in X$):

$$K(x, x) > 0$$
$$E_K(x) \mathfrak{H}_K = \mathfrak{K}_x \tag{5.28}$$
$$E_K(x)^* v_x = 0 \Leftrightarrow v_x = 0. \tag{5.29}$$

In view of Theorem 5.3.3, from now on, we shall write Φ for elements $[\Phi] \in \mathfrak{H}_K$, it being understood that

$$\Phi(x) = E_K(x) \Phi = \sigma([\Phi])(x) \in \mathfrak{K}_x. \tag{5.30}$$

With this convention, elements in $\mathcal{S}_K$ will be written

$$\xi[v_x] = E_K(x)^* v_x = K(\cdot, x) v_x. \tag{5.31}$$

Definition 5.3.4 (Reproducing kernel Hilbert space): *The Hilbert space $\mathfrak{H}_K$, constructed in the above manner, using a measurable field of Hilbert spaces, is called a reproducing kernel Hilbert space, and the operator function $K(x, y)$ is called a reproducing kernel for $\mathfrak{H}_K$. The map $E_K(x) : \mathfrak{H}_K \to \mathfrak{K}_x$ is called the evaluation map at $x \in X$.*

It is readily proven that, given a Hilbert space of equivalence classes of measurable vector fields (taken from a measurable field of Hilbert spaces $\{\mathfrak{K}_x \mid x \in X\}$), a reproducing kernel, when it exists, is unique. Furthermore, given such a Hilbert space, the defining condition for a reproducing kernel is that it constitute a map that associates with each pair $(x, y) \in X \times X$ an operator $K(x, y) \in \mathcal{L}(\mathfrak{K}_y, \mathfrak{K}_x)$, such that, for fixed x and v_x, the vector field $K(\cdot, x) v_x$ is measurable and , the *reproducing condition* (5.23) is satisfied. Alternatively, in view of (5.31), the reproducing condition may also be written as

$$\langle K(\cdot, x) v_x | \Phi \rangle_K = \langle v_x | \Phi(x) \rangle_x, \tag{5.32}$$

for all $x \in X$, $v_x \in \mathfrak{K}_x$ and $\Phi \in \mathfrak{H}_K$.

5.3.2 Coherent states and POV functions

Let us write for the range of $E_K(x)^*$

$$\mathfrak{H}_K(x) = E_K(x)^* \mathfrak{K}_x \subset \mathfrak{H}_K. \tag{5.33}$$

Then, (5.29) shows that $\mathfrak{H}_K(x)$ is an n-dimensional subspace of $\mathfrak{H}_K$ and $E_K(x)^*$ is a vector space isomorphism. Let $\mathbb{P}_K(x)$ be the projection operator

$$\mathbb{P}_K(x)\mathfrak{H}_K = \mathfrak{H}_K(x). \tag{5.34}$$

We claim that

$$\mathbb{P}_K(x) = E_K(x)^* K(x,x)^{-1} E_K(x); \tag{5.35}$$

that is, for any $\Phi \in \mathfrak{H}_K$,

$$\mathbb{P}_K(x)\Phi = E_K(x)^* K(x,x)^{-1} \Phi(x), \tag{5.36}$$

and, in particular,

$$\Phi = E_K(x)^* K(x,x)^{-1} \Phi(x), \quad \forall \Phi \in \mathfrak{H}_K(x). \tag{5.37}$$

Eq. (5.35) indeed defines a projection operator. The relations $\mathbb{P}_K(x)^* = \mathbb{P}_K(x)^2 = \mathbb{P}_K(x)$ are immediate to verify, noting the equality $K(x,x) = E_K(x)E_K(x)^*$ that follows from (5.21).

Define the map $V_K(x) : \mathfrak{K}_x \to \mathfrak{H}_K(x)$, by

$$V_K(x) = E_K(x)^* K(x,x)^{-\frac{1}{2}}, \tag{5.38}$$

where $K(x,x)^{-\frac{1}{2}}$ denotes the inverse of the (bounded) positive square root operator $K(x,x)^{\frac{1}{2}} \in \mathcal{L}(\mathfrak{K}_x)$. Then $V_K(x)^* = K(x,x)^{-\frac{1}{2}} E_K(x)$.

The map $V_K(x)$ is unitary. Indeed, by inspection, $V_K(x)^* V_K(x) = I_{\mathfrak{K}_x}$ (the unit operator on $\mathfrak{K}_x$), and, for all $v_x \in \mathfrak{K}_x$, one has

$$V_K(x)V_K(x)^* E_K(x)^* v_x = E_K(x)^* K(x,x)^{-1} E_K(x)E_K(x)^* v_x = E_K(x)^* v_x.$$

Defining the operator

$$F_K(x) = V_K(x) K(x,x) V_K(x)^{-1} \tag{5.39}$$

as the image of $K(x,x)$ on $\mathfrak{H}_K(x)$ under $V_K(x)$, we find that

$$F_K(x) E_K(x)^* v_x = E_K(x)^* K(x,x) v_x. \tag{5.40}$$

Extend $F_K(x)$ to all of $\mathfrak{H}_K$ by defining it to be zero on the orthogonal complement of $\mathfrak{H}_K(x)$, and, for simplicity of notation, denote both this extended operator and its restriction to $\mathfrak{H}_K(x)$ by the same symbol. Then, we obtain

$$F_K(x) = E_K(x)^* E_K(x). \tag{5.41}$$

Thus, $F_K(x)$ is a bounded positive operator on $\mathfrak{H}_K$ of norm

$$\|F_K(x)\| = \|K(x,x)\|. \tag{5.42}$$

The projection operator $\mathbb{P}_K(x)$ is the *support* of $F_K(x)$, which implies that $\mathbb{P}_K(x)F_K(x)\mathbb{P}_K(x) = F_K(x)$. Furthermore, since the square root operator is $F_K(x)^{\frac{1}{2}} = V_K(x)K(x,x)^{\frac{1}{2}}V_K(x)^{-1}$, the following is easily seen to hold:

$$F_K(x)^{\frac{1}{2}}\Phi = V_K(x)\Phi(x) \in \mathfrak{H}_K(x), \quad \forall \Phi \in \mathfrak{H}_K(x), \tag{5.43}$$

whence,

$$E_K(x)^*v_x = F_K(x)^{\frac{1}{2}}V_K(x)v_x \tag{5.44}$$

and

$$F_K(x)^{\frac{1}{2}}E_K(x)^*v_x = E_K(x)^*K(x,x)^{\frac{1}{2}}v_x. \tag{5.45}$$

Since, for arbitrary $\Phi, \Psi \in \mathfrak{H}_K$, (5.41) implies

$$\langle \Phi | F_K(x)\Psi \rangle_K = \langle \Phi(x) | \Psi(x) \rangle_x, \tag{5.46}$$

and Φ, Ψ are elements of the measurable field of vectors $\mathfrak{M}$, the positive operator-valued function $x \mapsto F_K(x)$ is measurable which, in view of the example worked out in Section 5.2, also implies that $\{\mathfrak{H}_K(x) \mid x \in X\}$ is a measurable field of Hilbert spaces. The associated set $\mathfrak{M}_K$ of measurable vector fields consists of all elements in $\prod_{x \in X} \mathfrak{H}_K(x)$ of the type $x \mapsto V_K(x)\Phi(x)$, where $\Phi \in \mathfrak{M}$. Additionally, the rank of the operator $F_K(x)$ is finite and equal to n, for each $x \in X$, and the set of vectors $\{F_K(x)^{\frac{1}{2}}\Phi \mid x \in X, \ \Phi \in \mathfrak{H}_K\}$ is total in $\mathfrak{H}_K$.

Definition 5.3.5: *The function F_K is said to be the POV function canonically associated with the reproducing kernel Hilbert space $\mathfrak{H}_K$.*

The operator $F_K(x)$, being a rank n positive operator, can be written in the form

$$F_K(x) = \sum_{i=1}^{n} |\xi_x^i\rangle\langle\xi_x^i|, \tag{5.47}$$

where the vectors $\xi_x^i \in \mathfrak{H}_K(x)$, $i = 1, 2, \ldots, n$, are linearly independent (and nonzero). Moreover, if $v_x^i = V_K(x)^{-1}\xi_x^i \in \mathfrak{K}_x$, $i = 1, 2, \ldots, n$, is the corresponding (linearly independent) set of vectors in $\mathfrak{K}_x$, then

$$K(x,x) = \sum_{i=1}^{n} |v_x^i\rangle\langle v_x^i|, \tag{5.48}$$

and we infer from (5.38) that

$$\xi_x^i = K(\cdot,x)K(x,x)^{-\frac{1}{2}}v_x^i = E_K(x)^*K(x,x)^{-\frac{1}{2}}v_x^i. \tag{5.49}$$

Definition 5.3.6 (Coherent states): *The set of vectors*

$$\mathfrak{S}_K = \{\xi_x^i \mid x \in X, \ i = 1, 2, \ldots, n\} \subset \mathcal{S}_K \tag{5.50}$$

is called a family of coherent states (CS), associated with the reproducing kernel K.

The above definition of CS is the most general one employed in this book. In later chapters, we shall look at more specific families of CS, which satisfy additional properties, usually arising from group symmetries. Clearly, the set of CS is total in $\mathfrak{H}_K$. Moreover, the measurability properties of $K(x,y)$ and an application of Lemma 5.2.3 demonstrate that, for each i, $x \mapsto \xi_x^i \in \mathfrak{H}_K(x)$ can be chosen as a measurable vector field. A point ought to be made here about the choice of the linearly independent set of vectors $\{\xi_x^i\}_{i=1}^n$ used to represent $F_K(x)$. If we were to choose for these vectors the eigenvectors of $F_K(x)$, then (5.47) would simply be its spectral representation and the ξ_x^i its eigenvectors. On the other hand, starting with this representation, if we define a second set of linearly independent vectors $\eta_x^i = \sum_{j=1}^n \mathcal{U}(x)_{ji} \xi_x^j$, $i = 1, 2, \ldots, n$, where the $\mathcal{U}(x)_{ji}$ are the elements of an $n \times n$ unitary matrix, then once again $F_K(x) = \sum_{i=1}^n |\eta_x^i\rangle\langle\eta_x^i|$. The vectors η_x^i, however, are no longer necessarily mutually orthogonal.

5.3.3 Some isomorphisms, bases, and ν-selections

Given the reproducing kernel Hilbert space, constructed above, it is immediate to verify that the quantity

$$\widehat{K}(x,y) = F_K(x)^{\frac{1}{2}} F_K(y)^{\frac{1}{2}} \tag{5.51}$$

is an admissible, positive-definite ν-kernel on the measurable field of Hilbert spaces $\{\mathfrak{H}_K(x) \mid x \in X\}$. Thus, using this kernel, we could construct a second Hilbert space $\mathfrak{H}_{\widehat{K}}$, for which $\widehat{K}$ would be the reproducing kernel. Canonically associated with this new Hilbert space, one would have a POV function $x \mapsto F_{\widehat{K}}(x)$. As we show in the next lemma, however, this new Hilbert space $\mathfrak{H}_{\widehat{K}}$ is not sensibly different from $\mathfrak{H}_K$.

Lemma 5.3.7: *The mapping*

$$W : \mathfrak{H}_K \to \mathfrak{H}_{\widehat{K}}, \qquad (W\Phi)(x) = F_K(x)^{\frac{1}{2}} \Phi, \tag{5.52}$$

is unitary and maps $F_K(x)$ onto $F_{\widehat{K}}(x)$, for each $x \in X$; i.e.,

$$F_{\widehat{K}}(x) = W F_K(x) W^{-1}. \tag{5.53}$$

Furthermore, the respective evaluation maps and reproducing kernels are related by

$$E_{\widehat{K}}(x) = V_K(x) E_K(x) W^{-1} = F_K(x)^{\frac{1}{2}} W^{-1} \tag{5.54}$$

$$\widehat{K}(x,y) = V_K(x) K(x,y) V_K(y)^{-1} = E_{\widehat{K}}(x) E_{\widehat{K}}(y)^*, \tag{5.55}$$

while the coherent states $\widehat{\xi}_x^i$, $i = 1, 2, \ldots, n$, satisfying

$$F_{\widehat{K}}(x) = \sum_{i=1}^n |\widehat{\xi}_x^i\rangle\langle\widehat{\xi}_x^i|, \tag{5.56}$$

are the vectors
$$\widehat{\xi}_x^i = W\xi_x^i = WV_K(x)v_x^i. \tag{5.57}$$

Proof. Clearly W is a linear map, and since the vectors $\Phi \in \mathfrak{H}_K(x)$, as x ranges through X, are total in $\mathfrak{H}_K$, it is enough to prove the unitarity of W on such vectors. Let $F_K(x)^{-\frac{1}{2}} : \mathfrak{H}_K(x) \to \mathfrak{H}_K(x)$ be the inverse of the operator $F_K(x)^{\frac{1}{2}}$ restricted to $\mathfrak{H}_K(x)$. For such a vector Φ,
$$(W\Phi)(y) = F_K(y)^{\frac{1}{2}}\Phi = \widehat{K}(y,x)F_K(x)^{-\frac{1}{2}}\Phi,$$
and, hence, for all $\Phi \in \mathfrak{H}_K(x)$,
$$W\Phi = \widehat{K}(\cdot,x)F_K(x)^{-\frac{1}{2}}\Phi. \tag{5.58}$$
Thus, using the reproducing property (5.32) in $\mathfrak{H}_{\widehat{K}}$, we obtain, for $\Phi, \Psi \in \mathfrak{H}_K(x)$,
$$\begin{aligned}
\langle W\Phi | \Psi \rangle_{\widehat{K}} &= \langle \widehat{K}(\cdot,x)F_K(x)^{-\frac{1}{2}}\Phi | \Psi \rangle_{\widehat{K}} \\
&= \langle F_K(x)^{-\frac{1}{2}}\Phi | \Psi \rangle_{\mathfrak{H}_K(x)} \\
&= \langle \Phi | F_K(x)^{-\frac{1}{2}}\Psi \rangle_{\mathfrak{H}_K(x)} \\
&= \langle \Phi | W^*\Psi \rangle_K,
\end{aligned}$$
which yields $W^*\Psi = F_K(x)^{-\frac{1}{2}}\Psi$, for each $\Psi \in \mathfrak{H}_K(x)$. With this and (5.52), we obtain immediately that $WW^* = W^*W = I_{\mathfrak{H}_K(x)}$, thus proving the unitarity of W.

Next, for $\Phi \in \mathfrak{H}_K$, (5.52) and (5.43) imply
$$E_{\widehat{K}}(x)(W\Phi) = (W\Phi)(x) = F_K(x)^{\frac{1}{2}}\Phi = V_K(x)\Phi(x) = V_K(x) \circ E_K(x)\Phi,$$
from which (5.53), (5.54), and (5.55) follow. Finally, (5.57) results from (5.49), upon noting that $F_{\widehat{K}}(x) = E_{\widehat{K}}(x)^* E_{\widehat{K}}(x)$. □

It emerges from the above discussion that a reproducing kernel Hilbert space is always associated to a POV function. This is made precise in the following theorem, as a direct consequence of the foregoing lemma.

Theorem 5.3.8: Let $\mathfrak{H}$ be an abstract Hilbert space, $x \mapsto F(x)$ a weakly measurable POV function on $\{X, \nu\}$, with values in $\mathcal{L}(\mathfrak{H})^+$, satisfying:

1. rank $F(x) = n < \infty$, $n \neq 0$, for all $x \in X$;

2. the set of vectors
$$\mathcal{S} = \{F(x)^{\frac{1}{2}}\phi \mid \phi \in \mathfrak{H} \quad x \in X\} \tag{5.59}$$
is total in $\mathfrak{H}$.

Then, $\{\mathfrak{K}_x \mid x \in X\}$, where $\mathfrak{K}_x \subset \mathfrak{H}$ is the range of $F(x)$, is a measurable field of Hilbert spaces on which
$$K(x,y) = F(x)^{\frac{1}{2}} F(y)^{\frac{1}{2}} \tag{5.60}$$

5.3. Reproducing kernel Hilbert spaces

is an admissible ν-kernel. The associated reproducing kernel Hilbert space $\mathfrak{H}_K$ and the canonical POV function $x \mapsto F_K(x)$ are related by the unitary map

$$W_K : \mathfrak{H} \to \mathfrak{H}_K, \qquad (W_K\phi)(x) = F(x)^{\frac{1}{2}}\phi, \tag{5.61}$$
$$W_K F(x) W_K^{-1} = F_K(x). \tag{5.62}$$

If

$$F(x) = \sum_{i=1}^{n} |\eta_x^i\rangle\langle\eta_x^i|, \tag{5.63}$$

where $\eta_x^i \in \mathfrak{K}_x$, $i = 1, 2, \ldots, n$, are linearly independent and $x \mapsto \eta_x^i$ are measurable vector fields, then the coherent states

$$\mathfrak{S}_K = \{\xi_x^i = W_K \eta_x^i \mid i = 1, 2, \ldots, n, \quad x \in X\} \tag{5.64}$$

form a total set in $\mathfrak{H}_K$. In general, a vector $\phi \in \mathfrak{K}_x$ maps to

$$W_K\phi = E_K(x)^* F(x)^{-\frac{1}{2}}\phi = K(\cdot,x) F(x)^{-\frac{1}{2}}\phi \in \mathfrak{H}_K, \tag{5.65}$$

where $F(x)^{-\frac{1}{2}}$ is the inverse of the square root operator $F(x)^{\frac{1}{2}}$ restricted to $\mathfrak{K}_x$.

(Again, in the definition of the kernel in (5.60), the restriction of the operator $F(x)^{\frac{1}{2}}$ to $\mathfrak{K}_x$ is implied.)

Thus, we have shown that having a reproducing kernel Hilbert space is completely equivalent to having a POV function on a Hilbert space satisfying the conditions of this theorem. Note that, in the statement of this theorem, we could have replaced the hypothesis on the existence of a measurable POV function $x \mapsto F(x)$ and conditions 1 and 2 by the following equivalent hypothesis.

There exists a set of vectors

$$\mathfrak{S} = \{\eta_x^i \mid i = 1, 2, \ldots, n, \quad x \in X\},$$

which is total in $\mathfrak{H}$ and satisfies:

1. *for each $x \in X$, the vectors η_x^i $i = 1, 2, \ldots, n$, are linearly independent;*

2. *for each $\phi \in \mathfrak{H}$, the functions $x \mapsto \langle\eta_x^i|\phi\rangle$, $i = 1, 2, \ldots, n$, are measurable.*

The above construction of a reproducing kernel Hilbert space can also be carried out in terms of the set of basis vectors η_x^i $i = 1, 2, \ldots, n$, $x \in X$, $F(x) = \sum_{i=1}^{n} |\eta_x^i\rangle\langle\eta_x^i|$, and indeed, for explicit computations, it is more useful to do it this way: The $n \times n$ matrix-valued function $\mathbf{K}(x,y)$, with matrix elements

$$\mathbf{K}(x,y)_{ij} = \langle\eta_x^i|\eta_y^j\rangle, \qquad i,j = 1, 2, \ldots, n, \tag{5.66}$$

defines a positive-definite kernel on the measurable field of Hilbert spaces $\{\mathfrak{K}_x \mid x \in X\}$, with $\mathfrak{K}_x = \mathbb{C}^n$ for each x. The resulting reproducing kernel Hilbert space consists of measurable maps $\Phi : X \to \mathbb{C}^n$. In particular,

$$(E_K(x)^* \mathbf{v})(y) = \mathbf{K}(y, x)\mathbf{v}, \qquad \mathbf{v} \in \mathbb{C}^n. \tag{5.67}$$

The unitary map $W_K : \mathfrak{H} \to \mathfrak{H}_K$ is now given, componentwise, by

$$(W_K \phi)^i(x) = \Phi^i(x) = \langle \eta_x^i | \phi \rangle. \tag{5.68}$$

The matrix $\mathbf{K}(x, y)$ is related to the operator $K(x, y)$ in (5.60) by

$$K(x, y) = \sum_{i,j=1}^n |u_x^i\rangle \mathbf{K}(x, y)_{ij} \langle u_y^j|, \tag{5.69}$$

where u_x^i is the vector $F(x)^{-1} \eta_x^i$ $(= K(x,x)^{-\frac{1}{2}} v_x^i)$, considered as an element in $\mathfrak{H}_K(x)$ [the restriction of $F(x)$ to $\mathfrak{H}_K(x)$ being again implied here] and $\langle u_y^j |$ is the dual of $|u_y^j\rangle$ as a vector in $\mathfrak{H}_K(y)$.

Suppose now that $\widetilde{\mathfrak{H}}$ is a Hilbert space of (equivalence classes) of ν-measurable vector fields from $\{\mathfrak{K}_x \mid x \in X\}$ and $\mathfrak{H}_K$ is a reproducing kernel Hilbert space that is a subspace of $\widetilde{\mathfrak{H}}$. Let $\mathbb{P}_K$ be the projection operator

$$\mathbb{P}_K \widetilde{\mathfrak{H}} = \mathfrak{H}_K. \tag{5.70}$$

Then, for arbitrary $[\Phi] \in \widetilde{\mathfrak{H}}$, $\mathbb{P}_K[\Phi]$ is the unique element in $\mathfrak{H}_K$, such that $(\mathbb{P}_K \Phi)(x) \in \mathfrak{K}_x$ is defined by

$$\langle v_x | (\mathbb{P}_K \Phi)(x) \rangle_x = \langle \xi[v_x] | \Phi \rangle_K = \langle K(\cdot, x) v_x | \Phi \rangle_K, \tag{5.71}$$

for all $v_x \in \mathfrak{K}_x$ [see (5.32)]. On the other hand, if again $\mathfrak{H}$ is a Hilbert space of (equivalence classes) of ν-measurable vector fields from $\{\mathfrak{K}_x \mid x \in X\}$, we can ask under what conditions $\mathfrak{H}$ itself would be a reproducing kernel Hilbert space. This is settled by the following theorem (see [88] for proof).

Theorem 5.3.9: *Let $\mathfrak{H}$ be a Hilbert space of equivalence classes of ν-measurable vector fields taken from the measurable field of Hilbert spaces $\{\mathfrak{K}_x \mid x \in X\}$. Then the following are equivalent:*

1. *there exists an admissible, positive-definite ν-kernel K on $\{\mathfrak{K}_x \mid x \in X\}$ such that $\mathfrak{H} = \mathfrak{H}_K$;*

2. *there exists a ν-selection σ on $\mathfrak{H}$, such that for each $x \in X$, the evaluation map, $E_K(x) : \mathfrak{H} \to \mathfrak{K}_x$, $E_K(x)[\Phi] = \sigma([\Phi])(x)$, is continuous (in the norms of $\mathfrak{H}$ and $\mathfrak{K}_x$).*

5.3.4 A reconstruction problem: Example of a holomorphic map (♣)

An immediate application of the theory of reproducing kernel Hilbert spaces is in the reconstruction of certain types of functions from a knowledge of

5.3. Reproducing kernel Hilbert spaces

their values at countable sets of points. Let $\mathfrak{H}_K$ be a reproducing kernel Hilbert space of ν-measurable vector fields from $\{\mathfrak{K}_x \mid x \in X\}$. In general, X is not a discrete space and the set of coherent states (5.50) contains more vectors than is required to form a basis (in physicists' jargon, they form an overcomplete set of vectors). Thus, it should be possible to choose a countable subset of points $\{x_j\}_{j=1}^{N}$, $N \geq \dim \mathfrak{H}_K$, (note that the dimension of $\mathfrak{H}_K$ is in general infinite), such that the set of CS $\xi_{x_j}^{i}$, $i = 1, 2, \ldots, n$, $j = 1, 2, \ldots, N$, is total in $\mathfrak{H}_K$. Choosing such a set of CS, if, for $\Phi \in \mathfrak{H}_K$, the set of complex numbers

$$\Phi^i(x_j) = \langle \xi_{x_j}^i | \Phi \rangle, \quad i = 1, 2, \ldots, n, \quad j = 1, 2, \ldots, N, \tag{5.72}$$

are known, it is possible to determine Φ itself. In other words, it is possible to reconstruct $\Phi \in \mathfrak{H}_K$ from a knowledge of its values $\Phi(x_j) \in \mathfrak{K}_{x_j}$ at the points $x_j \in X$, $j = 1, 2, \ldots, N$.

A simple but powerful example of the construction of reproducing kernel Hilbert spaces and discretizations in the above sense is provided by a class of holomorphic maps [236]. Once again, let $\mathcal{D} = \{z \in \mathbb{C} \mid |z| < 1\}$ be the open unit disc in the complex plane and $\mathfrak{H}$ be an abstract Hilbert space. Let $\{c_n\}_{n=0}^{\infty}$ be an infinite complex sequence satisfying

$$c_n \neq 0, \quad \lim_{n \to \infty} \left| \frac{c_n}{c_{n+1}} \right| = 1, \quad n = 0, 1, 2, \ldots \infty, \tag{5.73}$$

and $\{e_n\}_{n=0}^{\infty}$ be an orthonormal basis in $\mathfrak{H}$. It is then readily checked that the series

$$\zeta_{\bar{z}} = \sum_{n=0}^{\infty} \overline{c_n} e_n \bar{z}^n \tag{5.74}$$

is strongly convergent, so that $\zeta_{\bar{z}} \in \mathfrak{H}$. Furthermore, the convergence is uniform on every compact subset of $\mathcal{D}$. Thus, for each $\phi \in \mathfrak{H}$, the function

$$\Phi(z) = \langle \zeta_{\bar{z}} | \phi \rangle = \sum_{n=0}^{\infty} c_n \langle e_n | \phi \rangle z^n \tag{5.75}$$

is analytic in $\mathcal{D}$. From this, it also follows that the set of vectors, $\zeta_{\bar{z}}$, $z \in \mathcal{D}$, is total in $\mathfrak{H}$. The function $K : \mathbb{C} \times \mathbb{C} \to \mathbb{C}$ [see (5.66)],

$$K(z, \bar{z}') = \langle \zeta_{\bar{z}} | \zeta_{\bar{z}'} \rangle,$$

is then a positive-definite kernel on the space $\mathfrak{H}_K(\mathcal{D})$ of all holomorphic functions of the type (5.75). Thus, $\mathfrak{H}_K(\mathcal{D})$ is a reproducing kernel Hilbert space and the map

$$W_K : \mathfrak{H} \to \mathfrak{H}_K(\mathcal{D}), \quad (W_K \phi)(z) = \Phi(z) = \langle \zeta_{\bar{z}} | \phi \rangle, \tag{5.76}$$

is a unitary isomorphism.

If $\{z_j\}_{j=1}^{\infty}$ is a sequence of points in $\mathcal{D}$ converging to $z_0 \in \mathcal{D}$, then the values of the analytic function Φ in (5.75) at the points $\{z_j\}_{j=0}^{\infty}$ determine

it completely on all of $\mathcal{D}$. This implies that the countable set of vectors
$$\xi_{\bar{z}_j} = W_K \zeta_{\bar{z}_j}, \qquad \xi_{\bar{z}_j}(z) = \langle \zeta_{\bar{z}} | \zeta_{\bar{z}'} \rangle, \qquad j = 0, 1, 2, \ldots, \infty,$$
is total in $\mathfrak{H}_K(\mathcal{D})$, and, hence, any vector in it can be written in terms of these. We ought to point out, however, that it is in general not possible to build frames using the vectors $\xi_{\bar{z}_j}$, since the operator represented by $\sum_{j=0}^{\infty} |\xi_{\bar{z}_j}\rangle\langle\xi_{\bar{z}_j}|$ is in general unbounded. On the other hand, it can be verified that, if λ_j, $j = 0, 1, 2, \ldots, \infty$, are positive nonzero numbers, satisfying $\sum_{j=0}^{\infty} \lambda_j < \infty$, then the operator defined by the weak sum $\sum_{j=0}^{\infty} \lambda_j |\xi_{z_j}\rangle\langle\xi_{z_j}|$ is bounded with finite trace.

Actually, our reconstruction problem is rendered particularly simple if the vectors $\xi_{x_j}^i$ form a *(discrete) frame* in $\mathfrak{H}_K$, in the sense of Chapter 3, Section 3.4. Then, indeed,
$$\sum_{i=1}^{n} \sum_{j=1}^{N} |\xi_{x_j}^i\rangle\langle\xi_{x_j}^i| = A, \tag{5.77}$$
where both A and A^{-1} are bounded operators on $\mathfrak{H}_K$. Let $\widetilde{\xi}_{x_j}^i$, $i = 1, 2, \ldots, n$, $j = 1, 2, \ldots, N$, constitute the *dual frame* $\mathcal{F}\{\widetilde{\xi}_{x_j}^i, A^{-1}, n\}$. Then, by (3.36), we see that
$$\Phi = \sum_{i=1}^{n} \sum_{j=1}^{N} \Phi^i(x_j) \, \widetilde{\xi}_{x_j}^i. \tag{5.78}$$

The kth component of $\Phi(x) \in \mathfrak{K}_x$, in the basis v_x^i, $i = 1, 2, \ldots, n$, [see (5.48)] is, therefore,
$$\Phi^k(x) = \sum_{i=1}^{n} \sum_{j=1}^{N} \Phi^i(x_j) \, \langle \xi_x^k | A^{-1} \xi_{x_j}^i \rangle \tag{5.79}$$
and
$$\Phi(x) = \sum_{k=1}^{n} \Phi^k(x) K(x,x)^{-1} v_x^k. \tag{5.80}$$

Clearly, the important quantity here, as always, is the *frame kernel* $K_A(x,y)$, with matrix elements
$$K_A(x,y)_{ij} = \langle \xi_x^i | A^{-1} \xi_y^j \rangle. \tag{5.81}$$

5.4 Some properties of reproducing kernel Hilbert spaces

Reproducing kernel Hilbert spaces have an array of interesting and useful properties, some of which we now derive. The assumptions and setting are as for Theorem 5.3.3. We start out with a minimum norm problem.

5.4. Some properties of reproducing kernel Hilbert spaces

Theorem 5.4.1: *If $\mathfrak{H}_K$ is a reproducing kernel Hilbert space and $\mathfrak{H}_K(x) \subset \mathfrak{H}_K$, as in (5.33), the vectors $\Phi \in \mathfrak{H}_K(x)$ are determined by their values at x. Of all $\Phi \in \mathfrak{H}_K$, satisfying $\|\Phi(x)\|_x = \alpha$ (fixed constant), the vector $\Phi_{\min} \subset \mathfrak{H}_K(x)$, such that $\|\Phi_{\min}(x)\|_x = \alpha$ and $\Phi_{\min}(x)$ is an eigenvector of $K(x,x)$ corresponding to the highest eigenvalue $\lambda_{\max}$, has minimum norm in $\mathfrak{H}_K$, and*

$$\Phi_{\min} = \xi[\Phi_{\min}(x)/\lambda_{\max}] = \frac{1}{\lambda_{\max}} K(\cdot, x)\Phi_{\min}(x). \tag{5.82}$$

Proof. The first statement is contained in (5.37) and (5.31). For the proof of the second statement, let us rewrite (5.26) as

$$\|\Phi\|_K \geq \|K(x,x)\|^{-\frac{1}{2}} \|\Phi(x)\|_x, \quad \Phi \in \mathfrak{H}_K. \tag{5.83}$$

On the other hand, $\xi[v_x](x) = K(x,x)v_x \Rightarrow \|\xi[v_x](x)\|_x = \|K(x,x)v_x\|_x$. Suppose that v_x is an eigenvector of $K(x,x)$ with eigenvalue λ [we have $\lambda > 0$, since $K(x,x)$ is a strictly positive operator]. Thus, $K(x,x)^{\frac{1}{2}}v_x = \lambda^{\frac{1}{2}}v_x$ which implies $K(x,x)v_x = \lambda^{\frac{1}{2}}K(x,x)^{\frac{1}{2}}v_x$. Hence, for this vector $\|\xi[v_x](x)\|_x = \lambda^{\frac{1}{2}}\|K(x,x)^{\frac{1}{2}}v_x\|_x = [\langle v_x | K(x,x)v_x \rangle_x]^{\frac{1}{2}} = \lambda^{\frac{1}{2}}\|\xi[v_x]\|_K$, by (5.31) and (5.23). Thus,

$$\|\xi[v_x]\|_K = \lambda^{-\frac{1}{2}}\|\xi[v_x](x)\|_x.$$

If v_x is the eigenvector for the highest eigenvalue $\lambda_{\max}$ of $K(x,x)$, then since $\|K(x,x)\| = \lambda_{\max}$, we have for this vector

$$\|\xi[v_x]\|_K = \|K(x,x)\|^{-\frac{1}{2}} \|\xi[v_x](x)\|_x,$$

which thus minimizes (5.83). Writing $\Phi_{\min} = \xi[v_x]$, with v_x normalized so as to ensure $\|\Phi_{\min}\| = \alpha$, using (5.37), we get

$$v_x = \frac{\Phi_{\min}(x)}{\lambda_{\max}},$$

from which (5.82) results. □

The next theorem gives a kind of internal condition for ascertaining when a Hilbert space of measurable vector fields is a reproducing kernel Hilbert space.

Theorem 5.4.2: *If $\{u_i\}_{i=1}^{\dim \mathfrak{H}}$ is an orthonormal basis in $\mathfrak{H}_K$, then*

$$K(x,y) = \sum_{i=1}^{\dim \mathfrak{H}} |u_i(x)\rangle\langle u_i(y)|. \tag{5.84}$$

If $\mathfrak{H} \subset [\mathfrak{M}]$ is a Hilbert space of ν-equivalence classes of measurable vector fields in $\{\mathfrak{K}_x \mid x \in X\}$, then $\mathfrak{H}$ is a reproducing kernel Hilbert space if there exists a sequence of vector fields $\{u_i\}_{i=1}^{\dim \mathfrak{H}} \subset \mathfrak{M}$, such that the sequence of ν-equivalence classes $\{[u_i]\}_{i=1}^{\dim \mathfrak{H}}$ is an orthonormal basis of $\mathfrak{H}$ and:

1. for each $x \in X$,

$$\sum_{i=1}^{\dim \mathfrak{H}} \|u_i(x)\|_x < \infty; \qquad (5.85)$$

2. the set of vectors $\{u_i(x) \mid i = 1, 2, \ldots, \dim \mathfrak{H}\}$ is total in $\mathfrak{K}_x$, for each $x \in X$.

In this case, the reproducing kernel is constructed using (5.84), and

$$\xi[v_x] = \sum_{i=1}^{\dim \mathfrak{H}} \langle u_i(x)|v_x\rangle_x \, u_i. \qquad (5.86)$$

Furthermore, every orthonormal basis of vectors in a reproducing kernel Hilbert space, with K satisfying the strict positivity condition (5.17), fulfils conditions 1 and 2 above.

The proof of this theorem is an adaptation of a standard proof, given for example, in [Mes], and we omit it. A number of useful norm estimates hold for reproducing kernel Hilbert spaces, which can be derived using relations such as (5.26), (5.31), (5.60); and (5.61).

Theorem 5.4.3: Let $\mathfrak{H}_K$ be a reproducing kernel Hilbert space (of ν-measurable vector fields taken from a measurable field of Hilbert spaces $\{\mathfrak{K}_x \mid x \in X\}$), F_K be the canonically associated POV function and $E_K(x) : \mathfrak{H}_K \to \mathfrak{K}_x$ be the evaluation map at $x \in X$. Then,

$$\|K(x,y)\| = \|F_K(x)^{\frac{1}{2}} F_K(y)^{\frac{1}{2}}\| \leq [\|F_K(x)\| \, \|F_K(y)\|]^{\frac{1}{2}},$$
$$x, y \in X, \qquad (5.87)$$

$$\|E_K(x)\| = \|E_K(x)^*\| = \|F_K(x)\|^{\frac{1}{2}}, \quad x \in X, \qquad (5.88)$$

$$\|\Phi(x)\|_x \leq \|K(x,x)\|^{\frac{1}{2}} \|\Phi\|_K, \quad x \in X, \ \Phi \in \mathfrak{H}_K. \qquad (5.89)$$

The vectors $\xi[v_x]$, $x \in X$, $v_x \in \mathfrak{K}_x$, in (5.31) satisfy

$$\|\xi[v_x]\|_K = \|K(x,x)^{\frac{1}{2}} v_x\|_x \quad \text{and} \quad \|\xi[v_x](x)\|_x = \|K(x,x) v_x\|_x. \qquad (5.90)$$

6
Square Integrable and Holomorphic Kernels

In this chapter, we study two special types of reproducing kernel Hilbert spaces, which are probably the most widely occurring types in the physical literature. While the very general reproducing kernel Hilbert spaces, constructed in the last chapter, were spaces of vector-valued functions, they were not assumed to be Hilbert spaces of square integrable functions, with respect to any measure. Most reproducing kernel Hilbert spaces that arise in physics and in group representation theory do, on the other hand, turn out to be spaces of square integrable fuctions. Another widely occurring variety of reproducing kernel Hilbert spaces are spaces of holomorphic or square integrable holomorphic functions. We look at these two situations more closely in this chapter. Recall from the discussion in Chapter 2 that the family of canonical CS arise from a reproducing kernel Hilbert space of square integrable functions, and, indeed, they may also be associated to a space of analytic functions (the Bargmann space).

6.1 Square integrable kernels

The two examples of reproducing kernels, appearing in (2.66) in Chapter 2 and (4.91) in Chapter 4, respectively [the former coming from the Bargmann space for the representation of the canonical CS and the latter coming from spaces of holomorphic functions on the unit disc, used to describe representations of $SU(1,1)$ from the discrete series] are both *holomorphic kernels*; i.e., they arise in Hilbert spaces of holomorphic functions.

110   6. Square Integrable and Holomorphic Kernels

They are also *square integrable* kernels, in that they both satisfy an integral relation of the type:

$$\int_X K(x,z)K(z,y)\, d\nu(z) = K(x,y). \tag{6.1}$$

Motivated by this observation, we introduce a somewhat more general notion of square integrability for reproducing kernels.

Definition 6.1.1 (Square integrable kernel): *Let $\{\mathfrak{K}_x \mid x \in X\}$ be a measurable field of Hilbert spaces (with respect to the Borel measure ν), K be an admissible, positive-definite ν-kernel, and $\mathfrak{H}_K$ be the corresponding reproducing kernel Hilbert space. The kernel K is said to be square integrable if there exists a bounded operator $A_K \in \mathfrak{H}_K$, with bounded inverse, such that, for each $v_x \in \mathfrak{K}_x$ and $v_y \in \mathfrak{K}_y$,*

$$\int_X \langle v_x | K(x,z)K(z,y) v_y \rangle \, d\nu(z) = \langle \xi[v_x] | A_K \xi[v_y] \rangle \tag{6.2}$$

[here we are using the notation $\xi[v_x]$ introduced in (5.31)]. An equivalent way of expressing the above condition for square integrability is

$$\int_X F_K(x)\, d\nu(x) = \sum_{i=1}^n \int_X |\xi_x^i\rangle\langle\xi_x^i|\, d\nu(x) = A_K, \quad A_K, A_K^{-1} \in \mathcal{L}(\mathfrak{H}_K), \tag{6.3}$$

the integral being assumed to converge weakly. This, in turn, is equivalent to the *frame condition*, already given in (3.11),

$$\mathsf{m}(A)\|\Phi\|_K^2 \leq \int_X \langle \Phi | F_K(x)\Phi \rangle_K \, d\nu(x) \leq \mathsf{M}(A)\|\Phi\|_K^2, \quad \Phi \in \mathfrak{H}_K, \tag{6.4}$$

where $\mathsf{m}(A)$ and $\mathsf{M}(A)$ are positive real numbers, such that $0 < \mathsf{m}(A) \leq \mathsf{M}(A) < \infty$. In the special case of $A_K = I_K$ (the identity operator on $\mathfrak{H}_K$), the condition (6.2) is exactly the relation (6.1).

For square integrable kernels, a useful reconstruction formula for vectors $\Phi_K \in \mathfrak{H}_K$ can be obtained in terms of the values $\Phi_K^i(x) = \langle \xi_x^i | \Phi_K \rangle$ of their components Φ_K^i evaluated at the points x. Indeed, it is not hard to verify that, for any $\Phi_K \in \mathfrak{H}_K$,

$$\Phi_K = \sum_{i=1}^n \int_X \Phi_K^i(x) A_K^{-1} \xi_x^i \, d\nu(x). \tag{6.5}$$

Let $\Phi, \Psi \in \mathfrak{H}_K$. Then,

$$\langle \Phi | \Psi \rangle_K = \langle \Phi | [\int_X F_K(x)\, d\nu(x)] (A_K^{-1}\Psi) \rangle_K,$$

which, by virtue of (5.41) and the weak convergence of (6.3), implies that

$$\langle \Phi | \Psi \rangle_K = \int_X \langle \Phi | E_K(x)^* E_K(x)(A_K^{-1}\Psi) \rangle_K \, d\nu(x) = \int_X \langle \Phi(x) | (A_K^{-1}\Psi)(x) \rangle_x \, d\nu(x) \tag{6.6}$$

6.1. Square integrable kernels

Thus, the scalar product in $\mathfrak{H}_K$ is realized as an L^2-product with an operator density. Again, when $A_K = I_K$,

$$\langle \Phi | \Psi \rangle_K = \int_X \langle \Phi(x) | \Psi(x) \rangle_x \, d\nu(x), \tag{6.7}$$

and comparing with (5.11), we see that in this case the reproducing kernel Hilbert space $\mathfrak{H}_K$ becomes a subspace of the direct integral Hilbert space $\widetilde{\mathfrak{H}}$ in (5.12). Also, now the operators

$$a_K(\Delta) = \int_\Delta F_K(x) \, d\nu(x), \qquad \Delta \in \mathcal{B}(X), \tag{6.8}$$

constitute a normalized POV measure, and following the discussion in Section 3.2, the PV measure

$$(\widetilde{P}(\Delta)\Psi)(x) = \chi_\Delta(x)\Psi(x), \qquad \Psi \in \widetilde{\mathfrak{H}},$$

is its minimal Naĭmark extension to $\widetilde{\mathfrak{H}}$. The projection operator $\mathbb{P}_K : \widetilde{\mathfrak{H}} \to \mathfrak{H}_K$ now acts in the manner [see (5.70) and (5.71)]:

$$(\mathbb{P}_K \Psi)(x) = \int_X K(x,y)\Psi(y) \, d\nu(y), \qquad \Psi \in \widetilde{\mathfrak{H}}, \tag{6.9}$$

while (5.32) appears in the integral form:

$$E_K(x)\Psi = \int_X K(x,y)\Psi(y) \, d\nu(y) = \Psi(x), \qquad \Psi \in \mathfrak{H}_K. \tag{6.10}$$

In general, however (i.e., when A_K is not necessarily equal to I_K), the relation (5.32) assumes the form:

$$E_K(x)\Psi = \int_X K(x,y)(A_K^{-1}\Psi)(y) \, d\nu(y) = \Psi(x), \qquad \Psi \in \mathfrak{H}_K. \tag{6.11}$$

Although in the above definition of square integrability we assumed both A_K and A_K^{-1} to be bounded operators, a more general situation, where A_K is bounded but A_K^{-1} is only assumed to be densely defined, can also be envisaged. This involves, however, having to deal with certain functional analytic technicalities related to unbounded operators (see [12] for details).

Since we are assuming A_K^{-1} to be bounded, quite generally, we may rewrite (6.3) as

$$\int_X F_{\overline{K}}(x) \, d\nu(x) = I_K, \quad \text{where} \quad F_{\overline{K}}(x) = A_K^{-\frac{1}{2}} F_K(x) A_K^{-\frac{1}{2}}. \tag{6.12}$$

Since $F_{\overline{K}}(x)$ is again a positive operator-valued function of x, satisfying all conditions of Theorem 5.3.8, there exists another reproducing kernel Hilbert space $\mathfrak{H}_{\overline{K}}$, with kernel

$$\overline{K}(x,y) = F_{\overline{K}}(x)^{\frac{1}{2}} F_{\overline{K}}(y)^{\frac{1}{2}}. \tag{6.13}$$

This is an admissible, positive-definite ν-kernel, defined on the measurable field of Hilbert spaces $\{\overline{\mathfrak{K}}_x \mid x \in X\}$, where

$$\overline{\mathfrak{K}}_x = \{A_K^{-\frac{1}{2}}\Phi \mid \Phi \in \mathfrak{H}_K(x)\} \tag{6.14}$$

[$\mathfrak{H}_K(x)$ being as in (5.33)], and it satisfies the square integrability condition

$$\int_X \overline{K}(x,z)\overline{K}(z,y)\,d\nu(z) = \overline{K}(x,y). \tag{6.15}$$

Following (5.61), we could construct the unitary map,

$$W_{\overline{K}} : \mathfrak{H}_K \to \mathfrak{H}_{\overline{K}}, \qquad (W_{\overline{K}}\Phi)(x) = F_{\overline{K}}(x)^{\frac{1}{2}}\Phi. \tag{6.16}$$

This map does make the two Hilbert spaces $\mathfrak{H}_K$ and $\mathfrak{H}_{\overline{K}}$ isometrically isomorphic, but it does not map the coherent states of these spaces into one another, unless $A_K = I_K$. Indeed, let ξ_x^i, $i = 1, 2, \ldots, n$, $x \in X$, be the coherent states of $\mathfrak{H}_K$ [see (5.47)–(5.50)] and $\overline{\xi}_x^i$, $i = 1, 2, \ldots, n$, $x \in X$, those of $\mathfrak{H}_{\overline{K}}$. Then it is easily verified that

$$W_{\overline{K}}(A_K^{-\frac{1}{2}}\xi_x^i) = \overline{\xi}_x^i, \qquad i = 1, 2, \ldots, n, \quad x \in X. \tag{6.17}$$

6.2 Holomorphic kernels

We proceed now to a consideration of a class of holomorphic kernels, which, as it will turn out, are all square integrable. Let $\mathbb{D} \subset \mathbb{C}$ be a *domain*, i.e., an open connected set,

$$d\nu(z,\overline{z}) = \frac{dz \wedge d\overline{z}}{2\pi i} = \frac{1}{\pi}dy \wedge dx, \qquad z = x + iy, \tag{6.18}$$

be the Lebesgue measure on $\mathbb{D}$ and $d\mu(z,\overline{z}) = \rho(z,\overline{z})d\nu(z,\overline{z})$ be any other measure, equivalent to ν, where ρ is a continuous, positive function, which does not vanish anywhere on $\mathbb{D}$. Let $\mathfrak{H} = L^2(\mathbb{D}, d\mu)$, and denote the norm in it by $\|\ldots\|_{hol}$. Suppose that there exists a nonempty subset of vectors in $\mathfrak{H}$, which can be identified with functions analytic in z. Let $L^2_{hol}(\mathbb{D}, d\mu) \subset \mathfrak{H}$ denote this subset. Note that, if, for example, $\mathbb{D} = \mathbb{C}$ and $\mu = \nu$, then no nonvanishing analytic functions exist at all in $\mathfrak{H}$. On the other hand, with μ as in (2.62), $L^2_{hol}(\mathbb{D}, d\mu)$ is the Bargmann space of entire analytic functions discussed in Chapter 2, Section 2.4. Similarly, when $\mathbb{D} = \mathcal{D}$, the unit disc, we have the entire class of Hilbert spaces of holomorphic functions $\mathfrak{H}^j_{hol}$ discussed in Section 4.2.2. We now have the following important result (proof adapted from [Sug]).

Lemma 6.2.1: $L^2_{hol}(\mathbb{D}, d\mu)$ *is a closed Hilbert subspace of* $\mathfrak{H}$, *on which the evaluation map*

$$E_{hol}(z) : L^2_{hol}(\mathbb{D}, d\mu) \to \mathbb{C}, \qquad E_{hol}(z)f = f(z), \tag{6.19}$$

6.2. Holomorphic kernels

is bounded and linear for all $z \in \mathbb{D}$, and, moreover, for any compact subset $C \subset \mathbb{D}$, there exists a constant $k(C) > 0$, such that

$$|f(z)| \leq k(C)\|f\|_{hol}, \qquad (6.20)$$

for all $f \in L^2_{hol}(\mathbb{D}, d\mu)$ and $z \in \mathbb{C}$.

Proof. The linearity of $E_{hol}(z)$ is obvious, and its boundedness would follow directly once (6.20) is proved. Let us therefore prove this relation. Let $f \in L^2_{hol}(\mathbb{D}, d\mu)$ and $z \in \mathbb{D}$. Choose $\varepsilon \in (0,1)$, such that $V_\varepsilon(z) = \{w \mid |w - z| < \varepsilon\} \subset \mathbb{D}$. Taylor expanding f around z in $V_\varepsilon(z)$, we may write

$$f(w) = \sum_{k=0}^{\infty} a_k (w - z)^k, \qquad a_k \in \mathbb{C}.$$

Setting $f_k(w) = (w - z)^k$,

$$\begin{aligned}
\langle f_k | f_\ell \rangle &= \int_{V_\varepsilon(z)} \overline{f_k(w)} f_\ell(w)\, d\nu(w, \overline{w}) = \frac{1}{\pi} \int_0^\varepsilon r\, dr \int_0^{2\pi} r^{k+\ell} e^{-i(k-\ell)\theta}\, d\theta \\
&= \frac{2\varepsilon^{k+\ell+2}}{k+\ell+2}\, \delta_{k\ell}.
\end{aligned}$$

Thus,

$$\begin{aligned}
\|f\|_\varepsilon^2 &= \langle f | f \rangle_\varepsilon = \sum_{k=0}^{\infty} |a_k|^2 \|f_k\|_\varepsilon^2 \\
&= \sum_{k=0}^{\infty} |a_k|^2 \frac{\varepsilon^{2(k+1)}}{k+1},
\end{aligned}$$

and, since $a_0 = f(z)$, this implies

$$\int_{V_\varepsilon(z)} |f(w)|^2\, d\nu(w, \overline{w}) \geq |f(z)|^2 \varepsilon^2. \qquad (6.21)$$

Now, let $\varepsilon < 1$ be chosen so that the compact set

$$C' = \{w \in \mathbb{C} \mid \mathrm{dist}(C, w) \leq \varepsilon\}$$

is contained in C. [Here $\mathrm{dist}(C, w)$ is the infimum of $|z - w|$, over all $z \in C$.] Then, for any $z \in C$, $V_\varepsilon(z) \subset C'$. Going back to the measure $d\mu(w, \overline{w}) = \rho(w, \overline{w})\, d\nu(w, \overline{w})$, let

$$r(C) = \inf_{w \in C'} \rho(w, \overline{w}).$$

Hence, for all $z \in C$,

$$\begin{aligned}
\|f\|_{hol} &= \int_{\mathbb{D}} |f(w)|^2 \rho(w, \overline{w})\, d\nu(w, \overline{w}) \geq \int_{V_\varepsilon(z)} |f(w)|^2 \rho(w, \overline{w})\, d\nu(w, \overline{w}) \\
&\geq \varepsilon^2 r(C) |f(z)|^2,
\end{aligned}$$

by (6.21); so that taking $k(C) = [r(C)]^{-\frac{1}{2}} \varepsilon^{-1}$, we obtain (6.20).

We still have to prove that $L^2_{hol}(\mathbb{D}, d\mu)$ is closed. Let $\{f_m\}_{m=0}^\infty$ be a Cauchy sequence in $L^2_{hol}(\mathbb{D}, d\mu)$. Since $L^2_{hol}(\mathbb{D}, d\mu) \subset \mathfrak{H}$, there is an $f \in \mathfrak{H}$ such that $\lim_{m\to\infty} \|f_m - f\|_{hol} = 0$. By virtue of (6.21), it follows that the complex sequence $\{f_m(z)\}_{m=0}^\infty$ converges to some function $g(z)$, and this convergence is uniform on every compact subset C of $\mathbb{D}$. Being the uniform limit of holomorphic functions, g must then also be holomorphic and, as in the standard proof of the completeness of L^2-spaces, we infer that $f(z) = g(z)$ almost everywhere. Hence, $g \in L^2_{hol}(\mathbb{D}, \mu)$, which implies that $L^2_{hol}(\mathbb{D}, \mu)$ is closed. □

Combining the various results in this section with this lemma, we arrive at our main theorem on holomorphic kernels, which in fact is a special case of a more general result proved in [198]:

Theorem 6.2.2 (Holomorphic kernel-space theorem): *The subspace $L^2_{hol}(\mathbb{D}, d\mu)$ of $\mathfrak{H} = L^2(\mathbb{D}, d\mu)$ is a reproducing kernel Hilbert space with square integrable kernel $K_\mathbb{D} : \mathbb{D} \times \mathbb{D} \to \mathbb{C}$,*

$$K_\mathbb{D}(z, \overline{z}') = E_{hol}(z) E_{hol}(\overline{z}')^*, \tag{6.22}$$

such that

$$\int_\mathbb{D} K_\mathbb{D}(z, \overline{w}) K_\mathbb{D}(w, \overline{z}') \, d\mu(w, \overline{w}) = K_\mathbb{D}(z, \overline{z}'). \tag{6.23}$$

For fixed $w \in \mathbb{D}$, the kernel $K_\mathbb{D}(z, \overline{w})$ is holomorphic in z.

From the proof of Lemma 6.2.1, it is also clear that, if $\mathbb{D}$ is a bounded domain, then $L^2_{hol}(\mathbb{D}, d\nu)$ (i.e., with respect to the Lebesgue measure) is always nonempty. [Indeed, the identity function $\mathbb{I}(z) = 1$, $\forall z \in \mathbb{D}$, is always in $L^2_{hol}(\mathbb{D}, d\nu)$.] In this case, the reproducing kernel $K_\mathbb{D}$ is called the *Bergman kernel* of the domain $\mathbb{D}$. As an example, the kernel K^j_{hol} in (4.91), with $j = 1$, is the Bergman kernel of $\mathbb{D} = \mathcal{D}$, the unit disc. In general, the kernel K_{hol} is called the μ-Bergmann kernel of $\mathbb{D}$.

The above theorem admits several generalizations. First, $\mathbb{D}$ could be taken to be a domain in $\mathbb{C}^k$, so that we would be considering Hilbert spaces of holomorphic functions of k complex variables, $z_1, z_2, \ldots, z_k$. Writing $\mathbf{z} = (z_1, z_2, \ldots, z_k)$, the measure ν in (6.18) would now be replaced by

$$d\nu(\mathbf{z}, \overline{\mathbf{z}}) = \frac{1}{(2\pi i)^k} \prod_{i=1}^k dz_i \wedge d\overline{z}_i. \tag{6.24}$$

Furthermore, the density ρ in the definition of μ,

$$d\mu(\mathbf{z}, \overline{\mathbf{z}}) = \rho(\mathbf{z}, \overline{\mathbf{z}}) \, d\nu(\mathbf{z}, \overline{\mathbf{z}}), \tag{6.25}$$

could in general be an *admissible weight* [242]. To understand the meaning of this concept, let $L^2(\mathbb{D}, d\mu)$ be, as before, the Hilbert space of all complex-valued functions on $\mathbb{D} \subset \mathbb{C}^k$, which are square integrable with respect to

(6.25), and, again, let $L^2_{hol}(\mathbb{D}, d\mu)$ be the set of vectors in $L^2(\mathbb{D}, d\mu)$ that can be identified with holomorphic functions. We assume that this set is nonempty. Let $E_{hol}(\mathbf{z}) : L^2_{hol}(\mathbb{D}, d\mu) \to \mathbb{C}$, $E_{hol}(\mathbf{z})f = f(\mathbf{z})$, be the evaluation map at $\mathbf{z} \in \mathbb{D}$. Then, ρ is said to be an admissible weight if, (1) it is Lebesgue measurable, positive and nonzero on all of $\mathbb{D}$; and (2) for any $\mathbf{z} \in \mathbb{D}$, there exist a neighborhood $V(\mathbf{z})$ and a constant $k(\mathbf{z})$, such that, for all $\mathbf{w} \in V(\mathbf{z})$, $\|E_{hol}(\mathbf{w})\| \leq k(\mathbf{z})$. It can be shown that, if ρ is an admissible weight, then $L^2_{hol}(\mathbb{D}, d\mu)$ is a closed subspace of $L^2(\mathbb{D}, d\mu)$, admitting a reproducing kernel $K_{hol}(\mathbf{z}, \overline{\mathbf{w}}) = E_{hol}(\mathbf{z})E_{hol}(\mathbf{w})^*$, which is holomorphic in $\mathbf{z}$ and square integrable in the sense of (6.23).

The second generalization is a little more sophisticated, in that it involves holomorphic vector bundles [243]. Let $\mathbf{E} = (E, \pi, M)$ be a holomorphic vector bundle, with base space M (which is a complex analytic manifold), total space E, and canonical projection map $\pi : E \to M$. Suppose that M has complex dimension k and that the fiber $\pi^{-1}(x)$ over the point $x \in M$ is isomorphic to $\mathbb{C}^n$. Let h be a hermitian structure on $\mathbf{E}$; i.e., it associates a scalar product $\langle \cdot | \cdot \rangle_x$ to each fiber $\pi^{-1}(x)$, $x \in M$, such that, if s and t are smooth sections of $\mathbf{E}$, then $h_x(s,t) = \langle s(x)|t(x)\rangle_x$, and the map $x \mapsto \langle s(x)|t(x)\rangle_x$ is smooth. Let μ be a volume form on M, and consider the Hilbert space $L^2(\mathbf{E}, h, d\mu)$. This is the completion of the space of all smooth, compactly supported sections of $\mathbf{E}$, with respect to the scalar product

$$\langle s|t\rangle_\mu = \int_M \langle s(x)|t(x)\rangle_x \, d\mu)(x). \tag{6.26}$$

Let $L^2_{hol}(\mathbf{E}, h, d\mu)$ be the set of all vectors in $L^2(\mathbf{E}, h, d\mu)$ that can be identified with holomorphic sections of $\mathbf{E}$ and assume that this set is nonempty. If $\mathbf{E}^*$ denotes the dual bundle of $\mathbf{E}$ and $\pi' : \mathbf{E}^* \to M$ is the corresponding projection map, consider, for any $v^* \in \mathbf{E}^*$, the linear map

$$\mathcal{E}_{v^*} : L^2_{hol}(\mathbf{E}, h, d\mu) \to \mathbb{C}, \qquad \mathcal{E}_{v^*}(s) = \langle v^*; s(\pi'(v^*))\rangle, \tag{6.27}$$

$\langle \cdot; \cdot \rangle$ denoting the dual pairing between $\mathbf{E}$ and $\mathbf{E}^*$. Then, it can be shown that there exist a neighborhood $V(v^*)$ of v^* and a constant $k(v^*) > 0$, such that, for all $w^* \in V(v^*)$,

$$|\mathcal{E}_{w^*}(s)| \leq k(v^*)\|s\|_\mu. \tag{6.28}$$

From this, it follows that the evaluation map

$$E_{hol}(x) : L^2_{hol}(\mathbf{E}, h, d\mu) \to \pi^{-1}(x), \qquad E_{hol}(x)s = s(x), \tag{6.29}$$

is continuous and $L^2_{hol}(\mathbf{E}, h, d\mu)$ is a closed subspace of $L^2(\mathbf{E}, h, d\mu)$. The maps $\mathcal{E}_{v^*}$ and $E_{hol}(x)$ are related by

$$\mathcal{E}_{v^*}(s) = \langle v^*; E_{hol}(x)s\rangle, \tag{6.30}$$

where $x \in \pi^*(v^*)$. The Hilbert space $L^2_{hol}(\mathbf{E}, h, d\mu)$ admits the reproducing kernel

$$K_{hol}(x,y) = E_{hol}(x)E_{hol}(y)^* : \pi^{-1}(y) \to \pi^{-1}(x), \tag{6.31}$$

which again satisfies the square integrability condition (6.1), and is holomorphic in x. The earlier constructions of μ-Bergman kernels all become special cases of this general result.

6.3 Coherent states: The holomorphic case

As we have seen in the previous chapter, given a reproducing kernel Hilbert space $\mathfrak{H}_K$ of measurable vector fields on a locally compact space X, there always exists an overcomplete set of vectors ξ_x^i, $i = 1, 2, \ldots, n$, $x \in X$ in $\mathfrak{H}_K$, defined by the kernel K [see (5.49)–(5.50)], which constitutes its coherent states. In concrete situations, such states may not be related to any group representation at all. In this section, we shall discuss examples of such CS. The first one pertains to spaces of holomorphic functions, and the second one is a nonholomorphic extension of the first. Another example is the set of CS on the unit circle, first obtained in [103]. These are derived from canonical CS on the line by a direct integral decomposition (of Bloch type) of $L^2(\mathbb{R}, dx)$ into infinitely many copies of $L^2(S^1, d\theta)$. This yields a tight frame, but the CS cannot be obtained as an orbit of a fixed vector under some group action. That example was reexamined in [199], including the holomorphic aspects discussed here, and a complete and rigorous treatment has finally appeared in [151]. It is interesting to note that the unitary transformation mapping the Hilbert space $L^2(\mathbb{R})$ of the canonical CS onto the one of the new CS, namely, $L^2(S^1 \times S^{1*})$, where S^{1*} denotes the dual of S^1, is the Zak transform, familiar in solid-state theory. We shall meet this transform explicitly again in the discussion of Gabor frames in Chapter 16, Section 16.1. As an extreme example, of course, one finds the discrete frames discussed in Chapter 3, Section 3.4.

Although most of the subsequent chapters will deal with CS originating from a group representation, CS are also often associated to the complex structures possessed by certain types of spaces, unrelated to any group representation. The Hilbert spaces of holomorphic functions $L_{hol}^2(\mathbb{D}, d\mu)$ discussed in Section 6.1 admit such CS. Similarly, for the class of holomorphic maps mentioned in the example of Section 5.3.4, the vectors ζ_z in (5.74) or (6.32), or, equivalently, the vectors ξ_z in (5.76), are CS that are not necessarily related to any group representation.

In such cases, the reproducing kernel can often be realized as a square integrable kernel in the sense of Definition 6.1.1; i.e., it may be possible to find a Borel measure ν for which (6.2) holds. In the case of the holomorphic maps of the unit disc $\mathcal{D} \in \mathbb{C}$ of Section 5.3.4, it is particularly interesting to examine this point since it can be related [236] to the classical *moment problem* [Akh, Sho]. To begin with, note that, if we take $c_n = \sqrt{n+1}$ and $e_n = u_n$, $n = 0, 1, \ldots, \infty$, in (5.74) [or (6.32)], then the vectors $\zeta_{\overline{z}}$ are precisely the vectors $\zeta_{\overline{z}}^j$ appearing in (4.95)–(4.97), for $j = 1$, in the holomorphic discrete series representation of $SU(1,1)$. Thus, the Hilbert space $\mathfrak{H}_K$ in (5.76) is now the L^2-space of (4.77), and the kernel K is the same as that in (4.91) and (4.92) for $j = 1$. Of course, this kernel is square integrable since it satisfies (4.98). Generally, going back to the vectors

$$\zeta_z = \sum_{n=0}^{\infty} c_n e_n z^n \qquad (6.32)$$

defined in (5.74), assume, for simplicity, that the coefficients c_n are all real, and consider the formal operator

$$A = \int_{\mathcal{D}} |\zeta_z\rangle\langle\zeta_z| \, d\mu(z, \bar{z}), \qquad (6.33)$$

where the measure μ is defined as [see (6.18) and (6.25)]:

$$d\mu(z, \bar{z}) = \frac{1}{K(z, \bar{z})} d\nu(z, \bar{z}) = \frac{1}{K(z, \bar{z})} \frac{dz \wedge d\bar{z}}{2\pi i},$$

$$K(z, \overline{z'}) = \langle \zeta_{z'} | \zeta_z \rangle, \quad K(z, \bar{z}) = \|\zeta_{\bar{z}}\|^2, \qquad z, z' \in \mathcal{D}. \quad (6.34)$$

For arbitrary $\phi, \psi \in \mathfrak{H}$,

$$\left| \int_{\mathcal{D}} \langle\phi|\zeta_z\rangle\langle\zeta_z|\psi\rangle \, d\mu(z, \bar{z}) \right| \leq \|\phi\| \, \|\psi\| \int_{\mathcal{D}} d\nu(z, \bar{z}) = \|\phi\| \, \|\psi\|.$$

Thus, as a weak integral, the right-hand side of (6.33) defines A as a bounded positive operator, with $\|A\| \leq 1$. Since the vectors $\zeta_{\bar{z}}$ are overcomplete in $\mathfrak{H}$, $\langle\phi|A\phi\rangle = 0$ implies $\phi = 0$, so that A^{-1} exists, although it may not be bounded. The spectrum of A is, however, easily computed. Indeed,

$$\langle e_m | A e_n \rangle = c_m c_n \int_{\mathcal{D}} \bar{z}^m z^n \, d\mu(z, \bar{z}), \qquad (6.35)$$

as is readily inferred using (5.74). Introducing polar coordinates (r, θ) and noting that

$$K(z, \bar{z}) = \sum_{k=0}^{\infty} c_k^2 |z|^{2n} = \sum_{k=0}^{\infty} c_k^2 \, r^{2n}$$

depends on r^2 only, the integration with respect to θ may be completed to obtain

$$\langle e_m | A e_n \rangle = \left[c_n^2 \int_0^1 r^{2n+1} \rho(r^2) \, dr \right] \delta_{mn}, \qquad (6.36)$$

where we have written

$$\rho(r^2) = \left[\sum_{k=0}^{\infty} c_k^2 r^{2k} \right]^{-1}. \qquad (6.37)$$

118   6. Square Integrable and Holomorphic Kernels

Thus,

$$A = \sum_{n=0}^{\infty} \lambda_n |e_n\rangle\langle e_n|, \qquad \lambda_n = c_n^2 \int_0^1 r^{2n+1} \rho(r^2)\, dr = \frac{c_n^2}{2} \int_0^1 r^n \rho(r)\, dr. \tag{6.38}$$

The boundedness of A ensures that, for finite n, $\lambda_n < \infty$. Thus, A has a pure point spectrum. Since λ_n could go to zero as $n \to \infty$, however, A^{-1} could be unbounded. In any case,

$$A^{-1} = \sum_{n=0}^{\infty} \lambda_n^{-1} |e_n\rangle\langle e_n|, \quad \text{and} \quad A^{-\frac{1}{2}} = \sum_{n=0}^{\infty} \lambda_n^{-\frac{1}{2}} |e_n\rangle\langle e_n|. \tag{6.39}$$

If we wanted A to be the identity operator, (6.36) would impose the constraints

$$\int_0^1 r^n \rho(r)\, dr = \frac{2}{c_n^2}, \qquad n = 0, 1, 2, \ldots, \infty. \tag{6.40}$$

For a given set of constants c_n, these cannot in general be satisfied. However, if we replace $\rho(r^2)\, dr$ in the above equation by an arbitrary measure $dF(r)$, (6.40) becomes just the moment problem for determining this measure, given the set of constants c_n [Sho]. A solution to this problem would then yield a measure

$$d\widehat{\mu}(z,\bar{z}) = \frac{1}{\pi} r\, dF(r)\, d\theta, \tag{6.41}$$

for which

$$\int_{\mathcal{D}} |\zeta_z\rangle\langle\zeta_z|\, d\widehat{\mu}(z,\bar{z}) = I. \tag{6.42}$$

Thus, with respect to this measure, the kernel $K(z,\bar{z}') = \langle\zeta_{z'}|\zeta_z\rangle$ would become square integrable in the sense of Definition 6.1.1; i.e.:

$$\int_{\mathcal{D}} K(z,\bar{z}'') K(z'',\bar{z}')\, d\widehat{\mu}(z'',\bar{z}'') = K(z,\bar{z}'),$$

and $\mathfrak{H}_K = L^2_{hol}(\mathcal{D}, d\widehat{\mu})$. Recall that $\mathfrak{H}_K$ is the Hilbert space consisting of all holomorphic functions of the form $\Phi(z) = \langle\zeta_{\bar{z}}|\phi\rangle$, $\phi \in \mathfrak{H}$, and these now become square integrable with respect to $d\widehat{\mu}$. While this is an attractive procedure, the measure $dF(r)$ may turn out to be highly singular, and, generally, $d\widehat{\mu}$ would not be absolutely continuous with respect to the Lebesgue measure $d\nu$ of $\mathcal{D}$.

Considering the operator $A^{-\frac{1}{2}}$, which has the eigenvectors e_n and eigenvalues $\lambda_n^{-\frac{1}{2}}$, $n = 0, 1, 2, \ldots \infty$, we see that the weighted CS

$$\eta_{\bar{z}} = A^{-\frac{1}{2}} \zeta_{\bar{z}} = \sum_{n=0}^{\infty} b_n e_n \bar{z}^n, \qquad b_n = \left[\frac{1}{2} \int_0^1 r^n \rho(r)\, dr\right]^{-\frac{1}{2}}, \tag{6.43}$$

6.3. Coherent states: The holomorphic case   119

are well-defined vectors in $\mathfrak{H}$. Indeed, it is not hard to see that $b_{n+1}/b_n \to 1$ as $n \to \infty$, and, hence, the coefficients b_n, $n = 0, 1, 2, \ldots, \infty$, also define a holomorphic map on the unit disc, in the sense that each one of the functions $\Psi(z) = \langle \eta_{\bar{z}} | \phi \rangle$, $\phi \in \mathfrak{H}$, is holomorphic on $\mathcal{D}$. Furthermore,

$$\int_{\mathcal{D}} |\eta_z\rangle\langle\eta_z| \, d\mu(z,\bar{z}) = I, \tag{6.44}$$

and the corresponding kernel $K_\eta(z, \bar{z}') = \langle \eta_{z'} | \eta_z \rangle$ is square integrable with respect to the original measure $d\mu$. By Theorem 6.2.2, the functions $\Psi(z)$ constitute the Hilbert space $L^2_{hol}(\mathcal{D}, d\mu)$ of holomorphic functions square integrable with respect to $d\mu$, and K_η is the reproducing kernel for this space. The norm of $L^2_{hol}(\mathcal{D}, d\mu)$, however, is different from the norm of the original reproducing kernel Hilbert space $\mathfrak{H}_K$, having the kernel $K(z, \bar{z}') = \langle \zeta_{z'} | \zeta_z \rangle = \langle \eta_{z'} | A \eta_z \rangle$. Since $\langle \zeta_z | \phi \rangle = \langle \eta_z | A^{\frac{1}{2}} \phi \rangle$, for any $\phi \in \mathfrak{H}$, we see that, with $\Phi(z) = \langle \zeta_{\bar{z}} | \phi \rangle$,

$$\|\Phi\|^2_{L^2_{hol}(\mathcal{D},d\mu)} = \int_{\mathcal{D}} |\Phi(z)|^2 \, d\mu(z,\bar{z}) = \langle \phi | A \phi \rangle_{\mathfrak{H}},$$

while $\|\Phi\|^2_K = \|\phi\|^2_{\mathfrak{H}}$, where $\|\cdots\|_K$ denotes the norm in $\mathfrak{H}_K$ (see [5.76]). Thus, while $\mathfrak{H}_K \hookrightarrow L^2_{hol}(\mathcal{D}, d\mu)$, and this inclusion is continuous, if $A^{-\frac{1}{2}}$ is unbounded, $\mathfrak{H}_K$ only forms an open dense set in $L^2_{hol}(\mathcal{D}, d\mu)$. As sets, the two Hilbert spaces $\mathfrak{H}_K$ and $L^2_{hol}(\mathcal{D}, d\mu)$ become equal if $A^{-\frac{1}{2}}$ is bounded.

6.3.1 An example of a holomorphic frame (♣)

We illustrate this section with an example of a sequence $\{c_n\}$ that leads to a frame $\mathcal{F}\{\zeta_z, A\}$. Suppose the orthonormal basis $e_n \equiv |e_n\rangle$ in $\mathfrak{H}$ consists of eigenvectors of a positive self-adjoint operator

$$X|e_n\rangle = x_n|e_n\rangle, \tag{6.45}$$

where the nondegenerate eigenvalues x_n form a strictly increasing sequence of positive numbers starting with $x_0 = 0$:

$$x_0 = 0 < x_1 < x_2 < \ldots < x_n < x_{n+1} < \ldots. \tag{6.46}$$

We associate to the number x_n another one, which represents its "factorial." Mimicking the "q-language" (see [96], [236], and references therein), we shall put

$$[x_n]! \equiv x_1 x_2 \ldots x_n \ (n \geq 1), \quad [x_0]! \equiv 1. \tag{6.47}$$

Next, we take for the sequence $\{c_n\}$ in (6.32) the value $c_n = 1/\sqrt{[x_n]!}$. The corresponding CS,

$$|\zeta_z\rangle = \sum_{n=0}^{\infty} \frac{z^n}{\sqrt{[x_n]!}} |e_n\rangle, \tag{6.48}$$

are labeled by complex numbers z within the disk $\mathcal{D}_R = \{z \in \mathbb{C} \mid |z| < R\}$, with radius R given by

$$R = \limsup_{n \to \infty} \sqrt[n]{[x_n]!}. \tag{6.49}$$

The reproducing kernel or inner product between two CS equals

$$K(z, \overline{z'}) = \langle \zeta_{z'} | \zeta_z \rangle = \sum_{n=0}^{\infty} \frac{(z\overline{z'})^n}{[x_n]!} \equiv \exp_F(z\overline{z'}). \tag{6.50}$$

Here, we have introduced the notation $\exp_F$ for the "exponential" function associated to the sequence $\{x_n\}$:

$$\exp_F(t) = \sum_{n=0}^{\infty} \frac{t^n}{[x_n]!}. \tag{6.51}$$

Its radius of convergence is precisely R. Associated to the family $\{\zeta_z\}$ are the operator

$$A = \int_{\mathcal{D}_R} |\zeta_z\rangle\langle\zeta_z| \, d\mu(z, \overline{z}), \tag{6.52}$$

where

$$d\mu(z, \overline{z}) = \frac{1}{\exp_F(|z|^2)} \frac{dz \wedge d\overline{z}}{2\pi i}, \tag{6.53}$$

and its spectral decomposition

$$A = \sum_{n=0}^{\infty} \lambda_n |n\rangle\langle n|, \tag{6.54}$$

with,

$$\lambda_n = \frac{1}{[x_n]!} \int_0^{R^2} \frac{t^n}{\exp_F(t)} \, dt \equiv \frac{\Gamma_F(n+1)}{[x_n]!}, \tag{6.55}$$

in a self-evident notation.

Let us now specialize [143] this CS machinery to the case in which $\{x_n\}$ is the sequence $\mathbb{Z}_\tau^+$ of positive τ-integers, where τ denotes the golden mean $\tau = \frac{1}{2}(1+\sqrt{5})$, a notion we shall fully develop in Chapter 13, Section 13.4.1. These τ-integers form a quasiperiodic sequence, in which the difference $x_{n+1} - x_n$ takes only two values, 1 or $\tau^{-1} = \tau - 1 = \frac{1}{2}(\sqrt{5} - 1)$. The first few elements of the sequence are the following:

$$0, 1, \tau, \tau^2, \tau^2 + 1, \tau^3, \tau^3 + 1, \tau^3 + \tau, \ldots. \tag{6.56}$$

This counting system can be built from the representation of natural numbers in the Fibonacci numeration system, where the Fibonacci numbers are defined by

$$f_{n+2} = f_{n+1} + f_n, \quad f_0 = 1, \quad f_1 = 2. \tag{6.57}$$

6.3. Coherent states: The holomorphic case   121

To the (unique) representation [83] of a natural number,

$$n = \sum_{i=0}^{j_n} \xi_i f_i, \quad \xi_i \in \{0,1\}, \; \xi_i \xi_{i+1} = 0, \tag{6.58}$$

corresponds the $(n+1)$st positive τ-integer

$$x_n = \sum_{i=0}^{j_n} \xi_i \tau^i. \tag{6.59}$$

Note that the sequence $\{x_n, n \geq 0\} \equiv \mathbb{Z}_\tau^+$ is "almost" like the natural numbers. In connection with the present problem, it is interesting to have first some insight on the associated exponential

$$\exp_F(t) = \sum_{n=0}^{\infty} \frac{t^n}{[x_n]!} = 1 + t + \frac{t^2}{\tau} + \frac{t^3}{\tau^3} + \frac{t^4}{\tau^3(\tau^2+1)} + \ldots \tag{6.60}$$

and on its inverse $1/\exp_F(t)$. The first few terms of the latter are given by

$$\frac{1}{\exp_F(t)} = 1 - t + \frac{t^2}{\tau^2} - \frac{t^4}{\tau^3(\tau^2+1)} - \frac{t^5}{\tau^6(\tau^2+1)} + \ldots \tag{6.61}$$

Here, the radius of convergence is infinite : $\mathcal{D}_R = \mathbb{C}$. We show in Figure 6.1 the graph of the corresponding Gaussian

$$G_F(t) \equiv 1/\exp_F(t^2). \tag{6.62}$$

We have computed numerically the Gaussian integral

$$\sqrt{[\pi]} = \int_{-\infty}^{+\infty} \frac{1}{\exp_F(t^2)} \, dt, \tag{6.63}$$

where the notation used is a reminder of the standard case. We find (with $\doteq$ denoting an approximation of the above quantity),

$$\sqrt{[\pi]} \doteq 2.9155. \tag{6.64}$$

Next, consider the spectrum $\{\lambda_n, n \in \mathbb{N}\}$ of the operator A defined by (6.53)–(6.55). The first eigenvalues λ_n, computed using (6.55) read

$$\lambda_0 \doteq 0.8107, \; \lambda_1 \doteq 0.8771, \; \lambda_2 \doteq 0.8728, \; \lambda_3 \doteq 0.8347, \; \lambda_4 \doteq 0.8494, \; \ldots \tag{6.65}$$

The numerical computation of the limit of the eigenvalues gives

$$\lim_{n\to\infty} \lambda_n \doteq 0.8539. \tag{6.66}$$

The striking fact is that this limit is found to be equal to $\frac{1}{2}\sqrt{[\pi]}$:

$$\lim_{n\to\infty} \frac{\Gamma_F(n+1)}{[x_n]!} = \lim_{n\to\infty} \frac{1}{x_1 x_2 \ldots x_n} \int_0^\infty \frac{t^n}{\exp_F(t)} \, dt = \int_0^\infty \frac{1}{\exp_F(t)} \, dt. \tag{6.67}$$

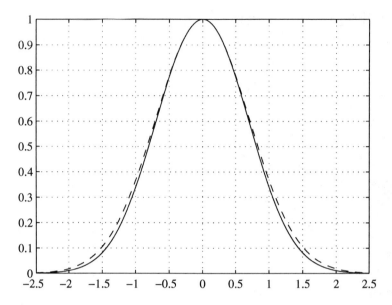

Figure 6.1. The τ-Gaussian $G_F(t) \equiv 1/\exp_F(t^2)$ (solid line), compared to the usual Gaussian (dashed line).

Unfortunately, no rigorous proof of this remarkable fact exists at the moment. Finally, we see from the behavior of λ_n that A indeed defines a (nontight) frame.

6.3.2 A nonholomorphic excursion (♣)

In fact, good physical reasons exist to go beyond the holomorphic class of CS. Recent explorations of nonholomorphic extensions of (6.32) [142, 192, 193, 194] stem from the following observation about the familiar canonical CS (2.14),

$$|z\rangle = e^{-\frac{1}{2}|z|^2} \sum_{n=0}^{\infty} \frac{z^n}{\sqrt{n!}} |n\rangle, \quad (6.68)$$

namely, the (phase space) complex parameter z should *also* be interpreted in terms of the action-angle variables (J, γ) for the harmonic oscillator

$$z = \sqrt{J}\, e^{-i\gamma}. \quad (6.69)$$

Considering the expansion (6.68) from this point of view, we rewrite it as

$$|z\rangle \equiv |J, \gamma\rangle = e^{-\frac{J}{2}} \sum_{n=0}^{\infty} \frac{J^{\frac{n}{2}} e^{-in\gamma}}{\sqrt{n!}} |n\rangle, \quad (6.70)$$

6.3. Coherent states: The holomorphic case

and then new CS features become apparent. In particular, they suggest a number of requirements that the *normalized* "action-angle" CS,

$$|J,\gamma\rangle, \quad 0 \leq J, \quad -\infty < \gamma < \infty, \tag{6.71}$$

ought to fulfill, indicating yet another close connection between the classical and quantum formulations of a given (1-D) system with Hamiltonian H:

1. Continuity:

$$(J',\gamma') \to (J,\gamma) \implies |J',\gamma'\rangle \to |J,\gamma\rangle;$$

2. Resolution of the identity:

$$\int |J,\gamma\rangle\langle J,\gamma| \, d\mu(J,\gamma) = I$$

(even if we keep the freedom to weaken this requirement by dealing with a nontrivial frame);

3. Temporal stability:

$$e^{-iHt}|J,\gamma\rangle = |J,\gamma+t\rangle;$$

4. Action identity:

$$\langle J,\gamma|H|J,\gamma\rangle = J.$$

We observe that properties 3 and 4 simply mimic the classical behavior of action-angle variables. In particular, Condition 3 tells us that, under time evolution, any coherent state always remains a coherent state.

Now choose a Hamiltonian H with a discrete spectrum bounded from below and adjusted so that $H \geq 0$. We assume also that all eigenvalues of H are nondegenerate,

$$H|e_n\rangle = x_n|e_n\rangle, \tag{6.72}$$

where the eigenvalues x_n form, as above, a strictly increasing sequence of positive numbers, starting from $x_0 = 0$

$$x_0 = 0 < x_1 < x_2 < \ldots < x_n < x_{n+1} < \ldots < x_\infty, \tag{6.73}$$

where x_∞ may be finite or infinite. It is then straightforward to check that the states

$$|J,\gamma\rangle = N(J)^{-1} \sum_{n=0}^{\infty} \frac{J^{\frac{n}{2}} e^{-ix_n\gamma}}{\sqrt{[x_n]!}} |n\rangle \tag{6.74}$$

exist and are normalized if $0 \leq J < R$, where R is the radius of convergence of the normalization factor

$$N(J)^2 = \sum_{n=0}^{\infty} \frac{J^n}{\sqrt{[x_n]!}}. \tag{6.75}$$

Moreover, they satisfy the four requirement listed above: 1, 3 and 4 are immediate, whereas 2 leads to a classical moment problem:

$$\begin{aligned}I &= \int |J,\gamma\rangle\langle J,\gamma| \, d\mu(J,\gamma) \\ &= \int_0^R \left[\lim_{\Gamma\to\infty} \frac{1}{2\Gamma} \int_{-\Gamma}^{\Gamma} |J,\gamma\rangle\langle J,\gamma| \, d\gamma \right] N(J)^2 \rho(J) \, dJ,\end{aligned} \quad (6.76)$$

where $\rho(J)$ is a probability distribution, such that

$$\int_0^R u^n \rho(u) \, du = [x_n]!. \quad (6.77)$$

As a first example, let us examine the case of a particle in an infinite well of width π. The adjusted Hamiltonian

$$H = -\frac{d^2}{dx^2} - 1 \quad (6.78)$$

(in units where $\hbar/2m = 1$) has the spectrum

$$x_n = n(n+2), \quad n = 0, 1, 2, \ldots$$

and eigenstates

$$\langle x|n\rangle = \sqrt{\frac{2}{\pi}} \sin(n+1)x. \quad (6.79)$$

The corresponding CS are

$$|J,\gamma\rangle = N(J)^{-1} \sum_{n=0}^{\infty} \frac{J^{\frac{n}{2}} e^{-in(n+2)\gamma}}{n!\sqrt{\frac{1}{2}(n+1)(n+2)}} |n\rangle. \quad (6.80)$$

Here, we have

$$N(J)^2 = 2 \sum_{n=0}^{\infty} \frac{J^n}{n!(n+2)!} = \frac{2}{J} I_2(2\sqrt{J}) = {}_0F_1(3, J), \quad (6.81)$$

where I_ν is the modified Bessel function. The moment problem

$$\frac{1}{2} n!(n+2)! = \int_0^{\infty} u^n \rho(u) \, du$$

has for solution the density $\rho(u) = u K_2(2\sqrt{u})$, where K_ν is again a modified Bessel function [Mag].

Next we consider an example for which $\lim_{n\to\infty} x_n = x_\infty < \infty$. We choose a Coulomb-like discrete spectrum

$$x_n \equiv 1 - \frac{1}{(n+1)^2}, \quad n = 0, 1, 2 \ldots. \quad (6.82)$$

In this case,
$$[x_n]! = \frac{n+2}{2(n+1)}. \tag{6.83}$$

Therefore, $\lim_{n\to\infty} x_n = 1$ and $\lim_{n\to\infty}[x_n]! = \frac{1}{2}$. This means that the radius of convergence of $N(J)^2$, as given in (6.75), is $R = 1$. In fact, we can get the normalization factor in closed form, for $0 \leq J < 1$,
$$N(J)^2 = \frac{2}{J}\left[\frac{1}{1-J} + \frac{\ln(1-J)}{J}\right]. \tag{6.84}$$

The moment problem
$$[x_n]! = \frac{n+2}{2(n+1)} = \int_0^1 u^n \rho(u)\, du \tag{6.85}$$

is solved by
$$\rho(u) = \frac{1}{2}[1 + \delta(u - 1^-)], \quad 0 \leq u < 1. \tag{6.86}$$

Thus, we have obtained a set of CS of the form
$$|J,\gamma\rangle = N(J)^{-1}\sum_{n=0}^{\infty}\sqrt{\frac{2n+2}{n+2}}\, J^{\frac{n}{2}}\, e^{-(1-\frac{1}{(n+1)^2})\gamma}\, |n\rangle. \tag{6.87}$$

The extension of this result to the actual Coulomb problem in space is a purely technical matter. Taking into account the angular degrees of freedom would lead, for instance, to CS $|J, \gamma, \Omega\rangle$, where Ω denotes a direction in space, for which one has the so-called spin CS $|\Omega\rangle$ (see [194] and Section 7.2).

We should not conclude this section without mentioning the extension of this "coherent" action-angle formalism from systems with continuous spectrum to systems that have both a discrete and a continuous spectrum [142].

7
Covariant Coherent States

By now, we have encountered several examples of coherent states — the canonical CS, associated to the Weyl–Heisenberg group, discussed in detail in Chapter 2; vector CS, introduced in Chapter 4, Section 4.2.1; CS associated to the discrete series representations of $SU(1,1)$, in Section 4.2.2 of the same chapter; and general coherent states, associated with any reproducing kernel Hilbert space, in Chapter 5, Section 5.3.2. At this point, certain common features are already seen to emerge, as follows.

1. Coherent states form an overcomplete family of vectors in a Hilbert space and define a reproducing kernel on it.

2. The existence of the reproducing kernel enables one to map the Hilbert space unitarily to a space of (vector-valued) functions — in quantum mechanical language, the CS become wave functions on some appropriate space (usually, the phase space of the underlying classical system).

3. Group representations often facilitate the construction of such families, the individual CS being now labeled by the points of some homogeneous space of the group — thereby inheriting covariance properties under the action of the group.

4. Square integrability of the reproducing kernel, when it is present, reflects special properties of the functions constituting the reproducing kernel Hilbert space and of the group representation.

128   7. Covariant Coherent States

In this and the next few chapters, we embark on a systematic study of CS associated with group representations. We begin by giving a general definition of such CS and indicate how it is broad enough to encompass all examples of group-related CS found in the literature. Later, in Chapter 8, we pass on to a detailed treatment of square integrable representations of groups — or representations emanating from the discrete series — and CS associated with them. In Chapter 9, we specialize to a class of groups of the semidirect product type, admitting square integrable representations, and for which many of the quantities appearing in the general theory can be explicitly calculated.

7.1   Covariant coherent states

At the expense of being pedantic, we wish to reiterate that, generally, coherent states are simply a preferred overcomplete family of vectors in a reproducing kernel Hilbert space. There is no assumption of group covariance or square integrability in the general setting. Most coherent states arising in physics, however, are associated with symmetry groups of the system. We introduce, therefore, a general definition to encompass all such group-related coherent states.

7.1.1   A general definition

As before, let G be a locally compact group, μ be the left Haar measure on it, H be a closed subgroup of G, and $X = G/H$ be the corresponding left coset space, with a quasi-invariant measure ν. For a global Borel section $\sigma : X \to G$ of the group, let ν_σ be the unique quasi-invariant measure defined in (4.19):

$$d\nu_\sigma(x) = \lambda(\sigma(x), x) d\nu(x), \quad \text{where} \quad \lambda(g, x) d\nu(x) = d\nu(g^{-1}x), \ (\forall g \in G).$$

Definition 7.1.1 (Covariant coherent states): *(1) Let U be a unitary representation of G in a Hilbert space $\mathfrak{H}$ and η^i, $i = 1, 2, \ldots, n$, be a set of linearly independent vectors in $\mathfrak{H}$. Then, if the set of vectors*

$$\mathfrak{S}_\sigma = \{\eta^i_{\sigma(x)} = U(\sigma(x))\eta^i \mid i = 1, 2, \ldots, n, \ x \in X\}, \qquad (7.1)$$

is total in $\mathfrak{H}$, we call it a family of covariant coherent states for U.
(2) Let $F = \sum_{i=1}^n |\eta^i\rangle\langle\eta^i|$, $F_\sigma(x) = U(\sigma(x))FU(\sigma(x))^*$, and suppose that

$$\int_X F_\sigma(x) \, d\nu_\sigma(x) = A_\sigma, \qquad A_\sigma, \ A_\sigma^{-1} \in \mathcal{L}(\mathfrak{H}), \qquad (7.2)$$

the integral converging weakly. Then, the coherent states $\mathfrak{S}_\sigma$ are called square integrable and the representation U is said to be square integrable mod(H, σ). In this case, the vectors η^i, $i = 1, 2, \ldots, n$, are called admissible mod(H, σ) and the operator F is called a *resolution generator* for the representation.

Unless otherwise stated, the number n of vectors η^i (which are often called *fiducial vectors* [Kl2, 189]) will be assumed to be finite. Given a family of covariant CS, square integrable or not, there is an associated reproducing kernel

$$K_\sigma(x, y) = F_\sigma(x)^{\frac{1}{2}} F_\sigma(y)^{\frac{1}{2}} \tag{7.3}$$

and a reproducing kernel Hilbert space $\mathfrak{H}_{K_\sigma}$, to which $\mathfrak{H}$ is unitarily equivalent as a consequence of Theorem 5.3.8 (see, in particular, (5.61)–(5.65); see also [200]).

Consider now the sections $\sigma_g : X \to G$, $g \in G$, which are covariant translates of σ under g:

$$\sigma_g(x) = g\sigma(g^{-1}x) = \sigma(x)h(g, g^{-1}x), \tag{7.4}$$

where h is the cocycle defined in (4.37). Let ν_{σ_g} be the measure $d\nu_{\sigma_g}(x) = \lambda(\sigma_g(x), x)\, d\nu$, again constructed using (4.19), and $F_{\sigma_g}(x) = U(\sigma_g(x))FU(\sigma_g(x))^*$. If U is square integrable mod(H, σ), there is a general covariance property enjoyed by the positive operators defined by the weak integrals,

$$a_{\sigma_g}(\Delta) = \int_\Delta F_{\sigma_g}(x)\, d\nu_{\sigma_g}(x), \qquad g \in G, \Delta \in \mathcal{B}(X). \tag{7.5}$$

Indeed, it is easily verified [using (4.8)] that the covariance condition

$$U(g)a_\sigma(\Delta)U(g)^* = a_{\sigma_g}(g\Delta), \qquad U(g)A_\sigma U(g)^* = A_{\sigma_g}, \tag{7.6}$$

holds. In the case in which X admits a left invariant measure m and one takes $\nu = m$, and

$$U(h)FU(h)^* = F, \qquad h \in H, \tag{7.7}$$

it is an immediate consequence of (7.4) that $U(g)A_\sigma U(g)^* = A_\sigma$, for all $g \in G$. If the representation U is assumed to be irreducible, Schur's lemma (see Chapter 4, Section 4.3) then implies that $A = \lambda I$, where λ is a positive real number. Moreover, in this case, the operators $a_\sigma(\Delta)$ do not depend on σ — a situation that has already been encountered for vector CS in Chapter 4, Section 4.2.1 — and thus, the pair $\{U, a\}$ forms a system of covariance in the sense of Definition 4.2.1. We now call the representation U simply *square integrable* mod(H) and the vectors η^i, $i = 1, 2, \ldots, n$, *admissible* mod(H).

7.1.2 The Gilmore–Perelomov CS and vector CS

A ready example of a family of covariant coherent states is obtained for any unitary representation U of the group G, which admits a *cyclic vector* η (a *fortiori* if it is irreducible). In this case, the set of vectors $\eta_g = U(g)\eta$, $g \in G$, is dense in $\mathfrak{H}$ and hence satisfies Condition 1 of Definition 7.1.1, with $H = \{e\}$, the trivial subgroup consisting of the identity element of G, (so that $X = G$), the trivial section $\sigma(g) = g$, and $n = 1$. Thus, the orbit of any cyclic vector is a family of coherent states, which do not in general have to be square integrable. Furthermore, square integrability in this case would imply that

$$\int_G |\eta_g\rangle\langle\eta_g| \, d\mu(g) = \lambda I, \tag{7.8}$$

for some real number λ (and a similar relation would hold if the right invariant measure were to be substituted for $d\mu$).

On the other hand, consider the closed subgroup H of G that stabilizes the cyclic vector η up to a multiplicative factor; i.e.,

$$U(h)\eta = \exp[i\omega(h)]\eta, \qquad h \in H, \tag{7.9}$$

where ω is a real-valued function on H (the fact that H is a subgroup is trivial to verify.) It follows that $\chi(h) = \exp[i\omega(h)]$ is a character of H, since clearly

$$\chi(h_1 h_2) = \chi(h_1)\chi(h_2), \; h_1, h_2 \in H, \quad \text{and} \quad \chi(e) = 1. \tag{7.10}$$

Letting $X = G/H$ and taking any Borel section $\sigma : X \to G$, we may define another family of covariant CS

$$\eta_{\sigma(x)} = U(\sigma(x))\eta, \qquad x \in X, \tag{7.11}$$

since this set of vectors is also total in $\mathfrak{H}$. Such CS will be referred to as the *Gilmore–Perelomov coherent states* [Per, 144, 246]. Note that, if now the representation is also square integrable in the sense of (7.8) and the measure ν_σ is actually *invariant* under the action of G on X, then H is necessarily a compact subgroup of G, although just square integrability mod(H, σ) will not ensure this. From a mathematical point of view, it is natural to consider the notion of square integrability of a representation of a group G modulo its center $Z(G)$. While this is commonly done in the mathematical literature [Bor], it is, however, unduly restrictive. The more general setting mentioned above was introduced, independently, by Gilmore [144] and Perelomov [246]. It was anticipated, to some extent, by Klauder [189], using the argument that coherent states arise in quantum theories, and there, as is well known, states are represented not by individual vectors in a Hilbert space $\mathfrak{H}$, but rather by rays (one-dimensional subspaces of $\mathfrak{H}$). Upon normalization, a quantum state is thus represented by a family $\{\eta = \exp[i\alpha]\eta_o \mid \alpha \in [0, 2\pi)\}$. Take now a coherent state $\eta_g = U(g)\eta, g \in G$, of the Gilmore–Perelomov type; the corresponding quantum state does not depend on g, but rather on its equivalence class modulo H. Although a section is not explicitly used in either [144] or [246], it appears implicitly in the form of a particular factorization of the elements of G. Indeed, take

$g \in G$, and let $x = gH$ be the coset to which it belongs (or, equivalently, its canonical projection in the principal bundle $G \to X = G/H$). Then, any Borel section $\sigma : X \to G$ defines a factorization $g = \sigma(x)h(x)$, with $h(x) \in H$, and, conversely, one can always find a factorization $g = g_o(y)h(g)$, with $h(g) \in H$, which then defines a Borel section $\sigma(x) = g_o(g)$. Before passing to some concrete examples, we might point out that the formulations of Gilmore [144] and Perelomov [Per, 246] differ in the assumptions they make concerning G, U, and η. Gilmore considers a general symmetry group G, but restricts U to be a square integrable representation and η a highest (or lowest) weight vector of U, while Perelomov requires G to be a Lie group, which allows for the use of the Lie algebra, e.g., in the choice of the basic vector η (see Section 4.5). Coherent states are then defined through (7.11), without any square integrability assumptions initially, thus allowing U to be any unitary irreducible representation (including those of the principal or the complementary series), and η to be any normalized vector in the representation space.

If the representation U is not square integrable, and if by analogy with (7.2), we formally write

$$\int_X |\eta_{\sigma(x)}\rangle\langle\eta_{\sigma(x)}| \, d\nu(x) = A_\sigma, \tag{7.12}$$

(for simplicity, we take here $n = 1$ and assume that the measure ν is invariant), then A_σ is not a bounded operator — perhaps not even to a bona fide operator on the Hilbert space $\mathfrak{H}$. It may be possible, however, to give a meaning to A_σ as a sesquilinear form on $\mathfrak{K} \times \mathfrak{K}$, where $\mathfrak{K}$ is some dense domain in $\mathfrak{H}$. An interesting possibility would be to reformulate this statement in the language of *rigged Hilbert spaces* [Gel]. This means that one would consider the triplet of spaces

$$\mathfrak{K} \subset \mathfrak{H} \subset \mathfrak{K}', \tag{7.13}$$

where $\mathfrak{K}'$ denotes the topological dual of $\mathfrak{K}$, in a suitable locally convex topology. Then, a sesquilinear form over $\mathfrak{K} \times \mathfrak{K}$ may be identified with a continuous map from $\mathfrak{K}$ into $\mathfrak{K}'$. This kind of an approach has become familiar, both in the theory of representations of noncompact groups [82, 226] and in quantum mechanics [25]. In the present context, however, this is still largely speculative, although some steps have already been taken in this direction in [240]. (A similar reformulation will be considered in Section 7.3, leading to unbounded frames).

Both the canonical coherent states (2.37) and the CS arising from the discrete series representations of $SU(1,1)$ [see (4.99)] are examples of Gilmore–Perelomov coherent states. Indeed, the form of η^s in (2.6) and of the representation $U^\lambda(\theta, q, p) = e^{i\lambda\theta} U^\lambda(q, p)$ of the Weyl–Heisenberg group in (2.33) tell us that, with $H = \Theta$, the phase subgroup,

$$U^\lambda(\theta)\eta^s = e^{i\lambda\theta}\eta^s, \qquad \theta \in \Theta. \tag{7.14}$$

Thus, using the section $\sigma(q,p) = (0,q,p)$ in (2.36), we identify the canonical CS, $\eta^s_{\sigma(q,p)} = U^\lambda(q,p)\eta^s$ in (2.37) as being precisely of the Gilmore–Perelomov type. If, as noted in Chapter 2, Section 2.2, we derive the canonical CS from a unitary representation of the oscillator group $H(4)$, then, choosing again for η the ground state $|0\rangle$, one finds that $H = U(1) \times U(1)$, and the construction proceeds as above, with $X \simeq \mathbb{C}$ [289].

In the case of the discrete series representation U^j of $SU(1,1)$ given in (4.86), if we take $H = K \simeq U(1)$, the maximal compact subgroup [see (4.65)], and the vector

$$\eta = (2j-1)^{\frac{1}{2}} u_0, \quad \text{with} \quad u_0(z) = 1, \, z \in \mathcal{D}, \tag{7.15}$$

then, once again,

$$U^j(k)\eta = e^{-ij\phi}\eta, \quad k = \text{diag}(e^{i\frac{\phi}{2}}, e^{-i\frac{\phi}{2}}) \in K, \tag{7.16}$$

by (4.88). Hence, the CS, $\eta_{\sigma(z)} = U^j(\sigma(z))\eta$ in (4.99), generated by the vector η, are again of the Gilmore–Perelomov type.

CS associated with square integrable representations of semisimple Lie groups, which are of the Gilmore–Perelomov type, will be discussed more systematically in Section 7.2.

Vector coherent states, of which we examined a square integrable version in Chapter 4, Section 4.2.1, are an obvious generalization of the Gilmore–Perelomov CS. Going back to the definition of covariant CS, let $\mathfrak{K}$ be the n-dimensional subspace of $\mathfrak{H}$ generated by the vectors η^i, $i = 1,2,\ldots,n$. Assume that $\mathfrak{K}$ is stable under the action of the operators $U(h)$, $h \in H$, i.e., $U(h)\eta \in \mathfrak{K}$, for all $h \in H$ and $\eta \in \mathfrak{K}$. Then, the functions [see (5.68), $\Phi : X \to \mathbb{C}^n$, with components $\Phi^i(x) = \langle \eta^i_{\sigma(x)} | \phi \rangle$, $i = 1,2,\ldots,n$, for all $\phi \in \mathfrak{H}$, constitute a reproducing kernel Hilbert space with matrix kernel [see (5.66)] $\mathbf{K}(x,y)$ having matrix elements

$$\mathbf{K}(x,y)_{ij} = \langle \eta^i_{\sigma(x)} | \eta^j_{\sigma(y)} \rangle, \quad i,j = 1,2,\ldots,n. \tag{7.17}$$

The mapping

$$W_\eta : \mathfrak{H} \to \mathfrak{H}_K, \quad (W_\eta \phi)^i(x) = \Phi^i(x) = \langle \eta^i_{\sigma(x)} | \phi \rangle, \, i = 1,2,\ldots,n, \, x \in X, \tag{7.18}$$

is a Hilbert space isometry. Denoting the image of $U(g)$ under W_η (as an operator on $\mathfrak{H}_K$) by $U_\eta(g)$, we find, as in (4.60)–(4.62),

$$(U_\eta(g)\Phi)(x) = V(h(g^{-1},x))^* \Phi(g^{-1}x), \quad \Phi \in \mathfrak{H}_K, \, x \in X, \tag{7.19}$$

where, as defined in (4.54), $V(h)$, $h \in H$, constitutes an n-dimensional unitary representation of H on $\mathbb{C}^n$. Since in Section 4.2.1 we had assumed [see (4.58)] square integrability of the kernel $\mathbf{K}(x,y)$, and consequently there was the associated system of covariance (4.59), it was possible to extend the representation U_η to an induced representation $\widetilde{U}$ on the Hilbert space $\mathbb{C}^n \otimes L^2(X,d\nu)$, of which $\mathfrak{H}_K$ then became a subspace. The conditions for ensuring square integrability will be taken up again in Section 9.2. In general, however, one does not have square integrability, although the form of the representation U_η in (7.19) resembles that of an induced representation. The vectors $\eta^i_{\sigma(x)}$, $i = 1, 2, \ldots, n$, $x \in X$, are called *vector coherent states (VCS)*, and the representation U_η is called a *VCS representation* [105, 252, 260, 261, 262, 263]. In [263], the subspace $\mathfrak{K}$ is not assumed to be finite dimensional, nor is one necessarily restricted to representations on Hilbert spaces.

Let us emphasize again that the Gilmore–Perelomov CS are a special case of the VCS when the dimension of the subspace $\mathfrak{K}$ is unity. Moreover, while the VCS arise from representations of the subgroup H on the subspace $\mathfrak{K}$, for general covariant CS, Definition 7.1.1 given above does not require $\mathfrak{K}$ to carry a representation of H.

7.1.3 A geometrical setting

It is possible to give an invariant geometrical description of vector coherent states [5] using vector bundles, which generalizes a similar construction proposed in [254] for the Gilmore–Perelomov type CS, employing complex line bundles. In order to present this construction, it is necessary to use some elementary notions from the theory of principal bundles and their associated vector bundles. (An easy introduction to the subject may, for instance, be found in [vWe].)

The point of this construction is to obtain the VCS as sections in a vector bundle associated with the principal bundle $\pi : G \to X$, i.e., the bundle giving the natural fibration of G over the homogeneous space G/H. [Here π denotes the projection map $\pi(g) = gH$.] For the purposes of this discussion, we shall use the symbol $[g]$ to denote the left coset $gH = x \in X = G/H$ and assume that U is a unitary representation of G on $\mathfrak{H}$, admitting a *finite dimensional cyclic subspace* $\mathfrak{K}$, which is stable under U when the latter is restricted to H; i.e., the set $\{U(g)\eta \mid \eta \in \mathfrak{K}, g \in G\}$ is total in $\mathfrak{H}$ and,

$$U(h)\eta \in \mathfrak{K}, \quad \text{for all } \eta \in \mathfrak{K}, h \in H. \tag{7.20}$$

Let $\mathbb{P}_\mathfrak{K}$ be the projection operator for the subspace $\mathfrak{K}$, and write

$$U(h)\mathbb{P}_\mathfrak{K} = V(h), \quad h \in H. \tag{7.21}$$

We also assume that the the representation $h \mapsto V(h)$, of the subgroup H, acts *irreducibly* on $\mathfrak{K}$.

Looking at G as a principal bundle, fibered over X and with structure group H, we consider the G-homogeneous associated bundle $B = G \times_{V(H)} \mathfrak{K}$. This bundle, also fibered over X, is obtained by identifying elements $(g, \eta), (g', \eta') \in G \times \mathfrak{K}$, whenever $g' = gh$ and $\eta' = V(h)^{-1}\eta$ for some $h \in H$. Thus, elements in B can be put into one-to-one correspondence with vectors $\eta_g = U(g)\eta \in \mathfrak{H}$, for all $g \in G$ and $\eta \in \mathfrak{K}$. In the terminology of [254], these vectors may be called a family of *coherent vectors*. In view of the cyclicity of the subspace $\mathfrak{K}$ with respect to the representation U, it should be possible to extract a suitable subset of vectors from among this set, which would be the CS of this representation, satisfying the conditions of Definition 7.1.1, and we shall do so momentarily. Identifying the bundle $B = G \times_{V(H)} \mathfrak{K}$ with this family of coherent vectors, the canonical projection $\pi_B : B \to X$ in it is, very simply, $\pi_B(\eta_g) = [g]$. Sections in this bundle can be put into one-to-one correspondence with functions $\mathbf{f} : G \to \mathfrak{K}$, which satisfy

$$\mathbf{f}(gh) = V(h)^{-1}\mathbf{f}(g), \qquad h \in H. \tag{7.22}$$

Consider next the functions $\Phi : G \to \mathfrak{K}$, defined by

$$\Phi(g) = \mathbb{P}_{\mathfrak{K}} U(g)^{-1}\phi, \qquad \phi \in \mathfrak{H}. \tag{7.23}$$

Clearly these functions satisfy (7.22) and hence describe sections of B. We denote this class of sections (for all $\phi \in \mathfrak{H}$) by $\mathfrak{H}_K$, and then it is clear that $\mathfrak{H}_K$ has the natural structure of a reproducing kernel Hilbert space, with reproducing kernel $K : G \times G \to \mathcal{L}(\mathfrak{K})$ ($=$ bounded operators on $\mathfrak{K}$),

$$K(g, g') = \mathbb{P}_{\mathfrak{K}} U(g)^{-1} U(g') \mathbb{P}_{\mathfrak{K}}. \tag{7.24}$$

Indeed, since $K(g, g') = K(g', g)^*$ and as an operator on $\mathfrak{K}$, $K(g, g)$ is strictly positive, K is a strictly positive-definite kernel, and, therefore, following Theorem 5.3.3, the relationship

$$\langle K(\,\cdot\,, g)\eta | K(\,\cdot\,, g')\eta' \rangle_K = \langle \eta | K(g, g')\eta' \rangle_{\mathfrak{H}}, \tag{7.25}$$

can be used to define a scalar product [see (5.23)], $\langle \cdot | \cdot \rangle_K$, on the linear span of sections of the type $K(\,\cdot\,, g)\eta$, for all $g \in G$ and $\eta \in \mathfrak{K}$, which will then extend to all sections of the type (7.23). Note that the section $K(\,\cdot\,, g)\eta$ assumes the value

$$K(g', g)\eta = \mathbb{P}_{\mathfrak{K}} U(g)^* \eta_{g'} \in \mathfrak{K}, \tag{7.26}$$

at the point $g' \in G$. Note also, that the measurable field of Hilbert spaces, postulated by Theorem 5.3.3, is in this case $\{\mathfrak{K}_g \mid g \in G\}$, with $\mathfrak{K}_g = \mathfrak{K}$ for all $g \in G$. It then follows that $\mathfrak{H}_K$ is complete in the above scalar product, hence, a reproducing kernel Hilbert space, and the map $W : \mathfrak{H} \to \mathfrak{H}_K$, such that

$$(W_K \phi)(g) = \Phi(g) = \mathbb{P}_{\mathfrak{K}} U(g)^{-1}\phi, \qquad g \in G, \phi \in \mathfrak{H}, \tag{7.27}$$

is unitary. Clearly,

$$W_K \eta_g = K(\,\cdot\,, g)\eta, \qquad g \in G, \tag{7.28}$$

meaning that the coherent vectors are mapped to sections in the associated bundle B and the linear span of this family of sections defines a reproducing kernel Hilbert space $\mathfrak{H}_K$, which is unitarily equivalent to the original space $\mathfrak{H}$ of the representation U. The reproducing property of the kernel $\mathfrak{H}_K$ is reflected in the evaluation condition for functions Φ in $\mathfrak{H}_K$ at points $g \in G$:

$$\langle K(\,\cdot\,,g)\eta|\Phi\rangle_K = \langle \eta|\Phi(g)\rangle_{\mathfrak{H}}, \qquad \eta \in \mathfrak{H},\, g \in G. \tag{7.29}$$

Let $\{\eta^i\}_{i=1}^n$ (n = dimension of $\mathfrak{K}$) be an orthonormal basis for $\mathfrak{K}$ and $\sigma: X \to G$ be a (global) Borel section of G. Then, the set of vectors,

$$\mathfrak{S}_\sigma = \{\eta^i_{\sigma([g])} \mid i = 1,2,\ldots n,\ [g] \in X\}, \tag{7.30}$$

is total in $\mathfrak{H}$ and forms a family of vector coherent states for the representation U. This then is the differential geometric interpretation of vector CS. They arise as (in general measurable) sections in a vector bundle, naturally associated with the group representation in question.

We shall also be interested in conditions for the square integrability mod(H) of these VCS, i.e., conditions to ensure a resolution of the identity of the type:

$$\sum_{i=1}^n \int_X |\eta^i_{\sigma([g])}\rangle\langle \eta^i_{\sigma([g])}|\, d\nu([g]) = \int_X \langle \phi|\mathbb{P}_{\mathfrak{K}}([g])\phi\rangle\, d\nu([g]) = \lambda I < \infty,$$

$$\lambda > 0, \tag{7.31}$$

for all $\phi \in \mathfrak{H}$, where we have written, with obvious justification,

$$\mathbb{P}_{\mathfrak{K}}([g]) = U(g)\mathbb{P}_{\mathfrak{K}}U(g)^*. \tag{7.32}$$

In this case, however, it will be necessary to impose the condition that the measure ν be invariant on X. We shall return to this point in Chapter 9, Section 9.2, but let us note here that, from our discussion on induced representations in Section 4.2 of Chapter 4, we know that the set of sections **f** in (7.22), which are square integrable with respect to the invariant measure ν, forms a Hilbert space $L^2_{\mathfrak{K}}(X, d\nu)$, on which the representation $g \mapsto {}^V U(g)$, $g \in G$,

$$({}^V U(g)\mathbf{f})(g') = \mathbf{f}(g^{-1}g'), \qquad g' \in G, \tag{7.33}$$

is unitary (by virtue of the invariance of ν). Furthermore, this is just the standard induced representation (4.45) of G (induced from the UIR V of the subgroup H).

Suppose now that the reproducing kernel Hilbert space $\mathfrak{H}_K$ forms a subspace of $L^2_{\mathfrak{K}}(X, d\nu)$. Then, the unitary map W_K in (7.27) will clearly intertwine the representation U (of G on $\mathfrak{H}$) with the induced representation ${}^V U$, and thus, U would be a subrepresentation of an induced representation. Additionally, if all sections Φ of the type (7.23) are elements of $L^2_{\mathfrak{K}}(X, d\nu)$, then, for any two sections $\Phi, \Phi' \in \mathfrak{H}_K$,

$$\int_X \langle \Phi(g)|\Phi'(g)\rangle_{\mathfrak{H}}\, d\nu([g]) < \infty, \tag{7.34}$$

and, in particular, using sections of the type (7.28),

$$\int_X \langle K(g'',g)\eta|K(g'',g')\eta'\rangle_{\mathfrak{H}}\, d\nu([g'']) =$$

$$= \int_X \langle \eta | U(g^{-1}) \mathbb{P}_{\mathfrak{K}}([g'']) U(g) \eta' \rangle \, d\nu([g''])$$
$$= \int_X \langle \eta | \mathbb{P}_{\mathfrak{K}}([g'']) \eta' \rangle \, d\nu([g'']) < \infty, \tag{7.35}$$

for arbitrary $\eta, \eta' \in \mathfrak{K}$, the last equality following from the invariance of the measure ν. Thus, we arrive at the condition

$$I(\eta) \equiv \int_X \langle \eta | \mathbb{P}_{\mathfrak{K}}([g]) \eta \rangle \, d\nu([g]) < \infty, \tag{7.36}$$

for all $\eta \in \mathfrak{K}$. It will be shown in Theorem 9.2.5 in Chapter 9 that this condition is both necessary and sufficient for the square integrability mod(H) of the representation U of G.

7.2 Example: The classical theory of coherent states

The classical theory of coherent states, largely due to Gilmore [144] and Perelomov [246], applies primarily to semisimple and nilpotent Lie groups. We will briefly survey it in this section, referring the reader to the monograph of Perelomov [Per] for a more detailed presentation and concrete examples, together with original references.

7.2.1 CS of compact semisimple Lie groups

Let G be a compact semisimple Lie group, with Lie algebra $\mathfrak{g}$. Then, any unitary irreducible representation of G is square integrable and finite dimensional. So, the only question for constructing CS is: What is the best choice for the basic vector η in a given physical situation?

An obvious requirement is to try and find a vector η that yields CS which are 'closest to the classical states', in other words, minimal uncertainty states. A criterion for that was found independently by Perelomov [Per, 246] and by Delbourgo and Fox [106, 107], and generalized further by Spera [271]. If η is a highest (or lowest) weight vector in the chosen representation, then the corresponding coherent states have minimal uncertainty. This means that they minimize the dispersion ΔC_2 of the quadratic Casimir operator $C_2 = \sum_{i,k=1}^n g_{ik} X^i X^k$, that is, the quantity

$$\Delta C_2 = \langle C_2 \rangle - \sum_{i,k=1}^n g_{ik} \langle X^i \rangle \langle X^k \rangle, \tag{7.37}$$

where $\langle . \rangle$ denotes the mean value in the state η, namely, $\langle A \rangle = \langle \eta | A | \eta \rangle$.

We illustrate this again on the simplest example, namely, $SU(2)$, already mentioned in Chapter 4, Section 4.5.1. In the unitary spin-j irreducible representation $D^{(j)}$, the vector $|j,-j\rangle$ is a lowest weight vector; i.e. $J_-|j,-j\rangle = 0$) and has an isotropy subgroup $U(1)$. Then, the corresponding CS, $D^{(j)}(g)|j,\pm j\rangle$, have minimal uncertainty and are indexed by points of $SU(2)/U(1) \simeq S^2$, the Bloch sphere of quantum optics. These are the spin CS or Bloch states, widely used in atomic physics and quantum optics [Scu, 47, 253].

In fact, this choice of a highest weight vector is a particular case of a general criterion introduced by Perelomov [Per, 246], which points towards some deep geometric properties of CS. Namely, if the isotropy subalgebra $\mathfrak{b} \subset \mathfrak{g}^c$ of η is maximal in the sense of Perelomov (here $\mathfrak{g}^c$ denotes the complexified Lie algebra of $\mathfrak{g}$, see Section 4.5.1), that is, $\mathfrak{b} \oplus \overline{\mathfrak{b}} = \mathfrak{g}^c$, then η generates minimal uncertainty CS.

Coming back to $SU(2)$, it is easy to see that $\mathfrak{b}_\pm = \{J_0, J_\pm\}$, $G^c = SL(2,\mathbb{C})$, $H^c, Z_\pm$ are the one-parameter subgroups generated by $J_0, J_\pm$, respectively, and $X_+ = G^c/B_- \simeq X_- = B_+\backslash G^c \simeq \mathbb{C} \simeq S^2$ (the last homeomorphism being the stereographic projection from the Riemann sphere onto the complex plane). Then, the Gauss decomposition (4.148) of elements of G^c, $g = z_+ h z_-$, $z_\pm \in Z_\pm$, $h \in H^c$, yields a corresponding factorization for the representation operators $D^{(j)}(g)$:

$$D^{(j)}(g) = \exp(zJ_+)\exp(xJ_0)\exp(z'J_-). \tag{7.38}$$

Choosing the lowest weight vector $|j,-j\rangle$, with isotropy subalgebra $\mathfrak{b}_- = \{J_0, J_-\}$, one gets for the corresponding CS

$$|z\rangle = D^{(j)}(g)|j,-j\rangle = c_j \exp(zJ_+)|j,-j\rangle, \quad z \in \mathbb{C}, \tag{7.39}$$

wher c_j is a suitable constant. The same technique is widely used in nuclear physics, for the unitary groups $SU(n)$, in particular, $SU(3)$ and $U(3)$ (which, while not semisimple, can be treated in the same way). We refer to [Kl2] for original references.

An example: Spin CS and the Hepp-Lieb model (♣)

As discussed in Chapter 2, Section 2.7.2, a system of N two-level atoms interacting with a radiation field exhibits a phase transition, radiance → superradiance, at a certain critical temperature T_c, when the coupling is strong enough. This result was obtained in the simplest approximation by Wang and Hioe [283] with the help of canonical CS, and derived rigorously by Hepp and Lieb by standard methods [168]. In a further paper [169], the last two authors obtained a far-reaching generalization, this time using Bloch atomic or spin CS [Scu, 47, 253]. The difference between the two approaches may be caricatured by saying that canonical CS make the radiation field classical, while the atomic CS make the atomic variables classical.

The mathematical problem, as usual, resides in the calculation of the partition function. One starts with the Hamiltonian (2.92)–(2.94), written for N two-level atoms interacting with M photon modes ($k = 1, \ldots, M$). As in (2.98), the quantity to be estimated is $Z = \operatorname{Tr} e^{-\beta H}$, where the trace is over the tensor product of the Fock space $\mathfrak{F}$ of the photon modes and the Hilbert space $\mathfrak{H}$ of the N spin-1/2 atoms. The key point is that the total "spin," J, is a constant of the motion, so that

$$Z = \sum_{J=0}^{N/2} Y(N, J) Z(J), \quad Z(J) = \operatorname{Tr} e^{-\beta H(J)}, \tag{7.40}$$

where $H(J)$ is the Hamiltonian for a spin, $\vec{S}$, of magnitude J, and the combinatorial factor $Y(N, J) = N!(2J+1)[(J+1+\frac{1}{2}N)!(\frac{1}{2}N - J)!]^{-1}$ is the number of ways to construct an angular momentum J from N spin-1/2 particles. At this point, one uses the following bounds, obtained in [206] with atomic CS,

$$\widetilde{Z}(J) \leq (2J+1)^{-1} Z(J) \leq \widetilde{Z}(J+1), \tag{7.41}$$

where

$$\widetilde{Z}(J) = (4\pi)^{-1} \operatorname{Tr}_{\mathfrak{F}} \int e^{-\beta \widetilde{H}(J,\Omega)} \, d\Omega,$$

and $\widetilde{H}(J, \Omega)$ is the Hamiltonian $H(J)$ with $\vec{S}$ replaced by $J\vec{\Omega}$ ($\vec{\Omega}$ is a point on the unit sphere). Calculating the trace explicitly, one then finds that, for large N and fixed M, the free energy $-\beta \ln \widetilde{Z}(J)$ coincides with that obtained for a single mode in the rotating-wave approximation, but with an effective coupling constant

$$\lambda^2 = \max_\phi \sum_{k=1}^{M} |2\lambda_k \cos\phi|^2 \nu_k^{-1},$$

that is, the result of Wang and Hioe [283] described in Section 2.7.2, see (2.106).

Next, atomic CS may be used to understand the thermodynamical stability of a system with infinitely many modes. The result is that the system is stable; that is, $Z \leq \exp NA$, for some constant A, iff the atoms have very repulsive cores. Otherwise, a lower bound may be derived, namely,

$$Z > e^{CN^2}, \quad C > 0.$$

This is true irrespective of whether the atomic coordinates are treated classically or quantum mechanically. A final result of [169] is to extend the treatment to a finite-mode case, in which the atoms have translational degrees of freedom. In each of these various situations, the reasoning always rests on estimates obtained in [206] using atomic CS.

7.2.2 CS of noncompact semisimple Lie groups

If G is a noncompact semisimple Lie group, the square integrable representations are exactly those of the *discrete series* For such a representation, all that we have said for compact groups goes through, except for the definition of minimal uncertainty states. Indeed, the quadratic Casimir operator C_2 is not positive definite for a noncompact group, so that the minimization of ΔC_2 can no longer be interpreted as "optimal localization" of the state, which is the crucial semiclassical feature of CS [104]. In the case of $SU(1,1)$, however, it turns out that the correct CS not only minimize ΔC_2, but they also saturate another inequality directly derived from the commutation relations, of the general form

$$\langle A \rangle \langle B \rangle \geq \frac{1}{2} \langle [A, B] \rangle, \tag{7.42}$$

and which fully justifies their quasi-classical characterization. As a matter of fact, they share this property with the Barut–Girardello CS [65], mentioned in Chapter 1. We shall examine these CS of $SU(1,1)$ in great detail in Chapter 11, Section 11.3, in the study of the contraction process from an anti-de Sitter to a Minkowskian space–time [$SU(1,1) \simeq SO(1,2)$ is the anti-de Sitter group in 1+1 space–time dimensions] [104, 255].

Applications of CS, coming from noncompact semisimple Lie groups, are again numerous. For instance, besides the example just mentioned, $SU(1,1)$ CS have been used for deriving a consistent path integral formalism [Ino]; CS for $Sp(3, \mathbb{R})$ and other noncompact symplectic groups underlie the detailed structure of many nuclei [252], [260]–[263] (see below). Another striking application is the long-standing open problem of the hydrogen atom CS. The question was originally asked by Schrödinger whether one could find quantum states of the atom sufficiently well localized on the Kepler orbits, such that they would converge to their classical counterparts in the semiclassical limit $\hbar \to 0$. Schrödinger's idea was to find states analogous to the corresponding states of the harmonic oscillator, namely, the canonical CS, but he was not successful.

Now, it is well known that $SO_o(4,2)$ is the dynamical group of the hydrogen atom [64]. By this, one means that the group is generated by the observables of the system and has the property that a single UIR of it yields the entire energy spectrum. In the present case, the H-atom, both classical and quantum, has an accidental degeneracy, due to the existence of two conserved vector observables, the angular momentum (as for any central potential) and the Runge–Lenz vector. As a consequence, the group that explains the degeneracy of the spectrum is $SO(4)$, for the discrete spectrum, and $SO_o(3,1)$, for the continuous spectrum [24, 61]. More generally, $SO_o(4,1)$ may be taken as the invariance group of the system, in the sense that a single UIR of the latter describes the whole discrete spectrum. On top of that, if one adds ladder operators, describing electromagnetic dipole transitions between different energy levels, one obtains the group $SO_o(4,2)$.

Then, the relevant UIR of the latter remains irreducible when restricted to $SO_o(4,1)$. On that basis, various CS associated to $SO_o(4,2)$ or its subgroup $SU(1,1)$ have been proposed as candidates for the states sought by Schrödinger [66], but no really satisfactory solution has been found. It was even suggested that the problem has no solution, because the $SO_o(4,2)$ CS cannot be well localized [290]. Thus the problem remains open. It is worth mentioning here that a different solution to this problem, using CS not related to any group, has been presented recently in [142].

An example: CS for the nuclear collective model (♣)

The microscopic theory of nuclear collective structure, a refinement of the traditional Bohr–Mottelson collective model of nuclei, has a long history and much literature surrounds it. A comprehensive survey and the original references may be found in the review paper [261]. Group theory plays a central role in these models.

The starting point is to realize that the relevant observables for a nucleus contain both positions and momenta, which obey the CCR (2.1). Therefore, the symmetry group necessarily contains canonical transformations in $\mathbb{R}^3$, that is, the group $Sp(3,\mathbb{R})$ [sometimes called $Sp(6,\mathbb{R})$]. More precisely, one starts with an independent particle Hamiltonian

$$H = \frac{1}{2m} \sum_{s=1}^{A-1} \sum_{i=1}^{3} P_{si}^2 + \frac{1}{2} m\omega^2 \sum_{s=1}^{A-1} \sum_{i=1}^{3} Q_{si}^2 + V(Q), \qquad (7.43)$$

where A is the number of nucleons (the center of mass motion has been removed). The first two terms in (7.43) simply constitute a harmonic oscillator Hamiltonian, and $V(Q)$ is a collective potential. One then notes that the model is defined in terms of the following observables: The six quadrupole moments Q_{ij}, the generators of deformations and rotations S_{ij}, L_{ij}, and the six components of the quadrupole flow tensor K_{ij}, where

$$\begin{aligned} Q_{ij} &= \sum_s Q_{si} Q_{sj}, \\ S_{ij} &= \sum_s (Q_{si} P_{sj} + P_{si} Q_{sj}), \\ L_{ij} &= \sum_s (Q_{si} P_{sj} - Q_{sj} P_{si}), \\ K_{ij} &= \sum_s P_{si} P_{sj}. \end{aligned} \qquad (7.44)$$

Using again the CCR, one finds that the operators (7.44) generate the Lie algebra $\mathfrak{sp}(3,\mathbb{R})$, and the Hamiltonian (7.43) is a polynomial in these generators.

The next step is to build an irreducible representation of this Lie algebra. The standard technique, pioneered by Rowe [260]–[263] (see also [105]), is to use the vector coherent state method, which we have sketched in Chapter 4, Section 4.2.1. Using this approach, one is then able to build an orthonormal basis in the representation Hilbert space, adapted to the subgroup chain

$$Sp(3,\mathbb{R}) \supset U(3) \supset U(1) \times SU(3) \supset SO(3) \qquad (7.45)$$

[$SU(3)$ is the invariance group of the Elliott model, a precursor of the present collective model]. Basis vectors of this kind can be chosen as eigenstates of the Hamiltonian and with definite angular momentum. In that way, CS of $Sp(3,\mathbb{R})$ are indeed the building blocks of a host of different collective models of the nuclei. Further details may be found in [261].

7.2.3 CS of non-semisimple Lie groups

For non-semisimple groups, the emphasis has been put on the existence of CS systems based on a coset space $X = G/H$ (see [Per, Section 10.2] for a review). When G is a *nilpotent* Lie group, rather complete results have been obtained in [223], making extensive use of Kirillov's method of orbits [Kir]. Some of these results remain valid when G is a *solvable* or a *reductive* Lie group [224].

Let G be a Lie group of one of these types. Under the coadjoint action, the space $\mathfrak{g}^*$ is foliated into orbits, as usual, but, in addition, each orbit is associated to a single UIR of G. In addition, every coadjoint orbit is a symplectic manifold, making it a natural candidate for a classical phase space and, hence, a starting point for a quantization procedure. This is the basis of the method of geometric quantization [Woo], within the framework of which much work on CS has been done [207, 234, 235, 254, 271]. Coadjoint orbits have been used also for the construction of CS for inhomogeneous groups, such as the Euclidean, Galilei, and Poincaré groups [6, 29, 102], where, in addition to quantization, they have also been studied in connection with notions of quantum localizability and covariance [267]. Actually, each of these examples is the semidirect product of a translation group $\mathbb{R}^n$ by the corresponding homogeneous group. Accordingly, we will discuss them in detail, albeit mostly by the method of induced representations, in Chapter 10. An exception is the affine Weyl–Heisenberg group, studied in [185] by the method of coadjoint orbits. We will describe this example briefly in Chapter 15, Section 15.2.

7.3 Square integrable covariant CS: The general case

It is true that the Gilmore–Perelomov CS cover a large number of interesting situations, of which Perelomov's monograph [Per] offers a vast choice of examples. Yet this class is not general enough for all physical purposes. In

142   7. Covariant Coherent States

practice, the index space $X = G/H$ is often specified *a priori*, on physical grounds, and it may be that H is not the stability subgroup of any vector η in the representation space. For instance, in (geometric) quantization [Woo], one takes for X a *phase space*, which could be a coadjoint orbit of G or, more generally, a symplectic G-space [102]. This is the situation for the kinematical or relativity groups (Euclidean, Galilei, or Poincaré), which we will analyze in detail in Chapter 10. Interestingly enough, there are phase spaces that do lead to Gilmore–Perelomov CS. A notable example is the theory of wavelets, which we shall study closely in Chapters 12 and 14.

In the more general situations, one has to exploit the theory of covariant CS in almost its full generality, starting from Definition 7.1.1. The only restriction is that the unitary representation U is usually assumed to be irreducible and square integrable mod (H,σ), for some Borel section σ. In principle, all relevant results may be read off Chapters 5 and 6, in particular, Theorem 5.3.8 and Section 6.1. It may be useful, however, to summarize these results in the form in which we will use them in Chapters 10, 14, and 15. This has the additional advantage of facilitating the comparison with the existing literature.

To make things slightly simpler, we shall assume here that the measure ν on X is invariant. The general case of a quasi-invariant measure is then easily deduced from our previous analysis.

Theorem 7.3.1: *Let G be a locally compact group, H be a closed subgroup of G, and $X = G/H$, equipped with an invariant measure ν. Let $\sigma : X \to G$ be a global Borel section. Let U be a unitary irreducible representation of G in a Hilbert space $\mathfrak{H}$.*

Assume that the representation U is square integrable $\mathrm{mod}(H,\sigma)$; that is, there exist n linearly independent vectors η^i, $i = 1, 2, \ldots, n$ in $\mathfrak{H}$ such that the integral

$$\sum_{i=1}^{n} \int_X U(\sigma(x))|\eta^i\rangle\langle\eta^i|U(\sigma(x))^* \, d\nu(x) \tag{7.46}$$

converges weakly to a bounded, positive self-adjoint operator A_σ, with bounded inverse A_σ^{-1}. Then, the family of CS

$$\mathfrak{S}_\sigma = \{\eta^i_{\sigma(x)} = U(\sigma(x))\eta^i \mid i = 1, 2, \ldots, n, \, x \in X\} \tag{7.47}$$

has the following properties.

1. *The set $\mathfrak{S}_\sigma$ is total in $\mathfrak{H}$.*

2. *The range of the linear map $W_K : \mathfrak{H} \to \mathbb{C}^n \otimes L^2(X, d\nu)$, given componentwise by*

$$(W_K \phi)^i(x) = \langle \eta^i_{\sigma(x)} | \phi \rangle, \quad \phi \in \mathfrak{H}, \tag{7.48}$$

7.3. Square integrable covariant CS: The general case 143

is a (closed) subspace $\mathfrak{H}_K$ of $\mathbb{C}^n \otimes L^2(X, d\nu)$. It is complete with respect to the new scalar product

$$\langle \Phi | \Psi \rangle_K \equiv \langle \Phi | W_K A_\sigma^{-1} W_K^{-1} \Psi \rangle_{\mathbb{C}^n \otimes L^2(X, d\nu)}, \qquad (7.49)$$

and then $W_K : \mathfrak{H} \to \mathfrak{H}_K$ is unitary.

3. The projection $\mathbb{P}_K = W_K W_K^*$ from $\mathbb{C}^n \otimes L^2(X, d\nu)$ onto $\mathfrak{H}_K$ is an integral operator, and $\mathfrak{H}_K$ is a reproducing kernel Hilbert space of functions:

$$\Phi(x) = \int_X \mathbf{K}_\sigma(x,y) \Phi(y)\, d\nu(y),\ \forall\, \Phi \in \mathfrak{H}_K. \qquad (7.50)$$

The kernel of $\mathbb{P}_K$ is given explicitly by the matrix $\mathbf{K}_\sigma(x,y)_{ij} = \langle \eta^i_{\sigma(x)} | A_\sigma^{-1} \eta^j_{\sigma(y)} \rangle$, and it is square integrable:

$$\sum_{j=1}^n \int_X \mathbf{K}_\sigma(x,y)_{ij} \mathbf{K}_\sigma(y,z)_{jk}\, d\nu(y) = \mathbf{K}_\sigma(x,z)_{ik}. \qquad (7.51)$$

4. Being unitary, the map W_K may be inverted on its range by the adjoint operator, $W_K^{-1} = W_K^*$ on $\mathfrak{H}_K$, to obtain the reconstruction formula [see (6.5)]:

$$\phi = W_K^{-1} \Phi = \sum_{i=1}^n \int_X \Phi^i(x) A_\sigma^{-1} \eta^i_{\sigma(x)}\, d\nu(x),\ \Phi \in \mathfrak{H}_K. \qquad (7.52)$$

For practical purposes, it may be useful to rewrite the square integrability condition in the form of the familiar frame condition

$$\mathsf{m}(A_\sigma) \|\phi\|^2 \leq \sum_{i=1}^n \int_X |\langle \eta^i_{\sigma(x)} | \phi \rangle|^2\, d\nu(x) \leq \mathsf{M}(A_\sigma) \|\phi\|^2,\ \forall\, \phi \in \mathfrak{H}, \qquad (7.53)$$

where the frame bounds have been denoted $\mathsf{m}(A_\sigma), \mathsf{M}(A_\sigma)$, because they are in fact the infimum and the supremum, respectively, of the self-adjoint operator A_σ.

Suppose that the Hilbert space $\mathfrak{H}$ corresponds to the description of a quantum system, which admits G as a symmetry group. Then, in more familiar terms, Properties 1 and 2 mean that the CS family $\mathfrak{S}_\sigma$ yields a new representation of the system under consideration, unitarily equivalent to the original one. By Property 3, the Hilbert space of this CS representation is a space of (usually well behaved) functions. Finally, Property 4 means that the vector $\phi = W_K^{-1} \Phi$ may be expanded in (or reconstructed from) the CS $\eta^i_{\sigma(x)}$. This is the interpretation in terms of which the reconstruction formula is most often used, for instance, in signal processing (see Chapter 12). Notice that part of this construction is standard in noncommutative harmonic analysis, where the CS map (7.48) is sometimes called the Poisson transform [222].

7.3.1 Some further generalizations

Before concluding this section, it may be worthwhile to make contact with several notions related to CS that have cropped up in the literature.

1. *Unbounded frames:*
   In the whole analysis so far, we have assumed that the operator A_σ has a *bounded* inverse A_σ^{-1}, but this is in fact not necessary. The important point is that A_σ be invertible. If A_σ^{-1} is unbounded (it is necessarily densely defined, as the inverse of a self-adjoint operator), the whole development goes through, with some minor modifications. First, the CS map W_K defined in (7.48), restricted to the dense domain $D(A_\sigma^{-1})$, may again be shown to be an isometry onto $W_K[D(A_\sigma^{-1})]$ when the latter is equipped with the new scalar product (7.49). Then, W_K extends by continuity to a unitary map from $\mathfrak{H}$ onto $\mathfrak{H}_K$, which therefore is a subspace (though not necessarily closed) of $\mathbb{C}^n \otimes L^2(X, d\nu)$. The rest is as before, provided all vectors $\eta^i_{\sigma(x)}$, $i = 1, 2, \ldots, n$, $x \in X$, belong to the domain $D(A_\sigma^{-1})$. Otherwise, some of the relations, for instance, the reproducing kernel, have to be taken in a distributional sense. For further details, see [12], where the analysis is made in full generality.
   When A_σ^{-1} is unbounded, it is also possible to recast the construction in the language of rigged Hilbert spaces [Gel], mentioned earlier. The basic structure is a triplet of Hilbert spaces,
   $$\mathfrak{H}_K \subset \overline{\mathfrak{H}_K} \subset \mathfrak{H}_K'. \tag{7.54}$$
   In this relation, $\mathfrak{H}_K$ denotes, as before, the completion of $W_K[D(A_\sigma^{-1})]$ with respect to the norm $\|\cdot\|_K$, which is the graph norm of the self-adjoint operator $W_K A_\sigma^{-1/2} W_K^{-1}$ acting in $\mathbb{C}^n \otimes L^2(X, d\nu)$. Then, $\overline{\mathfrak{H}_K}$ is the closure of $\mathfrak{H}_K$ in $\mathbb{C}^n \otimes L^2(X, d\nu)$, and $\mathfrak{H}_K'$ is the dual of $\mathfrak{H}_K$, that is, the completion of $\mathfrak{H}_K$ in the norm $\langle \cdot | W_K A_\sigma^{-1} W_K^{-1} \cdot \rangle_{\mathbb{C}^n \otimes L^2(X, d\nu)}$. If $\eta^i_{\sigma(x)} \in D(A_\sigma^{-1})$, $\forall i = 1, 2, \ldots, n$, $x \in X$, then all three spaces $\mathfrak{H}_K, \overline{\mathfrak{H}_K}, \mathfrak{H}_K'$ are reproducing kernel Hilbert spaces, with the *same* reproducing kernel $\mathbf{K}_\sigma(x, y)$. One may view the space $\mathfrak{H}_K'$ as carrying the unbounded version of the dual frame.

2. *Weighted CS:*
   In the discussion at the end of Section 6.1, we have seen that it is always possible to transform a nontight frame, with $A_K \neq I_K$, into a tight one, by mapping the Hilbert space $\mathfrak{H}_K$ unitarily onto another one $\mathfrak{H}_{\overline{K}}$ with help of the well-defined operator $A_K^{-1/2}$. This corresponds to the replacement of (6.3) by (6.12). The CS on both sides, however, *do not* map into each other. When applied to the present situation, this move corresponds to the definition of *weighted*

7.3. Square integrable covariant CS: The general case

CS introduced in [13], namely, the vectors

$$\widetilde{\eta}_x^i = A_K^{-1/2} \eta_{\sigma(x)}^i. \tag{7.55}$$

While these vectors do constitute a tight frame, they are not covariant CS, since they have been "weighted" along the orbit $\{U(\sigma(x)) \mid x \in X\}$ by the operator $A_K^{-1/2}$. This procedure becomes more suggestive and useful if, and probably only if, the operator A_K is diagonal in $\mathfrak{H}_k$, that is, it is a multiplication operator by a function $\mathcal{A}_K(x)$. This indeed happens for a number of physically interesting cases, such as the Poincaré group or the affine Weyl–Heisenberg group, which we shall analyze in subsequent chapters.

3. *Quasi-coherent states:*
   It may happen that the square integrability condition (7.2) is not satisfied as it stands, but that the integral does converge when the translated measure ν_σ is replaced by some other quasi-invariant measure. Then the formalism goes through, as developed in Chapter 5, but one loses covariance properties with respect to translation of the section, such as (7.6) [19].

4. *Quasi-sections:*
   Finally, it may happen that no section is admissible, but that the square integrability condition is satisfied when the base space X is suitably reparametrized, i.e., when σ is replaced by the composition $\sigma_\varphi \equiv \sigma \circ \varphi : X \to G$, where $\varphi : X \to X$ is a homeomorphism (reparametrization) of the base space. This map, called a *quasi-section* of the principal bundle $\pi : G \to G/H = X$, satisfies the condition $\pi \circ \sigma_\varphi = \varphi$. Here again, the whole machinery of Chapter 5 applies and CS may be defined. A case in which this more general approach is needed is that of the massless representations of the Poincaré group $\mathcal{P}_+^\uparrow(1,1)$ in one space and one time dimensions (see [29] and Chapter 10, Section 11.1).

Of course, all of these variants are particular cases of the general theory developed here, and of unequal importance *per se*. From a historical point of view, however, they have contributed to expand the scope of the theory, thus paving the way to its present status. Note, however, that weighted CS are fundamental if we insist on a probabilistic interpretation of the POV measure (e.g., in quantum mechanics). Also, the lack of covariance, which leads to quasi-coherent states, may sometimes be very meaningful from a physical point of view (e.g., for Poincaré CS, which we will discuss in Chapter 11, Section 11.1).

8
Coherent States from Square Integrable Representations

This chapter is devoted to a fairly detailed development of the theory of square integrable group representations. We have already seen examples of such representations. Indeed, the discrete series representations, U^j of $SU(1,1)$, discussed in Chapter 4, Section 4.2.2, are square integrable. To make this clear, observe that the discussion in that section, leading up to (4.73), tells us that the invariant Haar measure on $SU(1,1)$ can be written as

$$d\mu(g) = \frac{1}{4\pi} d\nu(z,\bar{z})\, d\phi, \tag{8.1}$$

where $d\nu(z,\bar{z})$ is the invariant measure (4.73) on the open unit disc. This follows from the fact that an arbitrary group element, $g \in SU(1,1)$, can be written as $g = \sigma(z)k$ [see (4.65) and (4.66)]. Moreover, the measure $d\mu$, so defined, is both left and right invariant, the group being unimodular. On the other hand, from the definition of the CS $\eta_{\sigma(z)}$ in (4.99) and the resolution of the identity in (4.100), we find that the vectors $\eta_g = U^j(g)\eta$, with $\eta = (2j-1)^{\frac{1}{2}} u_0$ [see (4.93)], satisfy the resolution of the identity (on the Hilbert space $\mathfrak{H}_{hol}$ of analytic functions on the open unit disc),

$$\int_{SU(1,1)} |\eta_g\rangle\langle\eta_g|\, d\mu(g) = I_{hol}. \tag{8.2}$$

A resolution of the identity of the above type, with respect to the entire group, forms the basis for the square integrability of the representation — a notion that we now make precise.

8.1 Square integrable group representations

Group representations U that are square integrable mod($\{e\},\sigma$), where $\sigma(g) = g$, for all $g \in G$ and $n = 1$, in Definition 7.1.1 of covariant CS, are simply called *square integrable*. In this section and the next, we give a detailed account of the properties of such representations, following mainly [Gaa], [85], [159], [160], and [248]. An algebraic treatment of the subject may be found in [114]; the case in which the group G is unimodular is treated in detail in [Dix2]. Square integrable representations of a group G are also said to belong to the *discrete series* of its representations. We emphasize again that square integrability in this setting implies the existence of a vector $\eta \in \mathfrak{H}$

$$\int_G |\eta_g\rangle\langle\eta_g| \, d\mu(g) = I, \qquad \eta_g = U(g)\eta, \tag{8.3}$$

(μ = left Haar measure). For a compact group, the above relation would hold for any (suitably normalized) vector η in the representation space $\mathfrak{H}$. Hence, every representation of a compact group is square integrable. For noncompact groups, however, there do exist representations that are not square integrable, and we have already encountered this situation. For example, the representation U^λ of the Weyl–Heisenberg group, given in (2.33), is not square integrable, for the Hilbert space does not contain any nontrivial vector η that would satisfy (8.3). The reason for this is that the invariant Haar measure is now $d\theta \, dq \, dp$, and the integration with respect to θ would clearly diverge.

CS of semisimple Lie groups, compact and noncompact, have been discussed in Sections 7.2.1 and 7.2.2, respectively, the simplest cases being $SU(2)$ and $SU(1,1)$.

For the remainder of this section, and in the next section, U will be taken to be a unitary, irreducible representation of G on the Hilbert space $\mathfrak{H}$.

Definition 8.1.1 (Admissible vector): *A vector $\eta \in \mathfrak{H}$ is said to be admissible if*

$$I(\eta) = \int_G |\langle U(g)\eta|\eta\rangle|^2 \, d\mu(g) < \infty. \tag{8.4}$$

Note that since $d\mu_r(g) = d\mu(g^{-1})$, [see (4.5)] and since $U(g)$ is unitary,

$$I(\eta) = \int_G |\langle U(g^{-1})\eta|\eta\rangle|^2 \, d\mu_r(g) = \int_G |\langle \eta|U(g)\eta\rangle|^2 \, d\mu_r(g).$$

Hence,

$$I(\eta) = \int_G |\langle U(g)\eta|\eta\rangle|^2 \, d\mu_r(g), \tag{8.5}$$

so that it is immaterial whether the left or the right invariant Haar measure is used in the definition of admissibility. Note also that, if $\eta \neq 0$, then $I(\eta) \neq 0$. Indeed, since $g \mapsto \langle U(g)\eta|\eta\rangle$ is a continuous function, and the measure $d\mu$ is invariant under left translations, $I(\eta) = 0$ implies $\langle U(g)\eta|\eta\rangle = 0$, for all $g \in G$. Since $U(g)\eta$, $g \in G$, is a dense set of vectors in $\mathfrak{H}$, this implies that $\eta = 0$.

Lemma 8.1.2: *If $\eta \in \mathfrak{H}$ is an admissible vector, then so also is $\eta_g = U(g)\eta$, for all $g \in G$.*

Proof. Indeed,

$$\begin{aligned}
I(\eta_g) &= \int_G |\langle U(g')\eta_g|\eta_g\rangle|^2 \, d\mu(g') = \int_G |\langle U(g^{-1}g'g)\eta|\eta\rangle|^2 \, d\mu(g'), \\
&= \int_G |\langle U(g'g)\eta|\eta\rangle|^2 \, d\mu(g'), \quad \text{by the left invariance of } d\mu, \\
&= \int_G |\langle U(g')\eta|\eta\rangle|^2 \Delta(g^{-1}) \, d\mu(g'), \quad \text{by (4.5),} \\
&= \frac{1}{\Delta(g)} \int_G |\langle U(g')\eta|\eta\rangle|^2 \, d\mu(g'), \quad \text{by (4.4).}
\end{aligned}$$

Thus,

$$I(\eta_g) = \frac{1}{\Delta(g)} I(\eta) < \infty. \tag{8.6}$$

□

Let $\mathcal{A}$ denote the set of all admissible vectors. Then, as a consequence of this lemma, $\mathcal{A}$ is stable under $U(g)$, $g \in G$. Since U is irreducible, either $\mathcal{A} = \{0\}$, i.e., it consists of the zero vector only, or $\mathcal{A}$ is *total* in $\mathfrak{H}$. Furthermore, it turns out (see the proof of Theorem 8.1.3) that

$$\eta \in \mathcal{A} \quad \text{iff} \quad \int_G |\langle U(g)\eta|\phi\rangle|^2 \, d\mu(g) < \infty, \ \forall \phi \in \mathfrak{H}, \tag{8.7}$$

and this in turn implies that $\eta_1 + \eta_2$ is admissible if η_1, η_2 are, i.e., $\mathcal{A}$ is a vector subspace of $\mathfrak{H}$. Therefore, either $\mathcal{A} = \{0\}$ or $\mathcal{A}$ is *dense* in $\mathfrak{H}$. For $\eta \in \mathcal{A}$, $\eta \neq 0$, we shall write

$$c(\eta) = \frac{I(\eta)}{\|\eta\|^2}. \tag{8.8}$$

The condition for the admissibility $\mathrm{mod}(H, \sigma)$, of a set of vectors η^i, $i = 1, 2, \ldots, n$, was given in Definition 7.1.1 (see also Theorem 7.3.1). The next theorem shows that in the present context that definition is equivalent to the one given above; the theorem also captures practically all of the important properties of square integrable representations. Recall the definitions of the left and right regular representations, U_ℓ and U_r, of G, given in Chapter 3, Section 4.2.3.

Theorem 8.1.3: *The UIR $g \mapsto U(g)$ of the locally compact group G is square integrable if and only if $\mathcal{A} \neq \{0\}$. In this case, for any $\eta \in \mathcal{A}$, the*

mapping
$$W_\eta : \mathfrak{H} \to L^2(G, d\mu), \qquad (W_\eta \phi)(g) = [c(\eta)]^{-\frac{1}{2}} \langle \eta_g | \phi \rangle, \quad \phi \in \mathfrak{H},\ g \in G, \tag{8.9}$$
is a linear isometry onto a (closed) subspace $\mathfrak{H}_\eta$ of $L^2(G, d\mu)$. On $\mathfrak{H}$, the resolution of the identity
$$\frac{1}{c(\eta)} \int_G |\eta_g\rangle\langle \eta_g| \, d\mu(g) = I \tag{8.10}$$
holds. The subspace $\mathfrak{H}_\eta = W_\eta \mathfrak{H} \subset L^2(G, d\mu)$ is a reproducing kernel Hilbert space, so that the corresponding projection operator
$$\mathbb{P}_\eta = W_\eta W_\eta^*, \qquad \mathbb{P}_\eta L^2(G, d\mu) = \mathfrak{H}_\eta, \tag{8.11}$$
has the reproducing kernel K_η,
$$(\mathbb{P}_\eta \widetilde{\Phi})(g) = \int_G K_\eta(g, g') \widetilde{\Phi}(g') \, d\mu(g'), \qquad \widetilde{\Phi} \in L^2(G, d\mu),$$
$$K_\eta(g, g') = \frac{1}{c(\eta)} \langle \eta_g | \eta_{g'} \rangle, \tag{8.12}$$
as its integral kernel. Furthermore, W_η intertwines U and the left regular representation U_ℓ,
$$W_\eta U(g) = U_\ell(g) W_\eta, \qquad g \in G. \tag{8.13}$$
Of course, since here $n = 1$, we have identified the operator $K_\eta(g, g')$ and its matrix kernel $\mathbf{K}_\eta(g, g')$.

Before proving this theorem, we observe that an entirely analogous result holds with the right regular representation U_r. Thus, for each $\eta \in \mathcal{A}$, there exists a linear isometry,
$$W_\eta^r : \mathfrak{H} \to L^2(G, d\mu_r), \qquad (W_\eta^r \phi)(g) = [c(\eta)]^{-\frac{1}{2}} \langle \eta_{g^{-1}} | \phi \rangle, \quad \phi \in \mathfrak{H},\ g \in G. \tag{8.14}$$
The corresponding reproducing kernel is
$$K_\eta^r(g, g') = \frac{1}{c(\eta)} \langle \eta_{g^{-1}} | \eta_{g'^{-1}} \rangle = K_\eta(g^{-1}, g'^{-1}). \tag{8.15}$$

Proof. The domain $\mathcal{D}(W_\eta)$ of W_η is the set of all vectors $\phi \in \mathfrak{H}$ such that
$$\frac{1}{c(\eta)} \int_G |\langle \eta_g | \phi \rangle_\mathfrak{H}|^2 \, d\mu(g) < \infty.$$
But, for any $\phi \in \mathcal{D}(W_\eta)$ and $g' \in G$, we have
$$\frac{1}{c(\eta)} \int_G |\langle \eta_g | U(g') \phi \rangle_\mathfrak{H}|^2 \, d\mu(g) = \frac{1}{c(\eta)} \int_G |\langle \eta_{g'^{-1}g} | \phi \rangle_\mathfrak{H}|^2 \, d\mu(g)$$
$$= \frac{1}{c(\eta)} \int_G |\langle \eta_g | \phi \rangle_\mathfrak{H}|^2 \, d\mu(g),$$
by the invariance of μ.

Thus, $\mathcal{D}(W_\eta)$ is stable under U, and, hence, dense in $\mathfrak{H}$, since U is irreducible. Moreover, on $\mathcal{D}(W_\eta)$, the intertwining property (8.13) follows easily from (8.9) and the definition of the left regular representation in (4.105).

We prove next that, as a linear map, W_η is closed. Let $\{\phi_n\}_{n=1}^\infty \subset \mathcal{D}(W_\eta)$ be a sequence converging to $\phi \in \mathfrak{H}$, and let the corresponding sequence $\{W_\eta \phi_n\}_{n=1}^\infty \subset L^2(G, d\mu)$ converge to $\Phi \in L^2(G, d\mu)$. Then, by the continuity of the scalar product in $\mathfrak{H}$,

$$\lim_{n\to\infty} W_\eta \phi_n(g) = \lim_{n\to\infty} \langle \eta_g | \phi_n \rangle = \langle \eta_g | \phi \rangle. \tag{8.16}$$

Thus, since $W_\eta \phi_n \to \Phi$ in $L^2(G, d\mu)$ and $W_\eta \phi_n(g) \to \langle \eta_g | \phi \rangle$ pointwise,

$$\langle \eta_g | \phi \rangle = \Phi(g),$$

almost everywhere (with respect to μ), whence,

$$\int_G |\langle \eta_g | \phi \rangle|^2 \, d\mu(g) < \infty,$$

implying that $\phi \in \mathcal{D}(W_\eta)$ and $W_\eta \phi = \Phi$; i.e., W_η is closed.

Using Lemma 4.3.3 (the extended Schur's lemma), we establish the boundedness of $W_\eta : \mathcal{D}(W_\eta) \to L^2(G, d\mu)$. Hence, $\mathcal{D}(W_\eta) = \mathfrak{H}$, and, furthermore, W_η is a multiple of an isometry,

$$\|W_\eta \phi\|_{L^2(G,d\mu)}^2 = \lambda \|\phi\|_{\mathfrak{H}}^2, \qquad \phi \in \mathfrak{H}, \tag{8.17}$$

where $\lambda \in \mathbb{R}^+$. To fix λ, take $\phi = \eta$. Then,

$$\lambda = \frac{\|W_\eta \eta\|_{L^2(G,d\mu)}^2}{\|\eta\|^2} = \frac{I(\eta)}{c(\eta)\|\eta\|^2} = 1, \tag{8.18}$$

by (8.4), (8.8), and (8.9).

Thus, W_η is an isometry; i.e., $W_\eta^* W_\eta = I$, which implies that (8.10) holds. Therefore, the range of W_η is a closed subspace of $L^2(G, d\mu)$, and the projection on it is $\mathbb{P}_\eta = W_\eta W_\eta^*$. Then, (8.12) and (8.13) follow from a direct computation. $\square$

An immediate consequence of this theorem is the following important result.

Corollary 8.1.4: *Every square integrable representation of a locally compact group G is unitarily equivalent to a subrepresentation of its left regular representation (and, hence, of its right regular representation as well).*

The proof of this corollary consists simply in showing that the projection $\mathbb{P}_\eta$ on the range of W_η commutes with the left regular representation. Indeed:

$$\begin{aligned}\mathbb{P}_\eta U_\ell(g) &= W_\eta W_\eta^* U_\ell(g) = W_\eta (U_\ell(g^{-1}) W_\eta)^* = W_\eta (W_\eta U(g^{-1}))^* \\ &= W_\eta U(g) W_\eta^* = U_\ell(g) W_\eta W_\eta^* = U_\ell(g) \mathbb{P}_\eta.\end{aligned}$$

Since W_η is an isometry, its inverse is equal to its adjoint on its range; i.e., $W_\eta^{-1} = W_\eta^*$ on $\mathfrak{H}_\eta$. Then, applying both sides of (8.10) to an arbitrary vector $\phi \in \mathfrak{H}$, we obtain the *reconstruction formula* [see (6.5)]

$$\phi = W_\eta^* \Phi = \frac{1}{[c(\eta)]^{\frac{1}{2}}} \int_G \Phi(g) \eta_g \, d\mu(g). \qquad (8.19)$$

In (8.52), we shall obtain a generalized version of this reconstruction formula using two different admissible vectors.

The outcome of Theorem 8.1.3 is a total set of CS indexed by the points of the group G itself. While this does happen in simple examples, it is rather rare. A case in point are the 1-D wavelets, which we shall discuss in great detail in Chapter 12. Another, rather exotic, example has been given in [255], for the group $SU(1,1)$. In general, however, the CS systems of physical interest are supported by a quotient manifold $X = G/H$, where H may be the stability subgroup of the generating vector η (the Gilmore–Perelomov situation), but need not be so. Yet, CS on a group have a number of interesting mathematical properties that warrant a thorough analysis, which is the purpose of the present chapter.

We note that a vector $\Phi \in W_\eta \mathfrak{H} = \mathfrak{H}_\eta = \mathbb{P}_\eta L^2(G, d\mu)$, if and only if there exists a vector $\phi \in \mathfrak{H}$ such that $\Phi(g) = [c(\eta)]^{-\frac{1}{2}} \langle \eta_g | \phi \rangle$ for *almost* all $g \in G$ (with respect to the measure μ). This also means, in view of the strong continuity of the representation $g \mapsto U(g)$, that $\Phi(g)$ can be identified with the *bounded continuous function* of G,

$$g \mapsto [c(\eta)]^{-\frac{1}{2}} \langle \eta_g | \phi \rangle = \langle U(g)\eta | \phi \rangle$$
$$\sup_{g \in G} |\langle \eta_g | \phi \rangle| \leq \sup_{g \in G} \|U(g)\eta\| \, \|\phi\| = \|\eta\| \, \|\phi\|. \qquad (8.20)$$

Hence, the reproducing kernel subspace $\mathfrak{H}_\eta$ can be identified with a space of bounded, continuous functions on the group G.

In addition, the reproducing kernel $K_\eta(g, g')$ is in the present case a *convolution* kernel on G: $K_\eta(g, g') = \langle \eta | U(g^{-1}g') \eta \rangle$. This implies that K_η has a regularizing effect. For instance, if G is a Lie group, and η is appropriately chosen, the elements of $\mathfrak{H}_\eta$ can be made to be infinitely differentiable functions, which extend to holomorphic functions on the complexified group G^c [238]. This gives rise to some of the attractive holomorphic properties of CS, and their geometrical implications. Another consequence of the convolution character of K_η is that the kernel, and hence the elements of $\mathfrak{H}_\eta$, have interpolation properties that prove useful in practical computations [159].

Finally, it should be emphasized that the reproducing kernel K_η is the main tool for computing the *efficiency* or *resolving power* of the transform W_η, in wavelet analysis (see Chapter 14, Section 14.3.2). Notice that each admissible vector η determines its own reproducing kernel K_η and reproducing kernel subspace $\mathfrak{H}_\eta$.

Example: The connected affine group (♣)

We have encountered the connected affine group of the line, G_+, in Section 4.1. As one of the simplest examples of an application of Theorem 8.1.3, let us explicitly display the square integrability properties of a UIR of this group. Recall that G_+, which is also called the $ax+b$ group, consists of all transformations of $\mathbb{R}$ of the form $x \mapsto ax+b$, $x \in \mathbb{R}$, with $a > 0$, $b \in \mathbb{R}$ and a group element is given by a pair $(b,a) \in \mathbb{R} \times \mathbb{R}_*^+$. The invariant measures on G_+ are easily computed to be

$$d\mu(b,a) = \frac{db\,da}{a^2}, \quad \text{and} \quad d\mu_r(b,a) = \frac{db\,da}{a}, \qquad (8.21)$$

so that the modular fuction is

$$\Delta(b,a) = \frac{1}{a}. \qquad (8.22)$$

There are only two nontrivial, unitary irreducible representations of G_+. In order to examine these, consider the Hilbert space $L^2(\mathbb{R}, dx)$ and on it the representation $U(b,a)$,

$$(U(b,a)f)(x) = a^{-\frac{1}{2}} f(\frac{x-b}{a}), \quad f \in L^2(\mathbb{R}, dx), \quad g = (b,a) \in G_+. \qquad (8.23)$$

This representation is unitary, but not irreducible. In order to isolate its irreducible components, let us go over to the Fourier-transformed Hilbert space, which we denote by $L^2(\widehat{\mathbb{R}}, d\xi)$. On this space, the above representation transforms to $\widehat{U}(b,a)$, with

$$(\widehat{U}(b,a)\widehat{f})(\xi) = a^{\frac{1}{2}} \widehat{f}(a\,\xi) e^{-ib\xi}, \quad \widehat{f} \in L^2(\widehat{\mathbb{R}}, d\xi). \qquad (8.24)$$

It is now clear that the two subspaces,

$$\widehat{\mathfrak{H}}^\pm = L^2(\widehat{\mathbb{R}}^\pm, d\xi), \qquad (8.25)$$

are stable under the action of the $\widehat{U}(b,a)$, and in fact constitute irreducible subsaces under this action. We denote the restrictions of $\widehat{U}$ to these two subspaces by $\widehat{U}^+$ and $\widehat{U}^-$, respectively. The two subrepresentations are then inequivalent, but both are square integrable, and we only look at $\widehat{U}^+$. Let $\widehat{\eta} \in \widehat{\mathfrak{H}}^+$ be an admissible vector, in the sense of Definition 8.1.1. An easy computation then yields

$$I(\eta) = \int_{\mathbb{R}\times\mathbb{R}_*^+} |\langle \widehat{U}^+(b,a)\widehat{\eta}|\widehat{\eta}\rangle|^2 \frac{db\,da}{a^2} = 2\pi \,\|\widehat{\eta}\|^2 \int_0^\infty \left|\frac{\widehat{\eta}(\xi)}{\xi}\right|^2 d\xi. \qquad (8.26)$$

The constant $c(\widehat{\eta})$ in (8.8) now assumes the form,

$$c(\widehat{\eta}) = 2\pi \int_0^\infty \left|\frac{\widehat{\eta}(\xi)}{\xi}\right|^2 d\xi, \qquad (8.27)$$

which means that the vector $\widehat{\eta} \in \widehat{\mathfrak{H}}^+$ is admissible if and only if it satisfies the condition

$$\int_0^\infty \frac{2\pi}{\xi} |\widehat{\eta}(\xi)|^2 \, d\xi < \infty. \tag{8.28}$$

In other words, for admissibility, $\widehat{\eta}$ has to lie in the domain of the positive, unbounded operator C:

$$(C\widehat{\eta})(\xi) = \left[\frac{2\pi}{\xi}\right]^{\frac{1}{2}} \widehat{\eta}(\xi), \qquad \xi \geq 0. \tag{8.29}$$

We shall see in Theorem 8.2.1, that this is the generic situation, in the sense that admissible vectors are always characterized by being in the domain of a certain positive operator, determined uniquely by the group representation itself. Coherent states for this representation are then the families of vectors,

$$\widehat{\eta}_{ba} = \widehat{U}^+(b,a)\widehat{\eta}, \qquad (b,a) \in G_+, \tag{8.30}$$

with $\widehat{\eta}$ being any admissible vector. The CS satisfy the resolution of the identity (on $\widehat{\mathfrak{H}}^+$):

$$\frac{1}{c(\widehat{\eta})} \int_{\mathbb{R} \times \mathbb{R}_*^+} |\widehat{\eta}_{ba}\rangle\langle\widehat{\eta}_{ba}ba| \, \frac{db \, da}{a^2} = I. \tag{8.31}$$

The corresponding reproducing kernel is

$$K(b,a;\, b',a') = \frac{1}{c(\widehat{\eta})} \int_0^\infty e^{i\xi(b-b')} \, [aa']^{\frac{1}{2}} \, \overline{\widehat{\eta}(a\,\xi)} \, \widehat{\eta}(a'\,\xi) \, d\xi. \tag{8.32}$$

This example is of great interest for applications, since it is the group-theoretical backbone of the theory of wavelets, which we shall discuss at length in Chapter 12, Section 12.2 [the vectors (8.30) are exactly the 1-D wavelets]. Before leaving it, let us note that, in the case of the $SU(1,1)$ CS, mentioned at the beginning of this chapter, the operator C is $(2j+1)^{\frac{1}{2}} I_{hol}$, or just a multiple of the identity.

In our discussion of square integrable representations so far, the representation U has been assumed to be irreducible. This requirement may be weakened in several ways. A first possibility is to take a direct sum of square integrable representations (actually *every* square integrable representation is of this type [Gaa]). In this case one may prove the following result [125].

Theorem 8.1.5: *Let G be a locally compact group, with left Haar measure μ. Let U be a strongly continuous unitary representation of G into a Hilbert space $\mathfrak{H}$, and assume that U is a direct sum of disjoint square integrable representations U_i:*

$$U = \bigoplus_i U_i, \quad \text{in} \quad \mathfrak{H} = \bigoplus_i \mathfrak{H}_i. \tag{8.33}$$

Let η be an admissible vector, in the sense of Definition 8.1.1. Then,

$$\int_G |\langle U(g)\eta|\phi\rangle|^2 \, d\mu(g) = \sum_i c_i \|\mathbb{P}_i\phi\|^2, \qquad \phi \in \mathfrak{H}, \tag{8.34}$$

where $\mathbb{P}_i$ is the projection on $\mathfrak{H}_i$ and

$$c_i = \|\mathbb{P}_i\eta\|^{-2} \int_G |\langle U_i(g)\mathbb{P}_i\eta|\mathbb{P}_i\eta\rangle|^2 \, d\mu(g). \tag{8.35}$$

If, in addition, all constants c_i are equal, then the map $W_\eta : \phi \mapsto \langle U(g)\eta|\phi\rangle$ is an isometry (up to a constant) from $\mathfrak{H}$ into $L^2(G, d\mu)$.

Thus, when the conditions of this theorem are satisfied, CS may be built in the usual way. By similar arguments, the same is true if some of the components U_i are mutually unitarily equivalent.

Another generalization is to take for U a *cyclic* representation, with η a cyclic vector [86]. In this case, assuming the admissibility condition (8.4), all of Theorem 8.1.3 can be recovered. The relationship between cyclicity and square integrability has been studied systematically in some recent work [128].

A more radical situation has been considered recently, in the case of the Euclidean [182] and Poincaré [195] groups. The idea is to take for U a direct integral of irreducible, Wigner-type representations $U^{(m)}$ over the mass parameter m, on an interval $[m_1, m_2]$, and for such a reducible representation, CS may be built. Taking *all* positive values of m, however, results in an *irreducible* representation of the corresponding affine or similitude group, to which we shall return in Chapter 14, Section 15.3. The same idea was exploited to a certain extent in [71].

Finally, although the framework used here is that of square integrable representations, a parallel theory may be derived for *integrable representations*, for which many of the above results extend to Banach spaces [114, 123].

8.2 Orthogonality relations

If G is a compact group and U a unitary irreducible representation of G, then, according to the Peter–Weyl theorem [Gaa], the matrix elements $\langle U(g)\psi|\phi\rangle$ of U satisfy certain *orthogonality relations*, and one may construct an orthonormal basis of $L^2(G, d\mu)$ consisting of such matrix elements. When G is only locally compact, square integrable representations have the same property. Thus, among all UIR's, the square integrable representations are the direct generalizations of the irreducible representations of compact groups. These orthogonality relations, well known when G is unimodular [Dix2, 148], extend to non-unimodular groups as well [85, 114, 159].

Theorem 8.2.1 (Orthogonality relations): *Let G be a locally compact group and U be a square integrable representation of G on the Hilbert space $\mathfrak{H}$. Then there exists a unique positive, self-adjoint, invertible operator C in $\mathfrak{H}$, the domain $\mathcal{D}(C)$ of which is dense in $\mathfrak{H}$ and is equal to $\mathcal{A}$, the set of all admissible vectors; if η and η' are any two admissible vectors and ϕ, ϕ' are arbitrary vectors in $\mathfrak{H}$, then*

$$\int_G \overline{\langle \eta'_g | \phi' \rangle} \langle \eta_g | \phi \rangle \, d\mu(g) = \langle C\eta | C\eta' \rangle \, \langle \phi' | \phi \rangle. \tag{8.36}$$

Furthermore, $C = \lambda I$, $\lambda > 0$, if and only if G is unimodular.

Proof. Let $\eta, \eta' \in \mathcal{A}$, and consider the corresponding isometries $W_\eta, W_{\eta'}$, defined as in (8.9). With $W_\eta^* : L^2(G, d\mu) \to \mathfrak{H}$ denoting, as before, the adjoint of the linear map $W_\eta : \mathfrak{H} \to L^2(G, d\mu)$, the operator $W_{\eta'}^* W_\eta$ is bounded on $\mathfrak{H}$. Next, for all $g \in G$,

$$\begin{aligned} W_{\eta'}^* W_\eta U(g) &= W_{\eta'}^* U_\ell(g) W_\eta, \quad \text{by (8.13)}, \\ &= [U_\ell(g^{-1}) W_{\eta'}]^* W_\eta = [W_{\eta'} U(g^{-1})]^* W_\eta \\ &= U(g) W_{\eta'}^* W_\eta. \end{aligned}$$

By Schur's lemma, $W_{\eta'}^* W_\eta$ is therefore a multiple of the identity on $\mathfrak{H}$:

$$W_{\eta'}^* W_\eta = \lambda(\eta, \eta') I, \qquad \lambda(\eta, \eta') \in \mathbb{C} \tag{8.37}$$

$[\lambda(\eta, \eta')$ is antilinear in η and linear in $\eta']$. Applying Theorem 8.1.3, we find, for $\eta = \eta'$,

$$\lambda(\eta, \eta) = 1, \qquad \eta \in \mathcal{A}. \tag{8.38}$$

Set

$$q(\eta, \eta') = [c(\eta) c(\eta')]^{\frac{1}{2}} \lambda(\eta, \eta'), \tag{8.39}$$

with $c(\eta)$ as in (8.8). Using (8.9), we obtain

$$\begin{aligned} \int_G \overline{\langle \eta'_g | \phi' \rangle} \langle \eta_g | \phi \rangle \, d\mu(g) &= [c(\eta) c(\eta')]^{\frac{1}{2}} \int_G \overline{(W_{\eta'} \phi')(g)} (W_\eta \phi)(g) \, d\mu(g) \\ &= [c(\eta) c(\eta')]^{\frac{1}{2}} \langle W_{\eta'} \phi' | W_\eta \phi \rangle_{L^2(G, d\mu)} \\ &= [c(\eta) c(\eta')]^{\frac{1}{2}} \langle \phi' | W_{\eta'}^* W_\eta \phi \rangle_{\mathfrak{H}}, \end{aligned}$$

for all $\eta, \eta' \in \mathcal{A}$, and $\phi, \phi' \in \mathfrak{H}$. Inserting (8.37) and (8.39) into this, we get

$$\int_G \overline{\langle \eta'_g | \phi' \rangle} \langle \eta_g | \phi \rangle \, d\mu(g) = q(\eta, \eta') \langle \phi' | \phi \rangle_{\mathfrak{H}}. \tag{8.40}$$

From (8.37) and (8.40), we also find that

$$W_{\eta'}^* W_\eta = \frac{1}{[c(\eta) c(\eta')]^{\frac{1}{2}}} \int_G |\eta'_g\rangle \langle \eta_g| \, d\mu(g). \tag{8.41}$$

8.2. Orthogonality relations

Now, (8.40) and (8.37) together imply that $q : \mathcal{A} \times \mathcal{A} \to \mathbb{C}$ is a positive, symmetric, sesquilinear form on the dense domain $\mathcal{A}$. Moreover, since q is independent of ϕ, ϕ', taking $\phi = \phi' \neq 0$, we obtain

$$q(\eta, \eta') = \frac{1}{\|\phi\|^2} \int_G \overline{\langle U(g)\eta'|\phi\rangle} \langle U(g)\eta|\phi\rangle \, d\mu(g). \tag{8.42}$$

We next prove that the sesquilinear form q is closed on its form domain $\mathcal{A}$. Indeed, consider on $\mathcal{A}$ the scalar product and associated norm:

$$\langle \eta|\eta'\rangle_q = \langle \eta|\eta'\rangle_{\mathfrak{H}} + q(\eta, \eta'), \quad \|\eta\|_q^2 = \|\eta\|_{\mathfrak{H}}^2 + q(\eta, \eta), \qquad \eta, \eta' \in \mathcal{A}. \tag{8.43}$$

Let $\{\eta_k\}_{k=1}^\infty \subset \mathcal{A}$ be a Cauchy sequence in the $\|\ldots\|_q$-norm. Clearly, $\{\eta_k\}_{k=1}^\infty$ is also a Cauchy sequence in the norm of $\mathfrak{H}$, implying that there exists a vector $\eta \in \mathfrak{H}$ such that $\lim_{k\to\infty} \|\eta_k - \eta\|_\mathfrak{H} = 0$. Also, since the sequence is Cauchy in the $\|\ldots\|_q$-norm, $q(\eta_j - \eta_k, \eta_j - \eta_k) \to 0$ for $j, k \to \infty$. From (8.42), we infer that the sequence of functions,

$$\{\widetilde{\Phi}_k\}_{k=1}^\infty \subset L^2(G, d\mu), \qquad \widetilde{\Phi}_k(g) = \langle U(g)\eta_k|\phi\rangle_\mathfrak{H},$$

is a Cauchy sequence in $L^2(G, d\mu)$. Thus there exists a vector $\widetilde{\Phi} \in L^2(G, d\mu)$ satisfying

$$\lim_{k\to\infty} \|\widetilde{\Phi}_k - \widetilde{\Phi}\|_{L^2(G, d\mu)} = 0, \tag{8.44}$$

and, therefore, the sequence $\{\widetilde{\Phi}_k\}_{k=1}^\infty$ also converges to $\widetilde{\Phi}$ weakly, with the sequence of norms $\{\|\widetilde{\Phi}_k\|_{L^2(G, d\mu)}\}_{k=1}^\infty$ remaining bounded. Moreover, for any $g \in G$,

$$\lim_{k\to\infty} \langle U(g)\eta_k|\phi\rangle_\mathfrak{H} = \langle U(g)\eta|\phi\rangle_\mathfrak{H} \quad \Rightarrow \quad \lim_{k\to\infty} |\widetilde{\Phi}_k(g) - \widetilde{\Phi}(g)| = 0.$$

Thus, by (8.44), $\widetilde{\Phi}(g) = \langle U(g)\eta|\phi\rangle_\mathfrak{H}$, for all $g \in G$ and all $\phi \in \mathfrak{H}$, so that $g \mapsto \langle U(g)\eta|\phi\rangle_\mathfrak{H}$ defines a vector in $L^2(G, d\mu)$. Taking $\phi = \eta$, we see that this implies $\eta \in \mathcal{A}$. Next,

$$\begin{aligned} \lim_{k\to\infty} \|\eta_k - \eta\|_q &= \lim_{k\to\infty} \|\eta_k - \eta\|_\mathfrak{H} + \lim_{k\to\infty} q(\eta_k - \eta, \eta_k - \eta) \\ &= 0 + \lim_{k\to\infty} \frac{1}{\|\phi\|^2} \|\widetilde{\Phi}_k - \widetilde{\Phi}\|_{L^2(G, d\mu)}, \quad \text{by (8.42)}, \\ &= 0, \quad \text{by (8.44)}. \end{aligned}$$

Consequently, $\mathcal{A}$ is complete in the $\|\ldots\|_q$-norm, so that q is closed (see, for example, [Ree]). Since q is a closed, symmetric, positive form, the well-known second representation theorem [Kat] implies that the existence of a unique positive self-adjoint operator C, with domain $\mathcal{A}$, such that

$$q(\eta, \eta') = \langle C\eta|C\eta'\rangle_\mathfrak{H}. \tag{8.45}$$

Next, if $\eta \neq 0$, then, by (8.37), (8.39), and (8.45),

$$\|C\eta\|^2 = c(\eta) = \frac{I(\eta)}{\|\eta\|^2} \neq 0 \tag{8.46}$$

[see (8.8)]. So, C is injective, and, consequently, it is invertible. Moreover, its inverse C^{-1} is densely defined, as the inverse of an invertible self-adjoint operator [indeed, it is easily seen that $\mathrm{Ran}(C)$ (the range of C) is dense in $\mathfrak{H}$].

It remains to prove the last statement. By (8.42), we have, for all $g \in G$,

$$\begin{aligned} q(U(g)\eta, U(g)\eta') &= \frac{1}{\|\phi\|^2} \int_G \overline{\langle U(g'g)\eta'|\phi\rangle} \langle U(g'g)\eta|\phi\rangle \, d\mu(g'), \\ &= \frac{\Delta(g^{-1})}{\|\phi\|^2} \int_G \overline{\langle U(g')\eta'|\phi\rangle} \langle U(g')\eta|\phi\rangle \, d\mu(g'), \\ &= q(\eta, \eta'). \end{aligned}$$

Comparing with (8.45), we find that, for all $\eta, \eta' \in \mathcal{A}$,

$$\langle CU(g)\eta|CU(g)\eta'\rangle_{\mathfrak{H}} = \frac{1}{\Delta(g)} \langle C\eta|C\eta'\rangle_{\mathfrak{H}}. \tag{8.47}$$

Now, C^2 is positive and densely defined in $\mathfrak{H}$. In addition, its domain is invariant under U. Indeed, let $\eta' \in \mathcal{D}(C^2)$, which implies that $\eta' \in \mathcal{D}(C)$, $C\eta' \in \mathcal{D}(C)$, and $\eta'_g \in \mathcal{D}(C)$. Then, (8.47) may be written as

$$\langle C\eta_g|C\eta'_g\rangle_{\mathfrak{H}} = \frac{1}{\Delta(g)} \langle \eta|C^2\eta'\rangle_{\mathfrak{H}},$$

which shows that $C\eta'_g \in \mathcal{D}(C)$ as well, i.e., $\eta'_g \in \mathcal{D}(C^2)$. Thus, (8.47) implies that, on the dense invariant domain $\mathcal{D}(C^2)$:

$$C^2 U(g) = \frac{1}{\Delta(g)} U(g) C^2. \tag{8.48}$$

This then shows, using the extended Schur's Lemma 4.3.3, with $U_1 = U_2$, that $\Delta(g) = 1$, for all $g \in G$; that is, G is unimodular if and only if $C = \lambda I$, $\lambda > 0$. □

The operator C is known in the mathematical literature as the *Duflo-Moore operator* [71, 114, 248], often denoted $C = K^{-1/2}$. Actually, it can be shown (see, for example, [Sug]), that, if G is compact, then

$$C = [\dim \mathfrak{H}]^{-\frac{1}{2}} I. \tag{8.49}$$

(Note that, with G compact and U irreducible, $\dim \mathfrak{H}$ is finite.) If G is not compact, but just unimodular, then [see (8.46)] with $\|\eta\| = 1$,

$$C = [c(\eta)]^{\frac{1}{2}} I, \tag{8.50}$$

so that the value of $c(\eta)$ does not depend of $\eta \in \mathcal{A}$. In that case, we call $d_U \equiv c(\eta)^{-1}$ the *formal dimension* of the representation U. In this terminology, when G is a non-unimodular group, the formal dimension of a square integrable representation U is the positive self-adjoint (possibly unbounded) operator C^{-2}.

To conclude this section, we derive a generalized version of the resolution of the identity (8.10).

Corollary 8.2.2: *Let U be a square integrable representation of the locally compact group G. If η and η' are any two nonzero admissible vectors, then, provided $\langle C\eta | C\eta' \rangle \neq 0$,*

$$\frac{1}{\langle C\eta | C\eta' \rangle} \int_G |\eta'_g\rangle\langle \eta_g| \, d\mu(g) = I. \tag{8.51}$$

Proof. This is mere restatement of the orthogonality relation (8.36), since the vectors ϕ and ϕ' are arbitrary. $\square$

From here, we get the reconstruction formula

$$\phi = \frac{[c(\eta')]^{\frac{1}{2}}}{\langle C\eta | C\eta' \rangle} \int_G \Phi(g) \eta'_g \, d\mu(g), \quad \phi \in \mathfrak{H}, \quad \Phi(g) = (W_\eta \phi)(g), \tag{8.52}$$

provided $\langle C\eta | C\eta' \rangle \neq 0$, which generalizes (8.19). Here, η is the analyzing vector and η' the vector used for reconstruction. This reconstruction formula can be further generalized to situations in which only one of the vectors η, η' is admissible, provided the other one satisfies a more stringent condition and the mutual compatibility condition

$$0 < |c_{\eta\eta'}| < \infty, \quad \text{where} \quad c_{\eta\eta'} = \frac{1}{\langle C\eta | C\eta' \rangle} \int_G \langle \eta' | \eta'_g \rangle \langle \eta_g | \eta \rangle \, d\mu(g) \tag{8.53}$$

is satisfied. If $\eta^\#$ denotes the admissible vector, then one requires that $W_{\eta^\#}\eta^\# \in L^1(G, d\mu)$. While the extended formula (8.52) has been known for a long time in wavelet analysis [98], where it is widely used in practice — for example, when η' is taken to be the δ-function, (8.52) yields the so-called Morlet reconstruction formula — its validity for a general group was proved in [174]. Another example, in two dimensions, is obtained by taking a singular analyzing wavelet — a delta function on a line — and this yields the inverse Radon transform [172].

An important consequence of (8.52) is that there are many kernels associated to a given η, namely, all the functions

$$K_{\eta\eta'}(g, g') = \frac{1}{\langle C\eta | C\eta' \rangle} \langle \eta_g | \eta'_{g'} \rangle, \tag{8.54}$$

each one of which defines the evaluation map on $\mathfrak{H}_\eta \in L^2(G, d\mu)$:

$$\int_G K_{\eta\eta'}(g, g') \Phi(g') \, d\mu(g') = \Phi(g), \quad \Phi \in \mathfrak{H}_\eta = W_\eta(\mathfrak{H}). \tag{8.55}$$

It ought to be noted, however, that, if $\eta \neq \eta'$, $K_{\eta\eta'}$ is not a positive-definite kernel, and, hence, not a reproducing kernel, although, as an

160   8. Coherent States from Square Integrable Representations

integral operator on $L^2(G, d\mu)$, it is *idempotent*, that is, square integrable:

$$\int_G K_{\eta\eta'}(g, g'') K_{\eta\eta'}(g'', g') \, d\mu(g'') = K_{\eta\eta'}(g, g'). \tag{8.56}$$

8.3 The Wigner map

An alternative perspective is obtained by recasting the orthogonality relations into a slightly different form. The latter is rooted in quantum physics and can be traced back to Wigner [284] and to Weyl (it was later rediscovered by the engineer Ville [280] in signal analysis; see also Chapter 12). The Wigner distribution is the Fourier transform of the product of the shifted wave function $\psi(x)$ with its complex conjugate shifted the other way around:

$$W(q, p) = \frac{1}{2\pi} \int e^{-ipx} \, \overline{\psi\left(q - \frac{x}{2}\right)} \, \psi\left(q + \frac{x}{2}\right) dx. \tag{8.57}$$

$W(q, p)$ is a real function on phase space, which is the quotient of the Weyl–Heisenberg group by its center, G_{WH}/Θ (see Chapter 2, Section 2.2). To a certain extent it may be viewed as a compromise between the probability densities $|\psi(q)|^2$ and $|\widehat{\psi}(p)|^2$ on position and momentum space, respectively. Indeed, it satisfies the marginality conditions:

$$\int W(q, p) \, dp = |\psi(q)|^2,$$
$$\int W(q, p) \, dq = |\widehat{\psi}(p)|^2. \tag{8.58}$$

The Wigner distribution function $W(q, p)$, however, is not everywhere positive, which prevents its use as a genuine probability distribution: it is only a quasiprobability distribution. Yet, it is extensively used in many domains of quantum physics, in particular, in quantum optics [Fla, Gro, Kim, Scu, 22, 286].

Let us now recast (8.57) into a form that brings out its group-theoretical content. Introducing the displacement operators of the UIR of the Weyl–Heisenberg group, already mentioned in Chapter 2, Section 2.2, namely,

$$U(q, p) = e^{i(pQ - qP)} = e^{-\frac{i}{2}pq} \, e^{ipQ} \, e^{-iqP} \tag{8.59}$$

and the parity operator

$$\Pi\psi(x) = \psi(-x), \tag{8.60}$$

we see that

$$W(q, p) = \frac{1}{\pi} \text{Tr} \left(U(q, p) \Pi U(q, p)^* |\psi\rangle\langle\psi| \right)$$

$$= \frac{1}{\pi}\text{Tr}\left(U(2q,2p)^*|\psi\rangle\langle\psi|\Pi\right), \qquad (8.61)$$

since

$$\Pi U(q,p) = U(q,p)^*\Pi. \qquad (8.62)$$

The importance of the parity operator in the context of the Weyl–Heisenberg group has been stressed in [155]. Looked at in this way, the Wigner distribution function appears as a function on the Weyl–Heisenberg group (actually a homogeneous space of it). For the purposes of generalizing it to other possible groups, it is useful to look at it somewhat differently. A simple computation shows that (8.57) can be rewritten in the form

$$W(q,p) = \frac{1}{2\pi}\int_{\mathbb{R}^2} e^{i(qp'-pq')}\text{Tr}\left(U(q',p')^*|\psi\rangle\langle\psi|\right)\,dq'\,dp'. \qquad (8.63)$$

In this form, the Wigner function appears essentially as a Fourier transform of a function of the type

$$(q',p') \mapsto \text{Tr}\left(U(q',p')^*|\psi\rangle\langle\phi|\right), \qquad (8.64)$$

defined on the group. The interpretation of $W(q,p)$ is then as a function on a *coadjoint orbit* of the group [22, 286] — an object that has the natural structure of a phase space. In any case, the primary object of interest is the function (8.64), and we now proceed to look at such functions in the context of general square integrable representations.

Once again, let U be a unitary square integrable representation of a group G in a Hilbert space $\mathfrak{H}$, and let $\mathcal{B}_2(\mathfrak{H})$ denote the Hilbert space of all Hilbert–Schmidt operators on $\mathfrak{H}$, with scalar product:

$$\langle \mathbf{X}_2|\mathbf{X}_1\rangle_{\mathcal{B}_2(\mathfrak{H})} = \text{Tr}\left(\mathbf{X}_2^*\mathbf{X}_1\right), \qquad \mathbf{X}_1, \mathbf{X}_2 \in \mathcal{B}_2(\mathfrak{H}). \qquad (8.65)$$

To every element $\mathbf{X} \in \mathcal{B}_2(\mathfrak{H})$, associate the function on G

$$(\mathcal{W}\mathbf{X})(g) = \text{Tr}(U(g)^*\mathbf{X}C^{-1}), \qquad (8.66)$$

where C is the Duflo–Moore operator of Theorem 8.2.1. Define the vectors $\zeta_i = C\eta_i \in \text{Ran}(C) = \mathcal{D}(C^{-1})$, $i=1,2$ [recall that $\text{Ran}(C)$ is dense in $\mathfrak{H}$]. Then, if $\mathbf{X}_i$ denotes the rank one operator $\mathbf{X}_i = |\phi_i\rangle\langle\zeta_i| \in \mathcal{B}_2(\mathfrak{H})$, one has

$$(\mathcal{W}\mathbf{X}_i)(g) = \text{Tr}(U(g)^*\mathbf{X}_i C^{-1}) = \langle U(g)C^{-1}\zeta_i|\phi_i\rangle, \quad i=1,2, \qquad (8.67)$$

which shows that the function $\mathcal{W}\mathbf{X}_i$ is square integrable on G. Furthermore, in this notation, the orthogonality relation (8.36) becomes

$$\int_G \overline{(\mathcal{W}\mathbf{X}_2)(g)}(\mathcal{W}\mathbf{X}_1)(g)\,d\mu(g) = \text{Tr}(\mathbf{X}_2^*\mathbf{X}_1) = \langle \mathbf{X}_2|\mathbf{X}_1\rangle_{\mathcal{B}_2(\mathfrak{H})}, \qquad (8.68)$$

which means that $\mathcal{W}$ is isometric from $\mathfrak{H}\otimes\mathcal{D}(C^{-1})^\dagger$ into $L^2(G,d\mu)$, where $\mathfrak{H}\otimes\mathcal{D}(C^{-1})^\dagger$ denotes the dense subspace of $\mathcal{B}_2(\mathfrak{H})$ generated by vectors of the form $|\phi\rangle\langle\psi|$, $\phi \in \mathfrak{H}$, $\psi \in \mathcal{D}(C^{-1})$. Thus, $\mathcal{W}$ may be extended by continuity to an isometry, called the *Wigner map* or *Wigner transform* and denoted by the same symbol, $\mathcal{W} : \mathcal{B}_2(\mathfrak{H}) \to L^2(G, d\mu)$. Comparing the definition (8.66) with (8.61), we notice two differences. First, C reduces to the identity operator for the Weyl–Heisenberg group, which is unimodular. Next, the parity operator Π has been inserted in order that the Wigner function be real-valued.

The map $\mathcal{W}$ induces a decomposition of $\text{Ran}(\mathcal{W})$, which is a closed subspace of $L^2(G, d\mu)$ denoted by $L^2_\mathcal{W}(G, d\mu)$, into a direct sum of *coherent sectors*. First the unitary irreducible representation U lifts to a unitary *reducible* representation $\mathbb{U}_\ell$ on $\mathcal{B}_2(\mathfrak{H})$

$$\mathbb{U}_\ell(g) = U(g) \vee I, \quad g \in G, \tag{8.69}$$

where we have introduced the notation

$$(A \vee B)\mathbf{X} = A X B^*, \quad \mathbf{X} \in \mathcal{B}_2(\mathfrak{H}), \quad A, B \in \mathcal{L}(\mathfrak{H}). \tag{8.70}$$

Clearly, $A \vee B$ defines a bounded linear operator on $\mathcal{B}_2(\mathfrak{H})$ for each pair of operators A and B in $\mathcal{L}(\mathfrak{H})$. Moreover, the Wigner map $\mathcal{W}$ intertwines $\mathbb{U}_\ell$ with the left regular representation U_ℓ on $L^2(G, d\mu)$:

$$\mathcal{W}\mathbb{U}_\ell(g) = U_\ell(g)\mathcal{W}, \quad g \in G. \tag{8.71}$$

Next, note that $\langle \zeta_1 | \zeta_2 \rangle = 0$ in $\mathfrak{H}$ implies $\langle \mathbf{X}_1 | \mathbf{X}_2 \rangle = 0$ in $\mathcal{B}_2(\mathfrak{H})$, for $\mathbf{X}_i = |\phi_i\rangle\langle\zeta_i|$ [by (8.65)], and this in turn implies $\langle \mathcal{W}\mathbf{X}_1 | \mathcal{W}\mathbf{X}_2 \rangle = 0$ in $L^2(G, d\mu)$. Now, let $\{\zeta_j, j = 1, 2, \ldots, N\}$, where $N = \dim \mathfrak{H} \leq \infty$, be an orthonormal basis of $\mathfrak{H}$ contained in $\mathcal{D}(C^{-1})$. For each j, the set $\mathfrak{H} \otimes \overline{\zeta_j} = \{|\phi\rangle\langle\zeta_j| \, | \, \phi \in \mathfrak{H}\}$ is a vector subspace of $\mathcal{B}_2(\mathfrak{H})$, invariant under $\mathbb{U}_\ell$. Taking the restriction from $\mathcal{B}_2(\mathfrak{H})$ to this subspace, we see that

$$\mathbb{U}_\ell(g)|_{\mathfrak{H} \otimes \overline{\zeta_j}} \sim U(g) \quad \text{(unitary equivalence)}$$

$$\mathcal{W}|_{\mathfrak{H} \otimes \overline{\zeta_j}} = \widetilde{W}_{\eta_j}, \quad \text{with} \quad \eta_j = C^{-1}\zeta_j, \tag{8.72}$$

where $\widetilde{W}_{\eta_j} : \mathfrak{H} \otimes \overline{\zeta_j} \to L^2(G, d\mu)$ is essentially the isometric map defined in (8.9), with η_j replacing η,

$$(\widetilde{W}_{\eta_j} |\phi\rangle\langle\zeta_j|)(g) = \text{Tr}(U(g)^*|\phi\rangle\langle\zeta_j|C^{-1}) = \langle U(g)C^{-1}\zeta_j|\phi\rangle = \langle U(g)\eta_j|\phi\rangle,$$

[since $\|\zeta_j\| = 1$ and (8.46) together imply that $c(\eta_j) = 1$]. It follows that the Wigner map $\mathcal{W}$ transforms the orthogonal decomposition

$$\mathcal{B}_2(\mathfrak{H}) = \bigoplus_{j=1}^{N} \mathfrak{H} \otimes \overline{\zeta_j} \tag{8.73}$$

into the decomposition of $L^2_\mathcal{W}(G, d\mu)$ into coherent sectors

$$L^2_\mathcal{W}(G, d\mu) = \mathbb{P}_\mathcal{W} L^2(G, d\mu) = \bigoplus_{j=1}^{N} \mathfrak{H}_{\eta_j}, \tag{8.74}$$

where $\mathbb{P}_\mathcal{W} = \mathcal{W}\mathcal{W}^*$ is the projection operator onto $L^2_\mathcal{W}(G, d\mu)$ and $\mathfrak{H}_{\eta_j}$ is the range of $\widetilde{W}_{\eta_j}$ in $L^2(G, d\mu)$. Correspondingly, denoting by U^j_ℓ the restriction of U_ℓ to $\mathfrak{H}_{\eta_j}$, we obtain the orthogonal decomposition

$$U_\ell \mathbb{P}_\mathcal{W} = \bigoplus_{j=1}^{N} U_\ell^j, \qquad (8.75)$$

from which we see that the unitary image of the irreducible, square integrable representation U, on the Hilbert space $\mathfrak{H}$, appears with multiplicity exactly equal to the dimension of $\mathfrak{H}$ in the left regular representation U_ℓ of G. In general, the operator $\mathbb{P}_\mathcal{W}$ projects only onto a proper subspace of $L^2(G, d\mu)$, since the group G may have several inequivalent square integrable representations, as well as other representations, not square integrable, but nevertheless contained (as subrepresentations) in the left regular representation. The case of groups for which the left regular representation decomposes into a direct sum of square integrable representations alone is of particular interest in signal analysis. In Section 9.1, we shall study a class of such groups.

Finally, let us compute the inverse of the map $\mathcal{W}$ on $L^2_\mathcal{W}(G, d\mu)$. Consider an element in $\mathcal{B}_2(\mathfrak{H})$ of the type $|\phi\rangle\langle\zeta|$, with $\phi \in \mathfrak{H}$ and $\zeta \in \mathcal{D}(C^{-1})$ and let $f = \mathcal{W}(|\phi\rangle\langle\zeta|)$. Then, the integral

$$\int_G U(g)C^{-1} f(g)\, d\mu(g)$$

converges weakly to $|\phi\rangle\langle\zeta|$. Indeed, if $\phi' \in \mathfrak{H}$ and $\zeta' \in \mathcal{D}(C^{-1})$, then

$$\int_G \langle \phi' | U(g) C^{-1} \zeta' \rangle f(g)\, d\mu(g) =$$

$$= \int_G \langle \phi' | U(g) C^{-1} \zeta' \rangle \mathrm{Tr}(U(g)^* |\phi\rangle\langle\zeta| C^{-1})\, d\mu(g)$$

$$= \int_G \overline{\langle \phi | U(g) C^{-1} \zeta \rangle} \langle \phi' | U(g) C^{-1} \zeta' \rangle\, d\mu(g)$$

$$= \langle \phi' | \phi \rangle \langle \zeta | \zeta' \rangle, \qquad (8.76)$$

by virtue of the orthogonality relations. Also,

$$\left| \int_G \langle \phi' | U(g) C^{-1} \zeta' \rangle f(g)\, d\mu(g) \right| \leq \|\phi'\|\,\|\zeta'\|\,\|\phi\|\,\|\zeta\|,$$

and, hence, the relation (8.76) holds for all $\phi' \in \mathfrak{H}$ and $\zeta' \in \mathcal{D}(C^{-1})$. Thus, as a weak integral,

$$|\phi\rangle\langle\zeta| = \int_G U(g) C^{-1} f(g)\, d\mu(g), \qquad \phi \in \mathfrak{H},\ \zeta \in \mathcal{D}(C^{-1}).$$

This implies that, on the dense set of vectors $f \in L^2_\mathcal{W}(G, d\mu)$, comprising the image of $\mathfrak{H} \otimes \mathcal{D}(C^{-1})^\dagger$ under $\mathcal{W}$,

$$\mathcal{W}^{-1} f = \int_G U(g) C^{-1} f(g)\, d\mu(g), \qquad (8.77)$$

the integral being defined weakly. Since $\mathcal{W}$ is an isometry, the above relation can be extended by continuity to all of $L^2_{\mathcal{W}}(G, d\mu)$.

8.4 Modular structures and statistical mechanics

There is a remarkable algebraic structure encapsulated in the orthogonality relations of Section 8.2, which, while not directly related to the theory of CS states, displays several interesting features when looked at in conjunction with the decomposition (8.74) of $L^2_{\mathcal{W}}(G, d\mu)$ into coherent sectors. Since this aspect of the theory is overlooked in most discussions of square integrable representations, we summarize here some related *modular algebraic* features of the above decomposition. The analysis given below follows [6],[17], [117], [118] and [StZ]. Considerable use is made of the theory of von Neumann algebras and modular Hilbert algebras in deriving the results; however, we try to present here a simplified and self-contained version. The reader wishing to go deeper could consult, for example, [Dix1], [Ta1], [Ta2] or [StZ].

Once again, let us fix the orthonormal basis $\{\zeta_j \,|\, j = 1, 2, \ldots, N\}$, ($N = \dim \mathfrak{H} \leq \infty$), of $\mathfrak{H}$, where each ζ_i is contained in $\mathcal{D}(C^{-1})$ and, as before, write $\eta_j = C^{-1}\zeta_j$. Let $f_{ij} \in L^2_{\mathcal{W}}(G, d\mu)$ be the vectors, in the range of $\mathcal{W}$,

$$f_{ij} = \mathcal{W}\left(|\zeta_i\rangle\langle\zeta_j|\right) = \mathcal{W}_{\eta_j}\zeta_i, \qquad i,j = 1, 2, \ldots, N. \tag{8.78}$$

Since the $|\zeta_i\rangle\langle\zeta_j|$ form an orthonormal basis of $\mathcal{B}_2(\mathfrak{H})$, the f_{ij} make up an orthonormal basis for $L^2_{\mathcal{W}}(G, d\mu)$. An easy computation then shows that

$$J f_{ij} = f_{ji}, \qquad i,j = 1, 2, \ldots, N, \tag{8.79}$$

where J is the antiunitary map defined in (4.112). Next, using the notation introduced in (8.70), define the operators $F_{ij,\,k\ell}$ on $L^2_{\mathcal{W}}(G, d\mu)$:

$$F_{ij,\,k\ell} = \mathcal{W}\left[(|\zeta_i\rangle\langle\zeta_j|) \vee (|\zeta_k\rangle\langle\zeta_\ell|)\right] \mathcal{W}^{-1}, \qquad i,j,k,\ell = 1, 2, \ldots, N. \tag{8.80}$$

We shall also need the "summed up versions" of these operators, for which the notation

$$F_{ij,\,\cdot\cdot} = \sum_{k=1}^{N} F_{ij,kk} = \mathcal{W}\left[(|\zeta_i\rangle\langle\zeta_j|) \vee I\right] \mathcal{W}^{-1}, \quad \text{etc.,}$$

$$F_{\cdot\cdot,ij} = \sum_{k=1}^{N} F_{kk,ij} = \mathcal{W}\left[I \vee (|\zeta_i\rangle\langle\zeta_j|)\right] \mathcal{W}^{-1}, \tag{8.81}$$

will be employed. Using the explicit forms of $\mathcal{W}$ and $\mathcal{W}^{-1}$, it is immediately verified that

$$F_{ij,\,k\ell} = |f_{i\ell}\rangle\langle f_{jk}|, \qquad i,j,k,\ell = 1, 2, \ldots, N. \tag{8.82}$$

8.4. Modular structures and statistical mechanics

Hence, define the projection operators

$$\mathbb{P}_{ij} = |f_{ij}\rangle\langle f_{ij}| = F_{ii,\,jj}, \qquad i,j = 1,2,\ldots,N, \tag{8.83}$$

on $L^2_{\mathcal{W}}(G, d\mu)$. The projection operators $\mathbb{P}_j$, obtained by taking $\eta - \eta_j$ in (8.11), now appear as

$$\mathbb{P}_j = \sum_{i=1}^{N} \mathbb{P}_{ij} = F_{..,\,jj}, \qquad j = 1,2,\ldots,N. \tag{8.84}$$

Similarly, defining

$$\widehat{\mathbb{P}}_i = \sum_{j=1}^{N} \mathbb{P}_{ij} = F_{ii,\,..}, \qquad i = 1,2,\ldots,N, \tag{8.85}$$

we verify that

$$\widehat{\mathbb{P}}_i = J\mathbb{P}_i J,$$
$$(\widehat{\mathbb{P}}_i f)(g) = \int_G \overline{K_{\eta_i}(g^{-1}, g'^{-1})}\,[\Delta(g)\Delta(g')]^{-\frac{1}{2}} f(g')\, d\mu(g'),$$
$$f \in L^2_{\mathcal{W}}(G, d\mu). \tag{8.86}$$

Every operator of the type $F_{ij,\,..}$ commutes with each one of the projection operators $\mathbb{P}_k$, and, similarly, the projection operators $\widehat{\mathbb{P}}_k$ commute with the operators $F_{..,\,ij}$:

$$[\mathbb{P}_k, F_{ij,\,..}] = [\widehat{\mathbb{P}}_k, F_{..,\,ij}] = 0, \qquad i,j,k = 1,2,\ldots,N. \tag{8.87}$$

Let $U_\ell^{\mathcal{W}}$ and $\overline{U}_r^{\mathcal{W}}$ be the restrictions of the representations U_ℓ and $\overline{U}_r$, respectively, to $L^2_{\mathcal{W}}(G, d\mu)$ [see (4.110)]. From (8.69) and (8.71), we know that

$$\mathcal{W}\,(U(g) \vee I)\,\mathcal{W}^{-1} = U_\ell^{\mathcal{W}}(g), \qquad g \in G, \tag{8.88}$$

and, similarly,

$$\mathcal{W}\,(I \vee U(g))\,\mathcal{W}^{-1} = \overline{U}_r^{\mathcal{W}}(g), \qquad g \in G. \tag{8.89}$$

Denote by $\mathfrak{A}_\ell^{\mathcal{W}}$ and $\mathfrak{A}_r^{\mathcal{W}}$ the (von Neumann) algebras of operators on $L^2_{\mathcal{W}}(G, d\mu)$ generated by these two sets of unitary operators; that is, $\mathfrak{A}_\ell^{\mathcal{W}}$ (respectively, $\mathfrak{A}_r^{\mathcal{W}}$) consists of all bounded operators on $L^2_{\mathcal{W}}(G, d\mu)$ that commute with the set of all operators that commute with the $U_\ell^{\mathcal{W}}$ (respectively, $\overline{U}_r^{\mathcal{W}}$), for all $g \in G$. The relation (4.112), between the operators of the left and right regular representations, in conjunction with (8.88) and (8.89) allows us to write

$$J\mathfrak{A}_\ell^{\mathcal{W}} J = \mathfrak{A}_r^{\mathcal{W}}. \tag{8.90}$$

At this point, we pause for a while and recall the basic concepts and definitions concerning von Neumann algebras that we will need later on (see

[Bra], [Dix1], or [Ta2] for a systematic discussion). Let $\mathfrak{H}$ be an arbitrary Hilbert space and $\mathfrak{B}$ be a *-invariant set of bounded operators on $\mathfrak{H}$. Then, as already mentioned in Chapter 3, Section 3.1, its *commutant*, denoted by $\mathfrak{B}'$, is the set of all bounded operators that commute with all elements of $\mathfrak{B}$, and its bicommutant is $\mathfrak{B}'' = (\mathfrak{B}')'$. A von Neumann algebra $\mathfrak{A}$ of operators on $\mathfrak{H}$ is a *-invariant set of bounded operators, which is invariant under linear combinations and products, and, moreover, satisfies the relation $\mathfrak{A} = \mathfrak{A}''$. Equivalently (and this is the content of the basic theorem of von Neumann), $\mathfrak{A}$ is closed in the weak operator topology. For a von Neumann algebra $\mathfrak{A}$, its commutant is again a von Neumann algebra, and in general $\mathfrak{A}$ and $\mathfrak{A}'$ are different algebras. The *center* of the algebra $\mathfrak{A}$ (or $\mathfrak{A}'$) is the von Neumann algebra $\mathfrak{Z} = \mathfrak{A} \cap \mathfrak{A}'$. If the center is trivial, $\mathfrak{Z} = \mathbb{C}I$, i.e., consists only of multiples of the identity operator I on $\mathfrak{H}$, the von Neumann algebra $\mathfrak{A}$ is called a *factor*. At the opposite extreme, the algebra $\mathfrak{A}$ is called *abelian* if $\mathfrak{A} \subset \mathfrak{A}'$ and *maximal abelian* if $\mathfrak{A} = \mathfrak{A}'$. A projection operator $\mathbb{P} \in \mathfrak{A}$ is called *minimal* if there is no other nontrivial projection operator $\mathbb{P}'$ in $\mathfrak{A}$ that is smaller (i.e., projecting onto a smaller subspace) than $\mathbb{P}$. It is an important result that a von Neumann algebra is always generated by (that is, is the bicommutant of) the set of its projections.

A *state* $\widehat{\phi}$ on the algebra is a normalized, bounded linear functional on it, and we shall use the notation $\widehat{\phi}(A)$, $A \in \mathfrak{A}$, to denote its value on the element A of the algebra. The condition of being normalized means that, for the identity operator I on $\mathfrak{H}$, $\widehat{\phi}(I) = 1$. The set of states on an algebra is a convex set. Extreme points of this set are called the *pure states*. If there is a vector $\phi \in \mathfrak{H}$, such that, for all $A \in \mathfrak{A}$, $\widehat{\phi}(A) = \langle\phi|A\phi\rangle_{\mathfrak{H}}$, then $\widehat{\phi}$ is called a *vector state*. The state $\widehat{\phi}$ is called *faithful* if $\widehat{\phi}(A^*A) = 0$, for $A \in \mathfrak{A}$, implies that $A = 0$, and is called *normal* if it is continuous in the *ultraweak* topology. This latter condition means that, given any discrete family of mutually orthogonal projection operators P_i in $\mathfrak{A}$, $\sum_i \widehat{\phi}(P_i) = \widehat{\phi}(\sum_i P_i)$. Given a state $\widehat{\phi}$ on the algebra $\mathfrak{A}$, the *centralizer* of $\mathfrak{A}$ with respect to $\widehat{\phi}$ is the von Neumann algebra $\mathfrak{M}_{\widehat{\phi}} = \{B \in \mathfrak{A} \mid \widehat{\phi}([A,B]) = 0 \text{ for all } A \in \mathfrak{A}\}$. A vector $\phi \in \mathfrak{H}$ is said to be *cyclic* for $\mathfrak{A}$ if the set of vectors $\{A\phi \mid A \in \mathfrak{A}\}$ is dense in $\mathfrak{H}$; it is called *separating* for $\mathfrak{A}$ if $A\phi = 0$, for any $A \in \mathfrak{A}$ implies that $A = 0$ [note that a vector is cyclic (respectively, separating) for $\mathfrak{A}$ iff it is separating (respectively, cyclic) for $\mathfrak{A}'$].

Let $\mathfrak{A}$ be a von Neumann algebra and ϕ be a cyclic and separating vector for it. On the dense set of vectors $\mathfrak{A}\phi \subset \mathfrak{H}$, define the antilinear map $S^o_\phi : A\phi \mapsto A^*\phi$. Then, the Tomita–Takesaki theory of modular Hilbert algebras [Ta1] asserts that the operator S^o_ϕ is closable. Let S_ϕ be the closure of S^o_ϕ. The operator S_ϕ has the polar decomposition $S_\phi = J_\phi[\triangle_\phi]^{\frac{1}{2}}$, where J_ϕ is antiunitary and satisfies $J^2_\phi = I$, $J_\phi \mathfrak{A} J_\phi = \mathfrak{A}'$, while $\triangle_\phi = S^*_\phi S_\phi$. For any $\beta > 0$, the map $A \mapsto \alpha_\phi(t)[A] = [\triangle_\phi]^{it/\beta} A [\triangle_\phi]^{it/\beta}$, $t \in \mathbb{R}$, then defines a continuous one-parameter group of automorphisms of $\mathfrak{A}$. Also, this is the

8.4. Modular structures and statistical mechanics

only one-parameter group of automorphisms of this algebra, such that, for any two elements $A, B \in \mathfrak{A}$, there exists a complex function $F_{A,B}$, analytic in the strip $\{\operatorname{Im} z \subset (0, \beta)\}$ and continuous on its boundaries, which satisfies $F_{A,B}(t) = \widehat{\phi}(A\alpha_\phi(t)[B])$, $t \in \mathbb{R}$, and

$$F_{A,B}(t + i\beta) = \widehat{\phi}(\alpha_\phi(t)[B]A), \tag{8.91}$$

where $\widehat{\phi}$ is the state defined by ϕ. In the algebraic theory of statistical mechanics, a state $\widehat{\phi}$ on the algebra of observables, satisfying (8.91), is said to obey the *KMS condition* at the "natural temperature" β [Tal, 163] and such states are called *KMS states*. These states are taken as (Gibbs) equilibrium states of the physical system at hand, and it is one of the major goals of the theory to try and identify its KMS states.

Now, we may come back to our algebras $\mathfrak{A}_\ell^\mathcal{W} = \{U_\ell^\mathcal{W}(g), g \in G\}''$ and $\mathfrak{A}_r^\mathcal{W} = \{U_r^\mathcal{W}(g), g \in G\}''$. They are both von Neumann algebras. As such, they are closed in the weak topology; for example, $\mathfrak{A}_\ell^\mathcal{W}$ could also be generated by taking all operators $U_\ell^\mathcal{W}(g)$, $g \in G$, forming all possible products and their (complex) linear combinations and closing the resulting set in the weak topology. It is then not hard to see that, for arbitrary $A \in \mathcal{L}(\mathfrak{H})$,

$$A_\ell^\mathcal{W} := \mathcal{W}(A \vee I)\mathcal{W}^{-1} \in \mathfrak{A}_\ell^\mathcal{W} \quad \text{and} \quad A_r^\mathcal{W} := \mathcal{W}(I \vee A)\mathcal{W}^{-1} \in \mathfrak{A}_r^\mathcal{W}, \tag{8.92}$$

and that $\mathfrak{A}_\ell^\mathcal{W}$ and $\mathfrak{A}_r^\mathcal{W}$ consist entirely of elements of this type.

We proceed to demonstrate below how KMS states arise naturally on the algebras $\mathfrak{A}_\ell^\mathcal{W}$ and $\mathfrak{A}_r^\mathcal{W}$ in connection with their decompositions into irreducible components [see (8.75)].

From (8.92) and the fact that U is an irreducible representation, it is not hard to see that

$$\{\mathfrak{A}_\ell^\mathcal{W}\}' = \mathfrak{A}_r^\mathcal{W}, \quad \text{and} \quad \mathfrak{A}_\ell^\mathcal{W} \cap \mathfrak{A}_r^\mathcal{W} = \mathbb{C}I_\mathcal{W}, \tag{8.93}$$

where $I_\mathcal{W}$ denotes the identity operator on $L_\mathcal{W}^2(G, d\mu)$. Thus, the two von Neumann algebras $\mathfrak{A}_\ell^\mathcal{W}$ and $\mathfrak{A}_r^\mathcal{W}$ are mutual commutants, and both are factors. Furthermore, in the light of (8.87),

$$\widehat{\mathbb{P}}_j \in \mathfrak{A}_\ell^\mathcal{W}, \quad \text{and} \quad \mathbb{P}_j \in \mathfrak{A}_r^\mathcal{W}, \quad j = 1, 2, \ldots, N, \tag{8.94}$$

and we also observe that $\widehat{\mathbb{P}}_j$ and $\mathbb{P}_j$ are minimal projectors in their respective algebras.

The relations (8.90) and (8.93) are said to constitute a *modular structure* for the triple $\{\mathfrak{A}_\ell^\mathcal{W}, \mathfrak{A}_r^\mathcal{W}, J\}$. Clearly, the decomposition of the algebra $\mathfrak{A}_\ell^\mathcal{W}$ (or $\mathfrak{A}_r^\mathcal{W}$) into irreducible sectors follows the decomposition of U_ℓ (or $\overline{U}_r$) in (8.75). In view of (8.94), this means that the irreducible subspaces of $L_\mathcal{W}^2(G, d\mu)$, for the algebra $\mathfrak{A}_\ell^\mathcal{W}$ (or $\mathfrak{A}_r^\mathcal{W}$), arise from minimal projectors in its commutant. We proceed to display some additional structure inherent in the modular property.

168   8. Coherent States from Square Integrable Representations

Let α_i, $i = 1, 2, \ldots, N$, be a sequence of nonzero, positive numbers satisfying $\alpha_i \neq \alpha_j$ for $i \neq j$, and $\sum_{i=1}^{N} \alpha_i = 1$. Define the vectors

$$\phi = \sum_{i=1}^{N} \alpha_i^{\frac{1}{2}} \zeta_i \in \mathfrak{H}, \qquad \Phi = \sum_{i=1}^{N} \alpha_i^{\frac{1}{2}} f_{ii} \in L^2_{\mathcal{W}}(G, d\mu), \qquad (8.95)$$

the density matrix ρ_ϕ and the associated self-adjoint operator $H_\phi \in \mathcal{L}(\mathfrak{H})$,

$$\rho_\phi = \sum_{i=1}^{N} \alpha_i |\zeta_i\rangle\langle\zeta_i| = \exp[-\beta H_\phi], \qquad (8.96)$$

where $\beta > 0$ is arbitrary. Explicitly, H_ϕ has the spectral representation:

$$H_\phi = -\frac{1}{\beta} \sum_{i=1}^{N} \log \alpha_i \, |\zeta_i\rangle\langle\zeta_i|, \qquad (8.97)$$

and, therefore, its spectrum is purely discrete and lies in the open interval $(0, \infty)$.

Using the above operators, let us construct the following operators on $L^2_{\mathcal{W}}(G, d\mu)$:

$$H_\phi^\ell = \mathcal{W}(H_\phi \vee I)\mathcal{W}^{-1}, \qquad H_\phi^r = \mathcal{W}(I \vee H_\phi)\mathcal{W}^{-1},$$
$$H_\phi^{\mathcal{W}} = H_\phi^\ell - H_\phi^r, \qquad \triangle_\phi^{\mathcal{W}} = \exp[-\beta H_\phi^{\mathcal{W}}]. \qquad (8.98)$$

From the fact that

$$H_\phi^\ell = -\frac{1}{\beta} \sum_{k=1}^{N} \log \alpha_k \, \widehat{\mathbb{P}}_k, \qquad H_\phi^r = -\frac{1}{\beta} \sum_{k=1}^{N} \log \alpha_k \, \mathbb{P}_k \qquad (8.99)$$

and

$$\widehat{\mathbb{P}}_k f_{ij} = \delta_{ik} f_{kj}, \quad \mathbb{P}_k f_{ij} = \delta_{kj} f_{ik}, \quad i, j, k = 1, 2, \ldots, N,$$

where

$$f_{ij} = \mathcal{W}(|\zeta_i\rangle\langle\zeta_j|),$$

we get the spectral form of the self-adjoint operators $H_\phi^{\mathcal{W}}$ and $\triangle_\phi^{\mathcal{W}}$,

$$H_\phi^{\mathcal{W}} = \frac{1}{\beta} \sum_{k,\ell}^{N} \log \frac{\alpha_k}{\alpha_\ell} \mathbb{P}_{k\ell}, \qquad \triangle_\phi^{\mathcal{W}} = \sum_{k,\ell}^{N} \frac{\alpha_k}{\alpha_\ell} \mathbb{P}_{k\ell}, \qquad (8.100)$$

showing that the spectrum of $H_\phi^{\mathcal{W}}$ is also purely discrete, symmetric about the origin, and contained in $(-\infty, \infty)$. Note that, if $N = \dim \mathfrak{H} = \infty$, then all three operators H_ϕ^ℓ, H_ϕ^r, and $H_\phi^{\mathcal{W}}$ are unbounded. Finally, let us define the operator

$$S_\phi^{\mathcal{W}} = J[\triangle_\phi^{\mathcal{W}}]^{\frac{1}{2}} \quad \Rightarrow \quad S_\phi^{\mathcal{W}} f_{ij} = \left[\frac{\alpha_i}{\alpha_j}\right]^{\frac{1}{2}} f_{ji}, \quad i, j = 1, 2, \ldots, N. \qquad (8.101)$$

8.4. Modular structures and statistical mechanics

Also, note that the set of vectors $\Phi_g = U_\ell^{\mathcal{W}}(g)\Phi$, $g \in G$, with $\Phi \in L^2_{\mathcal{W}}(G, d\mu)$, as in (8.95), is dense in $L^2_{\mathcal{W}}(G, d\mu)$. On these vectors, the action of $S_\phi^{\mathcal{W}}$ is readily computed to be

$$S_\phi^{\mathcal{W}} \Phi_g = \Phi_{g^{-1}} = U_\ell^{\mathcal{W}}(g)^* \Phi. \tag{8.102}$$

We are now in a position to summarize the properties of the algebra $\mathfrak{A}_\ell^{\mathcal{W}}$ (or, equivalently, $\mathfrak{A}_r^{\mathcal{W}}$), emanating from its modular structure, and display explicitly its attendant KMS properties. The proofs of all assertions made below become transparent as soon as the equivalent statements are considered (using the inverse isometry $\mathcal{W}^{-1}$) on $\mathcal{B}_2(\mathfrak{H})$, for the algebra generated by the $\mathbb{U}_\ell(g)$, $g \in G$ [see (8.69)].

1. The vector Φ is cyclic and separating for $\mathfrak{A}_\ell^{\mathcal{W}}$.

2. The action

$$S_\phi^{\mathcal{W}} A_\ell \Phi = A_\ell^* \Phi, \qquad A_\ell \in \mathfrak{A}_\ell^{\mathcal{W}}, \tag{8.103}$$

defines $S_\phi^{\mathcal{W}}$ as a closable antilinear map; the antiunitary map J and the operator $[\triangle_\phi^{\mathcal{W}}]^{\frac{1}{2}}$ are just the components in its polar decomposition.

3. The state $\widehat{\phi}$ on $\mathfrak{A}_\ell^{\mathcal{W}}$,

$$\langle \widehat{\phi}; A_\ell \rangle = \langle \Phi | A_\ell \Phi \rangle = \mathrm{Tr}[A\rho_\phi], \quad A_\ell = \mathcal{W}(A \vee I)\mathcal{W}^{-1} \in \mathfrak{A}_\ell^{\mathcal{W}}, \quad A \in \mathcal{L}(\mathfrak{H}), \tag{8.104}$$

is a faithful normal vector state. It is a *KMS state* for the one-parameter group of evolution $t \mapsto \alpha_\phi(t)$ on $\mathfrak{A}_\ell^{\mathcal{W}}$,

$$A_\ell \mapsto \alpha_\phi(t)[A_\ell] = [\triangle_\phi^{\mathcal{W}}]^{-it/\beta} A_\ell [\triangle_\phi^{\mathcal{W}}]^{it/\beta} = \exp[itH_\phi^\ell] A_\ell \exp[-itH_\phi^\ell], \tag{8.105}$$

at the "natural temperature" β. Thus, for arbitrary $A_\ell, B_\ell \in \mathfrak{A}_\ell^{\mathcal{W}}$, the function F_{A_ℓ, B_ℓ},

$$F_{A_\ell, B_\ell}(t) = \langle \widehat{\phi}; A_\ell \alpha_\phi(t)[B_\ell] \rangle = \langle \Phi | A_\ell \alpha_\phi(t)[B_\ell] \Phi \rangle, \tag{8.106}$$

is analytic in the strip $\{\Im z \in (0, \infty)\}$ and continuous on its boundaries, as can easily be verified using the relation

$$[\triangle_\phi^{\mathcal{W}}]^{-it/\beta} = \mathcal{W}(\exp[itH_\phi] \vee \exp[itH_\phi])\mathcal{W}^{-1} = \exp[itH_\phi^{\mathcal{W}}]. \tag{8.107}$$

The KMS condition

$$F_{A_\ell, B_\ell}(t + i\beta) = \langle \widehat{\phi}; \alpha_\phi(t)[B_\ell] A_\ell \rangle \tag{8.108}$$

is now trivial to verify.

4. The centralizer $\mathfrak{M}_\phi$ of $\mathfrak{A}_\ell^{\mathcal{W}}$ with respect to $\widehat{\phi}$ is the *abelian* von Neumann algebra generated by the projectors $\widehat{\mathbb{P}}_k$, $k = 1, 2, \ldots, N$. This algebra, being generated by minimal projectors, is actually *atomic*. Since

by (8.99), the $\widehat{\mathbb{P}}_k$ are also the spectral projectors of the self adjoint operator H_ϕ^ℓ, the algebra $\mathfrak{M}_\phi$ is generated by this single element, and, consequently,

$$\mathfrak{M}_\phi = \{A_\ell \in \mathfrak{A}_\ell^\mathcal{W} \mid \alpha_\phi(t)[A_\ell] = A_\ell\} = \{H_\phi^\ell\}''. \tag{8.109}$$

It is now clear that every faithful normal state on $\mathfrak{A}_\ell^\mathcal{W}$ is of the above type, i.e., arises from an orthonormal basis, ζ_i, $i = 1, 2, \ldots, N$, in $\mathfrak{H}$ and leads to a vector state. The centralizer $\mathfrak{M}_\phi$ carries information on the decomposition of the algebra into irreducibles. Indeed, the pure states on $\mathfrak{M}_\phi$ lead directly to the projectors $\mathbb{P}_k$ in (8.84). Finally, we ought to mention that it is not necessary to choose the vectors ζ_i from the domain of C^{-1} to get a faithful normal state $\widehat{\phi}$ and the attendant KMS structure. It is only when $\zeta_i \in \mathcal{D}(C^{-1})$, however, that a decomposition using the pure states of $\mathfrak{M}_\phi$ will consist of coherent sectors alone.

9
Some Examples and Generalizations

We have seen explicit examples of square integrable representations in the last chapter, related to the $SU(1,1)$ and the connected affine groups. Here, we work out a few more examples to both illustrate the general theory of square integrable representations better and get a deeper understanding of the nature of the Duflo–Moore operator C, appearing in the orthogonality relations. We then move on to deriving a generalization of the notion of square integrability to accomodate CS of the Gilmore–Perelomov type and vector CS.

9.1 A class of semidirect product groups

The particular type of square integrable representations that we examine here, following [71], arises from a class of semidirect product groups. The regular representation of every group in this class consists of square integrable representations only. This class of groups includes the affine group G_{aff}, on which the theory of wavelets is based. We have already looked at the connected component of this group in Section 8.1, and we shall take it up again, in detail, in Chapter 12, Section 12.2 [see also (4.25) and (4.26) in Section 4.1].

Consider $\mathbb{R}^n$ as an abelian group, and denote by $\widehat{\mathbb{R}}^n$ its dual group. We shall actually identify $\widehat{\mathbb{R}}^n$ with $\mathbb{R}^n$ and express duality through the usual scalar product, $\langle \mathbf{k} \, ; \, \mathbf{x} \rangle = \mathbf{k} \cdot \mathbf{x}, \ \mathbf{k} \in \widehat{\mathbb{R}}^n, \ \mathbf{x} \in \mathbb{R}^n$. The group $GL(n, \mathbb{R})$ of all real $n \times n$ nonsingular matrices has a natural action on $\mathbb{R}^n$: $(\mathbf{x}, M) \mapsto M\mathbf{x}$, where $\mathbf{x} \in \mathbb{R}^n$, $M \in GL(n,\mathbb{R})$, and $M\mathbf{x}$ is just the matrix M acting on the vector $\mathbf{x}$. The dual action on $\widehat{\mathbb{R}}^n$ is $(\mathbf{k}, M) \mapsto M^T\mathbf{k}$, $\mathbf{k} \in \widehat{\mathbb{R}}^n$, and M^T denoting, as before, the transpose of the matrix M. Let H be an n-dimensional closed subgroup of $GL(n,\mathbb{R})$, and assume that H has an *open free orbit* in $\widehat{\mathbb{R}}^n$. This means that there exists a $\mathbf{k}_0 \in \widehat{\mathbb{R}}^n$ such that its orbit under H,

$$\widehat{\mathcal{O}} = H^T\mathbf{k}_0 = \{M^T\mathbf{k}_0 \mid M \in H\}, \tag{9.1}$$

is an open set in $\widehat{\mathbb{R}}^n$ and, furthermore,

$$M^T\mathbf{k} = \mathbf{k} \quad \Rightarrow \quad M = \mathbb{I}_n, \tag{9.2}$$

for any $\mathbf{k} \in \widehat{\mathcal{O}}$, where $\mathbb{I}_n$ is the $n \times n$ identity matrix. In other words, the stabilizer of any $\mathbf{k} \in \widehat{\mathcal{O}}$ is trivial. Note that in (9.1) $\mathbf{k}_0$ can be chosen to be any point in the orbit. Also, for fixed $\mathbf{k} \in \widehat{\mathcal{O}}$, the map $M \mapsto M^T\mathbf{k}$, $M \in H$, is a *homeomorphism* between H and $\widehat{\mathcal{O}}$ [146].

Let $G = \mathbb{R}^n \rtimes H$ be the semidirect product group, with elements $g = (\mathbf{a}, M)$, $\mathbf{a} \in \mathbb{R}^n$, $M \in H$ and multiplication law:

$$(\mathbf{a}_1, M_1)(\mathbf{a}_2, M_2) = (\mathbf{a}_1 + M_1\mathbf{a}_2,\ M_1M_2).$$

It is this class of semidirect products with which we shall be working, i.e., semidirect products of $\mathbb{R}^n$ with closed n-dimensional subgroups of $GL(n,\mathbb{R})$ having open free orbits in $\widehat{\mathbb{R}}^n$. The affine group G_{aff}, mentioned above, is of this type, with $n = 1$ and $H = \mathbb{R}\backslash\{0\}$. Let $d\mu_G$ and $d\mu_H$ denote the left Haar measures on G and H, respectively. The Lebesgue measure $d\mathbf{a}$ on $\mathbb{R}^n$, when the latter is considered as an abelian group, is the invariant measure on it. If $\det M$ is the determinant of M, then it is not hard to see that

$$d\mu_G(\mathbf{a}, M) = \frac{1}{|\det M|}\, d\mathbf{a}\, d\mu_H(M). \tag{9.3}$$

Similarly, if Δ_G and Δ_H are the modular functions of G and H, respectively [see (4.3)–(4.5)], then

$$\Delta_G(\mathbf{a}, M) = \frac{\Delta_H(M)}{|\det M|}, \qquad (\mathbf{a}, M) \in \mathbb{R}^n \rtimes H. \tag{9.4}$$

Recall that, for any integrable function f on $\mathbb{R}^n$,

$$\int_{\mathbb{R}^n} f(M\mathbf{x})\, d\mathbf{x} = \frac{1}{|\det M|} \int_{\mathbb{R}^n} f(\mathbf{x})\, d\mathbf{x}$$

($d\mathbf{x}$ being the Lebesgue measure on $\mathbb{R}^n$).

Consider next the Hilbert space $\mathfrak{H} = L^2(\mathbb{R}^n, d\mathbf{x})$ and its Fourier-transformed space $\widetilde{\mathfrak{H}} = L^2(\widehat{\mathbb{R}}^n, d\mathbf{k})$, with the Fourier isometry $\mathcal{F} : \mathfrak{H} \to \widetilde{\mathfrak{H}}$ being given by

$$(\mathcal{F}f)(\mathbf{k}) = \widehat{f}(\mathbf{k}) = \frac{1}{(2\pi)^{\frac{n}{2}}} \int_{\mathbb{R}^n} \exp[-i\mathbf{k} \cdot \mathbf{x}] f(\mathbf{x})\, d\mathbf{x}.$$

9.1. A class of semidirect product groups

On $\mathfrak{H}$, define the unitary representation U of $G = \mathbb{R}^n \rtimes H$:

$$(U(\mathbf{a}, M)f)(\mathbf{x}) = \frac{1}{|\det M|^{\frac{1}{2}}} f(M^{-1}(\mathbf{x} - \mathbf{a})), \qquad (\mathbf{a}, M) \in G, \quad f \in \mathfrak{H}. \tag{9.5}$$

Since $\mathbb{R}^n \simeq G/H$, this representation resembles a regular representation, but is defined on a coset space, instead of on the group itself. Such a representation is also called a *quasi-regular representation*. The unitarily equivalent representation $\widehat{U}(\mathbf{a}, M) = \mathcal{F}U(\mathbf{a}, M)\mathcal{F}^{-1}$, on the Fourier-transformed space $\widehat{\mathfrak{H}}$, has the form:

$$(\widehat{U}(\mathbf{a}, M)\widehat{f})(\mathbf{k}) = |\det M|^{\frac{1}{2}} \exp[-i\mathbf{k} \cdot \mathbf{a}]\widehat{f}(M^T\mathbf{k}), \qquad \widehat{f} \in L^2(\widehat{\mathbb{R}}^n, d\mathbf{k}). \tag{9.6}$$

Referring to the discussion of induced representations in Chapter 4, Section 4.2, we realize that U is the representation of G induced from the trivial representation of the subgroup H. In general, this representation is reducible. Let $\widehat{\mathcal{O}} \subset \widehat{\mathbb{R}}^n$ be an open free orbit, and let $\widehat{\mathfrak{H}}_{\widehat{\mathcal{O}}} = L^2(\widehat{\mathcal{O}}, d\mathbf{k})$ be the Hilbert subspace of $L^2(\widehat{\mathbb{R}}^n, d\mathbf{k})$ consisting of elements with supports contained in $\widehat{\mathcal{O}}$. Denote by $\mathfrak{H}_{\widehat{\mathcal{O}}}$ the subspace of $L^2(\mathbb{R}^n, d\mathbf{x})$, which is the inverse Fourier transform of $\widehat{\mathfrak{H}}_{\widehat{\mathcal{O}}}$; i.e., if $f \in \mathfrak{H}_{\widehat{\mathcal{O}}}$, then its Fourier transform $\widehat{f} \in \widehat{\mathfrak{H}}_{\widehat{\mathcal{O}}}$. From (9.6), it is then clear that $\widehat{\mathfrak{H}}_{\widehat{\mathcal{O}}}$ is an invariant subspace for the representation $\widehat{U}$. In fact, we shall see in Theorem 9.1.2 that $\widehat{U}$ restricted to this subspace (or, equivalently, U restricted to $\mathfrak{H}_{\widehat{\mathcal{O}}}$) is irreducible.

Using the homeomorphism $M \mapsto M^T\mathbf{k}$ (for some fixed $\mathbf{k}$), let us transfer the measure μ_H from H to $\widehat{\mathcal{O}}$, and denote the resulting measure by ν. Then, for any Borel set $E \subset H$,

$$\mu_H(E) = \nu(E'), \quad \text{where} \quad E' = E^T\mathbf{k} = \{M^T\mathbf{k} \in \widehat{\mathcal{O}} \mid M \in E\}. \tag{9.7}$$

A simple argument using the left invariance of μ_H then shows that ν is independent of the point $\mathbf{k}$ chosen in the definition of the homeomorphism $M \mapsto M^T\mathbf{k}$. Furthermore, the Lebesgue measure $d\mathbf{k}$ of $\widehat{\mathbb{R}}^n$, when restricted to $\widehat{\mathcal{O}}$, can be shown to be equivalent to $d\nu$. Thus, there exists a positive, Lebesgue measurable function $\mathcal{C}$ on $\widehat{\mathcal{O}}$ such that

$$d\nu(\mathbf{k}) = \mathcal{C}(\mathbf{k}) \, d\mathbf{k}, \qquad 0 < \mathcal{C}(\mathbf{k}) < \infty, \tag{9.8}$$

the above relations holding for almost all $\mathbf{k} \in \widehat{\mathcal{O}}$.

Lemma 9.1.1: *The function $\mathcal{C} : \widehat{\mathcal{O}} \to \mathbb{R}^+$ satisfies the following relations:*

1. *for any $\mathbf{k} \in \widehat{\mathcal{O}}$ and integrable function $\widehat{f} : \widehat{\mathcal{O}} \to \mathbb{C}$,*

$$\int_H \widehat{f}(M^T\mathbf{k}) \, d\mu_H(M) = \int_{\widehat{\mathcal{O}}} \widehat{f}(\mathbf{k}') \, d\nu(\mathbf{k}') = \int_{\widehat{\mathcal{O}}} \widehat{f}(\mathbf{k}')\mathcal{C}(\mathbf{k}') \, d\mathbf{k}'; \tag{9.9}$$

2. *for any $M \in H$ and (ν-almost) all $\mathbf{k} \in \widehat{\mathcal{O}}$,*

$$\mathcal{C}(M^T\mathbf{k}) = \frac{\Delta_H(M)}{|\det M|} \mathcal{C}(\mathbf{k}). \tag{9.10}$$

Proof. The two equalities in (9.9) follow from the definition of ν in (9.7) and the relation between $d\nu$ and $d\mathbf{k}$ in (9.8). To prove (9.10), it is enough to note that, for any integrable $\widehat{f}$,

$$\int_{\widehat{\mathcal{O}}} \widehat{f}(\mathbf{k})\mathcal{C}(M^T\mathbf{k})\, d\mathbf{k} = \frac{1}{|\det M|} \int_{\widehat{\mathcal{O}}} \widehat{f}((M^T)^{-1}\mathbf{k})\mathcal{C}(\mathbf{k})\, d\mathbf{k}$$
$$= \frac{1}{|\det M|} \int_{\widehat{\mathcal{O}}} \widehat{f}((M^T)^{-1}\mathbf{k})\, d\nu(\mathbf{k}).$$

Whence, writing $\mathbf{k} = N^T \mathbf{k}_0$, for some fixed $\mathbf{k}_0 \in \widehat{\mathcal{O}}$, $N \in H$, and using (9.9), we arrive at

$$\int_{\widehat{\mathcal{O}}} \widehat{f}(\mathbf{k})\mathcal{C}(M^T\mathbf{k})\, d\mathbf{k} = \frac{1}{|\det M|} \int_{\widehat{\mathcal{O}}} \widehat{f}((NM^{-1})^T \mathbf{k}_0)\, d\mu_H(N)$$
$$= \frac{\Delta(M)}{|\det M|} \int_{\widehat{\mathcal{O}}} \widehat{f}(N^T \mathbf{k}_0)\, d\mu_H(N),$$
$$= \frac{\Delta(M)}{|\det M|} \int_{\widehat{\mathcal{O}}} \widehat{f}(\mathbf{k})\, d\nu(\mathbf{k}),$$

where the second equality results from using the property (4.5) of the modular function. $\square$

Combining (9.4) and (9.10), we see that $\mathcal{C}$ is essentially the image of the modular function on $\widehat{\mathcal{O}}$, and, furthermore, if G is unimodular, then $\mathcal{C}(\mathbf{k})$ is a constant function. In general, however, $\mathcal{C}$ could be an unbounded function. Define an operator C on $\mathfrak{H}_{\widehat{\mathcal{O}}} = L^2(\widehat{\mathcal{O}}, d\mathbf{k})$ as

$$(C\widehat{f})(\mathbf{k}) = (2\pi)^{\frac{n}{2}} \mathcal{C}(\mathbf{k})^{\frac{1}{2}} \widehat{f}(\mathbf{k}), \qquad (9.11)$$

and let $\mathcal{A}$ denote its domain; i.e., $\widehat{f} \in \mathcal{A}$ if and only if

$$\int_{\widehat{\mathcal{O}}} \mathcal{C}(\mathbf{k})|\widehat{f}(\mathbf{k})|^2\, d\mathbf{k} < \infty. \qquad (9.12)$$

Clearly, $\mathcal{A}$ is dense in $\mathfrak{H}_{\widehat{\mathcal{O}}}$. It will turn out that C is precisely the Duflo-Moore operator appearing in the orthogonality relations (8.36).

Theorem 9.1.2: *The representation $\widehat{U}$ of the group $G = \mathbb{R}^n \rtimes H$, restricted to the Hilbert space $\mathfrak{H}_{\widehat{\mathcal{O}}} = L^2(\widehat{\mathcal{O}}, d\mathbf{k})$, is irreducible and square integrable and every vector $\widehat{f} \in \mathcal{A}$ is admissible.*

Proof. Let $\widehat{g}_1, \widehat{g}_2 \in \mathcal{A}$ and $\widehat{f}_1, \widehat{f}_2 \in \mathfrak{H}_{\widehat{\mathcal{O}}}$, and assume these functions to be sufficiently smooth, so that integrations with respect to δ-measures and interchanges under integrals may be performed. Then, writing

$$I = \int_G \overline{\langle \widehat{U}(\mathbf{a}, M)\widehat{g}_1 | \widehat{f}_1 \rangle} \langle \widehat{U}(\mathbf{a}, M)\widehat{g}_2 | \widehat{f}_2 \rangle\, d\mu_G(\mathbf{a}, M),$$

we find that

$$I = \int_G \left[\int_{\widehat{\mathcal{O}}} \exp\left[-i\mathbf{k}\cdot\mathbf{a}\right] \widehat{g}_1(M^T\mathbf{k}) \overline{\widehat{f}_1(\mathbf{k})} \, d\mathbf{k} \right.$$
$$\left. \times \left[\int_{\widehat{\mathcal{O}}} \exp\left[i\mathbf{k}'\cdot\mathbf{a}\right] \overline{\widehat{g}_2(M^T\mathbf{k}')} \widehat{f}_2(\mathbf{k}') \, d\mathbf{k}' \right] d\mathbf{a} \, d\mu_H(M) \right.$$
$$= \int_H \left[\int_{\widehat{\mathcal{O}}} \int_{\widehat{\mathcal{O}}} \int_{\mathbb{R}^n} \exp\left[-i(\mathbf{k}-\mathbf{k}')\cdot\mathbf{a}\right] d\mathbf{a} \, \overline{\widehat{f}_1(\mathbf{k})} \widehat{f}_2(\mathbf{k}') \right.$$
$$\left. \times \widehat{g}_1(M^T\mathbf{k}) \overline{\widehat{g}_2(M^T\mathbf{k}')} \, d\mathbf{k} \, d\mathbf{k}' \right] d\mu_H(M).$$

Next, using the relation,

$$\frac{1}{(2\pi)^n} \int_{\mathbb{R}^n} \exp\left[-i(\mathbf{k}-\mathbf{k}')\cdot\mathbf{a}\right] d\mathbf{a} = \delta(\mathbf{k}-\mathbf{k}'),$$

which holds in the sense of distributions (i.e., when integrated with respect to smooth functions), and performing the integration in $\mathbf{k}'$, we obtain

$$I = (2\pi)^n \int_{\widehat{\mathcal{O}}} \overline{\widehat{f}_1(\mathbf{k})} \left[\int_H \widehat{g}_1(M^T\mathbf{k}) \overline{\widehat{g}_2(M^T\mathbf{k})} \, d\mu_H(M) \right] \widehat{f}_2(\mathbf{k}) \, d\mathbf{k}.$$

Using (9.9) and (9.11), we get, finally,

$$\int_G \overline{\langle \widehat{U}(\mathbf{a},M)\widehat{g}_1|\widehat{f}_1\rangle} \langle \widehat{U}(\mathbf{a},M)\widehat{g}_2|\widehat{f}_2\rangle \, d\mu_G(\mathbf{a},M) = \langle \widehat{f}_1|\widehat{f}_2\rangle \langle C\widehat{g}_2|C\widehat{g}_1\rangle. \quad (9.13)$$

A simple continuity argument now enables us to extend this relation to all $\widehat{g}_1, \widehat{g}_2 \in \mathcal{A}$ and $\widehat{f}_1, \widehat{f}_2 \in \widehat{\mathfrak{H}}_{\widehat{\mathcal{O}}}$. We have recovered in this way, in the present context, exactly the orthogonality relations appearing in (8.36) for a square integrable representation. Taking $\widehat{g}_1 = \widehat{g}_2 = \widehat{g}$, we obtain the admissibility condition (8.4) for $\widehat{g}$. To prove square integrability of the representation $\widehat{U}$, when restricted to $\widehat{\mathfrak{H}}_{\widehat{\mathcal{O}}}$, it only remains to establish its irreducibility, and then $\mathcal{A}$ would be the set of admissible vectors. Let $\widehat{g} \in \mathcal{A}$ be arbitrary but nonzero, and suppose that $\widehat{f} \in \widehat{\mathfrak{H}}_{\widehat{\mathcal{O}}}$ is a vector orthogonal to $\widehat{U}(\mathbf{a},M)\widehat{g}$ for every $(\mathbf{a},M) \in G$. Then, $\langle \widehat{U}(\mathbf{a},M)\widehat{g}|\widehat{f}\rangle = 0$, for all $(\mathbf{a},M) \in G$. Hence, by (9.13), $\|\widehat{f}\|^2 \|C\widehat{g}\|^2 = 0$, so that $\widehat{f} = 0$. Since $\mathcal{A}$ is dense in $\widehat{\mathfrak{H}}_{\widehat{\mathcal{O}}}$, this implies that every vector in $\widehat{\mathfrak{H}}_{\widehat{\mathcal{O}}}$ is cyclic. Thus, the representation $\widehat{U}$ restricted to $\widehat{\mathfrak{H}}_{\widehat{\mathcal{O}}}$ is irreducible and hence square integrable. □

In order to facilitate the comparison of this result with the general formalism developed in Chapter 7, we may point out that the measure ν on $\widehat{\mathcal{O}}$ is invariant under H, whereas the Lebesgue measure is only quasi-invariant. Hence, we get a unitarily equivalent picture if we use as representation space $L^2(\widehat{\mathcal{O}}, d\nu)$ and the corresponding representation $U_{\widehat{\mathcal{O}}}$, which is defined exactly as in (9.5) and (9.6), but *without* the factor $|\det M|^{\pm\frac{1}{2}}$. As stated in (9.8), this factor is the Radon–Nikodym derivative and equals $\mathcal{C}(\mathbf{k})$, upon using the homeomorphism $M \mapsto M^T\mathbf{k}_o$, where M is the element of H that maps the fixed base point $\mathbf{k}_o$ to $\mathbf{k}$.

176   9. Some Examples and Generalizations

Given a group of the type $\mathbb{R}^n \rtimes H$, with H an n-dimensional subgroup of $GL(n, \mathbb{R})$, it is not always the case that open free orbits of H exist in $\widehat{\mathbb{R}}^n$. If one such orbit does exist, however, then $\widehat{\mathbb{R}}^n$ is in fact a disjoint union of such orbits. Indeed, the left regular representation is in this case a direct sum of irreducible subrepresentations, i.e., of square integrable representations [60]. In addition, the restriction to free orbits may be dropped, as we shall see in Section 9.1.2.

9.1.1 Three concrete examples (♣)

1. The similitude group in two dimensions

As our first example, we take $G = \mathbb{R}^2 \rtimes H$, where $H = \mathbb{R}_*^+ \times SO(2)$ consists of dilations and rotations of the plane. The group G is called the *similitude* group of $\mathbb{R}^2$ and denoted $SIM(2)$. This is the group that underlies the 2-D wavelet transform we shall study in Chapter 14.

For $a > 0$, $\theta \in [0, 2\pi)$, the element $M = (\lambda, \theta) \in SIM(2)$ acts on $\mathbb{R}^2$ as:

$$\mathbf{x} = \begin{pmatrix} x_1 \\ x_2 \end{pmatrix} \mapsto \lambda r(\theta)\mathbf{x} \equiv \lambda \begin{pmatrix} \cos\theta & -\sin\theta \\ \sin\theta & \cos\theta \end{pmatrix} \begin{pmatrix} x_1 \\ x_2 \end{pmatrix}. \quad (9.14)$$

The dual action on $\widehat{\mathbb{R}}^2$ is $\mathbf{k} \mapsto \lambda r(-\theta)\mathbf{k}$. Hence, there are two orbits, namely, $\widehat{\mathcal{O}}_o = \{0\}$ and $\widehat{\mathcal{O}}_1 = \widehat{\mathbb{R}}^2 \setminus \{0\}$. The latter is open and free, as follows from inspection.

Therefore, the corresponding unitary representation

$$(U(\mathbf{a}, M(\lambda, \theta))f)(\mathbf{x}) = \frac{1}{\lambda} f(\lambda^{-1} r(-\theta)(\mathbf{x} - \mathbf{a})), \quad f \in L^2(\mathbb{R}^2, d\mathbf{x}), \quad (9.15)$$

or, equivalently,

$$(\widehat{U}(\mathbf{a}, M(\lambda, \theta))\widehat{f})(\mathbf{k}) = \lambda e^{-i\mathbf{k}\cdot\mathbf{a}} \widehat{f}(\lambda r(-\theta)\mathbf{k}), \quad f \in L^2(\widehat{\mathbb{R}}^2, d\mathbf{k}), \quad (9.16)$$

is irreducible and square integrable. The group $H = \mathbb{R}_*^+ \times SO(2)$ is unimodular (being abelian), and its Haar measure is $d\mu_H(\lambda, \theta) = \lambda^{-1} d\lambda\, d\theta$. The homeomorphism $M \mapsto M^T \mathbf{k}_o$ between H and the open orbit $\widehat{\mathcal{O}}_1$ may be expressed by choosing the fixed vector $\mathbf{k}_o = (1, 0) \in \mathbb{R}^2$ and mapping $M = (\lambda, \theta)$ onto $\mathbf{k} = M^T \mathbf{k}_o \neq 0$, where $\lambda = |\mathbf{k}|$, $\theta = -\arg \mathbf{k}$. Then, the measure on $\widehat{\mathcal{O}}_1$, transferred from $d\mu_H$, is

$$d\nu(\mathbf{k}) = \frac{d\lambda}{\lambda} d\theta = \frac{d|\mathbf{k}|}{|\mathbf{k}|} d\theta = \frac{1}{|\mathbf{k}|^2} d\mathbf{k}.$$

Thus, the Duflo–Moore operator is

$$(C\widehat{f})(\mathbf{k}) = 2\pi |\mathbf{k}|^{-1} \widehat{f}(\mathbf{k}). \quad (9.17)$$

2. The affine Poincaré group in 1+1 dimensions

Let $G = \mathbb{R}^2 \rtimes H$, where $H = \mathbb{R}_*^+ \times \mathcal{L}_+^\uparrow(1,1)$ is the Lorentz group in one-time and one-space dimensions together with dilations. Elements in $\mathbb{R}_*^+ \times \mathcal{L}_+^\uparrow(1,1)$ are 2×2 real matrices M of the form

$$M = M(\lambda, \theta) = \lambda \begin{pmatrix} \cosh\theta & \sinh\theta \\ \sinh\theta & \cosh\theta \end{pmatrix}, \quad M = M^T, \lambda > 0, \ -\infty < \theta < \infty. \tag{9.18}$$

Thus G is the *affine Poincaré group* in one-space and one-time dimensions, i.e., the Poincaré group along with dilations. This group may also be called $SIM(1,1)$, since it is the similitude group of the $(1+1)$-dimensional Minkowski space. Although this group has square integrable representations, the Poincaré group itself, $\mathcal{P}_+^\uparrow(1,1) = \mathbb{R}^2 \rtimes \mathcal{L}_+^\uparrow(1,1)$ (i.e., without dilations), has no representations in the discrete series. (We shall return to this point later in Chapter 11, Section 11.1, and demonstrate that $\mathcal{P}_+^\uparrow(1,1)$ has nevertheless a representation square integrable mod(T, σ), for a certain class of sections σ, T being the time-translation subgroup.)

Following the physicists' convention, we denote vectors in $\mathbb{R}^2$ and $\widehat{\mathbb{R}}^2$ in this case by

$$x = \begin{pmatrix} x_0 \\ \mathbf{x} \end{pmatrix} \in \mathbb{R}^2, \quad k = \begin{pmatrix} k_0 \\ \mathbf{k} \end{pmatrix} \in \widehat{\mathbb{R}}^2, \tag{9.19}$$

and use the Minkowski inner product

$$x \cdot k = x_0 k_0 - \mathbf{x}\mathbf{k} \tag{9.20}$$

to define the dual pairing $\langle \mathbf{k} \, ; \, \mathbf{x} \rangle$. Also, instead of the usual Fourier transform, we shall use, as is customary in a relativistic theory, the *relativistic Fourier transform*:

$$\widetilde{f}(k) = \frac{1}{2\pi} \int_{\mathbb{R}^2} e^{ik\cdot x} f(x) \, dx, \quad f \in L^2(\mathbb{R}^2, dx),$$

$$f(x) = \frac{1}{2\pi} \int_{\widetilde{\mathbb{R}}^2} e^{-ik\cdot x} \widetilde{f}(k) \, dk \quad \widetilde{f} \in L^2(\widehat{\mathbb{R}}^2, dk), \tag{9.21}$$

(where, of course, $dx = dx_0 \, d\mathbf{x}$ and $dk = dk_0 \, d\mathbf{k}$). The representation (9.5) now assumes the form [note that $M(\lambda, \theta)^{-1} = M(\lambda^{-1}, -\theta)$]:

$$(U(a, M(\lambda, \theta))f)(x) = \frac{1}{\lambda} f(M(\lambda^{-1}, -\theta)(x-a)),$$
$$a = (a_0, \mathbf{a}) \in \mathbb{R}^2, \quad f \in L^2(\mathbb{R}^2, dx). \tag{9.22}$$

Using the identity,

$$M(\lambda^{-1}, \theta) \begin{pmatrix} 1 & 0 \\ 0 & -1 \end{pmatrix} M(\lambda, \theta) = \begin{pmatrix} 1 & 0 \\ 0 & -1 \end{pmatrix}, \tag{9.23}$$

the relativistic Fourier-transformed representation is seen to be

$$(\widetilde{U}(a, M(\lambda, \theta))\widetilde{f})(k) = \lambda \, e^{ik\cdot a} \widetilde{f}(M(\lambda, -\theta)k), \quad \widetilde{f} \in L^2(\widehat{\mathbb{R}}^2, dk), \tag{9.24}$$

which of course is unitarily equivalent to (9.22). The dual action of an element $M(\lambda, \theta) \in \mathbb{R}_*^+ \times \mathcal{L}_+^\uparrow(1,1)$ on $\widehat{\mathbb{R}}^2$ is found using $k \cdot (M(\lambda, \theta)x) = (M(\lambda, -\theta)k) \cdot x$, from which orbits under this action can be identified. Indeed, for the four vectors,

$$k_+ = \begin{pmatrix} 1 \\ 0 \end{pmatrix}, \quad k_- = \begin{pmatrix} -1 \\ 0 \end{pmatrix}, \quad k_r = \begin{pmatrix} 0 \\ 1 \end{pmatrix}, \quad k_\ell = \begin{pmatrix} 0 \\ -1 \end{pmatrix}, \quad (9.25)$$

we obtain the four orbits

$$\widehat{\mathcal{O}}_j = \{M(\lambda, -\theta)k_j \,|\, j = +, -, r, \ell;\ \lambda > 0,\ -\infty < \theta < \infty\}. \quad (9.26)$$

The orbit $\widehat{\mathcal{O}}_+$, for example, consists of all vectors $k \in \widehat{\mathbb{R}}^2$,

$$k = \begin{pmatrix} k_0 \\ \mathbf{k} \end{pmatrix} = \lambda \begin{pmatrix} \cosh\theta \\ -\sinh\theta \end{pmatrix}, \quad \lambda > 0,\ -\infty < \theta < \infty,$$

and is therefore the open cone in the $(k_0\ \mathbf{k})$-plane, lying above the lines $k_0 = \pm \mathbf{k}$, $k_0 > 0$. Similarly, $\widehat{\mathcal{O}}_-$ is the open cone lying below the lines $k_0 = \pm \mathbf{k}$, $k_0 < 0$, while $\widehat{\mathcal{O}}_r$ and $\widehat{\mathcal{O}}_\ell$ are the open right and left cones, i.e., lying to the right and left of the lines $\mathbf{k} = \pm k_0$, $\mathbf{k} > 0$ and $\mathbf{k} = \pm k_0$, $\mathbf{k} < 0$, respectively. Each one of these cones is an open free orbit, and their union is dense in $\widehat{\mathbb{R}}^2$:

$$\widehat{\mathcal{O}}_+ \cup \widehat{\mathcal{O}}_- \cup \widehat{\mathcal{O}}_r \cup \widehat{\mathcal{O}}_\ell = \widehat{\mathbb{R}}^2 \setminus \{(k_0, \mathbf{k}) \,|\, k_0 = \pm \mathbf{k}\}.$$

Besides these orbits, there are also the two null cones and the origin, corresponding to

$$k = \pm \begin{pmatrix} 1 \\ 1 \end{pmatrix}, \quad k = \pm \begin{pmatrix} 1 \\ -1 \end{pmatrix}, \quad k = \begin{pmatrix} 0 \\ 0 \end{pmatrix},$$

respectively, which are not open and free. Thus, $\widehat{\mathfrak{H}} = L^2(\widehat{\mathbb{R}}^2, dk)$ decomposes as the orthogonal direct sum

$$\widehat{\mathfrak{H}} = L^2(\widehat{\mathcal{O}}_+, dk) \oplus L^2(\widehat{\mathcal{O}}_-, dk) \oplus L^2(\widehat{\mathcal{O}}_r, dk) \oplus L^2(\widehat{\mathcal{O}}_\ell, dk),$$

and restricted to each one of these component Hilbert spaces, the representation $\widehat{U}(a, M)$ is irreducible and the four subrepresentations are mutually inequivalent. It is not hard to calculate the invariant measure on these orbits. Indeed, on any one of them,

$$d\nu(k) = \frac{dk_0\, d\mathbf{k}}{|k_0^2 - \mathbf{k}^2|}, \quad (9.27)$$

and consequently one gets, for the Duflo–Moore operator C in (9.11),

$$(C\widehat{f})(k) = 2\pi |k_0^2 - \mathbf{k}^2|^{-\frac{1}{2}} \widehat{f}(k), \quad (9.28)$$

on a dense set of vectors in the appropriate Hilbert space. (The square integrability of the affine Poincaré group in higher space–time dimensions has been studied in [77]. We shall look at these higher dimensional analogues in Chapter 14.)

As a final remark, notice that the square integrable representations considered here are the restrictions of certain square integrable representations of the (semisimple) conformal group $SO(2,2)$, of which the affine Poincaré group $\mathcal{P}_+^\uparrow(1,1)$ is a subgroup.

3. Dilations and translations in $\mathbb{R}^2$

For our third example, we take the group $G = \mathbb{R}^2 \rtimes H$, with H consisting of 2×2 matrices of the type

$$M = \begin{pmatrix} a_1 & 0 \\ 0 & a_2 \end{pmatrix}, \qquad a_1 a_2 \neq 0. \tag{9.29}$$

Then, $G = G_{\text{aff}} \times G_{\text{aff}}$, where G_{aff} is the full (disconnected) affine group [see Section 12.2 and also (4.25) and (4.26)]. For this group, there is only one open free orbit in $\widehat{\mathbb{R}}^2$:

$$\widehat{\mathcal{O}} = \{\mathbf{k} = M \begin{pmatrix} 1 \\ 1 \end{pmatrix} \mid M \in H\}. \tag{9.30}$$

Clearly, $\widehat{\mathcal{O}} = \widehat{\mathbb{R}}^2 \setminus \{\mathbf{0}\}$, which is dense in $\widehat{\mathbb{R}}^2$, and the invariant measure on it is easily computed to be

$$d\nu(\mathbf{k}) = \frac{d\mathbf{k}}{|k_1 k_2|}, \qquad \mathbf{k} = \begin{pmatrix} k_1 \\ k_2 \end{pmatrix}. \tag{9.31}$$

The representation

$$(\widehat{U}(\mathbf{a}, M)\widehat{f})(\mathbf{k}) = \exp[-i\mathbf{k} \cdot \mathbf{a}]\widehat{f}(M^{-1}\mathbf{k}), \tag{9.32}$$

of G on $L^2(\widehat{\mathcal{O}}, d\mathbf{k})$ is now irreducible and square integrable, and the operator C assumes the form

$$(C\widehat{f})(\mathbf{k}) = 2\pi |k_1 k_2|^{-\frac{1}{2}} \widehat{f}(\mathbf{k}), \tag{9.33}$$

for a dense set of vectors in $L^2(\widehat{\mathcal{O}}, d\mathbf{k})$.

9.1.2 A broader setting

The first example in the preceding section extends in a straightforward way to higher dimensions, that is, to the similitude group of $\mathbb{R}^n$,

$$SIM(n) = \mathbb{R}^n \rtimes (\mathbb{R}_*^+ \times SO(n)).$$

For $n > 2$, however, the theory outlined above no longer applies. Indeed, here too, there are only two orbits, namely $\widehat{\mathcal{O}}_o = \{\mathbf{0}\}$ and $\widehat{\mathcal{O}}_1 = \widehat{\mathbb{R}}^n \setminus \{\mathbf{0}\}$, and the unitary representation associated to the latter is irreducible. The orbit $\widehat{\mathcal{O}}_1$ is open in $\widehat{\mathbb{R}}^n$, but it is *not* free. For instance, the point $\mathbf{k}_o = (0, 0, \ldots, 1)$ is stabilized by $SO(n-1)$ and so does any other point of the orbit. Nevertheless, this representation is still square integrable, as a consequence of a more general result, proved in [127], which we now sketch.

With the same notation as in the preceding section, the following theorem holds.

Theorem 9.1.3: *Let $\widehat{U}$ be the unitary (quasi-regular) representation of $G = \mathbb{R}^n \rtimes H$ in $L^2(\widehat{\mathbb{R}}^n, d\mathbf{k})$. Let $\widehat{\mathcal{O}}$ be any measurable subset of $\widehat{\mathbb{R}}^n$, invariant under the action of H, and let $\widehat{U}_{\widehat{\mathcal{O}}}$ be the restriction of $\widehat{U}$ to $\mathfrak{H}_{\widehat{\mathcal{O}}} = L^2(\widehat{\mathcal{O}}, d\mathbf{k})$. Then:*
1. *$\widehat{U}_{\widehat{\mathcal{O}}}$ is irreducible iff the action of H on $\widehat{\mathcal{O}}$ is ergodic;*
2. *$\widehat{U}_{\widehat{\mathcal{O}}}$ is square integrable iff the stabilizer of any point of $\widehat{\mathcal{O}}$ is compact.*

Here, ergodicity of the action of H means that the closure of any H-invariant subset of $\widehat{\mathcal{O}}$ either is of measure zero or coincides with $\widehat{\mathcal{O}}$. In other words, $\widehat{\mathcal{O}}$ consists of the closure of a unique dense orbit under H, plus possibly some sets of measure zero (closures of orbits of smaller dimension). Note that this condition essentially rules out discrete sets of dilations. The proof of this theorem is almost the same as that of Theorem 9.1.2.

The Duflo–Moore operator (9.11), in this general situation, is again the operator of multiplication by the function $C(\mathbf{k})^{-1/2}$, where $C(\mathbf{k})$ is a Radon–Nikodym derivative, just as in (9.8). Identifying as before a point $\mathbf{k} \in \widehat{\mathcal{O}}_1$ with an element $h = h(\mathbf{k}) \in H$ by $\mathbf{k} = h(\mathbf{k})\mathbf{k}_o$, for some fixed base point $\mathbf{k}_o \in \widehat{\mathcal{O}}_1$, one has

$$C(\mathbf{k}) = \mathbf{\Delta}_H(h)\,|\det h|^{-1}, \qquad \mathbf{k} = h(\mathbf{k})\mathbf{k}_o.$$

The above theorem is tailormade to fit the *SIM(n)* group. Here, $H = \mathbb{R}_*^+ \times SO(n)$ is unimodular, and, for $\mathbf{k} \in \widehat{\mathcal{O}}_1$, one may choose $\mathbf{k}_o = (0, 0, \ldots, 1)$, so that $h(\mathbf{k}) = (|\mathbf{k}|, R(\mathbf{k}_o \to \mathbf{k}))$, where $R(\mathbf{k}_o \to \mathbf{k})$ denotes some rotation that maps $\mathbf{k}_o$ to $\mathbf{k}$. Thus, $\det h(\mathbf{k}) = |\mathbf{k}|^n$ and the Duflo-Moore operator is the operator of multiplication by $(2\pi)^{n/2}|\mathbf{k}|^{n/2}$.

An identical analysis can be carried out for the positive mass representations of the affine Poincaré group in 1+3 dimensions. The corresponding orbit is the interior of the future lightcone,

$$\widehat{\mathcal{O}}_+ = \{k \mid k_0 > 0,\ k_0^2 - \mathbf{k}^2 > 0\},$$

(generalizing the orbit $\widehat{\mathcal{O}}_+$ of the 1+1 dimensional case discussed in the previous section), and the stabilizer of any of its points is isomorphic to $SO(3)$. The same is true for the past lightcone $\widehat{\mathcal{O}}_-$, but *not* for the spacelike region $k^2 < 0$. For instance, the point $k^{(o)} = (0, \ldots, 0, 1)$, and, thus, every point of that orbit has a stabilizer isomorphic to $SO_o(1, 2)$. We shall come back to these higher dimensional examples in Chapter 15, Section 15.3.

Actually, both Theorem 9.1.2 and Theorem 9.1.3 follow from a more general result [197] (see also [23]) that completely settles the question of the square integrability of all induced representations of semidirect products of the type $\mathbb{R}^n \rtimes H$, where H is a locally compact topological group, acting continuously on $\mathbb{R}^n$. This result is stated in Theorem 10.3.1 in Chapter 10, where the induced representations of such groups are discussed in detail (see Section 10.2.4). Simply stated, according to this theorem, the quasi-regular representation $\widehat{U}$ is square integrable if and only if the orbit $\widehat{\mathcal{O}}$ has positive Lebesgue measure in $\widehat{\mathbb{R}^n}$.

9.2 A generalization: α- and V-admissibility

It is possible to generalize the notion of square integrability of a group representation, as envisaged in Section 8.1, to include CS of the Gilmore–Perelomov type. As a first example of such a generalization, we sketch a number of results, following [166], which extend Theorems 8.1.3 and 8.2.1 in an interesting way. Later, we go further and discuss the square integrability of vector CS as well in this context.

Let $H \subset G$ be a closed subgroup, and suppose that $\alpha : H \to \mathbb{C}$ is a unitary character of H; i.e., for all $h, h_1, h_2 \in H$,

$$\alpha(h) \in \mathbb{C}, \quad |\alpha(h)| = 1 \quad \text{and} \quad \alpha(h_1)\alpha(h_2) = \alpha(h_1 h_2). \tag{9.34}$$

Suppose that we are given a UIR, $g \mapsto U(g)$ of G on $\mathfrak{H}$, and assume that there are nonzero vectors $\eta \in \mathfrak{H}$, such that [see (7.9)]

$$U(h)\eta = \alpha(h)\eta, \quad h \in H. \tag{9.35}$$

Assume also that $X = G/H$ carries an *invariant measure* ν, and let $\sigma : X \to G$ be a Borel section.

Definition 9.2.1: *A nonzero vector $\eta \in \mathfrak{H}$ is said to be α-admissible for the UIR $g \mapsto U(g)$ if it satisfies (9.35) and*

$$I_\alpha(\eta) = \int_X |\langle U(\sigma(x))\eta | \eta \rangle|^2 \, d\nu(x) < \infty. \tag{9.36}$$

Obviously, this definition of α-admissibility is independent of the choice of the particular section σ. Denote by $\mathcal{A}_\alpha$ the set of all α-admissible vectors in $\mathfrak{H}$ and by $\mathfrak{H}^\alpha$ the closure of this set, which clearly is a subspace of $\mathfrak{H}$. If α_1 and α_2 are two different unitary characters of H and the vectors $\eta_1, \eta_2 \in \mathfrak{H}$ are, respectively, α_1- and α_2-admissible, then the fact that η_1 and η_2 are eigenvectors of the unitary operator $U(h)$ corresponding to different eigenvalues $\alpha_1(h)$ and $\alpha_2(h)$ implies that η_1 and η_2 are mutually orthogonal. From this, it follows that $\mathfrak{H}^{\alpha_1}$ and $\mathfrak{H}^{\alpha_2}$ are mutually orthogonal subspaces of $\mathfrak{H}$.

Next, let $h : G \times X \to H$ be the cocycle $h(g, x) = \sigma(gx)^{-1} g \sigma(x)$ [see (4.37)]. Then, since α defines a one-dimensional unitary representation of

H, the operators ${}^\alpha U(g)$, $g \in G$, acting on the Hilbert space $L^2(X, d\nu)$,

$$({}^\alpha U(g)\Phi)(x) = \alpha(h(g^{-1}, x))^{-1} \, \Phi(g^{-1}x), \qquad \Phi \in L^2(X, d\nu), \qquad (9.37)$$

realize the representation of G, which is induced from α, as is immediately verified by comparing with (4.39)–(4.41). If it can now be proved that (with an appropriate normalization N) the operators

$$a(\Delta) = \frac{1}{N} \int_\Delta |\eta_{\sigma(x)}\rangle\langle\eta_{\sigma(x)}| \, d\nu(x), \qquad \eta_{\sigma(x)} = U(\sigma(x))\eta,$$

where Δ runs through all Borel sets of X, and η is an α-admissible vector, define a normalized POV measure, then from our discussion of vector CS in Sections 4.2.1, 7.1.2, and 7.1.3 it would follow that the representation U is equivalent to a subrepresentation of the induced representation ${}^\alpha U$. This indeed is the case, and following the lines of the proof of Theorem 8.1.3, one arrives at the following generalization of it:

Theorem 9.2.2: *Let $g \mapsto U(g)$ be a UIR of the locally compact group G, and suppose that α is a unitary character of the closed subgroup H. Let $X = G/H$, and let $\sigma : X \to G$ be any Borel section. Assume that $\mathcal{A}_\alpha \neq \{0\}$ and $\eta \in \mathcal{A}_\alpha$. Then, the representation U is square integrable mod (H, σ) and the mapping $W_\eta : \mathfrak{H} \to L^2(X, d\nu)$ defined by*

$$(W_\eta \phi)(x) = [c_\alpha(\eta)]^{-\frac{1}{2}} \langle \eta_{\sigma(x)} | \phi \rangle, \qquad \phi \in \mathfrak{H}, \qquad (9.38)$$

where

$$c_\alpha(\eta) = \frac{I_\alpha(\eta)}{\|\eta\|^2} \quad \text{and} \quad \eta_{\sigma(x)} = U(\sigma(x))\eta \qquad (9.39)$$

is a linear isometry onto a (closed) subspace $\mathfrak{H}_\eta^\sigma$ of $L^2(X, d\nu)$. On $\mathfrak{H}$, the resolution of the identity,

$$\frac{1}{c_\alpha(\eta)} \int_X |\eta_{\sigma(x)}\rangle\langle\eta_{\sigma(x)}| \, d\nu(x) = I, \qquad (9.40)$$

holds. The subspace $\mathfrak{H}_\eta^\sigma = W_\eta \mathfrak{H} \subset L^2(X, d\nu)$ is a reproducing kernel Hilbert space with reproducing kernel

$$K_\eta^\sigma(x, y) = \frac{1}{c_\alpha(\eta)} \langle \eta_{\sigma(x)} | \eta_{\sigma(y)} \rangle. \qquad (9.41)$$

Furthermore, W_η intertwines U and the induced representation ${}^\alpha U$:

$$W_\eta U(g) = {}^\alpha U(g) W_\eta, \qquad g \in G. \qquad (9.42)$$

The above theorem settles the question of square integrability of representations admitting Gilmore–Perelomov type CS, for, if in the definition of these states in (7.9)–(7.11) the vector η is α-admissible with $\alpha(h) = \exp[i\omega(h)]$, then U is square integrable mod (H, σ). Note that, if $\nu(X)$, the total measure of the homogeneous space $X = G/H$, is finite, then any $\eta \in \mathfrak{H}$

satisfying (9.35) is α-admissible and, consequently, the representation U is square integrable.

Obvious examples of applications of the above theorem are the representations (2.32) of the Weyl–Heisenberg group, G_{WH}, giving rise to the canonical CS (2.37) and the discrete series representations (4.86) of $SU(1,1)$, leading to the CS (4.99). In the case of the Weyl–Heisenberg group, the character α is the representation, $\theta \mapsto \exp(i\lambda\theta)$, of the phase subgroup Θ and every vector in the representation space is α-admissible. For the $SU(1,1)$ representations, U^j, the representation $k \mapsto \exp(-ij\phi)$ of the maximal compact subgroup K plays the role of α, with $(2j+1)^{\frac{1}{2}} u_0$ [see (4.99)] being an α-admissible vector.

In general, $\mathcal{A}_\alpha$ is not dense in $\mathfrak{H}$, i.e., $\mathfrak{H}^\alpha \neq \mathfrak{H}$. However, the following version of the orthogonality relations can be proved, generalizing Theorem 8.2.1 [166]:

Theorem 9.2.3: *Let G be a locally compact group, H be a closed subgroup admitting the unitary character α, $X = G/H$, and $\sigma : X \to G$ any Borel section. Let U be a UIR of G on the Hilbert space $\mathfrak{H}$, $\mathcal{A}_\alpha \neq \{0\}$ and $\mathfrak{H}^\alpha \subset \mathfrak{H}$ be the closure of $\mathcal{A}_\alpha$. Then, there exists a unique positive, invertible, self-adjoint operator C_α on $\mathfrak{H}^\alpha$ such that, for arbitrary $\eta, \eta' \in \mathcal{A}_\alpha$, and $\phi, \phi' \in \mathfrak{H}$,*

$$\int_X \overline{\langle \eta_{\sigma(x)} | \phi \rangle} \langle \eta'_{\sigma(x)} | \phi' \rangle \, d\nu(x) = \langle C_\alpha \eta' | C_\alpha \eta \rangle \langle \phi | \phi' \rangle. \qquad (9.43)$$

If U' is another UIR of G, on the Hilbert space $\mathfrak{H}'$, which is not unitarily equivalent to U, $\sigma' : X \to G$ any Borel section, and ξ any α-admissible vector for the representation U', then, for arbitrary $\phi \in \mathfrak{H}$ and $\phi' \in \mathfrak{H}'$,

$$\int_X \overline{\langle \eta_{\sigma(x)} | \phi \rangle_{\mathfrak{H}}} \langle \xi_{\sigma'(x)} | \phi' \rangle_{\mathfrak{H}'} \, d\nu(x) = 0. \qquad (9.44)$$

Again, the obvious examples illustrating this theorem are the representations U^λ of the Weyl–Heisenberg group [see (2.32)]. Taking $\lambda = 1$, writing U for the corresponding representation and $U(\theta, q, p) = e^{i\theta} U(q, p)$, the orthogonality relations now assume the form

$$\frac{1}{2\pi} \int_{\mathbb{R}^2} \overline{\langle U(q,p)\phi_1 | \psi_1 \rangle} \langle U(q,p)\phi_2 | \psi_2 \rangle \, dq dp = \overline{\langle \phi_1 | \psi_1 \rangle} \langle \phi_2 | \psi_2 \rangle, \qquad (9.45)$$

for arbitrary vectors ϕ_1, ϕ_2, ψ_1, and ψ_2 in the Hilbert space and the operator $C_\alpha = (2\pi)^{\frac{1}{2}} I$.

An elegant generalization of Theorem 9.2.2, to include cases in wich the unitary character α is replaced by a representation of the subgroup H on a finite-dimensional Hilbert space — leading thereby to a square integrability condition for vector coherent states — can be derived based on the geometrical considerations of Section 7.1.3. We begin by generalizing the concept of α-admissibility. As before, we assume that the coset space $X = G/H$ carries an *invariant measure* $d\nu$.

184   9. Some Examples and Generalizations

Definition 9.2.4: *Let U be a UIR of G on the Hilbert space $\mathfrak{H}$ and H be a closed subgroup of G. Let V be a finite-dimensional UIR of H. A subspace $\mathfrak{K} \subset \mathfrak{H}$ is said to be V-admissible if $U(H)$ restricted to $\mathfrak{K}$ is unitarily equivalent to V and if there exists a nonzero vector $\eta \in \mathfrak{K}$ satisfying the condition (7.36)*

$$I(\eta) \equiv \int_X \langle \eta | \mathbb{P}_{\mathfrak{K}}([g])\eta \rangle \, d\nu([g]) < \infty,$$

where $\mathbb{P}_{\mathfrak{K}}$ is the projection operator for the subspace $\mathfrak{K} \subset \mathfrak{H}$ and $\mathbb{P}_{\mathfrak{K}}([g]) = U(g)\mathbb{P}_{\mathfrak{K}}U(g)^*$. If such a subspace $\mathfrak{K}$ exists, the representation U is said to be square integrable $\mathrm{mod}(H)$.

With the conditions of the above definition being satisfied, let us set

$$c_V(\eta) = \frac{I(\eta)}{\|\eta\|^2} \tag{9.46}$$

and denote by $\mathcal{D}$ the dense set in $\mathfrak{H}$ spanned by the vectors $U(g)\eta$, $\eta \in \mathfrak{K}$, $g \in G$. Then, in view of the second equality in (7.35), the map $W_\eta : \mathcal{D} \to L^2_{\mathfrak{K}}(X, d\nu)$,

$$(W_\eta \phi)(g) = [c_V(\eta)]^{-\frac{1}{2}} \mathbb{P}_{\mathfrak{K}} U(g^{-1})\phi, \tag{9.47}$$

is well defined. The set $\mathcal{D}$ is stable under U, and W_η [being a multiple of the map W_K in (7.27)] intertwines U and the induced representation $^V U$. Retracing the lines of the proof of Theorem 8.1.3, it is easily established that W_η is closed, while an application of Lemma 4.3.3 then shows that W_η is an isometry. Thus, we have proved the following result.

Theorem 9.2.5: *Let G be a locally compact group, H be a closed subgroup of G, U be a UIR of G on the Hilbert space $\mathfrak{H}$, and V be a finite-dimensional UIR of H. Let $\mathfrak{K}$ be a V-admissible subspace of $\mathfrak{H}$. Then, the mapping (9.47) extends to all of $\mathfrak{H}$ as a linear isometry, so that its range, $\mathfrak{H}_\eta$, is a closed subspace of $L^2_{\mathfrak{K}}(X, d\nu)$. On $\mathfrak{H}$, the resolution of the identity*

$$\frac{1}{c_V(\eta)} \int_X \mathbb{P}_{\mathfrak{K}}([g]) \, d\nu([g]) = I \tag{9.48}$$

holds. The subspace $\mathfrak{H}_\eta = W_\eta(\mathfrak{H}) \subset L^2_{\mathfrak{K}}(X, d\nu)$ is a reproducing kernel Hilbert space; the corresponding projection operator

$$\mathbb{P}_\eta = W_\eta W_\eta^*, \qquad \mathbb{P}_\eta L^2_{\mathfrak{K}}(X, d\nu) = \mathfrak{H}_\eta, \tag{9.49}$$

has the reproducing kernel K_η

$$\begin{aligned}(\mathbb{P}_\eta \Phi)(g) &= \int_X K_\eta(g, g') \mathbf{f}(g') \, d\nu([g']), \qquad \mathbf{f} \in L^2_{\mathfrak{K}}(X, d\nu), \\ K_\eta(g, g') &= \frac{1}{c(\eta)} \mathbb{P}_{\mathfrak{K}} U(g^{-1}) U(g') \mathbb{P}_{\mathfrak{K}},\end{aligned} \tag{9.50}$$

9.2. A generalization: α- and V-admissibility

as its integral kernel. Furthermore, W_η intertwines U and the induced representation $^V U$

$$W_\eta U(g) = {}^V U(g) W_\eta, \quad g \in G. \tag{9.51}$$

A few remarks are in order here, as follows.

- As a consequence of this theorem, the quantity $c_V(\eta)$ is independent of the particular $\eta \in \mathfrak{K}$ chosen to evaluate it.

- The condition (9.50) translates into the property

$$\int_X K_\eta(g, g'') \, K_\eta(g'', g') \, d\nu([g'']) = K_\eta(g, g'). \tag{9.52}$$

- The set of positive, bounded operators

$$a(\Delta) = \int_\Delta \mathbb{P}_{\mathfrak{K}}([g]) \, d\nu([g]),$$

for arbitrary Borel sets $\Delta \subset X$, form a normalized POV measure and $\{U, a\}$ is a system of covariance,

$$U(g) a(\Delta) U(g)^* = a(g\Delta), \tag{9.53}$$

which of course is just a restatement of the intertwining property (9.51).

- If we choose $H = \{e\}$, the subgroup consisting of only the identity element, and take for V (or α) the trivial representation, then the V-admissibility (or α-admissibility) of a vector $\eta \in \mathfrak{H}$, for the representation U, is the same as admissibility in the sense of Definition 8.1.1. Thus, if such a vector exists, U is square integrable.

In order to define coherent states, it is necessary to choose a section $\sigma : G/H \to G$ and an orthonormal basis, η^i, $i = 1, 2, \ldots, n$, in $\mathfrak{K}$ (n being its dimension). Then, $\mathbb{P}_{\mathfrak{K}} = \sum_{i=1}^n |\eta^i\rangle\langle\eta^i|$, and

$$\mathfrak{S}_\sigma = \{[c_V(\eta)]^{\frac{1}{2}} \, \eta^i_{\sigma(x)} = [c_V(\eta)]^{-\frac{1}{2}} \, U(\sigma(x))\eta^i \mid \\ i = 1, 2, \ldots, n, \ x \in X = G/H\},$$

is a family of (in general vector) coherent states for the representation U. The resolution of the identity (9.48), in terms of these CS, now has the familiar form

$$\frac{1}{[c_V(\eta)]} \sum_{i=1}^n \int_X |\eta^i_{\sigma(x)}\rangle\langle\eta^i_{\sigma(x)}| \, d\nu(x) = I,$$

which again is independent of the section σ chosen to define the CS.

9.2.1 Example of the Galilei group (♣)

Let us illustrate the results of Theorems 9.2.2 and 9.2.5 by examples drawn from certain representations of the *Galilei group*. The kinematics of a free nonrelativistic physical particle is governed by its invariance under the action of the Galilei group [Inö, 203, 204, 205], which is a ten-parameter group $\mathcal{G}$ of transformations of Newtonian space-time. An element $g \in \mathcal{G}$ is of the form

$$g = (b, \mathbf{a}, \mathbf{v}, R), \qquad \mathbf{a}, \mathbf{v} \in \mathbb{R}^3, \quad SO(3), \tag{9.54}$$

where b is a time and $\mathbf{a}$ a spatial translation, $\mathbf{v}$ is a velocity boost, and R is a spatial rotation. The action of g on a space-time point $(\mathbf{x}, t)$ is given by

$$g(\mathbf{x}, t) = (\mathbf{x}', t'), \quad \mathbf{x}' = R\mathbf{x} + \mathbf{v}t + \mathbf{a}, \quad t' = t + b. \tag{9.55}$$

Actually, in quantum mechanics, one needs to work with a *central extension* of $\mathcal{G}$, and, also, in order to accomodate particles with half-integral spins, it is necessary to replace the rotation group $SO(3)$ by its universal covering group $SU(2)$. Denoting this extended Galilei group by $\widetilde{\mathcal{G}}$, its elements will be written as $g = (\theta, b, \mathbf{a}, \mathbf{v}, \rho)$, $\theta \in \mathbb{R}$, $\rho \in SU(2)$. A general element of $SU(2)$ is a 2×2 complex matrix,

$$\rho = \begin{pmatrix} \alpha & \beta \\ -\bar{\beta} & \bar{\alpha} \end{pmatrix}, \quad \alpha, \beta \in \mathbb{C}, \quad |\alpha|^2 + |\beta|^2 = 1, \tag{9.56}$$

and the elements of the corresponding 3×3 rotation matrix $R(\rho)$ are then

$$R(\rho)_{ij} = \text{Tr}[\rho \sigma_j \rho^{-1} \sigma_i], \quad i, j = 1, 2, 3, \tag{9.57}$$

the σ_j's being the three Pauli matrices:

$$\sigma_1 = \begin{pmatrix} 0 & 1 \\ 1 & 0 \end{pmatrix}, \quad \sigma_2 = \begin{pmatrix} 0 & -i \\ i & 0 \end{pmatrix}, \quad \sigma_3 = \begin{pmatrix} 1 & 0 \\ 0 & -1 \end{pmatrix}. \tag{9.58}$$

The group multiplication in $\widetilde{\mathcal{G}}$ is

$$\begin{aligned} g_1 g_2 &= (\theta_1 + \theta_2 + \xi(g_1, g_2), b_1 + b_2, \mathbf{a}_1 + R(\rho_1)\mathbf{a}_2 + \mathbf{v}_1 b_2, \\ &\qquad \mathbf{v}_1 + R(\rho_1)\mathbf{v}_2, \rho_1 \rho_2), \\ g_i &= (\theta_i, b_i, \mathbf{a}_i, \mathbf{v}_i, \rho_i), \quad i = 1, 2, \end{aligned} \tag{9.59}$$

where $\xi : \mathcal{G} \times \mathcal{G} \to \mathbb{R}$ is a *multiplier*, which, up to equivalence (in the sense of cocycles), can be taken to be

$$\xi(g_1, g_2) = m[\frac{1}{2} \mathbf{v}_1^2 b_2 + \mathbf{v}_1 \cdot R(\rho_1) \mathbf{a}_2], \quad m = \text{const.} > 0. \tag{9.60}$$

Thus, there is a one-parameter family of multipliers (that is, of central extensions), indexed by the parameter m, which is interpreted as *mass* [203, 204].

9.2. A generalization: α- and V-admissibility

The multiplier ξ has the properties

$$\xi(e,e) = 0,$$
$$\xi(g_1, g_2) + \zeta(g_1 g_2, g_3) = \xi(g_1, g_2 g_3) + \xi(g_2, g_3),$$
$$g_1, g_2, g_3 \in \widetilde{\mathcal{G}}. \qquad (9.61)$$

The identity element of $\widetilde{\mathcal{G}}$ is $e = (0, 0, \mathbf{0}, \mathbf{0}, \mathbb{I}_2)$ (where $\mathbb{I}_2 = 2 \times 2$ identity matrix), while the inverse of $g = (\theta, b, \mathbf{a}, \mathbf{v}, \rho)$ is the element

$$g^{-1} = (-\theta - m[\frac{1}{2}\mathbf{v}^2 b - \mathbf{v} \cdot \mathbf{a}], -b, R(\rho^{-1})(\mathbf{v}b - \mathbf{a}), -R(\rho^{-1})\mathbf{v}, \rho^{-1}). \quad (9.62)$$

A number of subgroups of $\widetilde{\mathcal{G}}$ are of particular interest to us here for the construction of CS. The subroup $\widetilde{\mathcal{G}}'$ having elements $(\theta, 0, \mathbf{a}, \mathbf{v}, \rho)$, i.e., with the time part b set equal to zero, is called the *isochronous extended Galilei group*. Another useful subgroup of $\widetilde{\mathcal{G}}$ — one which is also a subgroup of $\widetilde{\mathcal{G}}'$ — consists of elements of the type $g = (\theta, 0, \mathbf{a}, \mathbf{v}, \mathbb{I}_2)$. We denote this group by $\widetilde{\mathcal{G}}_{WH}$ and note that it is isomorphic to the Weyl–Heisenberg group for three degrees of freedom. The subgroup $\mathcal{S}$ of space translations with elements $g = (0, 0, \mathbf{a}, \mathbf{0}, \mathbb{I}_2)$, and of pure Galilean translations (or velocity boosts) $\mathcal{V}$, with $g = (0, 0, \mathbf{0}, \mathbf{v}, \mathbb{I}_2)$, will also play important roles in the construction to follow. To these, we add the subgroup of time translations $\mathcal{T}$ with $g = (0, b, \mathbf{0}, \mathbf{0}, \mathbb{I}_2)$ and the phase subgroup Θ having elements $g = (\theta, 0, \mathbf{0}, \mathbf{0}, \mathbb{I}_2)$. Finally, we note that $\widetilde{\mathcal{G}}$ can be looked upon as the semidirect product,

$$\widetilde{\mathcal{G}} = (\Theta \times \mathcal{T} \times \mathcal{S}) \rtimes K, \qquad K = \mathcal{V} \rtimes SU(2), \qquad (9.63)$$

and observe that $\Theta \times \mathcal{T} \times \mathcal{S}$ is an abelian subgroup of $\widetilde{\mathcal{G}}$.

Following [204], we proceed now to explicitly compute a class of UIRs of $\widetilde{\mathcal{G}}$, before actually computing their CS. The reasons for this extended exercise are twofold. First, it will illustrate how the general technique of induced representations may be used to generate unitary *irreducible* representations of certain semidirect product groups [Ma1, Ma2, 210]. Second, in Chapter 10, we shall encounter representations of other semidirect product groups that are similarly obtained. The particular representations of $\widetilde{\mathcal{G}}$ that concern us here are the ones that describe free nonrelativistic quantum particles of mass $m > 0$ and spin $j = 0, 1/2, 1, 3/2, \ldots$. The construction of these representations exploits the semidirect product nature of $\widetilde{\mathcal{G}}$ displayed in (9.63). The *momentum space* realizations of these representations are carried by the Hilbert spaces $\mathfrak{K}^j \otimes L^2(\widehat{\mathbb{R}}^3, d\mathbf{k})$, where $\mathfrak{K}^j$ is a $(2j+1)$-dimensional spinor space (which we identify with $\mathbb{C}^{2j+1}$), $\widehat{\mathbb{R}}^3$ is the dual of $\mathbb{R}^3$ (which we again identify with $\mathbb{R}^3$), and $d\mathbf{k}$ is the Lebesgue measure on $\widehat{\mathbb{R}}^3$. In Chapter 11, we shall discuss the CS of $\widetilde{\mathcal{G}}$ related to these representations. Here, we use them to first obtain representations of the isochronous Galilei group $\widetilde{\mathcal{G}}'$ and then construct its CS.

188   9. Some Examples and Generalizations

The spaces $\mathfrak{K}^j$ carry the unitary irreducible representations of $SU(2)$, which we denote by $\mathcal{D}^j(\rho)$, $\rho \in SU(2)$. The abelian subgroup $\Theta \times \mathcal{T} \times \mathcal{S}$ of $\widetilde{\mathcal{G}}$ has unitary characters $\chi_{\gamma,E,\mathbf{p}}$ of the type,

$$\chi_{\gamma,E,\mathbf{p}}(\theta, b, \mathbf{a}) = \exp\left[i(\gamma\theta + Eb + \mathbf{p}\cdot\mathbf{a})\right], \qquad \gamma, E \in \mathbb{R}, \quad \mathbf{p} \in \mathbb{R}^3. \quad (9.64)$$

We determine the dual action of the subgroup K on such a character. Since

$$(0,0,\mathbf{0},\mathbf{v},\rho)(\theta,b,\mathbf{a},,\mathbb{I}_2) = (\theta + m[\frac{1}{2}\mathbf{v}^2 b + \mathbf{v}\cdot R(\rho)\mathbf{a}], b, R(\rho)\mathbf{a} + \mathbf{v}b, \mathbf{v}, \rho),$$

writing $\chi_{\gamma',E',\mathbf{p}'} = (\mathbf{v},\rho)^*[\chi_{\gamma,E,\mathbf{p}}]$ for the transformed character, under the dual action of $(\mathbf{v},\rho) \in K$, we find

$$\chi_{\gamma',E',\mathbf{p}'}(\theta, b, \mathbf{a}) =$$
$$= \chi_{\gamma,E,\mathbf{p}}(\theta + m[\frac{1}{2}\mathbf{v}^2 b + \mathbf{v}\cdot R(\rho)\mathbf{a}], b, R(\rho)\mathbf{a} + \mathbf{v}b, \mathbf{v}, \rho)$$
$$= \exp\left[i\{\gamma\theta + (E + \frac{m\gamma}{2}\mathbf{v}^2 + \mathbf{p}\cdot\mathbf{v})b + R(\rho^{-1})(m\gamma\mathbf{v} + \mathbf{p})\cdot\mathbf{a}\}\right],$$

and, thus,

$$\left.\begin{array}{rcl} \gamma' & = & \gamma, \\ E' & = & E + \dfrac{m\gamma\mathbf{v}^2}{2} + \mathbf{p}\cdot\mathbf{v}, \\ \mathbf{p}' & = & R(\rho^{-1})(m\gamma\mathbf{v}+\mathbf{p}). \end{array}\right\} \quad (9.65)$$

Next we observe that

$$E' - \frac{\mathbf{p}'^2}{2m\gamma} = E - \frac{\mathbf{p}^2}{2m\gamma} = E_0 = \text{const}. \quad (9.66)$$

Thus, the orbit of a character under the dual action of the subgroup K of $\widetilde{\mathcal{G}}$ is a paraboloid characterized by two constants, $E_0, \gamma \in \mathbb{R}$, and, without loss of generality, we choose $\gamma = 1$. Denoting an orbit by $\widehat{\mathcal{O}}_{E_0}$, the constant E_0 may be identified with the *internal energy* of the system. The character $\chi_{1,E_0,\mathbf{0}} \in \widehat{\mathcal{O}}_{E_0}$ is then a natural choice of a representative for the orbit. From (9.65), the stability subgroup of this character is seen to be $SU(2)$, so that

$$\widehat{\mathcal{O}}_{E_0} \simeq K/SU(2) \simeq \widehat{\mathbb{R}}^3. \quad (9.67)$$

The *invariant* measure on this orbit may therefore be simply taken to be the Lebesgue measure $d\mathbf{k}$ on $\widehat{\mathbb{R}}^3$. According to the general theory, the irreducible representations of $\widetilde{\mathcal{G}}$ can now be obtained by inducing from the UIRs, $V_{E_0,j}$ of the subgroup $(\Theta \times \mathcal{T} \times \mathcal{S}) \rtimes SU(2)$, where

$$V_{E_0,j}(\theta, b, \mathbf{a}, \rho) = \chi_{1,E_0,\mathbf{0}}(\theta, b, \mathbf{a})\,\mathcal{D}^j(\rho) = \exp\left[i(\theta + E_0 b)\right]\mathcal{D}^j(\rho). \quad (9.68)$$

Define the section $\lambda : K/SU(2) \simeq \widehat{\mathbb{R}}^3 \to \widetilde{\mathcal{G}}$:

$$\lambda(\mathbf{k}) = (0, 0, \mathbf{0}, \frac{\mathbf{k}}{m}, \mathbb{I}_2). \quad (9.69)$$

9.2. A generalization: α- and V-admissibility

Computing $g^{-1}\lambda(\mathbf{k})$, for $g = (\theta, b, \mathbf{a}, \mathbf{v}, \rho) \in \widetilde{\mathcal{G}}$, explicitly,

$g^{-1}\lambda(\mathbf{k}) =$
$$= (-\theta - m\lfloor\frac{1}{2}\mathbf{v}^2 b - \mathbf{v}\cdot\mathbf{a}], -b, -R(\rho^{-1})[\mathbf{a} - \mathbf{v}b], R(\rho^{-1})[\frac{\mathbf{k}}{m} - \mathbf{v}], \rho^{-1})$$
$$= \left(0, 0, \mathbf{0}, R(\rho^{-1})[\frac{\mathbf{k}}{m} - \mathbf{v}], \mathbb{I}_2\right)$$
$$\times \left(-\theta - \frac{\mathbf{k}^2 b}{2m} + \mathbf{k}\cdot\mathbf{a}, -b, -R(\rho^{-1})[\mathbf{a} - \frac{\mathbf{k}}{m}b], \mathbf{0}, \rho^{-1}\right),$$

which yields the cocycle $h : \widetilde{\mathcal{G}} \times \widehat{\mathbb{R}}^3 \to K$, $h(g, \mathbf{k}) = \lambda(g\mathbf{k})^{-1}g\lambda(\mathbf{k})$ [see (4.37) and (4.38)],

$$h(g^{-1}, \mathbf{k}) = (-\theta - \frac{\mathbf{k}^2 b}{2m} + \mathbf{k}\cdot\mathbf{a}, -b, -R(\rho^{-1})[\mathbf{a} - \frac{\mathbf{k}}{m}b], \mathbf{0}, \rho^{-1}),$$
$$h(g^{-1}, \mathbf{k})^{-1} = (\theta + \frac{\mathbf{k}^2 b}{2m} - \mathbf{k}\cdot\mathbf{a}, b, \mathbf{a} - \frac{\mathbf{k}}{m}b, \mathbf{0}, \rho). \tag{9.70}$$

Thus, finally,

$$V_{E_0, j}(h(g^{-1}, \mathbf{k}))^{-1} = \exp\left[i\{\theta + (E_0 + \frac{\mathbf{k}^2}{2m})b - \mathbf{k}\cdot\mathbf{a}\}\right] \mathcal{D}^j(\rho), \tag{9.71}$$

and inserting this into (4.41), we obtain the representation $\widetilde{U}^{E_0, j}$ of $\widetilde{\mathcal{G}}$ which is induced from the UIR $V_{E_0, j}$ of the subgroup $(\Theta \times \mathcal{T} \times \mathcal{S}) \rtimes SU(2)$. This representation acts on the Hilbert space $\mathfrak{K}^j \otimes L^2(\widehat{\mathbb{R}}^3, d\mathbf{k})$ in the manner:

$$(\widehat{U}^{E_0, j}(\theta, b, \mathbf{a}, \mathbf{v}, \rho)\widehat{\phi})(\mathbf{k}) = \exp\left[i\{\theta + (E_0 + \frac{\mathbf{k}^2}{2m})b - \mathbf{k}\cdot\mathbf{a}\}\right]$$
$$\times \mathcal{D}^j(\rho)\, \widehat{\phi}(R(\rho^{-1})(\mathbf{k} - m\mathbf{v})), \tag{9.72}$$

for all $\widehat{\phi} \in \mathfrak{K}^j \otimes L^2(\widehat{\mathbb{R}}^3, d\mathbf{k})$. This is the so-called *momentum space representation* for a nonrelativistic system of mass m, internal energy E_0, and spin-j. When dealing with a single system, the internal energy may be set equal to zero, simply by adjusting the ground state. We shall do this in what follows and write the representations as $\widehat{U}^j$. The representation on *configuration space* is obtained by going over to the Fourier-transformed space $\mathfrak{K}^j \otimes L^2(\mathbb{R}^3, d\mathbf{x})$ of vectors

$$\phi(\mathbf{x}) = (2\pi)^{-\frac{3}{2}} \int_{\widehat{\mathbb{R}}^3} e^{i\mathbf{k}\cdot\mathbf{x}}\widehat{\phi}(\mathbf{k})\, d\mathbf{k}.$$

The transformed representation U^j assumes the form:

$$(U^j(\theta, b, \mathbf{a}, \mathbf{v}, \rho)\phi)(\mathbf{x}) = \exp\left[i\{\theta + \frac{\mathbf{P}^2}{2m}b + m\mathbf{v}\cdot(\mathbf{x} - \mathbf{a})\}\right]$$
$$\times \mathcal{D}^j(\rho)\, \phi(R(\rho^{-1})(\mathbf{x} - \mathbf{a})), \tag{9.73}$$

190   9. Some Examples and Generalizations

where $\mathbf{P} = -i\nabla$, or, in terms of the time-translated functions,

$$\phi(\mathbf{x},t) = (\exp[-i\frac{\mathbf{P}^2}{2m}t]\phi)(\mathbf{x}),$$

the representation appears in the more familiar form:

$$(U^j(\theta, b, \mathbf{a}, \mathbf{v}, \rho)\phi)(\mathbf{x}, t) =$$
$$= \exp[i\{\theta - \frac{m\mathbf{v}^2}{2}(t-b) + m\mathbf{v}\cdot(\mathbf{x}-\mathbf{a})\}]$$
$$\times \mathcal{D}^j(\rho)\,\phi\left(R(\rho^{-1})(\mathbf{x}-\mathbf{a}-\mathbf{v}(t-b), t-b\right) \quad (9.74)$$

9.2.2 CS of the isochronous Galilei group (♣)

Unitary irreducible representations of the isochronous Galilei group $\widetilde{\mathcal{G}}'$ can now be read off from (9.72) after setting $b = 0$. These representations are also carried by $\mathfrak{K}^j \otimes L^2(\widehat{\mathbb{R}}^3, d\mathbf{k})$, and using the same notation as before (also taking $E_0 = 0$),

$$(\widehat{U}^j(\theta, \mathbf{a}, \mathbf{v}, \rho)\widehat{\phi})(\mathbf{k}) = \exp[i(\theta - \mathbf{k}\cdot\mathbf{a})]\,\mathcal{D}^j(\rho)\,\widehat{\phi}(R(\rho^{-1})(\mathbf{k}-m\mathbf{v})),$$
$$(\theta, \mathbf{a}, \mathbf{v}, \rho) \in \widetilde{\mathcal{G}}'. \quad (9.75)$$

We proceed to demonstrate the square integrability mod$(\Theta \times SU(2))$ of these representations in the sense of Theorems 9.2.2 and 9.2.5, following [7] and [166].

Denote by $\chi^{j\mu}$, $\mu = -j, -j+1, \ldots, j-1, j$, the canonical basis of $\mathfrak{K}^j$ (i.e., $\chi^{j\mu}$ is the vector with components $\delta_{j+\mu+1,\ell}$, $\ell = 1, 2, \ldots, 2j+1$), and let $\mathcal{D}^j(\rho)_{\mu\mu'}$ be the matrix elements of $\mathcal{D}^j(\rho)$ in this basis. Define a function $\varepsilon : \mathbb{R}^+ \to \mathbb{C}$, normalized in the manner

$$\int_0^\infty |\varepsilon(r)|^2\,r^2 dr = 1. \quad (9.76)$$

[This is an element of the radial Hilbert space $L^2(\mathbb{R}^+, r^2 dr)$.] Consider first the representations $\widehat{U}^j$ for integral values of the spin ($j = 0, 1, 2, \ldots$). For any $\mathbf{k} \in \widehat{\mathbb{R}}^3$, denote by $Y^{j\mu}(\mathbf{k})$, $\mu = -j, -j+1, \ldots, j-1, j$, the jth spherical harmonics in the angles of the three-vector $\mathbf{k}$, and let $d\Omega(\mathbf{k})$ be the area element on the unit sphere S^2. The following properties of the spherical harmonics are well known:

$$\int_{S^2} \overline{Y^{j\mu}(\mathbf{k})} Y^{j'\mu'}(\mathbf{k})\,d\Omega(\mathbf{k}) = \delta_{jj'}\delta_{\mu\mu'}, \quad (9.77)$$

$$Y^{j\mu}(R^{-1}\mathbf{k}) = \sum_{\mu'=-j}^{j} \mathcal{D}^j(R)_{\mu'\mu} Y^{j\mu'}(\mathbf{k}), \quad \mathbf{k} \in \widehat{\mathbb{R}}^3, \quad R \in SO(3) \quad (9.78)$$

9.2. A generalization: α- and V-admissibility

$$P^j(\cos\gamma) = \frac{4\pi}{2j+1} \sum_{\mu=-j}^{j} \overline{Y^{j\mu}(\mathbf{k})} Y^{j\mu}(\mathbf{k}'),$$

$$\text{where } \cos\gamma = \frac{\mathbf{k}\cdot\mathbf{k}'}{\|\mathbf{k}\|\,\|\mathbf{k}'\|}, \quad \mathbf{k},\mathbf{k}' \in \widehat{\mathbb{R}}^3, \quad (9.79)$$

and P^j is the Legendre polynomial of order j.

Consider finally the vector $\widehat{\eta} \in \mathfrak{K}^j \otimes L^2(\widehat{\mathbb{R}}^3, d\mathbf{k})$,

$$\widehat{\eta}(\mathbf{k}) = \frac{1}{(2j+1)^{\frac{1}{2}}} \sum_{\mu=-j}^{j} \chi^{j\mu}\, \overline{Y^{j\mu}(\mathbf{k})}\,\varepsilon(\|\mathbf{k}\|), \quad \mathbf{k} \in \widehat{\mathbb{R}}^3, \quad (9.80)$$

and with ε as in (9.76). Let α be the unitary character

$$\alpha(\theta, \rho) = e^{i\theta} \quad (9.81)$$

of the subgroup $H = \Theta \times SU(2)$ of $\widetilde{\mathcal{G}}'$. An application of (9.78) then yields

$$\widehat{U}^j(\theta, \mathbf{0}, \mathbf{0}, \rho)\widehat{\eta} = \alpha(\theta, \rho)\widehat{\eta}. \quad (9.82)$$

The quotient space $\Gamma = \widetilde{\mathcal{G}}'/H$ is isomorphic to $\mathbb{R}^6$, by virtue of the decomposition

$$(\theta, \mathbf{a}, \mathbf{v}, \rho) = (0, \mathbf{a}, \mathbf{v}, \mathbb{I}_2)(\theta, \mathbf{0}, \mathbf{0}, \rho). \quad (9.83)$$

We parametrize Γ by $(\mathbf{q}, \mathbf{p}) \in \mathbb{R}^6$, with $\mathbf{q} = \mathbf{a}$ and $\mathbf{p} = m\mathbf{v}$, in terms of which the invariant measure on it is just the Lebesgue measure $d\mathbf{q}d\mathbf{p}$. A natural choice of a section $\sigma : \Gamma \to \widetilde{\mathcal{G}}'$ is then

$$\sigma(\mathbf{q}, \mathbf{p}) = (0, \mathbf{q}, \frac{\mathbf{p}}{m}, \mathbb{I}_2), \quad (9.84)$$

and by a straightforward computation of the type used in the proof of Theorem 9.1.2, and using relations such as (9.77), we find that

$$I_\alpha(\widehat{\eta}) = \int_\Gamma |\langle \widehat{\eta}_{\sigma(\mathbf{q},\mathbf{p})} | \widehat{\eta}\rangle|^2 \, d\mathbf{q}d\mathbf{p} = (2\pi)^3. \quad (9.85)$$

We have thus shown that the vector $\widehat{\eta} \in \mathfrak{K}^j \otimes L^2(\widehat{\mathbb{R}}^3, d\mathbf{k})$ is α-admissible, and therefore $\widehat{U}^j$ is square integrable $\mod(\Theta \times SU(2))$. Since $\|\widehat{\eta}\|^2 = 1$, $c_\alpha = (2\pi)^3$ [see (9.39)], and hence the map $W_\eta : \mathfrak{K}^j \otimes L^2(\widehat{\mathbb{R}}^3, d\mathbf{k}) \to L^2(\Gamma, d\mathbf{q}d\mathbf{p})$,

$$(W_\eta\widehat{\phi})(\mathbf{q}, \mathbf{p}) = \frac{1}{(2\pi)^{\frac{3}{2}}} \langle \widehat{\eta}_{\sigma(\mathbf{q},\mathbf{p})} | \widehat{\phi}\rangle = \frac{1}{(2\pi)^{\frac{3}{2}}} \langle \widehat{U}^j(0, \mathbf{q}, \frac{\mathbf{p}}{m}, \mathbb{I}_2)\widehat{\eta} | \widehat{\phi}\rangle, \quad (9.86)$$

is an isometry. Furthermore, we have the resolution of the identity,

$$\frac{1}{(2\pi)^3} \int_\Gamma |\widehat{\eta}_{\sigma(\mathbf{q},\mathbf{p})}\rangle\langle\widehat{\eta}_{\sigma(\mathbf{q},\mathbf{p})}| \, d\mathbf{q}d\mathbf{p} = I, \quad (9.87)$$

for the *Galilei coherent states*, which we now define to be the vectors

$$\frac{1}{(2\pi)^{\frac{3}{2}}} \widehat{\eta}_{\sigma(\mathbf{q},\mathbf{p})}, \quad (\mathbf{q}, \mathbf{p}) \in \mathbb{R}^6,$$

in $\mathfrak{K}^j \otimes L^2(\widehat{\mathbb{R}}^3, d\mathbf{k})$. The image of $\mathfrak{K}^j \otimes L^2(\widehat{\mathbb{R}}^3, d\mathbf{k})$ in $L^2(\Gamma, d\mathbf{q}d\mathbf{p})$ under the isometry (9.86) is a reproducing kernel Hilbert space $\mathfrak{H}_\eta$, the elements of which are the *continuous* functions:

$$\Phi(\mathbf{q},\mathbf{p}) = \frac{1}{[(2j+1)(2\pi)^3]^{\frac{1}{2}}}$$
$$\times \sum_{m=-j}^{j} \int_{\widehat{\mathbb{R}}^3} e^{i\mathbf{k}\cdot\mathbf{q}} Y^{jm}(\mathbf{k}-\mathbf{p}) \overline{\varepsilon(\|\mathbf{k}-\mathbf{p}\|)} \, \widehat{\phi}^m(\mathbf{k}) \, d\mathbf{k}, \qquad (9.88)$$

where $\widehat{\phi}(\mathbf{k}) = (\widehat{\phi}(\mathbf{k})^{-j}, \widehat{\phi}(\mathbf{k})^{-j+1}, \ldots \widehat{\phi}(\mathbf{k})^{j-1}, \widehat{\phi}(\mathbf{k})^{j}) \in \mathfrak{K}^j \otimes L^2(\widehat{\mathbb{R}}^3, d\mathbf{k})$.
The reproducing kernel defining $\mathfrak{H}_\eta$ is

$$K_\eta(\mathbf{q},\mathbf{p};\mathbf{q}',\mathbf{p}') = \frac{1}{(2\pi)^3} \langle \widehat{\eta}_{\sigma(\mathbf{q},\mathbf{p})} | \widehat{\eta}_{\sigma(\mathbf{q}',\mathbf{p}')} \rangle$$
$$= \frac{1}{4\pi(2\pi)^3} \int_{\widehat{\mathbb{R}}^3} e^{i[\mathbf{k}\cdot(\mathbf{q}-\mathbf{q}')]} P^j\left(\frac{(\mathbf{k}-\mathbf{p})\cdot(\mathbf{k}-\mathbf{p}')}{\|\mathbf{k}-\mathbf{p}\|\,\|\mathbf{k}-\mathbf{p}'\|}\right)$$
$$\times \overline{\varepsilon(\|\mathbf{k}-\mathbf{p}\|)}\,\varepsilon(\|\mathbf{k}-\mathbf{p}'\|)\, d\mathbf{k}, \quad (9.89)$$

which can be obtained using (9.79). The coset decomposition (9.83) implies the following action of $g = (\theta, \mathbf{a}, \mathbf{v}, \rho) \in \widetilde{\mathcal{G}}'$ on Γ:

$$g^{-1}(\mathbf{q},\mathbf{p}) = (R(\rho^{-1})(\mathbf{q}-\mathbf{a}), R(\rho^{-1})(\mathbf{p}-m\mathbf{v})). \qquad (9.90)$$

Similarly, the cocycle $h : \widetilde{\mathcal{G}}' \times \Gamma \to \Theta \times SU(2)$ is found to be

$$h(g^{-1},(\mathbf{q},\mathbf{p})) = (-\theta - m\mathbf{v}\cdot(\mathbf{q}-\mathbf{a}), \rho^{-1}). \qquad (9.91)$$

Using these, we obtain the form of the representation

$$\mathbf{U}_\eta(g) = W_\eta\,\widehat{U}^j(g)\,W_\eta^{-1}, \qquad g \in \widetilde{\mathcal{G}}',$$

on $L^2(\Gamma, d\mathbf{q}\,d\mathbf{p})$, arising as the image of $\widehat{U}^j$ under the isometry (9.86):

$$(\mathbf{U}_\eta(\theta,\mathbf{a},\mathbf{v},\rho)\Phi)(\mathbf{q},\mathbf{p}) = \exp\left[i\{\theta + m\mathbf{v}\cdot(\mathbf{q}-\mathbf{a})\}\right]$$
$$\Phi(R(\rho^{-1})(\mathbf{q}-\mathbf{a}), R(\rho^{-1})(\mathbf{p}-m\mathbf{v})), \quad (9.92)$$

($\Phi \in \mathfrak{H}_\eta$), which extends to the whole of $L^2(\Gamma, d\mathbf{q}d\mathbf{p})$ as the representation of $\widetilde{\mathcal{G}}'$, which is induced from α.

To obtain a version of the orthogonality relations (recall that we are still considering representations for $j = 0, 1, 2, \ldots$), first note that any α-admissible vector must be of the form (9.80) with $\varepsilon \in L^2(\mathbb{R}^+, r^2 dr)$ (but not necessarily normalized). Then, if $\widehat{\eta}, \widehat{\eta}'$ are α-admissible vectors, a similar computation as that leading to (9.85) results in the orthogonality relation

$$\int_\Gamma \overline{\langle \widehat{\eta}_{\sigma(\mathbf{q},\mathbf{p})}|\widehat{\phi}\rangle} \langle \widehat{\eta}'_{\sigma(\mathbf{q},\mathbf{p})}|\widehat{\phi}'\rangle\, d\mathbf{q}d\mathbf{p} = (2\pi)^3 \langle \widehat{\eta}'|\widehat{\eta}\rangle\,\langle \widehat{\phi}|\widehat{\phi}'\rangle, \qquad (9.93)$$

9.2. A generalization: α- and V-admissibility

for arbitrary $\widehat{\phi}, \widehat{\phi}' \in \mathfrak{K}^j \otimes L^2(\widehat{\mathbb{R}}^3, d\mathbf{k})$, illustrating Theorem 9.2.3. This relation also shows that, in this case, the Duflo–Moore operator C_α, appearing in Theorem 9.2.3, is a multiple of the identity: $C_\alpha = (2\pi)^3 \, I$.

In order to take into account the representations for $j = 1/2, 3/2, 5/2, \ldots$, it is necessary to adopt a more general approach. In fact, using the notion of V-admissibility, we shall now derive a wide class of square integrable CS for *all* values of j, integral or half-odd integral. Let us first decompose the representation space $\mathfrak{K}^j \otimes L^2(\widehat{\mathbb{R}}^3, d\mathbf{k})$ into rotationally invariant subspaces. For any $\widehat{\phi} \in \mathfrak{K}^j \otimes L^2(\widehat{\mathbb{R}}^3, d\mathbf{k})$, and $\rho \in SU(2)$,

$$(\widehat{U}^j(0,\mathbf{0},\mathbf{0},\rho)\widehat{\phi} = \mathcal{D}^j(\rho)\widehat{\phi}(R(\rho^{-1})\mathbf{k}). \tag{9.94}$$

The space $L^2(\widehat{\mathbb{R}}^3, d\mathbf{k})$ can be written as the tensor product $L^2(S^2, d\Omega(\mathbf{k})) \otimes L^2(\mathbb{R}^+, r^2 dr)$, where S^2 is the unit sphere and $r = \|\mathbf{k}\|$. Let $\{\varepsilon_n\}_{n=1}^\infty$ be an orthonormal basis of $L^2(\mathbb{R}^+, r^2 dr)$:

$$\int_{\mathbb{R}^+} \overline{\varepsilon_n(r)} \varepsilon_{n'}(r) \, r^2 dr = \delta_{nn'}. \tag{9.95}$$

For each $\ell = 0, 1, 2, \ldots$, the spherical harmonics $Y^{\ell\mu}(\mathbf{k})$, $\mu = -\ell, -\ell+1, \ldots, \ell-1, \ell$, span a subspace of $L^2(S^2, d\Omega(\mathbf{k}))$ that is stable under rotations. Furthermore,

$$f_{n\ell\mu}(R^{-1}\mathbf{k}) = \sum_{\mu'} \mathcal{D}^\ell(R)_{\mu'\mu} f_{n\ell\mu'}(\mathbf{k}), \tag{9.96}$$

where

$$f_{n\ell\mu}(\mathbf{k}) = Y^{\ell\mu}(\mathbf{k}) \, \varepsilon_n(\|\mathbf{k}\|). \tag{9.97}$$

For fixed ℓ, n, let $\mathfrak{K}^{\ell n}$ be the subspace of $L^2(\widehat{\mathbb{R}}^3, d\mathbf{k})$ spanned by the unit vectors $f_{n\ell m}$, $m = -\ell, -\ell+1, \ldots, \ell-1, \ell$. Then,

$$L^2(\widehat{\mathbb{R}}^3, d\mathbf{k}) = \bigoplus_{n=1}^\infty \bigoplus_{\ell=0}^\infty \mathfrak{K}^{\ell n}, \tag{9.98}$$

and each subspace $\mathfrak{K}^{\ell n}$ carries a representation of $SU(2)$ unitarily equivalent to the UIR $\mathcal{D}^\ell$.

Next, in view of the well-known decomposition (e.g., for angular momenta)

$$\mathcal{D}^j(\rho) \otimes \mathcal{D}^\ell(\rho) \simeq \bigoplus_{J=|j-\ell|}^{j+\ell} \mathcal{D}^J(\rho) \tag{9.99}$$

($\simeq$ implying unitary equivalence), the space $\mathfrak{K}^j \otimes L^2(\widehat{\mathbb{R}}^3, d\mathbf{k})$ decomposes as

$$\mathfrak{K}^j \otimes L^2(\widehat{\mathbb{R}}^3, d\mathbf{k}) = \bigoplus_{n=1}^\infty \bigoplus_{\ell=1}^\infty \bigoplus_{J=|j-\ell|}^{j+\ell} \mathfrak{K}^{nJ\ell}, \tag{9.100}$$

in which each $\mathfrak{K}^{nJ\ell}$ is a $(2J+1)$-dimensional space, carrying a UIR of $SU(2)$, unitarily equivalent to $\mathcal{D}^J$. The vectors

$$\widehat{\eta}^{nJ\ell M}(\mathbf{k}) = \sum_\mu C(j,\mu;\ell, M-\mu \mid JM)\, \chi^{j\mu}\, Y^{\ell\, M-\mu}(\mathbf{k})\, \varepsilon(\|\mathbf{k}\|), \quad (9.101)$$

for $M = -J, -J+1, \ldots, J-1, J$, and where the $C(j,\mu;\ell, M-\mu \mid JM)$ are $SU(2)$ Clebsch–Gordan coefficients, form an orthonormal basis in $\mathfrak{K}^{nJ\ell}$ and satisfy

$$\widehat{U}^j(0,\mathbf{0},\mathbf{0},\rho)\widehat{\eta}^{nJ\ell M} = \mathcal{D}^J(\rho)\widehat{\eta}^{nJ\ell M}, \quad M = -J, -J+1, \ldots, J-1, J. \quad (9.102)$$

The sum over μ in (9.101) runs through a subset of $-|j-\ell|, -|j-\ell|+1, \ldots, j+\ell$, determined by the constraints $-j \leq \mu \leq j$ and $-\ell \leq M-\mu \leq \ell$. Let V^J denote the UIR of the subgroup $H = \Theta \times SU(2)$ of $\widetilde{\mathcal{G}}'$:

$$V^J(\theta, \rho) = e^{i\theta} \mathcal{D}^J(\rho). \quad (9.103)$$

Then, (9.102) implies that, for any vector $\widehat{\eta} \in \mathfrak{K}^{nJ\ell}$,

$$\widehat{U}^j(\theta, \mathbf{0}, \mathbf{0}, \rho)\widehat{\eta} = \widehat{V}^J(\theta, \rho)\widehat{\eta}. \quad (9.104)$$

This also means that the projection operator,

$$\mathbb{P}^{nJ\ell} = \sum_{M=-J}^{J} |\widehat{\eta}^{nJ\ell M}\rangle\langle\widehat{\eta}^{nJ\ell M}|, \quad (9.105)$$

for the subspace $\mathfrak{K}^{nJ\ell}$ is a minimal projector satisfying

$$\widehat{U}^j(0,\mathbf{0},\mathbf{0},\rho)\mathbb{P}^{nJ\ell}\widehat{U}^j(0,\mathbf{0},\mathbf{0},\rho)^* = \mathbb{P}^{nJ\ell}, \quad (9.106)$$

for all $\rho \in SU(2)$ [minimality implying that there is no projector $\mathbb{P} \subset \mathbb{P}^{nJ\ell}$ also satisfying (9.106)]. Furthermore, a straightforward computation using properties of spherical harmonics shows that

$$
\begin{aligned}
c_{V^J}(\widehat{\eta}^{nJ\ell M'}) &= \sum_{M=-J}^{J} \int_\Gamma |\langle \widehat{\eta}^{nJ\ell M}_{\sigma(\mathbf{q},\mathbf{p})} | \widehat{\eta}^{nJ\ell M'}\rangle|^2 \, d\mathbf{q}\, d\mathbf{p} \\
&= (2\pi)^3 \sum_{M=-J}^{J} \sum_\mu |C(j,\mu;\ell, M-\mu \mid JM)|^2 \\
&\quad \times |C(j,\mu;\ell, M'-\mu \mid JM')|^2. \quad (9.107)
\end{aligned}
$$

Thus, the subspace $\mathfrak{K}^{nJ\ell}$ of $\mathfrak{K}^j \otimes L^2(\widehat{\mathbb{R}}^3, d\mathbf{k})$ is V^J-admissible, and again the representation is square integrable mod$(\Theta \times SU(2))$, since the above integral is clearly independent of the section σ. This also implies an interesting relation among Clebsch-Gordan coefficients, namely, that the sum on the RHS of (9.107) is independent of M', and indeed, using the fact that the vectors $\widehat{\eta}^{nJ\ell M}$, $M = -J, -J+1, \ldots, J-1, J$, form an orthonormal

9.2. A generalization: α- and V-admissibility   195

basis for $\mathfrak{K}^{nJ\ell}$, it is not hard to show [by computing the integral in (9.107) directly] that
$$c_{V^J}(\widehat{\eta}^{nJ\ell M'}) = c_{V^J}(\widehat{\eta}) = (2\pi)^3(2J+1), \tag{9.108}$$
for any vector $\widehat{\eta} \in \mathfrak{K}^{nJ\ell}$.

Consequently, for each $n = 1, 2, \ldots$, and each $J = |j-\ell|, |j-\ell|+1, \ldots, j+\ell$, the map $W_{nJ\ell}: \mathfrak{K}^j \otimes L^2(\mathbb{R}^3, d\mathbf{k}) \to \mathfrak{K}^J \otimes L^2(\Gamma, d\mathbf{q}\, d\mathbf{p})$, given by
$$(W_{nJ\ell}\widehat{\phi})^M(\mathbf{q}, \mathbf{p}) = [(2\pi)^3(2J+1)]^{-\frac{1}{2}} \langle \widehat{\eta}^{nJ\ell M} | \widehat{\phi} \rangle,$$
$$M = -J, -J+1, \ldots, J-1, J, \tag{9.109}$$
is an isometry. On $\mathfrak{K}^j \otimes L^2(\mathbb{R}^3, d\mathbf{k})$, the resolution of the identity,
$$\frac{1}{(2\pi)^3(2J+1)} \sum_{M=-J}^{J} \int_\Gamma |\widehat{\eta}^{nJ\ell M}_{\sigma(\mathbf{q},\mathbf{p})}\rangle\langle \widehat{\eta}^{nJ\ell M}_{\sigma(\mathbf{q},\mathbf{p})}|\, d\mathbf{q}\, d\mathbf{p} = I, \tag{9.110}$$
is valid for the class of *Galilei coherent states*,
$$[(2\pi)^3(2J+1)]^{-\frac{1}{2}} \widehat{\eta}^{nJ\ell M}_{\sigma(\mathbf{q},\mathbf{p})}, \quad (\mathbf{q}, \mathbf{p}) \in \mathbb{R}^6, \quad M = -J, -J+1, \ldots, J-1, J. \tag{9.111}$$
Explicitly, the CS are the functions,
$$\frac{1}{[(2\pi)^3(2J+1)]^{\frac{1}{2}}} \widehat{\eta}^{nJ\ell M}_{\sigma(\mathbf{q},\mathbf{p})}(\mathbf{k}) =$$
$$= \frac{1}{[(2\pi)^3(2J+1)]^{\frac{1}{2}}} (\widehat{U}^j(0, \mathbf{q}, \frac{\mathbf{p}}{m}, \mathbb{I}_2)\widehat{\eta}^{nJ\ell M})(\mathbf{k})$$
$$= \frac{1}{[(2\pi)^3(2J+1)]^{\frac{1}{2}}} e^{-i\mathbf{k}\cdot\mathbf{q}} \widehat{\eta}^{nJ\ell M}(\mathbf{k} - \mathbf{p})$$
$$= \frac{1}{[(2\pi)^3(2J+1)]^{\frac{1}{2}}} e^{-i\mathbf{k}\cdot\mathbf{q}} \sum_\mu C(j, \mu; \ell, M-\mu \mid JM)\, \chi^{j\mu}$$
$$\times Y^{\ell\, M-\mu}(\mathbf{k}-\mathbf{p})\, \varepsilon_n(\|\mathbf{k}-\mathbf{p}\|), \tag{9.112}$$

($M = -J, -J+1, \ldots, J-1, J$), which in general are vector coherent states. The image of $\mathfrak{K}^j \otimes L^2(\mathbb{R}^3, d\mathbf{k})$ in $\mathfrak{K}^J \otimes L^2(\Gamma, d\mathbf{q}\, d\mathbf{p})$, under the isometry $W_{nJ\ell}$, is a reproducing kernel Hilbert space $\mathfrak{H}_{nJ\ell}$. Its elements are vector-valued functions $\mathbf{\Phi}$ with components,
$$\Phi^M(\mathbf{q}, \mathbf{p}) = \frac{1}{[(2\pi)^3(2J+1)]^{\frac{1}{2}}} \sum_\mu C(j, \mu; \ell, M-\mu \mid JM)$$
$$\times \int_{\mathbb{R}^3} e^{i\mathbf{k}\cdot\mathbf{q}}\, \overline{Y^{\ell\, M-\mu}(\mathbf{k}-\mathbf{p})}\, \overline{\varepsilon_n(\|\mathbf{k}-\mathbf{p}\|)}\widehat{\phi}^\mu(\mathbf{k})\, d\mathbf{k},$$
$$M = -J, -J+1, \ldots, J-1, J. \tag{9.113}$$

The reproducing kernel for $\mathfrak{H}_{nJ\ell}$ is matrix valued and has the components

$$[\mathbf{K}_{nJ\ell}(\mathbf{q},\mathbf{p};\, \mathbf{q}'\mathbf{p}')]_{MM'} =$$
$$= \frac{1}{(2\pi)^3(2J+1)} \sum_\mu \overline{C(j,\mu;\, \ell, M-\mu \mid JM)} C(j,\mu;\, \ell, M'-\mu \mid JM')$$
$$\times \int_{\mathbb{R}^3} \exp\left[i\mathbf{k}\cdot(\mathbf{q}-\mathbf{q}')\right] \overline{Y^{\ell\, M-\mu}(\mathbf{k}-\mathbf{p})} Y^{\ell\, M'-\mu}(\mathbf{k}-\mathbf{p}')$$
$$\overline{\varepsilon_n(\|\mathbf{k}-\mathbf{p}\|)}\, \varepsilon_n(\|\mathbf{k}-\mathbf{p}'\|)\, d\mathbf{k}, \qquad (9.114)$$

$M, M' = -J, -J+1, \ldots, J-1, J$. Finally, the image of $\widehat{U}^j$ in $\mathfrak{H}_{nJ\ell} \subset \mathfrak{K}^J \otimes L^2(\Gamma, d\mathbf{q}\, d\mathbf{p})$ under $W_{nJ\ell}$ is

$$\mathbf{U}_{nJ\ell}(g) = W_{nJ\ell} \widehat{U}^j(g) W_{nJ\ell}^{-1}, \qquad g \in \widetilde{\mathcal{G}}', \qquad (9.115)$$

$$(\mathbf{U}_{nJ\ell}(\theta, \mathbf{a}, \mathbf{v}, \rho)\Phi)(\mathbf{q}, \mathbf{p}) = \exp\left[i\{\theta + m\mathbf{v}\cdot(\mathbf{q}-\mathbf{a})\}\right]$$
$$\times \mathcal{D}^J(\rho)\Phi(R(\rho^{-1})(\mathbf{q}-\mathbf{a}), R(\rho^{-1})(\mathbf{p}-m\mathbf{v})), \quad (9.116)$$

which again extends to the whole of $\mathfrak{K}^J \otimes L^2(\Gamma, d\mathbf{q}\, d\mathbf{p})$ as the representation of $\widetilde{\mathcal{G}}'$ induced from V^J.

In conclusion, observe that, since the isochronous Galilei group $\widetilde{\mathcal{G}}'$ is essentially the Weyl–Heisenberg group for three degrees of freedom with rotations, the coherent states (9.111) are also the VCS of this latter group.

9.2.3 Atomic coherent states

The coherent states (9.111) and (9.112) are all possible CS arising from the (extended) Galilei group, that are labeled by phase space points, satisfy a resolution of the identity, and incorporate spin degrees of freedom. Let η denote the Fourier-transformed function corresponding to any one of the vectors $\widehat{\eta}_{\sigma(\mathbf{q},\mathbf{p})}^{nJ\ell M}$. Then, η is a quantum mechanical wave function in configuration space. Its time evolved version,

$$\eta(\mathbf{x}, t) = (e^{-iHt}\eta)(\mathbf{x}), \qquad (9.117)$$

for some quantum mechanical Hamiltonian H, obeys the Schrödinger equation. The *fiducial* vectors $\widehat{\eta}^{nJ\ell M}$, $M = -J, -J-1, \ldots, J$, from which the coherent states (9.111) and (9.112) are built, can be looked on as quantum mechanical wave functions for a system with spin j, orbital angular momentum ℓ and total angular momentum J. Thus, the associated coherent states (9.111) and (9.112) can be called atomic (or molecular) coherent states.

Actually, the resolution of the identity (9.110) is not so surprising and it is easy to check it by direct computation. In fact, a much more general expression can be obtained as follows. Let $\widehat{\eta}^i$, $i = 1, 2, \ldots, n$, be any set of orthonormal vectors in $\mathfrak{K}^j \otimes L^2(\widehat{\mathbb{R}}^3, d\mathbf{k})$, spanning any n-dimensional

9.2. A generalization: α- and V-admissibility

subspace $\mathfrak{K}$ having projection operator $\mathbb{P}_{\mathfrak{K}}$. Then, by direct computation,

$$\frac{1}{(2\pi)^3 n}\int_\Gamma \mathbb{P}_{\mathfrak{K}}(\mathbf{q},\mathbf{p})\,d\mathbf{q}\,d\mathbf{p} = I, \quad \mathbb{P}_{\mathfrak{K}}(\mathbf{q},\mathbf{p}) = \widehat{U}^j(\sigma(\mathbf{q},\mathbf{p}))\,\mathbb{P}_{\mathfrak{K}}\,\widehat{U}^j(\sigma(\mathbf{q},\mathbf{p}))^*, \tag{9.118}$$

for any one of the following sections [obtained from the basic section (9.84)]:

$$\sigma(\mathbf{q},\mathbf{p}) = (0,\mathbf{q},\frac{\mathbf{p}}{m},\mathbb{I}_2)\,(\theta(\mathbf{q},\mathbf{p}),\mathbf{0},\mathbf{0},\rho). \tag{9.119}$$

Here, $\rho \in SU(2)$ is fixed and $\theta(\mathbf{q},\mathbf{p})$ is an arbitrary, real-valued, measurable function of $(\mathbf{q},\mathbf{p})$. Taking the η^i to be the n eigenstates of some observable (e.g., the Hamiltonian of some atomic system in an external potential), the general coherent states $[(2\pi)^3 n]^{-\frac{1}{2}}\,\widehat{\eta}^i_{\sigma(\mathbf{q},\mathbf{p})}$, $i = 1, 2, \ldots, n$, $(\mathbf{q},\mathbf{p}) \in \Gamma$, can be used in atomic computations involving such n-level systems. These CS, however, would not, in general, display the covariance property (9.53). Even more generally, if F is any trace-class operator on $\mathfrak{K}^j \otimes L^2(\widehat{\mathbb{R}}^3, d\mathbf{k})$, then with σ, as in (9.119), and $c(F) = (2\pi)^3[\mathrm{Tr}\,F]$,

$$\frac{1}{c(F)}\int_\Gamma F(\mathbf{q},\mathbf{p})\,d\mathbf{q}\,d\mathbf{p} = I, \quad F(\mathbf{q},\mathbf{p}) = \widehat{U}^j(\sigma(\mathbf{q},\mathbf{p}))F\widehat{U}^j(\sigma(\mathbf{q},\mathbf{p}))^*. \tag{9.120}$$

10
CS of General Semidirect Product Groups

In Chapter 9, Section 9.1, we studied a class of semidirect product groups, the regular representations of which consisted entirely of a discrete sum of irreducible subrepresentations, all square integrable. These groups were of the general form $G = \mathbb{R}^n \rtimes H$, where H was an n-dimensional subgroup of $GL(n, \mathbb{R})$ and its action on the dual space $\widehat{\mathbb{R}}^n$ gave rise to open free orbits. In this chapter, we generalize this setting and consider semidirect products of the type $G = V \rtimes S$, where V is an n-dimensional real vector space (n is assumed to be finite) and S is usually a subgroup of $GL(V)$ (the group of all nonsingular linear transformations of V). We shall again examine the action of S on the dual vector space V^*, but now without the assumption that the orbits be open or free, and generally, the dimensions of these orbits could be lower than n, the dimension of the vector space V on which S acts. The analysis, however, will have features similar to those encountered in Section 9.1, although we shall no longer look for the sort of simple square integrability with respect to the entire group as was done there. An interesting interplay between the geometry of the orbits and the existence of CS, square integrable over the associated homogeneous spaces, will be uncovered, which will turn out to be intimately related to the theory of group representations emanating from coadjoint orbits [Kir]. Our exposition follows [13], [8], [7], and [102], and the main result on square integrability is formulated in Theorem 10.3.3.

In the spirit of Section 9.1, however, we shall also look at a somewhat more general result for the square integrability of a group representation, which does not necessarily require the presence of open free orbits. This result is stated (without proof) in Theorem 10.3.1, leading to the following hierarchy of generalizations: Theorem 9.1.2 proves the square integrability of the quasi-regular representation of G, when open free orbits are present; Theorem 9.1.3 generalizes this result to the case in which the orbit is open but not necessarily free; however, the stabilizing subgroup is compact; Theorem 10.3.1 generalizes further to the case of an arbitrary induced representation of G, under the restriction that the orbit have positive Lebesgue measure in V and that the representation of the subgroup of S_0 of S, from which the representation of G is induced, be square integrable; finally, Theorem 10.3.3 obtains conditions for the square integrability of an arbitrary induced representation of G, assuming only that the representation of the subgroup of S_0 be finite-dimensional.

We begin with a discussion of squeezed states as an illustration of the use of semidirect product groups for constructing CS.

10.1 Squeezed states (♣)

Squeezed states were introduced in (2.7) of Chapter 2 (see the discussion following that equation). We study here the origin of these states in a certain representation of the *metaplectic group*. The discussion is based on [268]. Recall that the general *gaussons*, or Gaussian pure states were defined in Chapter 2, Equation (2.7), as

$$\eta_{\mathbf{q},\mathbf{p}}^{U,V}(\mathbf{x}) = \pi^{-\frac{n}{4}}[\det U]^{\frac{1}{4}} \exp\left[i(\mathbf{x} - \frac{\mathbf{q}}{2}) \cdot \mathbf{p}\right]$$
$$\times \exp\left[-\frac{1}{2}(\mathbf{x} - \mathbf{q}) \cdot (U + iV)(\mathbf{x} - \mathbf{q})\right], \quad (10.1)$$

where $\mathbf{q}, \mathbf{p} \in \mathbb{R}^n$, U is a (strictly) positive definite $n \times n$ real, symmetric matrix and V is an arbitrary $n \times n$ real, symmetric matrix. In the case in which U is the identity matrix and $V = 0$, these states reduce to the canonical coherent states (for n degrees of freedom), while if $V = 0$ and U is not the identity matrix, we get squeezed states.

We begin by introducing the notation,

$$\mathbf{\mathfrak{x}} = (\mathbf{q}, \mathbf{p}) \in \mathbb{R}^{2n}, \qquad \mathcal{Z} = V - iU \in \mathcal{M}_n(\mathbb{C}), \quad (10.2)$$

to rewrite the $\eta_{\mathbf{q},\mathbf{p}}^{U,V}$ as $\eta_{\sigma(\mathbf{\mathfrak{x}}, \mathcal{Z})}$, in which the σ will presently be identified with a section of the metaplectic group. Note that, for each fixed $\mathcal{Z}$, we have a resolution of the identity on $L^2(\mathbb{R}^n, d\mathbf{x})$:

$$\frac{1}{(2\pi)^n} \int_{\mathbb{R}^6} |\eta_{\sigma(\mathbf{\mathfrak{x}},\mathcal{Z})}\rangle\langle\eta_{\sigma(\mathbf{\mathfrak{x}},\mathcal{Z})}| \, d\mathbf{\mathfrak{x}} = I, \qquad d\mathbf{\mathfrak{x}} = d\mathbf{q}\, d\mathbf{p}. \quad (10.3)$$

The metaplectic group is the semidirect product, $\text{Mp}(2n, \mathbb{R}) = G_{WH}(n) \rtimes \text{Sp}(2n, \mathbb{R})$, of the Weyl–Heisenberg group $G_{WH}(n)$ for n degrees of freedom and the real symplectic group $\text{Sp}(2n, \mathbb{R})$ of $\mathbb{R}^{2n}$. Elements $g \in \text{Mp}(2n, \mathbb{R})$

10.1. Squeezed states (♣)

are of the form
$$g = (\theta, \pmb{x}, M), \quad \text{where} \quad \theta \in \mathbb{R}, \quad (\theta, \pmb{x}) \in G_{WH}(n), \quad M \in \text{Sp}(2n, \mathbb{R}). \tag{10.4}$$

Recall that the product rule in $G_{WH}(n)$ is
$$(\theta_1, \pmb{x}_1)(\theta_2, \pmb{x}_2) = (\theta_1 + \theta_2 + \xi(\pmb{x}_1, \pmb{x}_2), \pmb{x}_1 + \pmb{x}_2), \tag{10.5}$$
where, of course, $\pmb{x}_i = (\mathbf{q}_i, \mathbf{p}_i)$, $i = 1, 2$, and
$$\xi(\pmb{x}_1, \pmb{x}_2) = \frac{1}{2}(\mathbf{p}_1 \cdot \mathbf{q}_2 - \mathbf{p}_2 \cdot \mathbf{q}_1). \tag{10.6}$$
The elements $M \in \text{Sp}(2n, \mathbb{R})$ are $2n \times 2n$ real matrices satisfying
$$M\beta M^T = \beta, \quad \det M = 1, \quad \text{where} \quad \beta = \begin{pmatrix} 0 & \mathbb{I}_n \\ -\mathbb{I}_n & 0 \end{pmatrix} \tag{10.7}$$
($\mathbb{I}_n = n \times n$ identity matrix). The product rule of $\text{Mp}(2n, \mathbb{R})$ is, thus,
$$(\theta_1, \pmb{x}_1, M_1)(\theta_2, \pmb{x}_2, M_2) = (\theta_1 + \theta_2 + \xi(\pmb{x}_1, M_1\pmb{x}_2), \pmb{x}_1 + M_1\pmb{x}_2, M_1 M_2). \tag{10.8}$$
The $n \times n$ complex matrices $\mathcal{Z} = V - iU$, where $V = V^T$ and $U = U^T$, $U > 0$, are $n \times n$ real matrices, define a certain coset space of $\text{Sp}(2n, \mathbb{R})$, as we shall now see.

Any element $M \in \text{Sp}(2n, \mathbb{R})$ can be written in block form as
$$M = \begin{pmatrix} A & B \\ C & D \end{pmatrix}, \quad \text{with} \quad AD^T - BC^T = \mathbb{I}_n, \quad AB^T = BA^T, \quad CD^T = DC^T, \tag{10.9}$$
where A, B, C, and D are $n \times n$ real matrices. The maximal compact subgroup of $\text{Sp}(2n, \mathbb{R})$ is $\text{Sp}(2n, \mathbb{R}) \cap SO(2n)$, a group that is isomorphic to $U(n)$ (the group of all $n \times n$ complex unitary matrices of unit determinant). This means that a general $M \in \text{Sp}(2n, \mathbb{R})$ has the polar decomposition:
$$M = p(M)^{\frac{1}{2}} K, \quad p(M) = MM^T, \quad K \in \text{Sp}(2n, \mathbb{R}) \cap SO(2n) \simeq U(n). \tag{10.10}$$
The matrix $p(M)^{\frac{1}{2}}$ is the unique positive square root of the positive-definite matrix $p(M)$. Moreover, as shown in [268], $p(M)$ can be conveniently written in the form
$$p(M) = M(U, V)M(U, V)^T,$$
$$\text{where} \quad M(U, V) = \begin{pmatrix} \mathbb{I}_n & 0 \\ -V & \mathbb{I}_n \end{pmatrix} \begin{pmatrix} U^{-\frac{1}{2}} & 0 \\ 0 & U^{\frac{1}{2}} \end{pmatrix}, \tag{10.11}$$
in which, once again, U is a (strictly) positive-definite, $n \times n$ real symmetric matrix and V is an $n \times n$ real symmetric matrix. Using the complex notation introduced above, we see that the coset space $\text{Sp}(2n, \mathbb{R})/[\text{Sp}(2n, \mathbb{R}) \cap SO(2n)]$ can be identified with the complex tubular domain $\mathbb{T}$ of matrices $\mathcal{Z} = V - iU$ (with V a real symmetric and U

a positive-definite, symmetric, $n \times n$ matrix). In view of (10.11), we shall write write $p(M) = p(\mathcal{Z})$. Then, for arbitrary $\mathcal{Z} \in \mathbb{T}$ and $M \in \mathrm{Sp}(2n, \mathbb{R})$, if we write

$$p(\mathcal{Z}') = M\, p(\mathcal{Z})\, M^T, \qquad (10.12)$$

a straightforward computation yields the transformation law

$$\mathcal{Z}' = M[\mathcal{Z}] = (D\mathcal{Z} - C)(A - B\mathcal{Z})^{-1}, \qquad (10.13)$$

for the action of $M \in \mathrm{Sp}(2n, \mathbb{R})$ on $\mathbb{T}$. Furthermore, it is clear that

$$\mathcal{Z} = M[\mathcal{Z}_0], \quad \text{where} \quad \mathcal{Z}_0 = -i\mathbb{I}_n, \quad M = p(\mathcal{Z})^{\frac{1}{2}} K, \qquad (10.14)$$

meaning that $\mathbb{T}$ is the orbit of $\mathcal{Z}_0 = -i\mathbb{I}_n$ under the action (10.13) of $\mathrm{Sp}(2n, \mathbb{R})$. It also means that the elements of the complex matrices $\mathcal{Z}$ provide a global coordinatization for the homogeneous space $\mathrm{Sp}(2n, \mathbb{R}) / [\mathrm{Sp}(2n, \mathbb{R}) \cap SO(2n)]$, which, as a geometrical object, is therefore a complex (Kähler) manifold.

The Lie algebra $\mathfrak{sp}(2n, \mathbb{R})$ of $\mathrm{Sp}(2n, \mathbb{R})$ consists of all $2n \times 2n$ real matrices X with the property that βX is a symmetric matrix. Thus, any $X \in \mathfrak{sp}(2n, \mathbb{R})$ has the block form,

$$X = \begin{pmatrix} \alpha & \beta \\ \gamma & -\alpha^T \end{pmatrix}, \qquad \beta = \beta^T, \ \gamma = \gamma^T, \qquad (10.15)$$

in terms of three $n \times n$ real matrices, α, β, and γ. If $\alpha = -\alpha^T$ and $\gamma = -\beta$, in the above, the corresponding X lies in the Lie algebra of the maximal compact subgroup $\mathrm{Sp}(2n, \mathbb{R}) \cap SO(2n)$. Clearly, β is such an element of the Lie algebra. Consider now the *adjoint action* of $M \in \mathrm{Sp}(2n, \mathbb{R})$ on β, as defined in (4.137):

$$\mathrm{Ad}_M(\beta) = M\beta M^{-1}. \qquad (10.16)$$

If the pairing between $\mathfrak{sp}(2n, \mathbb{R})$ and its dual $\mathfrak{sp}(2n, \mathbb{R})^*$ is given via the trace operation, β may itself be identified with an element of the dual space $\mathfrak{sp}(2n, \mathbb{R})^*$, and the *coadjoint action* of $\mathrm{Sp}(2n, \mathbb{R})$ on β is then identifiable with:

$$\mathrm{Ad}^{\#}_M(\beta) = M^{T^{-1}} \beta M^T. \qquad (10.17)$$

It is easily verified, using the decomposition (10.10), that the stability subgroup of β under both the coadjoint and adjoint actions is precisely the maximal compact subgroup $\mathrm{Sp}(2n, \mathbb{R}) \cap SO(2n)$. Thus, the complex domain $\mathbb{T}$ is in fact the orbit of β under the coadjoint (or adjoint) action.

In order to obtain the squeezed states $\eta^{U,V}_{q,p}$ in (10.1), it will be necessary to construct an appropriate representation of the metaplectic group $\mathrm{Mp}(2n, \mathbb{R})$. We do this by starting with a representation of the Weyl–Heisenberg group $G_{WH}(n)$, obtained by generalizing to n degrees of freedom the representation introduced in (2.33) for one degree of freedom. This extended UIR, denoted again by U^λ, $\lambda \in \mathbb{R}$, $\lambda \neq 0$, acts on the Hilbert space $L^2(\mathbb{R}^n, d\mathbf{x})$ in the manner

$$(U^\lambda(\theta, \mathbf{x})\psi)(\mathbf{x}) = e^{i\lambda\theta}e^{i\lambda\mathbf{p}\cdot(\mathbf{x}-\frac{\mathbf{q}}{2})}\phi(\mathbf{x}-\mathbf{q}), \qquad \mathbf{x} = (\mathbf{q}, \mathbf{p}), \quad \psi \in L^2(\mathbb{R}^n, d\mathbf{x}). \tag{10.18}$$

While, for different values of λ, the representations U^λ are unitarily inequivalent, for our purposes, it will be enough to choose $\lambda = 1$. We shall do so in the following and write the representation simply as U. Next, we note that the representation U can actually be extended to a UIR of the entire metaplectic group. Indeed, let us define the $2n$-component vector operator $\mathfrak{X}$ on $L^2(\mathbb{R}^n, d\mathbf{x})$, in terms of the position and momentum operators Q_i, P_i,

$$\mathfrak{X} = (Q_1, Q_2, \ldots Q_n, P_1, P_2, \ldots, P_n),$$
$$(Q_i\psi)(\mathbf{x}) = x_i\psi(\mathbf{x}), \quad (P_i\psi)(\mathbf{x}) = -i\frac{\partial}{\partial x_i}\psi(\mathbf{x}). \tag{10.19}$$

In terms of the operator $\mathfrak{X}$, the canonical commutation relations $[Q_k, P_\ell] = iI\delta_{ij}$ are conveniently rewritten as

$$[\mathfrak{X}_k, \mathfrak{X}_\ell] = iI(\beta)_{k\ell}, \qquad k, \ell = 1, 2, \ldots, 2n. \tag{10.20}$$

If M is a $2n \times 2n$ real matrix and $\mathfrak{X}' = M\mathfrak{X}$, then the components of $\mathfrak{X}'$ satisfy the CCR (10.20) if and only if $M \in \mathrm{Sp}(2n, \mathbb{R})$. Moreover, for each $M \in \mathrm{Sp}(2n, \mathbb{R})$, there exists a unitary operator $\mathcal{U}(M)$ on $L^2(\mathbb{R}^n, d\mathbf{x})$, such that $M \mapsto \mathcal{U}(M)$ is a representation of $\mathrm{Sp}(2n, \mathbb{R})$, which acts on the vector $\mathfrak{X}$ in the manner,

$$\mathcal{U}(M)\mathfrak{X}\,\mathcal{U}(M)^* = M^{-1}\mathfrak{X}. \tag{10.21}$$

Consequently, in this representation of $\mathrm{Sp}(2n, \mathbb{R})$, the elements X of the Lie algebra $\mathfrak{sp}(2n, \mathbb{R})$ [see (10.15)] are realized as the operators

$$\widehat{X} = -\frac{1}{2}\mathfrak{X}\cdot\beta X\mathfrak{X}, \tag{10.22}$$

on $L^2(\mathbb{R}^n, d\mathbf{x})$, which means that, if $M(t) = \exp[tX]$ is the one-parameter subgroup of $\mathrm{Sp}(2n, \mathbb{R})$ generated by X, then $\mathcal{U}(M(t)) = \exp[-i\widehat{X}t]$. Furthermore, for all $(\theta, \mathbf{x}) \in G_{WH}(n)$ and $M \in \mathrm{Sp}(2n, \mathbb{R})$, the relations

$$U(\theta, \mathbf{x})\mathfrak{X}U(\theta, \mathbf{x})^* = \mathfrak{X} - \mathbf{x}, \qquad \mathcal{U}(M)U(\theta, \mathbf{x})\mathcal{U}(M)^* = U(\theta, M\mathbf{x}), \tag{10.23}$$

are easily seen to hold. At the level of the generators $\widehat{X}$, one also has the equivalent relations

$$[\mathfrak{X}, \widehat{X}] = iX\mathfrak{X}, \qquad [\widehat{X}, \widehat{X}'] = i\widehat{[X, X']}. \tag{10.24}$$

The relations (10.20), (10.21), and (10.23) together [or, alternatively, the relations (10.24)] imply that the operators

$$\mathbf{U}(\theta, \mathbf{x}, M) = U(\theta, \mathbf{x})\mathcal{U}(M), \qquad (\theta, \mathbf{x}, M) \in \mathrm{Mp}(2n, \mathbb{R}), \tag{10.25}$$

define a UIR of $\mathrm{Mp}(2n, \mathbb{R})$ on $L^2(\mathbb{R}^n, d\mathbf{x})$. It is difficult to write explicitly the action of a general representation operator $\mathbf{U}(\theta, \mathbf{x}, M)$ on an arbitrary vector $\psi \in L^2(\mathbb{R}^n, d\mathbf{x})$. Identifying the coset $\mathrm{Mp}(2n,\mathbb{R})/[\Theta \times (\mathrm{Sp}(2n,\mathbb{R}) \cap SO(2n))]$, however, ($\Theta$ being the phase subgroup of G_{WH}) with $\mathbb{R}^{2n} \times \mathbb{T}$, let us define the section

$$\sigma : \mathbb{R}^{2n} \times \mathbb{T} \to \mathrm{Mp}(2n,\mathbb{R}), \quad \sigma(\boldsymbol{x}, \mathcal{Z}) = (0, \boldsymbol{x}, M(U,V)),$$
$$\boldsymbol{x} = (\mathbf{q},\mathbf{p}), \quad \mathcal{Z} = V - iU, \quad (10.26)$$

$M(U,V)$ being as in (10.11). It is then straightforward to verify that

$$\eta_{\sigma(\boldsymbol{x},\mathcal{Z})} = \eta_{\boldsymbol{x},\mathcal{Z}}^{U,V} = \mathbf{U}(\sigma(\boldsymbol{x},\mathcal{Z}))\eta, \quad \text{where}, \quad \eta(\mathbf{x}) = \pi^{-\frac{n}{4}} \exp[-\frac{1}{2}\mathbf{x}\cdot\mathbf{x}]. \quad (10.27)$$

Thus, finally, the general squeezed states or gaussons are all obtained from the ground state wave function of a system of n harmonic oscillators using a section in the metaplectic group. Moreover, writing

$$\mathbb{P} = |\eta\rangle\langle\eta|, \quad \text{and} \quad \mathbb{P}(\boldsymbol{x}, \mathcal{Z}) = |\eta_{\sigma(\boldsymbol{x},\mathcal{Z})}\rangle\langle\eta_{\sigma(\boldsymbol{x},\mathcal{Z})}|, \quad (10.28)$$

we find that

$$\mathbb{P}(\boldsymbol{x}, \mathcal{Z}) = \mathbf{U}(\theta, \boldsymbol{x}, M) \, \mathbb{P} \, \mathbf{U}(\theta, \boldsymbol{x}, M)^* \quad (10.29)$$

[assuming the decomposition (10.10) and (10.11) of $M \in \mathrm{Sp}(2n,\mathbb{R})$], showing that the corresponding projection operators form an orbit of the metaplectic group.

10.2 Geometry of semidirect product groups

For a general analysis of CS of semidirect product groups, it will now be necessary to examine some detailed features of the group $G = V \rtimes S$ and of the associated geometry of orbits. Much of the Unfortunately a proliferation of symbols and notation will occur in the course of the discussion. In order to make the reading easier, we have drawn up a table of the more important quantities at the end of Section 10.2.3 (Table 10.1).

10.2.1 A special class of orbits

A general group element in $G = V \rtimes S$ will be written as $g = (x,s)$, with $x \in V$ and $s \in S$; the group multiplication being given by $(x_1, s_1)(x_2, s_2) = (x_1 + s_1 x_2, s_1 s_2)$, in which the action of S on V is indicated by $x \mapsto sx$. If $k \in V^*$, the dual of V, and $<\,;\,>$ denotes the dual pairing between V^* and V, we shall indicate by $k \mapsto sk$ the action of S on V^*, defined in the manner

$$<sk\,;\,x> \;=\; <k\,;\,s^{-1}x>, \quad (10.30)$$

which we shall later identify with the *coadjoint action*.

10.2. Geometry of semidirect product groups

Let $k_0 \in V^*$ be a fixed vector and $\mathcal{O}^*$ be its orbit under S. Recall that $\mathcal{O}^*$ then consists of all elements $k \in V^*$ that are of the type $k = sk_0$, $s \in S$. Let S_0 be the stability subgroup of k_0; i.e., $s \in S_0$ iff $sk_0 = k_0$. Then, $S/S_0 \simeq \mathcal{O}^*$ via the map $gS_0 \mapsto gk_0$. Moreover, as a geometrical object, $\mathcal{O}^*$ is a smooth manifold, and, at any point $k \in \mathcal{O}^*$, we may consider the tangent space $T_k\mathcal{O}^*$. If we assume that the dimension of $\mathcal{O}^*$, as a real manifold, is m ($\leq n$, the dimension of V), then $T_k\mathcal{O}^*$ is a real vector space, also of dimension m. Moreover, $T_k\mathcal{O}^*$ can be identified in a natural way with a subspace of V^*. Indeed, let $\{e^{*\,i}\}_{i=1}^n$ be a basis of V^* and suppose that $k(t) = \sum_{i=1}^n k_i(t) e^{*\,i}$ is a smooth curve in $\mathcal{O}^*$, passing through $k = k(0)$ and defined for values of t lying in some open set $(-\varepsilon, \varepsilon) \subset \mathbb{R}$. The components $k_i(t)$ are some smooth functions of t. Then, differentiating with respect to t,

$$\dot{k}(0) = \sum_{i=1}^n \dot{k}_i(0) e^{*\,i} \in V^*$$

is a vector tangent to $\mathcal{O}^*$ at k. We shall always consider $T_k\mathcal{O}^*$ as being embedded in this manner: $T_k\mathcal{O}^* \subset V^*$. The space $T\mathcal{O}^* = \bigcup_{k \in \mathcal{O}^*} T_k\mathcal{O}^*$ of all tangent vectors at all points k is again a smooth manifold, of dimension $2m$, called the *tangent bundle* of $\mathcal{O}^*$. The cotangent space $T_k^*\mathcal{O}^*$ at the point $k \in \mathcal{O}^*$ is the dual space of $T_k\mathcal{O}^*$, which we therefore identify with a subspace of V; i.e., $T_k^*\mathcal{O}^* \subset V$. Correspondingly we have the *cotangent bundle* $T^*\mathcal{O}^* = \bigcup_{k \in \mathcal{O}^*} T_k^*\mathcal{O}^*$. As a manifold, $T^*\mathcal{O}^*$ has certain properties that make it resemble a classical phase space. It is a *symplectic manifold*, which means that it comes equipped with a nondegenerate, closed *two-form* — a point we shall return to a little later.

Along with the two vector bundles $T\mathcal{O}^*$ and $T^*\mathcal{O}^*$, there is a second pair of vector bundles that appear here simultaneously and that we also need to consider. For the tangent space $T_k\mathcal{O}^*$, let N_k be its *annihilator* in V; i.e.,

$$N_k = \{x \in V \mid\, <p\,;\,x> = 0,\ \forall p \in T_k\mathcal{O}^*\}. \tag{10.31}$$

Then,

$$N_k \oplus T_k^*\mathcal{O}^* = V, \tag{10.32}$$

and $N = \bigcup_{k \in \mathcal{O}^*} N_k$ is a manifold, referred to as the *normal bundle*. Analogously we define

$$N_k^* = \{p \in V^* \mid\, <p\,;\,x> = 0,\ \forall x \in T_k^*\mathcal{O}^*\}, \tag{10.33}$$

so that, again,

$$N_k^* \oplus T_k\mathcal{O}^* = V^*, \tag{10.34}$$

with N_k^* being identifiable with the dual of N_k and $N^* = \bigcup_{k \in \mathcal{O}^*} N_k^*$ with the dual bundle of N.

Consider again the coadjoint action [see (10.30)] of S on V^*, under which $k \mapsto k' = sk$. The derivative of this map at the point $k \in V^*$, which we denote by $D_k(s)$, is a linear map between the two tangent spaces $T_k V^*$ and $T_{k'} V^*$. The restriction of the map $k \mapsto sk$ to the set of points $k \in \mathcal{O}^*$, which we denote by $s|_{\mathcal{O}^*}$, leaves $\mathcal{O}^*$ invariant, and thus the restriction of the derivative $D_k(s)$ to $T_k \mathcal{O}^*$ is exactly $D_k(s|_{\mathcal{O}^*})$, the derivative of the restricted map $s|_{\mathcal{O}^*}$ at k, and this is a linear map between $T_k \mathcal{O}^*$ and $T_{k'} \mathcal{O}^*$. In particular, since, for $s \in S_0$, $sk_0 = k_0$, its derivative $D_{k_0}(s)$ at k_0 maps $T_{k_0} V^*$ to itself and, consequently, $D_{k_0}(s|_{\mathcal{O}^*})$, for $s \in S_0$, maps $T_{k_0} \mathcal{O}^*$ to itself. Since all tangent spaces $T_k V^*$ can be naturally identified with V^*, we identify the action of $D_k(s)$ on $T_k V^*$ with that of s itself, and it then follows that, for $s \in S_0$ and $p \in T_{k_0} \mathcal{O}^*$, $sp \in T_{k_0} \mathcal{O}^*$. (One can think of s as a matrix acting on the vector p. Then the derivative of this map, with respect to p, is simply s itself.) By duality, we then immediately obtain that S_0 leaves N_{k_0} invariant. Thus, $N_0 \rtimes S_0$ is a subgroup of G (for simplicity, we write $N_0 = N_{k_0}$). Let Γ denote the resulting left coset space,

$$\Gamma = G/H_0, \qquad H_0 = N_0 \rtimes S_0. \tag{10.35}$$

Also, set

$$V_0 = T_{k_0}^* \mathcal{O}^*, \qquad V_0^* = T_{k_0} \mathcal{O}^*, \tag{10.36}$$

so that, by (10.32) and (10.34), $N_0 \oplus V_0 = V$ and $N_0^* \oplus V_0^* = V^*$. Since any $x \in V$ can be written uniquely as $x = n + v$, with $n \in N_0$ and $v \in V_0$, it would seem plausible that Γ should be isomorphic, as a Borel space, to $V_0 \times \mathcal{O}^*$. In fact, as we explain below, Γ is isomorphic to $T^* \mathcal{O}^*$ as a symplectic manifold and to $V_0 \times \mathcal{O}^*$ as a Borel space.

10.2.2 The coadjoint orbit structure of Γ

We now present a reasonably self-contained and not-too-technical description of the coset space Γ in (10.35) as a *coadjoint orbit* of G and as a symplectic manifold, in a sense that we make precise. While no rigorous differential geometric proofs are given, the discussion can in fact be developed into exact proofs without too much difficulty. A more rigorous treatment may, for instance, be found in [Gui]. Let us rewrite the elements $g = (x, s)$ of G in matrix form,

$$g = \begin{pmatrix} s & x \\ 0 & 1 \end{pmatrix}, \tag{10.37}$$

with group multiplication being replicated by matrix multiplication. Denoting by $\mathfrak{g}, \mathfrak{v}$, and $\mathfrak{s}$ the Lie algebras of G, V, and S, respectively, an element $X = (a, J) \in \mathfrak{g}$, with $a \in \mathfrak{v}$ and $J \in \mathfrak{s}$, then appears as the matrix

$$X = \begin{pmatrix} J & a \\ 0 & 0 \end{pmatrix}. \tag{10.38}$$

Note that J can be identified with a linear transformation of V, and $\mathfrak{v}$ with V itself in an obvious fashion. The *adjoint action* of G on $\mathfrak{g}$ is then defined

10.2. Geometry of semidirect product groups

by

$$\text{Ad}_g(X) = gXg^{-1} = \begin{pmatrix} sJs^{-1} & sa - sJs^{-1}x \\ 0 & 0 \end{pmatrix}. \tag{10.39}$$

The resultant action on the dual space $\mathfrak{g}^*$ of the Lie algebra $\mathfrak{g}$ is now easily obtained. Indeed, let $\mathfrak{v}^*, \mathfrak{s}^*$ be the dual spaces of $\mathfrak{v}$ and $\mathfrak{s}$, respectively, and let $X^* = (a^*, J^*) \in \mathfrak{g}^*$, with $a^* \in \mathfrak{v}^*$ and $J^* \in \mathfrak{s}^*$. Also, the dual pairing in $(\mathfrak{g}^*, \mathfrak{g})$ has the form

$$< X^* \, ; \, X >_{\mathfrak{g}^*, \mathfrak{g}} \, = \, < a^* \, ; \, a >_{\mathfrak{v}^*, \mathfrak{v}} \, + \, < J^* \, ; \, J >_{\mathfrak{s}^*, \mathfrak{s}}. \tag{10.40}$$

We now define the *coadjoint action* of G on $\mathfrak{g}^*$:

$$< \text{Ad}_g^{\#}(X^*) \, ; \, X >_{\mathfrak{g}^*, \mathfrak{g}} \, = \, < X^* \, ; \, \text{Ad}_{g^{-1}}(X) >_{\mathfrak{g}^*, \mathfrak{g}}. \tag{10.41}$$

Writing $\text{Ad}_g^{\#}(X^*) = X^{*'} = (a^{*'}, J^{*'})$, using (10.39) and noting that $\mathfrak{v}^*$ can be identified with V^*, we easily derive,

$$\left. \begin{array}{rcl} a^{*'} & = & sa^* \\ J^{*'} & = & \text{Ad}_s^{\#S}(J^*) + (sa^*) \odot x \end{array} \right\}, \tag{10.42}$$

where the first equation reproduces the coadjoint action defined in (10.30),

$$< sa^* \, ; \, a >_{\mathfrak{v}^*, \mathfrak{v}} \, = \, < a^* \, ; \, s^{-1}a >_{\mathfrak{v}^*, \mathfrak{v}}, \quad a \in \mathfrak{v},$$

while

$$< \text{Ad}_s^{\#S}(J^*) \, ; \, J >_{\mathfrak{s}^*, \mathfrak{s}} \, = \, < J^* \, ; \, s^{-1}Js >_{\mathfrak{s}^*, \mathfrak{s}}, \quad J \in \mathfrak{s}, \tag{10.43}$$

and $(sa^*) \odot x$ denotes the element in $\mathfrak{s}^*$ for which

$$< (sa^*) \odot x \, ; \, J >_{\mathfrak{s}^*, \mathfrak{s}} \, = \, < sa^* \, ; \, Jx >_{\mathfrak{v}^*, \mathfrak{v}}, \quad J \in \mathfrak{s}. \tag{10.44}$$

Let us examine the orbit of $X_0^* = (k_0, 0) \in \mathfrak{g}^*$ under the coadjoint action. From (10.42)–(10.44),

$$\text{Ad}_g^{\#}(k_0, 0) = (sk_0, \, sk_0 \odot x) = (k, \, k \odot x), \quad g = (x, s). \tag{10.45}$$

Let M be the dimension of the Lie group S, considered as a real manifold. Since $\dim \mathcal{O}^* = m$, the dimension of the subgroup S_0 is $M - m$. Choose a basis $\{J_i\}_{i=1}^M$ for the Lie algebra $\mathfrak{s}$, in such a way that $J_{m+1}, J_{m+2}, \ldots, J_M$ is a basis for $\mathfrak{s}_0$, the Lie algebra of S_0. Recall that each J_i can be identified with a linear map from V to itself. Denote by $J_i^\dagger$ the dual map on V^* (now identified with $\mathfrak{v}^*$): $< J_i^\dagger k \, ; \, x > \, = \, < k \, ; \, J_i x >$, for all $x \in V$, $k \in V^*$. Then, since S_0 is the stability subgroup of k_0,

$$\left. \begin{array}{rcl} J_i^\dagger k_0 & \neq & 0, \quad i = 1, 2, \ldots, m, \\ J_i^\dagger k_0 & = & 0, \quad i = m+1, m+2, \ldots, M. \end{array} \right\} \tag{10.46}$$

Moreover, $J_i^\dagger k_0$, $i = 1, 2, \ldots, m$, is a basis for the tangent space $T_{k_0}\mathcal{O}^* = V_0^*$. Since $sk_0 = k$ implies that $s(T_{k_0}\mathcal{O}^*) = T_k\mathcal{O}^*$, it follows that $sJ_i^\dagger s^{-1}k$, $i = 1, 2, \ldots, m$, is a basis for $T_k\mathcal{O}^*$.

Consider now the element $k_0 \odot x \in \mathfrak{s}^*$. By (10.44),
$$< k_0 \odot x \, ; \, J >_{\mathfrak{s}^*,\mathfrak{s}} \; = \; < k_0 \, ; \, Jx >_{\mathfrak{v}^*,\mathfrak{v}},$$
and, in particular, for all J_i, $i = m+1, m+2, \ldots, M$,
$$< k_0 \odot x \, ; \, J_i >_{\mathfrak{s}^*,\mathfrak{s}} \; = \; < J_i^\dagger k_0 \, ; \, x >_{\mathfrak{v}^*,\mathfrak{v}} \; = \; 0, \quad \forall x \in V.$$
On the other hand, for $J_i = 1, 2, \ldots, m$, and $x \in N_0$ (the annihilator of $T_{k_0}\mathcal{O}^*$),
$$< k_0 \odot x \, ; \, J_i >_{\mathfrak{s}^*,\mathfrak{s}} \; = \; < J_i^\dagger k_0 \, ; \, x >_{\mathfrak{v}^*,\mathfrak{v}} \; = \; 0.$$
Thus, writing $x = n_{k_0} + v_{k_0}$, with $n_{k_0} \in N_0 = N_{k_0}$ and $v_{k_0} \in V_0 = T_{k_0}^* \mathcal{O}^*$, we get
$$k_0 \odot x = k_0 \odot v_{k_0},$$
which means that each $k_0 \odot x \in \mathfrak{s}^*$ can be identified with an element $v_{k_0} \in T_{k_0}^* \mathcal{O}^*$. Conversely, given any $v_{k_0} \in T_{k_0}^* \mathcal{O}^*$, we can define the element $k_0 \odot v_{k_0} \in \mathfrak{s}^*$ corresponding to it. Generally, using the decomposition
$$x = n_k + v_k, \qquad x \in V, \quad n_k \in N_k, \quad v_k \in T_k^* \mathcal{O}^*, \qquad (10.47)$$
we establish the correspondence
$$(k, k \odot x) \mapsto (k, v_k), \qquad (10.48)$$
between elements in the orbit $\mathcal{O}_{(k_0,0)}$ of $(k_0, 0) \in \mathfrak{g}^*$, under the coadjoint action of G, and elements in the cotangent bundle $T^*\mathcal{O}^*$. Moreover, it is clear from (10.45) that $H_0 = N_0 \rtimes S_0$ is the stability subgroup of $(k_0, 0) \in \mathfrak{g}^*$. Hence, finally,
$$\Gamma = G/H_0 \simeq \{\mathrm{Ad}_g^\#(k_0, 0) \mid g \in G\} = \mathcal{O}_{(k_0,0)} \simeq T^*\mathcal{O}^*. \qquad (10.49)$$

The action of G on $T^*\mathcal{O}^*$ is readily computed. On V, corresponding to the decomposition $V = N_k \oplus T_k^* \mathcal{O}^*$ [see (10.32)], let us introduce the two projection operators, $\mathbf{P}_k^n$ and $\mathbf{P}_k^v$, such that, for $x \in V$ and $x = n_k + v_k$ [see (10.47)],
$$\mathbf{P}_k^n x = n_k, \qquad \mathbf{P}_k^v x = v_k. \qquad (10.50)$$
Next, writing
$$\mathrm{Ad}_{(x,s)}^\#(k, k \odot v_k) = (k', k' \odot w_{k'}), \qquad (10.51)$$
and applying (10.42), we get
$$k' = sk, \qquad w_{k'} = \mathbf{P}_{k'}^v x + s v_k \in T_{sk}^* \mathcal{O}^*. \qquad (10.52)$$
(Note that $\mathbf{P}_{k'}^v s v_k = s v_k$.) Thus, under the action of $g = (x, s) \in G$, an element $(k, v_k) \in T^*\mathcal{O}^*$ transforms to
$$g(k, v_k) = (sk, \mathbf{P}_{sk}^v x + s v_k). \qquad (10.53)$$

10.2. Geometry of semidirect product groups

To make the correspondence between $T^*\mathcal{O}^*$ and G/H_0 more explicit, recall first that, as manifolds, $\mathcal{O}^* \simeq S/S_0$. Let $\Lambda : \mathcal{O}^* \to S$ be a global Borel section, such that

$$\begin{aligned}\Lambda(k_0) &= e = \text{identity element of } S,\\ \Lambda(k)k_0 &= k, \quad k \in \mathcal{O}^*,\end{aligned} \quad (10.54)$$

and which is a smooth map on some open dense set in $\mathcal{O}^*$. Then, any element $s \in S$ can be uniquely written as

$$s = \Lambda(k)s_0, \quad k \in \mathcal{O}^*, \quad s_0 \in S_0. \quad (10.55)$$

This section will be fixed once and for all. Although the specific families of coherent states we shall obtain using it will depend on the particular choice of this section, the results on square integrability, such as Theorems 10.3.3 and 10.3.5, will not, however. If $v_k \in T_k^*\mathcal{O}^*$, then $\Lambda(k)^{-1}v_k \in T_{k_0}^*\mathcal{O}^* = V_0$. For any $(x, s) \in G$, let $x = n_k + v_k$ be the decomposition (10.47) and $s = \Lambda(k)s_0$. Then, the coset $(x, s)H_0$ can conveniently be represented by the element $(v_k, \Lambda(k)) \in G$, following the coset decomposition,

$$(x, s) = (v_k, \Lambda(k))(n_0, s_0), \quad n_0 = \Lambda(k)^{-1}n_k \in N_0. \quad (10.56)$$

The transformation property of an arbitrary $(v_p, \Lambda(p))$ under the action of $g = (x, s)$ is then obtained as usual by computing $(x, s)(v_p, \Lambda(p))$ and decomposing as above. We get

$$(x, s) : (v_p, \Lambda(p)) \mapsto (\mathbf{P}_{sp}^v x + sv_p,\ \Lambda(sp)), \quad (10.57)$$

which is the same transformation rule as for $(p, v_p) \in T_p^*\mathcal{O}^*$ obtained in (10.53). Thus, an element $(p, v_p) \in T_p^*\mathcal{O}^*$ corresponds to the coset $(v_p, \Lambda(p))H_0 \in G/H_0$ and vice versa.

The following section $\sigma_\wp : G/H_0 \to G$ will be useful for the construction of CS later:

$$\sigma_\wp((v_p, \Lambda(p))H_0) = (v_p, \Lambda(p)). \quad (10.58)$$

Since any $v_p \in T_p^*\mathcal{O}^*$ can be written as $v_p = \Lambda(p)q$, for some $q \in V_0$, we shall actually write the above section as the mapping,

$$\sigma_\wp : V_0 \times \mathcal{O}^* \to G, \quad \sigma_\wp(q, p) = (\Lambda(p)q, \Lambda(p)). \quad (10.59)$$

We shall refer to $\sigma_\wp$ as the *principal section*.

10.2.3 Measures on Γ

Since the space $T^*\mathcal{O}^*$ is a cotangent bundle, it is a symplectic manifold [Gui] and consequently comes equipped with a nondegenerate two-form, which is invariant under the action of G and gives rise to an invariant measure on it. Alternatively, since $T^*\mathcal{O}^*$ is a coadjoint orbit, the discussion in Chapter 4, Section 4.5.2 [see, in particular (4.142)], also tells us that it carries an invariant measure. By virtue of the isomorphisms displayed in (10.49), these properties are also inherited by the coset space Γ. Locally, the two-form on $T^*\mathcal{O}^*$ may be constructed as follows. Let $p_1, p_2, \ldots, p_m$ be a set of local coordinates, defined on an open set $U \subset \mathcal{O}^*$. Then, there is a smooth map $\psi : U \to \mathbb{R}^m$, invertible on its range, with smooth inverse, and such that $p_i = \psi(p)_i$ is the ith component of $\psi(p)$ in $\mathbb{R}^m$, for $p \in U$. The corresponding tangent vectors, $\{\frac{\partial}{\partial p_i}\}_{i=1}^m$, span the tangent space $T_p\mathcal{O}^*$. Let $\{dp_i\}_{i=1}^m$ be the dual basis in $T_p^*\mathcal{O}^*$, for each $p \in U$. Then, a general cotangent vector $v_p \in T_p^*\mathcal{O}^*$ can be written as

$$v_p = \sum_{i=1}^m v_p^i \, dp_i, \qquad v_p^i \in \mathbb{R}.$$

The canonical two-form Ω, expressed locally on U in terms of these coordinates, is now

$$\Omega = \sum_{i=1}^m dv_p^i \wedge dp_i, \qquad (10.60)$$

which is easily checked to be invariant, closed, and nondegenerate. The associated left-invariant measure $d\omega$ on $T^*\mathcal{O}^*$ is then, locally,

$$d\omega = dv_p^1 \wedge dv_p^2 \wedge \ldots \wedge dv_p^m \wedge dp_1 \wedge dp_2 \wedge \ldots \wedge dp_m. \qquad (10.61)$$

Again, the invariance of this measure is easily established using the fact that the dv_p^i and dp_i transform contragrediently under S. We recognize $dv_p^1 \wedge dv_p^2 \wedge \ldots \wedge dv_p^m$ to be simply a version of the Lebesgue measure dv_p on $T_p^*\mathcal{O}^*$, expressed in these coordinates.

For our purposes, it will be useful to find a convenient Borel isomorphism between $T^*\mathcal{O}^*$ and $V_0 \times \mathcal{O}^*$ [$V_0 = T_{k_0}^*\mathcal{O}^*$, see (10.36)], and, eventually, to work with $V_0 \times \mathcal{O}^*$ as the underlying space for labeling coherent states, rather than with $T^*\mathcal{O}^*$. Recall that, as a manifold, $\mathcal{O}^*$ is isomorphic to the coset space S/S_0, and hence the map

$$c : T^*\mathcal{O}^* \to V_0 \times \mathcal{O}^*, \qquad c(v_p, p) = (\Lambda(p)^{-1} v_p, p) := (q, p), \qquad (10.62)$$

where $p \in \mathcal{O}^*$ and $v_p = \Lambda(p)q \in T_p^*\mathcal{O}^*$ is a Borel isomorphism. Denoting the Lebesgue measure on V_0 by dq, we may then write,

$$dv_p = f(p) \, dq, \qquad (10.63)$$

where

$$f(p) = \frac{r(p)}{|\det[\Lambda(p)^{-1}|_{T_p^*\mathcal{O}^*}]|} \qquad (10.64)$$

$\Lambda(p)^{-1}|_{T_p^*\mathcal{O}^*}$ denoting the restriction of $\Lambda(p)^{-1}$ to $T_p^*\mathcal{O}^*$ and $r(p)$ being a measurable function on $\mathcal{O}^*$, which is positive and nonzero on U, and can in fact be chosen to be smooth on it.

10.2. Geometry of semidirect product groups

At this point we make the further assumption that $\mathcal{O}^*$ carries an *invariant* measure $d\nu$ (under the action of S), which on U can be written in the form

$$d\nu(p) = m(p)\, dp_1 \wedge dp_2 \wedge \ldots \wedge dp_m, \qquad (10.65)$$

where again $m(p)$ is a measurable function (which also can be chosen to be smooth) on $\mathcal{O}^*$, which is positive and nonzero on the open set U, used to define the local coordinates p_i. It follows then that under the Borel isomorphism c, the invariant measure $d\omega$ on $T^*\mathcal{O}^*$ transforms locally on U to the measure

$$d\mu(q,p) = \frac{f(p)}{m(p)}\, dq\, d\nu(p). \qquad (10.66)$$

Hence, globally on $V_0 \times \mathcal{O}^*$, we may write

$$d\mu(q,p) = \rho(p)\, dq\, d\nu(p), \qquad (10.67)$$

where now $\rho(p)$ is a measurable function that is positive and nonzero almost everywhere on $\mathcal{O}^*$ (with respect to $d\nu$). Henceforth, we shall identify the coset space $\Gamma = G/H_0$, $H_0 = N_0 \rtimes S_0$, or the cotangent bundle $T^*\mathcal{O}^*$, as Borel spaces, with $V_0 \times \mathcal{O}^*$, equipped with the invariant measure $d\mu$ in (10.67). The invariance of $d\mu$ is with respect to the group transformation (10.52), transported to $V_0 \times \mathcal{O}^*$ via the isomorphism (10.62). Thus, under the action of the group element $g = (x,s)$, $(q,p) \mapsto (q',p') = g(q,p)$, with

$$q' = \Lambda(sp)^{-1}\mathbf{P}^v_{sp}x + h_0(s;p)q, \qquad p' = sp, \qquad (10.68)$$

as follows from (10.57) and (10.62), where $h_0(s;p) = \Lambda(sp)^{-1}s\Lambda(p) \in S_0$. Note also, that in view of this transformation property, the section $\sigma_\wp$ introduced in (10.59) transforms as:

$$g\sigma_\wp(q,p) = \sigma_\wp(g(q,p))\, (\Lambda(sp)^{-1}\mathbf{P}^n_{sp}x,\, h_0(s,p)). \qquad (10.69)$$

The following set of local coordinates for $\mathcal{O}^*$ will later turn out to be useful for computational purposes. Let $\{e_i\}_{i=1}^n$ be a basis of V, such that $e_1, e_2, \ldots, e_m$ is a basis for V_0. Let $\{e^{*\,i}\}_{i=1}^n$ be the dual basis for V^*; i.e.,

$$\langle e^{*\,i}\, ;\, e_j \rangle = \delta_{ij}, \qquad i,j = 1,2,\ldots,n,$$

and $e^{*\,1}, e^{*\,2}, \ldots, e^{*\,m}$ span V_0^*. Then, since V_0^* is the tangent space to $\mathcal{O}^*$ at the point k_0, it is always possible to find an open set $O(k_0) \subset \mathcal{O}^*$, containing k_0, such that the map, $\psi : O(k_0) \to \mathbb{R}^m$, with components,

$$\psi(k)_i := k_i = \langle k - k_0\, ;\, e_i \rangle, \qquad i = 1,2,\ldots,m, \qquad (10.70)$$

is a diffeomorphism. In this basis, an arbitrary $k \in O(k_0)$ has the expansion,

$$k = \sum_{i=1}^m (k_i + \alpha_i)e^{*\,i} + n^*, \qquad \alpha_i = \langle k_0\, ;\, e_i \rangle, \qquad i = 1,2,\ldots,m, \qquad (10.71)$$

where $n^* \in N_0^*$ and is completely determined by the k_i. If $q \in V_0$ and $q = \sum_{i=1}^{m} q^i e_i$, then

$$\langle k \, ; \, q \rangle = \sum_{i=1}^{m} (k_i + \alpha_i) q^i, \qquad (10.72)$$

a relation we shall need later.

As an aid to further reading, we have collected together in Table 10.1 some of the special symbols, defined quantities, and notations introduced in the last few pages.

Symbol	Definition	Relation
V	vector space, abelian group of dimension n	
V^*	dual space of V	
e_i $(e^{*\,i})$	basis of V (V^*)	
S	subgroup of $GL(V)$	
S_0	stabilizer of $k_0 \in V^*$	$S_0 \subset S$
G	semidirect product group	$G = V \rtimes S$
$\mathcal{O}^*$	orbit of k_0 in V^* under S, of dimension m	$\mathcal{O}^* \simeq S/S_0$
Λ	global Borel section, $\Lambda : \mathcal{O}^* \to S$	$s = \Lambda(k) s_0, \ s \in S, \ s_0 \in S_0$
$\mathfrak{g}, \mathfrak{v}, \mathfrak{s}$	Lie algebra of G, V, S	$X = (a, J) \in \mathfrak{g}, \ a \in \mathfrak{v}, J \in \mathfrak{s}$
$\mathfrak{g}^*, \mathfrak{v}^*, \mathfrak{s}^*$	dual spaces of $\mathfrak{g}, \mathfrak{v}, \mathfrak{s}$	$X^* = (a^*, J^*) \in \mathfrak{g}^*$
$O(k_0)$	open set around k_0, for local coordinates	$O(k_0) \subset \mathcal{O}^*$
ψ	coordinate chart, $\psi : O(k_0) \to \mathbb{R}^m$	$\psi(k)_i = k_i = \langle k - k_0 \, ; \, e_i \rangle$
$d\nu(k)$	invariant measure on $\mathcal{O}^*$	$d\nu(k) = m(k) \wedge_{i=1}^{m} dk_i$
$T_k \mathcal{O}^*$	tangent space of $\mathcal{O}^*$ at k, $V_0^* = T_{k_0} \mathcal{O}^*$	$T_k \mathcal{O}^* \subset V^*$
$T_k^* \mathcal{O}^*$	cotangent space of $\mathcal{O}^*$ at k, $V_0 = T_{k_0}^* \mathcal{O}^*$	$T_k^* \mathcal{O}^* \subset V$
N_k	annihilator of $T_k \mathcal{O}^*$ in V, $N_0 = N_{k_0}$	$V = N_k \oplus T_k^* \mathcal{O}^*$
N_k^*	annihilator of $T_k^* \mathcal{O}^*$ in V^*, $N_0^* = N_{k_0}^*$	$V^* = N_k^* \oplus T_k \mathcal{O}^*$
$\mathbf{P}_k^v$ $(\mathbf{P}_k^n)$	projection operator from V to $T_k^* \mathcal{O}^*$ (N_k)	
H_0	stabilizer of $(k_0, 0) \in \mathfrak{g}^*$ under coad-action	$H_0 = N_0 \rtimes S_0$
$T^* \mathcal{O}^*$	cotangent bundle of $\mathcal{O}^*$	$T^* \mathcal{O}^* \simeq G/H_0$
Γ	parameter space for labeling CS	$\Gamma = V_0 \times \mathcal{O}^* \stackrel{\text{Borel}}{\simeq} T^* \mathcal{O}^*$
$d\mu(q, p)$	invariant measure on Γ	$d\mu(q, p) = \rho(p) \, dq \, d\nu(p)$
$\sigma_\wp$	principal section, $\sigma_\wp : \Gamma \to G$	$\sigma_\wp(q, p) = (\Lambda(p) q, \ \Lambda(p))$

Table 10.1. Main notions and notations used in this chapter.

10.2.4 Induced representations of semidirect products

For semidirect product groups of the type we are considering here, all irreducible representations arise as induced representations and correspond to orbits $\mathcal{O}^*$ in V^* [Ma1]. Let us work out these representations in some detail, for we shall have to rely on their specific features to derive a condition for the existence of square integrable coherent states. It will turn out that the CS will be labeled by points in $\Gamma = G/H_0$ [see (10.35)], or equivalently, by points in $V_0 \times \mathcal{O}^*$ (recall that $V_0 = T^*_{k_0}\mathcal{O}^*$). These CS will be square integrable and the associated induced representations square integrable mod(H_0, σ) for appropriate sections σ.

Consider again the element $k_0 \in V^*$, of which $\mathcal{O}^*$ is the orbit under S. The associated unitary character χ of the abelian subgroup V,

$$\chi(x) = \exp[-i < k_0 \,;\, x >], \qquad x \in V, \tag{10.73}$$

defines a one-dimensional representation of V. Let $s \mapsto L(s)$ be a unitary irreducible representation of S_0, the stability subgroup of k_0, and carried by some Hilbert space $\mathfrak{K}$. Consider now the UIR, χL, of $V \rtimes S_0$ carried by $\mathfrak{K}$:

$$(\chi L)(x, s) = \exp[-i < k_0 \,;\, x >]\, L(s). \tag{10.74}$$

We need the representation of $G = V \rtimes S$, which is induced from χL. Clearly,

$$G/(V \rtimes S_0) \simeq \mathcal{O}^*, \tag{10.75}$$

and we shall need the section,

$$\lambda : \mathcal{O}^* \to G, \qquad \lambda(k) = (0, \Lambda(k)), \tag{10.76}$$

with Λ as defined in (10.54), which corresponds to the coset decomposition

$$(x, s) = (0, \Lambda(k))\, (\Lambda(k)^{-1}x, s_0), \qquad (x, s) \in G$$

[see (10.54)]. Also, since for an arbitrary $(x, s) \in G$ and $p \in \mathcal{O}^*$,

$$(x, s)(0, \Lambda(p)) = (0, \Lambda(sp))\, (\Lambda(sp)^{-1}x, \Lambda(sp)^{-1}s\Lambda(p)), \tag{10.77}$$

the action of G on $\mathcal{O}^*$ is also given by $k \mapsto (x, s)k = sk$ [see (10.30)]. Recall that we are assuming the measure $d\nu$ on $\mathcal{O}^*$ to be *invariant* under this action. (This will be the case if, for example, S_0 is a compact subgroup of S.) Denote by $h : G \times \mathcal{O}^* \to V \rtimes S_0$ and $h_0 : S \times \mathcal{O}^* \to S_0$ the two cocycles appearing in (10.77):

$$\begin{aligned} h((x, s), p) &= (\Lambda(sp)^{-1}x, h_0(s, p)), \\ h_0(s, p) &= \Lambda(sp)^{-1}s\Lambda(p). \end{aligned} \tag{10.78}$$

Then a straightforward computation yields

$$(\chi L)(h(x, s)^{-1}, p)) = \exp[-i < k \,;\, x >]\, L(h_0(s^{-1}, p)). \tag{10.79}$$

Following Chapter 4 Section 4.2 [see, in particular, (4.41)], we write the representation of G induced from χL and carried by the Hilbert space $^{\chi L}\mathfrak{H} = \mathfrak{K} \otimes L^2(\mathcal{O}^*, d\nu)$. Denoting this representation by $^{\chi L}U$, we obtain

$$(^{\chi L}U(x,s)\phi)(k) = \exp[i < k\,;\,x >] L(h_0(s^{-1},k))^{-1}\phi(s^{-1}k). \qquad (10.80)$$

This representation is irreducible.

10.3 CS of semidirect products

The foregoing discussion on semidirect product groups can be effectively used to derive general square integrability conditions for induced representations of the type in (10.80). We first state, without proof, a theorem on the square integrability of such a representation [197], which generalizes Theorems 9.1.2 and 9.1.3. The setting is the same as in the previous section.

Theorem 10.3.1: *The induced representation $^{\chi L}U$ of the semidirect product group $G = V \rtimes S$ is square integrable if and only if the orbit $\mathcal{O}^*$ has positive Lebesgue measure in V^* and the representation L is square integrable.*

Note that, if the orbit $\mathcal{O}^*$ has positive Lebesgue measure in V^*, then its dimension is necessarily n (i.e., the same as that of V^*) and, hence, the cotangent bundle $T^*\mathcal{O}^*$ has dimension $2n$.

We now derive a condition for the square integrability mod($H_0, \sigma_\wp$) of the representation $^{\chi L}U$, for the principal section $\sigma_\wp$, introduced in (10.59). We also assume that the Hilbert space $\mathfrak{K}$ is finite-dimensional. Let η^i, $i = 1, 2, \ldots, N < \infty$, be vectors in $^{\chi L}\mathfrak{H} = \mathfrak{K} \otimes L^2(\mathcal{O}^*, d\nu)$ that are smooth functions on $\mathcal{O}^*$ and have supports contained in the set $O(k_0)$, used to define the local coordinates k_i in (10.70). Introduce the vectors

$$\eta^i_{\sigma_\wp(q,p)} = {}^{\chi L}U(\Lambda(p)q, \Lambda(p))\,\eta^i, \qquad i=1,2,\ldots,N, \quad (q,p) \in V_0 \times \mathcal{O}^*,$$

and the positive bounded operator on $^{\chi L}\mathfrak{H}$:

$$F = \sum_{i=1}^{N} |\eta^i\rangle\langle\eta^i|. \qquad (10.81)$$

Later, we shall assume that it satisfies the condition,

$$^{\chi L}U(0,s_0)\,F\,{}^{\chi L}U(0,s_0)^* = F, \quad \text{for all} \quad s_0 \in S_0. \qquad (10.82)$$

The operator F has a kernel,

$$F(k,k') = \sum_{i=1}^{N} |\eta^i(k)\rangle_{\mathfrak{K}}\, {}_{\mathfrak{K}}\langle\eta^i(k')| \in \mathcal{L}(\mathfrak{K}), \qquad k,k' \in \mathcal{O}^* \qquad (10.83)$$

(the subscripts in the above expression indicating that the vectors come from the Hilbert space $\mathfrak{K}$), for which

$$\langle \phi | F \psi \rangle = \int_{\mathcal{O}^* \times \mathcal{O}^*} \langle \phi(k) | F(k,k') \psi(k') \rangle_{\mathfrak{K}} \, d\nu(k) \, d\nu(k'),$$

and if F also satisfies (10.82), then, using the explicit form of the representation $^{\chi^L}U$ in (10.80), this is seen to imply

$$L(s_0) F(k,k') L(s_0)^* = F(s_0 k, s_0 k'), \qquad s_0 \in S_0, \qquad (10.84)$$

almost everywhere.

For arbitrary $\phi, \psi \in \mathfrak{K} \otimes L^2(\mathcal{O}^*, d\nu)$, consider the formal integral,

$$\begin{aligned}
I_{\phi,\psi} &= \sum_{i=1}^{N} \int_{V_0 \times \mathcal{O}^*} \langle \phi | {}^{\chi^L}U(\sigma_\wp(q,p)) \, F \, {}^{\chi^L}U(\sigma_\wp(q,p))^* \, \psi \rangle \, d\mu(q,p) \\
&= \sum_{i=1}^{N} \int_{V_0 \times \mathcal{O}^*} \langle \phi | \eta^i_{\sigma_\wp(q,p)} \rangle \langle \eta^i_{\sigma_\wp(q,p)} | \psi \rangle \, d\mu(q,p), \qquad (10.85)
\end{aligned}$$

where $d\mu$ is the measure obtained in (10.66). Using the explicit form of the representation $^{\chi^L}U$ and the invariance of the measure $d\nu$, the above integral can be brought into the form,

$$\begin{aligned}
I_{\phi,\psi} = \sum_{i=1}^{N} \int_{\mathcal{O}^* \times \mathcal{O}^* \times V_0 \times \mathcal{O}^*} & \langle \phi(\Lambda(p)k) \,|\, L(h_0(\Lambda(p)^{-1}, \Lambda(p)k))^{-1} \eta^i(k) \rangle_{\mathfrak{K}} \\
& \times \langle \eta^i(k') \,|\, L(h_0(\Lambda(p)^{-1}, \Lambda(p)k)) \psi(\Lambda(p)k') \rangle_{\mathfrak{K}} \\
& \times e^{i\langle k - k' \,;\, q \rangle} \, d\nu(k) \, d\nu(k') \, d\mu(q,p).
\end{aligned}$$

At this point we introduce the coordinates (10.70) for the variables k and k', use the relation (10.72), and noting that the η^i are smooth functions, with supports contained in $O(k_0)$, we obtain a δ-measure type of integral with respect to q. Performing this integration and a second one over k', we obtain

$$\begin{aligned}
I_{\phi,\psi} = (2\pi)^m \sum_{i=1}^{N} \int_{\mathcal{O}^*} & d\nu(k) \int_{\mathcal{O}^*} d\nu(p) \, m(\Lambda(p)^{-1} k) \, \rho(p) \\
& \times \langle \phi(k) | L(h_0(\Lambda(p)^{-1}, k))^{-1} \, \eta^i(\Lambda(p)^{-1} k) \rangle_{\mathfrak{K}} \\
& \times \langle \eta^i(\Lambda(p)^{-1} k) | \, L(h_0(\Lambda(p)^{-1}, k)) \, \psi(k) \rangle_{\mathfrak{K}},
\end{aligned}$$

m being the density function [see (10.65)] that appears when the special coordinates (10.70) are introduced. Thus,

$$\begin{aligned}
I_{\phi,\psi} = (2\pi)^m \sum_{i=1}^{N} \int_{\mathcal{O}^*} & d\nu(k) \int_{\mathcal{O}^*} d\nu(p) \, m(\Lambda(p)^{-1} k) \, \rho(p) \\
& \times \langle \phi(k) | ({}^{\chi^L}U(0, \Lambda(p)) \eta^i)(k) \rangle_{\mathfrak{K}} \, \langle ({}^{\chi^L}U(0, \Lambda(p)) \eta^i)(k) | \psi(k) \rangle_{\mathfrak{K}}, \\
& \qquad (10.86)
\end{aligned}$$

Set

$$\mathcal{A}_{\sigma_\wp}(k) = (2\pi)^m \sum_{i=1}^{N} \int_{\mathcal{O}^*} d\nu(p)\ m(\Lambda(p)^{-1}k)\ \rho(p)$$
$$|({}^{x^L}U(0,\Lambda(p))\eta^i)(k)\rangle_\mathfrak{K}\ {}_\mathfrak{K}\langle({}^{x^L}U(0,\Lambda(p))\eta^i)(k)|,$$
(10.87)

the convergence of the integral in $\mathfrak{K}$ being in the weak sense. This is a measurable function on $\mathcal{O}^*$ that defines a formal operator $\mathcal{A}_{\sigma_\wp}$ on the Hilbert space $\mathfrak{K} \otimes L^2(\mathcal{O}^*, d\nu)$:

$$(\mathcal{A}_{\sigma_\wp}\phi)(k) = \mathcal{A}_{\sigma_\wp}(k)\phi(k),$$
(10.88)

for almost all $k \in \mathcal{O}^*$. If we make the assumption that $\mathcal{A}_{\sigma_\wp}(k)$ is a bounded operator (for almost all k, with respect to the invariant measure $d\nu$) and, furthermore, that the function $k \mapsto \|\mathcal{A}_{\sigma_\wp}(k)\|_\mathfrak{K}$ is (essentially) bounded, then $\mathcal{A}_{\sigma_\wp}$ becomes a bounded operator and we may write

$$I_{\phi,\psi} = \langle \phi | \mathcal{A}_{\sigma_\wp} \psi \rangle,$$
(10.89)

and, moreover,

$$\mathcal{A}_{\sigma_\wp} = \sum_{i=1}^{N} \int_{V_0 \times \mathcal{O}^*} |\eta^i_{\sigma_\wp(q,p)}\rangle\langle\eta^i_{\sigma_\wp(q,p)}|\ d\mu(q,p).$$
(10.90)

Suppose now that σ' is any other section that is related to the principal section $\sigma_\wp$ in the manner

$$\sigma'(q,p) = \sigma_\wp(q,p)\ (n(p), s_0(q,p)), \qquad (q,p) \in V_0 \times \mathcal{O}^*,$$
(10.91)

where $n(p)$ is a measurable function on $\mathcal{O}^*$, with values in $N_0 = T^*_{k_0}\mathcal{O}^*$, and $s_0(q,p)$ is a measurable function on $V_0 \times \mathcal{O}^*$, with values in S_0. Define the vectors, $\eta^i_{\sigma'(q,p)} = U(\sigma'(q,p))\eta^i$, $i = 1, 2, \ldots, N$, and the operator

$$\mathcal{A}_{\sigma'} = \sum_{i=1}^{N} \int_{V_0 \times \mathcal{O}^*} |\eta^i_{\sigma'(q,p)}\rangle\langle\eta^i_{\sigma'(q,p)}|\ d\mu(q,p).$$
(10.92)

Note that we have not yet imposed the invariance property (10.82) on F.

Lemma 10.3.2: *The operator $\mathcal{A}_{\sigma_\wp}$ is bounded if and only if there exists a constant $c > 0$, such that*

$$(2\pi)^m \sum_{i=1}^{N} \int_{\mathcal{O}^*} \|\eta^i(\Lambda(p)^{-1}k)\|_\mathfrak{K}^2\ m(\Lambda(p)^{-1}k)\rho(p)\ d\nu(p) < c,$$
(10.93)

for almost all $k \in \mathcal{O}^*$. If, furthermore, F satifies the invariance property (10.82), then for all other sections σ' of the type (10.91), $\mathcal{A}_{\sigma'} = \mathcal{A}_{\sigma_\wp}$.

10.3. CS of semidirect products

Proof. By virtue of the unitarity of the representation L, the condition in (10.93) is seen to be the same as the condition:

$$(2\pi)^m \sum_{i=1}^{N} \int_{\mathcal{O}^*} \|({}^{\chi L}U(0,\Lambda(p))\eta^i)(k)\|_{\mathfrak{K}}^2 \, m(\Lambda(p)^{-1}k)\rho(p) \, d\nu(p) < c.$$

Since $\mathcal{A}_{\sigma_\wp}(k)$ in (10.87) is a positive operator on $\mathfrak{K}$, the integral on the left-hand side of the expression above is exactly equal to $\|\mathcal{A}_{\sigma_\wp}(k)\|_{\mathfrak{K}}$, and, since $\mathcal{A}_{\sigma_\wp}$ is a (block) multiplication operator, $\|\mathcal{A}_{\sigma_\wp}\|$ is equal to the (essential) supremum of the function $k \mapsto \|\mathcal{A}_{\sigma_\wp}(k)\|_{\mathfrak{K}}$, which proves the first part of the lemma. The rest of the lemma is proved by first noting that, if F satisfies (10.82), then

$$\chi^L U(n(p), s_0(q,p)) \, F \, {}^{\chi L}U(n(p), s_0(q,p))^* = {}^{\chi L}U(n(p), e) \, F \, {}^{\chi L}U(n(p), e)^*,$$

and then repeating the same computations as were done to arrive at (10.87), but now using the vectors $\eta^i_{\sigma'(q,p)}$, instead of the $\eta^i_{\sigma_\wp(q,p)}$. □

We can now prove the main result of this section.

Theorem 10.3.3: *Suppose there exist vectors, $\eta^i \in {}^{\chi L}\mathfrak{H} = \mathfrak{K} \otimes L^2(\mathcal{O}^*, d\nu)$, $i = 1, 2, \ldots, N$, which (1) are smooth as functions on $\mathcal{O}^*$, having supports contained in the set $O(k_0)$ used to define the local coordinates k_i in (10.70), and (2) satisfy the conditions (10.82) and (10.93).*

Then, the induced representation ${}^{\chi L}U$ of the group $V \rtimes S$ on ${}^{\chi L}\mathfrak{H}$ is square integrable $\mathrm{mod}(H_0, \sigma_\wp)$. *The operator-valued function $k \mapsto \mathcal{A}_{\sigma_\wp}(k)$ is a constant,*

$$\mathcal{A}_{\sigma_\wp}(k) = c(\sigma_\wp) \, I_{\mathfrak{K}}, \tag{10.94}$$

where $c(\sigma_\wp) > 0$ and $I_{\mathfrak{K}}$ is the identity operator on $\mathfrak{K}$. The vectors

$$\mathfrak{S}_{\sigma_\wp} = \{ [c(\sigma_\wp)]^{-\frac{1}{2}} \eta^i_{\sigma_\wp(q,p)} = [c(\sigma_\wp)]^{-\frac{1}{2}} \, {}^{\chi L}U(\Lambda(p)q, \Lambda(p)) \, \eta^i \mid \\ i = 1, 2, \ldots, N, \ (q,p) \in V_0 \times \mathcal{O}^* \} \tag{10.95}$$

form a family of square integrable, covariant CS for this representation, and the resolution of the identity

$$\frac{1}{c(\sigma_\wp)} \sum_{i=1}^{N} \int_{V_0 \times \mathcal{O}^*} |\eta^i_{\sigma_\wp(q,p)}\rangle\langle\eta^i_{\sigma_\wp(q,p)}| \, d\mu(q,p) = I \tag{10.96}$$

holds on ${}^{\chi L}\mathfrak{H}$.

Proof. Going back to (10.90), by virtue of Lemma 10.3.2, the operator $\mathcal{A}_{\sigma_\wp}$ is seen to be bounded, once the vectors η^i satisfy conditions (1) and (2). For arbitrary $g \in G$, consider the operator ${}^{\chi L}U(g)\mathcal{A}_{\sigma_\wp}{}^{\chi L}U(g)^*$. Using the coset decomposition (10.69), we find

$${}^{\chi L}U(g) \, {}^{\chi L}U(\sigma_\wp(q,p)) = {}^{\chi L}U(\sigma_\wp(g(q,p))) \, {}^{\chi L}U(n(p), h_0(s,p)),$$

where $n(p) \in V_0$ and depends on p only. Thus,

$$^{\chi^L}U(g) \, ^{\chi^L}U(\sigma_\wp(q,p)) = \, ^{\chi^L}U(\sigma'(g(q,p))),$$

where σ' is a section of the type (10.91). From (10.90), we then obtain

$$^{\chi^L}U(g) \, A_{\sigma_\wp} \, ^{\chi^L}U(g)^* = \sum_{i=1}^{N} \int_{V_0 \times \mathcal{O}^*} |\eta^i_{\sigma'(g(q,p))}\rangle\langle\eta^i_{\sigma'(g((q,p))}| \, d\mu(q,p).$$

Invoking the invariance of the measure $d\mu$ and applying Lemma 10.3.2, we see that, for all $g \in G$,

$$^{\chi^L}U(g) \, A_{\sigma_\wp} \, ^{\chi^L}U(g)^* = A_{\sigma_\wp}.$$

The irreducibility of the representation $^{\chi^L}U(g)$ then implies that

$$A_{\sigma_\wp} = c(\sigma_\wp) \, I, \tag{10.97}$$

for some positive, nonzero constant $c(\sigma_\wp)$, and, hence, the rest of the theorem follows. □

Conditions (1) and (2) of this theorem constitute the *admissibility conditions* $\mod(H_0, \sigma_\wp)$ for the vectors η^i. Note also that, written out explicitly, (10.94) expresses the remarkable result,

$$c(\sigma_\wp) = (2\pi)^m \sum_{i=1}^{N} \int_{\mathcal{O}^*} |\langle (^{\chi^L}U(0,\Lambda(p))\eta^i)(k)|u\rangle_{\mathfrak{K}}|^2 \, m(\Lambda(p)^{-1}k) \, \rho(p) \, d\nu(p), \tag{10.98}$$

for any unit vector $u \in \mathfrak{K}n$ and, in particular,

$$c(\sigma_\wp) = (2\pi)^m \sum_{i=1}^{N} \int_{\mathcal{O}^*} \|\langle (^{\chi^L}U(0,\Lambda(p)) \, \eta^i)(k)\|_{\mathfrak{K}}^2 \, m(\Lambda(p)^{-1}k) \, \rho(p) \, d\nu(p), \tag{10.99}$$

for almost all $k \in \mathcal{O}^*$.

Theorem 10.3.3 is readily seen to include Theorem 9.1.2 as a special case. Indeed, if $\mathcal{O}^*$ is an open free orbit in V, it is in fact an open subset of V. Also, $S_0 = \{e\}$, and $T_p^*\mathcal{O}^*$ can be identified with V, for all $p \in \mathcal{O}^*$. In the coset decomposition (10.55), $\Lambda(k) = s$. The open set $O(k_0)$ is $\mathcal{O}^*$ itself. Moreover, comparing with (9.3) and noting that we have used the Borel isomorphism (10.62), $d\mu(q,p) = dq \, d\nu(p)$, with dq being exactly the Lebesgue measure on V. Thus, $f(p) = 1$, for all $p \in \mathcal{O}^*$. Furthermore, m is defined on all of $\mathcal{O}^*$ and can be identified with the function $\mathcal{C}$ appearing in (9.8). Thus, any vector η in the representation Hilbert space satisfies Condition (1) of Theorem 10.3.3 and (10.82), while the condition (10.93) simply becomes

$$(2\pi)^n \int_{\mathcal{O}^*} |\eta(s^{-1}k)|^2 \, \mathcal{C}(s^{-1}k) \, d\nu(p) < c. \tag{10.100}$$

Comparing with (9.9), this is equivalent to

$$(2\pi)^n \int_{\mathcal{O}^*} |\eta(p)|^2\, \mathcal{C}(p)\, d\nu(p) < c, \qquad (10.101)$$

which is the admissibility condition postulated in (9.11), whence Theorem 10.3.3 follows.

A few remarks are in order here, as follows.

- While Theorem 10.3.3 has been proved specifically for the section $\sigma_\wp$, and the definition of the CS in (10.95) depends also on the specific choice of the section $\Lambda(p)$ [see (10.54)], the admissibility condition (10.93) itself is independent of the choice of $\Lambda(p)$, as are the constant $c(\sigma_\wp)$ and the resolution of the identity (10.96), as can be easily verified. The same holds true for the quantity

$$F_{\sigma_\wp}(q,p) = \frac{1}{c(\sigma_\wp)} \sum_{i=1}^{N} |\eta^i_{\sigma_\wp(q,p)}\rangle\langle \eta^i_{\sigma_\wp(q,p)}|.$$

- While Theorem 10.3.1 is not quite a special case of Theorem 10.3.3, the former can in fact be proved using similar arguments and the square integrability of L.

- If the assumption of invariance of F under the subgroup S_0 in (10.82) is dropped, but (10.93) is retained, the square integrability of the representation $\chi^L U$ would still hold, $\mathrm{mod}(H_0, \sigma_\wp)$, but the resolution of the identity (10.96) would have to be replaced by the more general resolution of the operator $A_{\sigma_\wp}$ given in (10.90).

An example: The Euclidean group $E(n)$ (♣)

An easy example of an application of Theorem 10.3.3 is provided by the Euclidean group $E(2) = \mathbb{R}^2 \rtimes SO(2)$ [102]. This group has elements $g = (v, \vartheta), v = (v_1, v_2) \in \mathbb{R}^2, \vartheta \in [0, 2\pi]$. The orbits of $S = SO(2)$ on $V^* = \mathbb{R}^2$ are circles, so one chooses $\mathcal{O}^* = S^1$, the unit circle. The corresponding induced representation of $E(2)$ lives in $\mathfrak{H} = L^2(S^1, d\alpha)$ and reads as

$$(U(v,\vartheta)\psi)(\alpha) = e^{i(v_1 \cos\alpha + v_2 \sin\alpha)}\, \psi(\alpha - \vartheta). \qquad (10.102)$$

Writing $x = (x_1, x_2) = (\cos\alpha, \sin\alpha)$ for points on the unit circle, the cotangent bundle is

$$T^*S^1 = \{(v, x) \in \mathbb{R}^2 \times S^1 | v_1 x_1 + v_2 x_2 = 0\}. \qquad (10.103)$$

It carries coordinates $(a, \alpha) \in \mathbb{R} \times S^1$ defined by $v_1(a, \alpha) = -a\sin\alpha$, $v_2(a, \alpha) = a\cos\alpha$, and the corresponding invariant form is $da \wedge d\alpha$. A vector $\eta \in L^2(S^1, d\alpha)$ is admissible if it is a smooth function satisfying the following conditions:

1. the support of η is contained in the half-circle $(-\frac{\pi}{2}, \frac{\pi}{2})$;

2. η is even: $\eta(-\alpha) = \eta(\alpha)$;

3. $\displaystyle\int_{-\pi/2}^{\pi/2} \frac{|\eta(\alpha)|^2}{\cos\alpha} < \infty.$

The coherent states are $\eta_{a,\alpha} = U(v(a,\alpha),\alpha)\eta$, with $v(a,\alpha) = (v_1(a,\alpha), v_2(a,\alpha))$, $a \in \mathbb{R}$, $0 \le \alpha < 2\pi$, and with proper normalization of η,

$$\int_{-\infty}^{\infty}\int_{0}^{2\pi} |\eta_{a,\alpha}\rangle\langle\eta_{a,\alpha}|\, da\, d\alpha = I. \tag{10.104}$$

Finally, if η is admissible, the isometric map $W_\eta : \mathfrak{H} \to L^2(\Sigma, da d\alpha)$ reads as

$$(W_\eta \psi)(a,\alpha) = \int_{S^1} e^{-ia\sin(\alpha'-\alpha)}\eta(\alpha'-\alpha)d\alpha'. \tag{10.105}$$

Again, the characteristic feature here is the necessity of imposing a restriction on the support of η to guarantee its admissibility, in addition to the symmetry Condition 2 and the growth Condition 3.

The same method applies to the Euclidean group in n dimensions $E(n) = \mathbb{R}^n \rtimes SO(n)$. The orbits $\mathcal{O}^*$ are now spheres S^{n-1}, and the representation lives in $L^2(S^{n-1})$. Furthermore, the vector η has to be chosen $SO(n-1)$-invariant, this condition replacing the symmetry Condition 2. The method also applies to the Galilei and Poincaré groups. On the other hand, as stated in Section 8.1, CS of the Euclidean group have also been obtained using a reducible representation [182].

10.3.1 Admissible affine sections

The principal section $\sigma_\wp$, introduced in (10.59) and used to derive Theorem 10.3.3, is not the only section for which the representation ^{xL}U could be square integrable. Indeed, it is possible to find conditions for square integrability — although not necessarily with a resolution of identity — for a wide class of other sections, which we now discuss. Explicit examples, using the Poincaré and Galilei groups, will be worked out in the next chapter.

Since $V = N_0 \oplus V_0$ and $H_0 = N_0 \rtimes S_0$, any other section σ can be expressed in terms of $\sigma_\wp$ in the following manner:

$$\sigma(q,p) = \sigma_\wp(q,p)\,(n(q,p),\, s_0(q,p)), \tag{10.106}$$

where $n : V_0 \times \mathcal{O}^* \to N_0$ and $s_0 : V_0 \times \mathcal{O}^* \to S_0$ are Borel functions. In particular, we shall isolate a class of sections, called *affine sections*. These are characterized by the following forms for the functions n and s_0:

$$\begin{aligned} n(q,p) &= \Theta(p)q + \Phi(p), \\ s_0(q,p) &= s_0(p), \quad (q,p) \in V_0 \times \mathcal{O}^*, \end{aligned} \tag{10.107}$$

10.3. CS of semidirect products

where, for fixed p, $\Theta(p) : V \to V$ is a linear map such that $\mathrm{Ker}(\Theta(p)) = \mathrm{Ran}(\Theta(p)) = N_0$ and all three functions $\Theta : \mathcal{O}^* \to \mathcal{L}(V)$, $\Phi : \mathcal{O}^* \to N_0$ and $s_0 : \mathcal{O}^* \to S_0$ depend on p only. Additionally, they are assumed to be smooth on the open dense set on which $\Lambda(p)$ is smooth. Then,

$$\sigma(q,p) = (F(p)q + \Lambda(p)\Phi(p), \Lambda(p)s_0(p)), \tag{10.108}$$

where $F(p) : V \to V$ is the linear map

$$F(p) = \Lambda(p)(I_V + \Theta(p)), \quad (I_V = \text{ identity operator on } V). \tag{10.109}$$

Let $\Theta(p)^* : V^* \to V^*$ be the adjoint map to $\Theta(p)$ and I_{V^*} be the identity operator on V^*.

Definition 10.3.4: *The section σ is called an* admissible affine *section if, for each $p \in \mathcal{O}^*$, $I_{V^*} + \Theta(p)^*$ maps the set $O(k_0)$ used to define the local coordinates k_i in(10.70) into itself and*

$$\det [\mathcal{J}(p,k)] \neq 0, \quad p \in \mathcal{O}^*, \quad k \in O(k_0), \tag{10.110}$$

where $\mathcal{J}(p,k)$ is the Jacobian of the map $I_{V^*} + \Theta(p)^*$ restricted to $O(k_0)$.

To understand the nature of the admissible affine section $\sigma(q,p)$ better, let us write

$$\sigma(q,p) = (\widehat{q}, \Lambda(p)s_0(p)), \quad \widehat{q} = F(p)q + \Lambda(p)\Phi(p) \in V. \tag{10.111}$$

For fixed $p \in \mathcal{O}^*$, the set of all vectors $\widehat{q}$ span an m-dimensional affine subspace Σ_p of V. Let $\{e_i^*\}_{i=1}^n$ be a basis of V^*, such that the vectors $e_1^*, e_2^*, \ldots, e_m^*$ span V_0^* and the vectors $e_{m+1}^*, e_{m+2}^*, \ldots, e_n^*$ span N_0^*. For each $p \in \mathcal{O}^*$, define the basis,

$$b_i(p)^* = F(p)^{*-1}e_i^*, \quad i = 1, 2, \ldots, n, \tag{10.112}$$

in V^* [where, of course, $F(p)^* : V^* \to V^*$ is the adjoint of the operator $F(p)$]. Then, since $q \in V_0$ if and only if $< e_i^* ; q > = 0$, $i = m+1, m+2, \ldots, n$, we obtain from (10.111) the defining equations for the affine subspace Σ_p,

$$< b_i(p)^* ; \widehat{q} > = \tau_i(p), \quad i = m+1, m+2, \ldots, n, \tag{10.113}$$

where the constant term $\tau_i(p)$ (for fixed p), which could be zero, is given by

$$\tau_i(p) = < b_i(p)^* ; \Lambda(p)\Phi(p) > . \tag{10.114}$$

Thus, $\widehat{q} \in \Sigma_p$ if and only if it satisfies the system of linear equations (10.113), i.e., if and only if $\widehat{q}$ lies in the common intersection of the $n-m$ affine hyperplanes in V, having normals $b_i(p)^*$ and constants $\tau_i(p)$. For appropriate choices of Φ and F, it may turn out that the set $\bigcup_{p\in\mathcal{O}^*} \Sigma_p$ has the structure of a vector bundle. In particular, if $\tau_i(p) = 0$, $i = m+1, m+2, \ldots, n$, and the vectors $b_i(p)^*$, $i = m+1, m+2, \ldots, n$, span the subspace N_p^* (the annihilator of the cotangent space $T_p^*\mathcal{O}^*$ at p) of V^*, it would follow that $V = N_p \oplus \Sigma_p$, for each p, and, in this case, $\bigcup_{p \in \mathcal{O}^*} \Sigma_p$ is called a *parallel bundle*.

Note, finally, that once the section $\Lambda : \mathcal{O}^* \to S$ and the basis vectors $\{e_i^*\}_{i=1}^n \subset V^*$ are fixed, the vectors $b_i(p)^*$ and the scalars $\tau_i(p)$, $i = m+1, m+2, \ldots, n$, are uniquely determined as soon as Φ and Θ are known.

The class of affine admissible sections can be shown to be stable under the group $G = H \rtimes S$; i.e., if $\sigma(q, p)$ is an admissible affine section, then so also is the section $\sigma'(q, p) = g\sigma(g^{-1}(q, p))$, for each $g = (x, s) \in G$.

Using arguments similar to those which led to the proof of Theorem 10.3.3, it is now possible to prove the more general result on the square integrability of $^{x^L}U$ mod (H_0, σ):

Theorem 10.3.5: *Let σ be any admissible affine section, and suppose that there exist vectors, $\eta^i \in {}^{x^L}\mathfrak{H} = \mathfrak{K} \otimes L^2(\mathcal{O}^*, d\nu)$, $i = 1, 2, \ldots, N$, which (1) are smooth as functions on $\mathcal{O}^*$, having supports contained in the set $O(k_0)$ used to define the local coordinates k_i in (10.70), and (2) satisfy the invariance condition (10.82) under the subgroup S_0. Let*

$$\mathfrak{S}_\sigma = \{\, \eta^i_{\sigma(q,p)} = {}^{x^L}U(\Lambda(p)q, \Lambda(p))\, \eta^i \mid i = 1, 2, \ldots, N,\ (q,p) \in V_0 \times \mathcal{O}^* \,\}. \tag{10.115}$$

Then, the representation $^{x^L}U$ is square integrable mod (H_0, σ), with coherent states $\mathfrak{S}_\sigma$ and the expression

$$A_\sigma = \sum_{i=1}^N \int_{V_0 \times \mathcal{O}^*} |\eta^i_{\sigma(q,p)}\rangle\langle\eta^i_{\sigma(q,p)}|\, d\mu(q, p), \tag{10.116}$$

defining a rank-N frame $\mathcal{F}\{\eta^i_{\sigma(q,p)}, A_\sigma, N\}$, if and only if there exist two nonzero, positive numbers, a and b, $a < b$, for which

$$a < (2\pi)^m \sum_{i=1}^N \int_{\mathcal{O}^*} \|\eta^i(\Lambda(p)^{-1}k)\|_\mathfrak{K}^2\, \frac{m(\Lambda(p)^{-1}k)}{|\det[\mathcal{J}(p,k)]|}\, \rho(p)\, d\nu(p) < b. \tag{10.117}$$

In this case, A_σ is the operator on $^{x^L}\mathfrak{H}$,

$$(A_\sigma \phi)(k) = \mathcal{A}_\sigma(k)\phi(k), \tag{10.118}$$

with

$$\mathcal{A}_\sigma(k) = (2\pi)^m \sum_{i=1}^N \int_{\mathcal{O}^*} |({}^{x^L}U(0, \Lambda(p))\eta^i)(k)\rangle_\mathfrak{K}$$

$$\times {}_\mathfrak{K}\langle({}^{x^L}U(0, \Lambda(p))\eta^i)(k)|\, \frac{m(\Lambda(p)^{-1}k)}{|\det[\mathcal{J}(p,k)]|}\, \rho(p)\, d\nu(p). \tag{10.119}$$

Generally, admissible affine sections may exist, for which the operator A_σ can never be a multiple of the identity. We shall encounter such cases, for the Poincaré group, in the next chapter. The principal section $\sigma_\wp$ is in a sense the most obvious choice of a section, in that it is the one naturally associated to the cotangent bundle $T^*\mathcal{O}^*$.

11
CS of the Relativity Groups

In this chapter, we examine a few of the various relativity groups, that are of great importance in physics. The discussion will also illustrate the use of the results on semidirect products obtained in the last chapter.

11.1 The Poincaré groups $\mathcal{P}_+^\uparrow(1,3)$ and $\mathcal{P}_+^\uparrow(1,1)$

Among the relativity groups, the most important are the Poincaré and the Galilei groups in various space–time dimensions. These groups are semidirect products of a translation group $\mathbb{R}^n$ by the corresponding isometry group of space–time. Thus, they fall entirely within the scope of the preceding discussion.

11.1.1 The Poincaré group in 1+3 dimensions, $\mathcal{P}_+^\uparrow(1,3)$ (♣)

We begin with the most fundamental relativity group, namely, the Poincaré group $\mathcal{P}_+^\uparrow(1,3)$ in one time and three space dimensions. Accordingly, we treat it in a rather detailed fashion, except for some proofs, which may be found in [20].

The full Poincaré group $\mathcal{P}_+^\uparrow(1,3)$ is the twofold covering group, $\mathcal{P}_+^\uparrow(1,3) = \mathbb{R}^{1,3} \rtimes SL(2,\mathbb{C})$, where $\mathbb{R}^{1,3}$ is the group of space–time translations, with Minkowski metric. Elements of $\mathcal{P}_+^\uparrow(1,3)$ will be denoted by (a,A), $a = (a_0,\mathbf{a}) \in \mathbb{R}^{1,3}$, $A \in SL(2,\mathbb{C})$. The multiplication law is $(a,A)(a',A') = (a+\Lambda a', AA')$, where $\Lambda \in \mathcal{L}_+^\uparrow(1,3) = SO_o(1,3)$ (the proper, orthochronous Lorentz group) is the Lorentz transformation corresponding to A

$$\Lambda^\mu{}_\nu = \frac{1}{2}\mathrm{Tr}[A\sigma_\nu A^\dagger \sigma_\mu], \qquad \mu,\nu = 0,1,2,3, \tag{11.1}$$

$\sigma^1 = -\sigma_1 = \sigma_x$, $\sigma^2 = -\sigma_2 = \sigma_y$ and $\sigma^3 = -\sigma_3 = \sigma_z$ are the Pauli matrices [see (9.58)], $\sigma^0 = \sigma_0 = \mathbb{I}_2$, and the metric tensor is $g_{00} = 1 = -g_{11} = -g_{22} = -g_{33}$. For typographical simplicity, we will often write $\underline{p} \equiv \mathbf{p}$ for the spatial part of a four-vector p.

Let

$$\mathcal{V}_m^+ = \{k = (k_0, \mathbf{k}) \in \mathbb{R}^{1,3} \mid k^2 = k_0^2 - \mathbf{k}^2 = m^2,\ k_0 > 0\} \tag{11.2}$$

be the *forward mass hyperboloid*. Then,

$$k' = \Lambda k \Rightarrow \sigma \cdot k' = A\sigma \cdot k A^\dagger, \tag{11.3}$$

with $\sigma \cdot k = \sigma^\mu k_\mu = k_0 \mathbb{I}_2 - \mathbf{k} \cdot \boldsymbol{\sigma}$, $\boldsymbol{\sigma} = (\sigma_x, \sigma_y, \sigma_z)$.

The group $\mathcal{P}_+^\uparrow(1,3)$ is a semidirect product of the type described in Section 10.2; hence, its unitary irreducible representations may be obtained by the Mackey method of induced representations, as described in Section 10.2.4. In the Wigner realization, the unitary irreducible representation U_W^s of $\mathcal{P}_+^\uparrow(1,3)$ for a particle of mass $m > 0$ and spin $s = 0, \frac{1}{2}, 1, \frac{3}{2}, 2, \ldots$, is carried by the Hilbert space

$$\mathfrak{H}_W^s = \mathbb{C}^{2s+1} \otimes L^2(\mathcal{V}_m^+, \frac{d\mathbf{k}}{k_0}) \tag{11.4}$$

of $\mathbb{C}^{2s+1}$-valued functions ϕ on $\mathcal{V}_m^+$, which are square integrable:

$$\int_{\mathcal{V}_m^+} \phi(k)^\dagger \phi(k) \frac{d\mathbf{k}}{k_0} = \|\phi\|^2 = \langle \phi | \phi \rangle < \infty. \tag{11.5}$$

Explicitly, in agreement with (10.80), we have

$$(U_W^s(a,A)\phi)(k) = e^{ik\cdot a}\, \mathcal{D}^s(h(k)^{-1}Ah(\Lambda^{-1}k))\phi(\Lambda^{-1}k),$$
$$k \cdot a = k_0 a_0 - \mathbf{k} \cdot \mathbf{a}, \tag{11.6}$$

where $\mathcal{D}^s$ is the $(2s+1)$-dimensional, irreducible spinor representation of $SU(2)$ (carried by $\mathbb{C}^{2s+1}$) and

$$h(k) = \frac{m\mathbb{I}_2 + \sigma \cdot \overline{k}}{\sqrt{2m(k_0+m)}}, \qquad (\overline{k} = (k_0, -\mathbf{k})), \tag{11.7}$$

is the image in $SL(2,\mathbb{C})$ of the Lorentz boost Λ_k, which brings the four-vector $(m, \mathbf{0})$ to the four-vector k in $\mathcal{V}_m^+$ (σ is a four-vector of Pauli matrices).

The matrix form of the Lorentz boost is

$$\Lambda_k = \frac{1}{m}\begin{pmatrix} k_0 & \mathbf{k}^\dagger \\ \mathbf{k} & mV_k \end{pmatrix} = \Lambda_k{}^\dagger, \tag{11.8}$$

11.1. The Poincaré groups $\mathcal{P}_+^\uparrow(1,3)$ and $\mathcal{P}_+^\uparrow(1,1)$

where V_k is the 3×3 symmetric matrix

$$V_k = \mathbb{I}_3 + \frac{\mathbf{k} \otimes \mathbf{k}^\dagger}{m(k_0 + m)} = V_k{}^\dagger. \tag{11.9}$$

The Wigner representation (11.6) is not square integrable over $\mathcal{P}_+^\uparrow(1,3)$, as may be shown by a straightforward computation. Hence, we have to look for an appropriate quotient. The obvious choice is phase space, which, for a classical (spinless), relativistic particle, can be identified with

$$\Gamma = \mathcal{P}_+^\uparrow(1,3)/(T \times SU(2)), \tag{11.10}$$

T denoting the subgroup of time translations (as a further hint that T is a subgroup to quotient out, one may note that it is the integral over a_0 that diverges when checking for square integrability). For a particle with non-zero spin (treated as an additional classical degree of freedom), a geometric quantization program [Woo] would normally start with the phase space $\Gamma' = \mathcal{P}_+^\uparrow(1,3)/T \times SO(2)$. Since geometric quantization is not our objective here, however, we choose Γ in (11.10) as the phase space for a particle with arbitrary spin s. (In the terminology of geometric quantization, this means working on a $\mathbb{C}^{2s+1}$-bundle, rather than on a line bundle.) Of course, $T \times SU(2)$ cannot be the stability subgroup of any vector in $\mathfrak{H}_W^s$, since it contains translations. Hence, the CS we are going to construct will not be of the Gilmore–Perelomov type.

For $A \in SL(2,\mathbb{C})$, let

$$A = h(k)R(k), \quad R(k) \in SU(2), \tag{11.11}$$

be its Cartan decomposition (see Chapter 4, Section 4.5.2). An arbitrary element $(a, A) \in \mathcal{P}_+^\uparrow(1,3)$ has the left coset decomposition,

$$(a, A) = \left((0, \mathbf{a} - \frac{a_0 \mathbf{k}}{k_0}), h(k)\right)\left((\frac{m a_0}{k_0}, \mathbf{0}), R(k)\right), \tag{11.12}$$

according to (11.10). Thus, the elements in Γ have the global coordinatization, $(\mathbf{q}, \mathbf{p}) \in \mathbb{R}^6$

$$\mathbf{q} = \mathbf{a} - \frac{a_0 \mathbf{k}}{k_0}, \quad \mathbf{p} = \mathbf{k}. \tag{11.13}$$

In terms of these variables, the action of $\mathcal{P}_+^\uparrow(1,3)$ on Γ is given by $(\mathbf{q}, \mathbf{p}) \mapsto (\mathbf{q}', \mathbf{p}') = (a, A)(\mathbf{q}, \mathbf{p})$,

$$\begin{aligned}\mathbf{q}' &= \frac{1}{p_0'}\left(p_0'[\mathbf{a} + \Lambda(0,\mathbf{q})] - \mathbf{p}'[a_0 + \{\Lambda(0,\mathbf{q})\}_0]\right) \\ \mathbf{p}' &= \Lambda p, \quad p = (\sqrt{m^2 + \mathbf{p}^2}, \mathbf{p}),\end{aligned} \tag{11.14}$$

where $\Lambda \in \mathcal{L}_+^\uparrow(1,3)$ is related to A by (11.1) and $p_0' = (\Lambda p)_0$. It can be shown [4] that the measure $d\mathbf{q}\, d\mathbf{p}$ is invariant under this action, and, hence, represents the invariant measure ν on Γ in the variables $(\mathbf{q}, \mathbf{p})$.

Comparing with the general theory in Chapter 10, Section 10.2, the coset space Γ in this case is isomorphic to the cotangent bundle $T^*\mathcal{V}_m^+$ and $\mathcal{V}_m^+$ is the orbit of the vector $(m, \mathbf{0}) \in \mathbb{R}^{1,3}$ under $\mathcal{L}_+^\uparrow(1,3)$. The cotangent space $T^*_{(m,0)}\mathcal{V}_m^+$ is just $\mathbb{R}^3$, and, since $\mathcal{V}_m^+$ can be globally coordinatized using vectors in $\mathbb{R}^3$ [see (11.2)], the open set $U(k_0)$ introduced in the last chapter [see (10.70)], for defining local coordinates is now all of $\mathcal{V}_m^+$.

Although it would be possible, for constructing CS, to work with the principal section, introduced in (10.58) and (10.59), for reasons of physical transparency, it is better to begin, in the present case, with a different section. Accordingly, in terms of the variables $(\mathbf{q}, \mathbf{p})$, let us define the basic section, $\sigma_0 : \Gamma \to \mathcal{P}_+^\uparrow(1,3)$, by

$$\sigma_0(\mathbf{q}, \mathbf{p}) = ((0, \mathbf{q}), h(p)). \tag{11.15}$$

We call σ_0 the *Galilean section*. Later, we shall also link it to the principal section. Any other section $\sigma : \Gamma \to \mathcal{P}_+^\uparrow(1,3)$ is then related to σ_0 in the manner

$$\sigma(\mathbf{q}, \mathbf{p}) = \sigma_0(\mathbf{q}, \mathbf{p}) \, ((f(\mathbf{q}, \mathbf{p}), \mathbf{0}), R(\mathbf{q}, \mathbf{p})), \tag{11.16}$$

where $f : \mathbb{R}^6 \to \mathbb{R}$ and $R : \mathbb{R}^6 \to SU(2)$ are smooth functions. As explained in Section 10.3.1, we work with the particular class of *affine sections*, which now assume the form,

$$f(\mathbf{q}, \mathbf{p}) = \varphi(\mathbf{p}) + \mathbf{q} \cdot \boldsymbol{\vartheta}(\mathbf{p}), \tag{11.17}$$

where $\varphi : \mathbb{R}^3 \to \mathbb{R}$, $\boldsymbol{\vartheta} : \mathbb{R}^3 \to \mathbb{R}^3$ are smooth functions of $\mathbf{p}$ alone, and $R(\mathbf{q}, \mathbf{p}) = R(\mathbf{p})$ is a function of $\mathbf{p}$ alone. Moreover, we shall only deal here with sections for which $\varphi = 0$. Writing

$$\sigma(\mathbf{q}, \mathbf{p}) = (\hat{q}, h(p) R(\mathbf{p})), \qquad \hat{q} = (\hat{q}_0, \hat{\mathbf{q}}) \in \mathbb{R}^{1,3}, \tag{11.18}$$

we see that

$$\hat{q}_0 = \boldsymbol{\beta}(\mathbf{p}) \cdot \hat{\mathbf{q}}, \tag{11.19}$$

where $\boldsymbol{\beta}$ is the three-vector field

$$\boldsymbol{\beta}(\mathbf{p}) = \frac{p_0 \boldsymbol{\vartheta}(\mathbf{p})}{m + \mathbf{p} \cdot \boldsymbol{\vartheta}(\mathbf{p})}, \quad \text{so that} \quad \boldsymbol{\vartheta}(\mathbf{p}) = \frac{m \boldsymbol{\beta}(\mathbf{p})}{p_0 - \mathbf{p} \cdot \boldsymbol{\beta}(\mathbf{p})}. \tag{11.20}$$

We also introduce the *dual* vector field $\boldsymbol{\beta}^*$,

$$\boldsymbol{\beta}^*(\mathbf{p}) = \frac{\mathbf{p} - m V_p \boldsymbol{\beta}(\mathbf{p})}{p_0 - \mathbf{p} \cdot \boldsymbol{\beta}(\mathbf{p})}, \tag{11.21}$$

where V_p is the matrix defined in (11.9). Note that

$$\boldsymbol{\beta}^{**} = \boldsymbol{\beta} \quad \text{and} \quad \boldsymbol{\vartheta}(\mathbf{p}) = \frac{1}{m}[\mathbf{p} - m V_p \boldsymbol{\beta}^*(\mathbf{p})]. \tag{11.22}$$

The vector fields $\boldsymbol{\beta}(\mathbf{p})$ and $\boldsymbol{\beta}^*(\mathbf{p})$ have an interesting physical and geometrical interpretation, but that does not concern us here (see [20] for details).

11.1. The Poincaré groups $\mathcal{P}_+^\uparrow(1,3)$ and $\mathcal{P}_+^\uparrow(1,1)$

We now take an arbitrary affine section σ and, going back to the Hilbert space $\mathfrak{H}_W^s$ in (11.4), choose a set of vectors $\boldsymbol{\eta}^i$, $i = 1, 2, \ldots, 2s+1$, in it to define the formal operator [see (7.46) and (11.6)]:

$$A_\sigma = \sum_{i=1}^{2s+1} \int_{\mathbb{R}^6} |\boldsymbol{\eta}^i_{\sigma(\mathbf{q},\mathbf{p})}\rangle \langle \boldsymbol{\eta}^i_{\sigma(\mathbf{q},\mathbf{p})}| \, d\mathbf{q} \, d\mathbf{p}, \qquad \boldsymbol{\eta}^i_{\sigma(\mathbf{q},\mathbf{p})} = U_W^s(\sigma(\mathbf{q},\mathbf{p}))\boldsymbol{\eta}^i. \qquad (11.23)$$

From the general definition, in order for the set of vectors

$$\mathfrak{S}_\sigma = \{\boldsymbol{\eta}^i_{\sigma(\mathbf{q},\mathbf{p})} \mid (\mathbf{q},\mathbf{p}) \in \mathbb{R}^6, \ i = 1, 2, \ldots, 2s+1\} \subset \mathfrak{H}_W^s \qquad (11.24)$$

to constitute a family of coherent states for the representation U_W^s, the integral in (11.23) must converge weakly and define A_σ as a bounded invertible operator. In fact, it will be possible to choose vectors $\boldsymbol{\eta}^i$, such that, for each affine section σ, both A_σ and A_σ^{-1} are bounded; i.e., each family $\mathfrak{S}_\sigma$ of CS will define a rank-$(2s+1)$ frame.

To study the convergence properties of the operator integral in (11.23), we have to determine the convergence of the ordinary integral

$$I_{\phi,\psi} = \sum_{i=1}^{2s+1} \int_{\mathbb{R}^6} \langle \phi | \boldsymbol{\eta}^i_{\sigma(\mathbf{q},\mathbf{p})} \rangle \langle \boldsymbol{\eta}^i_{\sigma(\mathbf{q},\mathbf{p})} | \psi \rangle \, d\mathbf{q} \, d\mathbf{p} \qquad (11.25)$$

for arbitrary $\phi, \psi \in \mathfrak{H}_W^s$. In (11.18), set

$$\hat{A}(p) = h(p)R(\mathbf{p}) \quad \text{and} \quad \hat{\Lambda}(p) = \Lambda_p \rho(p), \qquad (11.26)$$

where $\hat{\Lambda}(p)$ and $\rho(p)$ are the matrices in the Lorentz group $\mathcal{L}_+^\uparrow(1,3)$ that correspond to $\hat{A}(p)$ and $R(\mathbf{p})$, respectively. Then,

$$\boldsymbol{\eta}^i_{\sigma(\mathbf{q},\mathbf{p})}(k) = \exp\{-i\mathbf{X}(\mathbf{k}) \cdot \mathbf{q}\} \mathcal{D}^s(v(k,p))\boldsymbol{\eta}^i(\hat{\Lambda}(p)^{-1}k), \qquad (11.27)$$

where

$$\mathbf{X}(\mathbf{k}) = \mathbf{k} - \frac{k \cdot p}{m}\boldsymbol{\vartheta}(\mathbf{p}) \qquad (11.28)$$

and

$$v(k,p) = h(k)^{-1}\hat{A}(p)h(\hat{\Lambda}(p)^{-1}k) \in SU(2). \qquad (11.29)$$

Substituting into (11.25) yields

$$I_{\phi,\psi} = \sum_{i=1}^{2s+1} \int_{\mathcal{V}_m^+ \times \mathcal{V}_m^+ \times \mathbb{R}^6} \exp\{-i[\mathbf{X}(\mathbf{k}) - \mathbf{X}(\mathbf{k}')] \cdot \mathbf{q}\} \phi(k)^\dagger \mathcal{D}^s(v(k,p))$$
$$\times \boldsymbol{\eta}^i(\hat{\Lambda}(p)^{-1}k)\boldsymbol{\eta}^i(\hat{\Lambda}(p)^{-1}k')^\dagger \mathcal{D}^s(v(k',p))^\dagger \psi(k') \frac{d\mathbf{k}}{k_0} \frac{d\mathbf{k}'}{k_0'} \, d\mathbf{q} \, d\mathbf{p}. \qquad (11.30)$$

In order to perform the k, k' integrations in (11.30), we need to change variables: $\mathbf{k} \mapsto \mathbf{X}(\mathbf{k})$. Computing the Jacobian $\mathcal{J}_{\mathbf{X}}(\mathbf{k})$ of this transformation from (11.28), we obtain for its determinant

$$\det[\mathcal{J}_\mathbf{X}(\mathbf{k})] = 1 + \frac{1}{mk_0}\boldsymbol{\vartheta}(\mathbf{p}) \cdot [k_0\mathbf{p} - \mathbf{k}p_0]. \tag{11.31}$$

Since at $\mathbf{k} = \mathbf{p} = \mathbf{0}$, $\det[\mathcal{J}_\mathbf{X}(\mathbf{k})] = 1$, and since we need $\det[\mathcal{J}_\mathbf{X}(\mathbf{k})] \neq 0$ in order to change variables, we must impose the condition that $\det[\mathcal{J}_\mathbf{X}(\mathbf{k})] > 0$, $\forall (\mathbf{k}, \mathbf{p})$. This, in turn, imposes restrictions on $\boldsymbol{\vartheta}$ and, hence, on the four-vector $\hat{q} = (\hat{q}_0, \hat{\mathbf{q}})$ in (11.18). Rewriting (11.31) in terms of $\boldsymbol{\beta}^*$, we get

$$\det[\mathcal{J}_\mathbf{X}(\mathbf{k})] = \frac{p_0(\Lambda_k\overline{p})_0}{mk_0}\left[1 + \frac{(\Lambda_k\overline{p})}{(\Lambda_k\overline{p})_0} \cdot \rho(k \to \overline{p})^\dagger \boldsymbol{\beta}^*(\mathbf{p})\right], \tag{11.32}$$

where we have introduced the rotation matrix $\rho(k \to \overline{p}) = \Lambda_p^{-1}\Lambda_k\Lambda_p\Lambda_k^{-1}$. Thus, the positivity of $\det[\mathcal{J}_\mathbf{X}(\mathbf{k})]$ would be ensured if the second term within the square brackets in (11.32) does not exceed 1 in magnitude, i.e., if $\|\boldsymbol{\beta}^*(\mathbf{p})\| < 1$, $\forall \mathbf{p}$. To ensure this, we need the following result.

Proposition 11.1.1: *The following conditions are equivalent.*

1. *The four-vector $\hat{q} = (\hat{q}_0, \hat{\mathbf{q}})$ is space-like, i.e., $|\hat{q}_0|^2 - \|\hat{\mathbf{q}}\|^2 < 0$.*

2. *For all $\mathbf{p} \in \mathbb{R}^3$, the three-vector field $\boldsymbol{\beta}$ obeys $\|\boldsymbol{\beta}(\mathbf{p})\| < 1$.*

3. *For all $\mathbf{p} \in \mathbb{R}^3$, the three-vector field $\boldsymbol{\beta}^*$ obeys $\|\boldsymbol{\beta}^*(\mathbf{p})\| < 1$.*

From this, we obtain a simple condition for the change of variables to be nonsingular.

Proposition 11.1.2: *The condition $\det[\mathcal{J}_\mathbf{X}(\mathbf{k})] > 0$ holds for all $\mathbf{k}, \mathbf{p} \in \mathbb{R}^3$ if and only if the four-vector $\hat{q} = (\hat{q}_0, \hat{\mathbf{q}})$ is space-like, i.e., if and only if any one of the equivalent conditions in Proposition 11.1.1 is satisfied.*

An affine section σ, for which the corresponding three-vector $\boldsymbol{\beta}$ satisfies any one of the equivalent conditions of Proposition 11.1.1 will be called a *space-like affine section*. This, in the present instance, is the equivalent of the admissibility condition on the section σ, postulated in Section 10.3.1.

We return now to the integral (11.30). For simplicity, we assume that the vectors $\boldsymbol{\eta}^i$, defining the CS, satisfy the condition of *rotational invariance*,

$$\mathcal{D}^s(R)\left(\sum_{i=1}^{2s+1}|\boldsymbol{\eta}^i\rangle\langle\boldsymbol{\eta}^i|\right)\mathcal{D}^s(R)^\dagger = \sum_{i=1}^{2s+1}|\boldsymbol{\eta}^i\rangle\langle\boldsymbol{\eta}^i|, \quad \forall R \in SU(2), \tag{11.33}$$

which implies that we may take

$$\boldsymbol{\eta}^i = \hat{e}_i \otimes \eta, \quad i = 1, 2, \ldots, 2s+1, \tag{11.34}$$

where $\hat{e}_i$ are the unit vectors (δ_{ij}), $j = 1, 2, \ldots, 2s+1$, in $\mathbb{C}^{2s+1}$ and $\eta \in L^2(\mathcal{V}_m^+, d\mathbf{k}/k_0)$. We assume, furthermore, that the function $|\eta(k)|^2$ itself is also rotationally invariant; i.e., $|\eta(\rho k)|^2 = |\eta(k)|^2$, $\forall \rho \in SO(3)$.

11.1. The Poincaré groups $\mathcal{P}_+^\uparrow(1,3)$ and $\mathcal{P}_+^\uparrow(1,1)$

Going back to the computation of $I_{\phi,\psi}$ in (11.30), we note that the $d\mathbf{q}$ integration yields a δ-measure in $\mathbf{X}$, and, hence, making the change of variables $\mathbf{k} \mapsto \mathbf{X}$, integrating and rearranging [using (11.32)], we obtain

$$I_{\phi,\psi} = \int_{\mathcal{V}_m^+ \times \mathcal{V}_m^+} \phi(k)^\dagger \mathcal{A}_\sigma(k,p) \psi(k) \, \frac{d\mathbf{p}}{p_0} \frac{d\mathbf{k}}{k_0}, \qquad (11.35)$$

where $\mathcal{A}_\sigma(k,p)$ is the $(2s+1) \times (2s+1)$-matrix kernel

$$\mathcal{A}_\sigma(k,p) = a_\sigma(k,p) \, |\eta(p)|^2 \, \mathbb{I}_{2s+1},$$

$$a_\sigma(k,p) = \frac{(2\pi)^3 m}{p_0 + \mathbf{p} \cdot \rho(k \to \Lambda_k^{-1} p)^\dagger \, \boldsymbol{\beta}^*(-\Lambda_k^{-1} p)}, \qquad (11.36)$$

$$= \frac{(2\pi)^3 m (\Lambda_k^{-1} p)_0}{m k_0 - [k_0(\Lambda_k^{-1} p) + \mathbf{k}(\Lambda_k^{-1} p)_0] \cdot \boldsymbol{\vartheta}(-\Lambda_k^{-1} p)}, \qquad (11.37)$$

and $\rho(k \to p)$ is the rotation matrix defined above. Assuming the integral (11.25) to exist for all $\phi, \psi \in \mathfrak{H}_W^s$, let us write

$$\mathcal{A}_\sigma(k) = \int_{\mathcal{V}_m^+} \mathcal{A}_\sigma(k,p) \, \frac{d\mathbf{p}}{p_0}. \qquad (11.38)$$

Then, the operator A_σ in (11.23) is a matrix-valued multiplication operator:

$$(A_\sigma \phi)(k) = \mathcal{A}_\sigma(k) \phi(k), \qquad \phi \in \mathfrak{H}_W^s. \qquad (11.39)$$

On $\mathfrak{H}_W^s$, define the operators $(P_0, \mathbf{P})$,

$$(P_\mu \phi)(k) = k_\mu \phi(k). \qquad (11.40)$$

We shall also denote the analogous operators on $L^2(\mathcal{V}_m^+, d\mathbf{k}/k_0)$ by the same symbols. P_0^{-1} is a bounded operator with spectrum $[0, \frac{1}{m}]$. With the above simplifications, (11.38) becomes

$$\mathcal{A}_\sigma(k) = \langle a_\sigma(k, P) \rangle_\eta \, \mathbb{I}_{2s+1}, \qquad (11.41)$$

where $\langle \cdot \rangle_\eta$ denotes the $L^2(\mathcal{V}_m^+, d\mathbf{k}/k_0)$ expectation value with respect to the vector η. Hence, for the operator A_σ [see (11.39)]

$$\|A_\sigma\| = \sup_{k \in \mathcal{V}_m^+} |\langle a_\sigma(k, P) \rangle_\eta|, \qquad (11.42)$$

provided this supremum exists. On the other hand, since we have $\|\boldsymbol{\beta}^*(-\Lambda_k^{-1} p)\| < 1$ and $\|\rho(k \to \Lambda_k^{-1} p)^\dagger\| = 1$, from (11.36), we get

$$\frac{1}{(2\pi)^3 m}(p_0 - \|\mathbf{p}\|) < \frac{1}{a_\sigma(k,p)} < \frac{1}{(2\pi)^3 m}(p_0 + \|\mathbf{p}\|). \qquad (11.43)$$

Thus, we have the following result.

Lemma 11.1.3: *If $\|\boldsymbol{\beta}(\mathbf{p})\| \leq 1$, $\forall \mathbf{p}$, then $a_\sigma(k,p)$ is a bounded function satisfying*

$$\frac{(2\pi)^3}{m}(p_0 - \|\mathbf{p}\|) \leq a_\sigma(k,p) \leq \frac{(2\pi)^3}{m}(p_0 + \|\mathbf{p}\|). \qquad (11.44)$$

Suppose now that η lies in the domain of $P_0^{1/2}$, i.e.,

$$\int_{V_m^+} |\eta(k)|^2 \, d\mathbf{k} < \infty, \qquad (11.45)$$

and set

$$\langle P_0 \pm \|\mathbf{P}\| \rangle_\eta = \int_{V_m^+} (p_0 \pm \|\mathbf{P}\|) |\eta(p)|^2 \, \frac{d\mathbf{p}}{p_0}. \qquad (11.46)$$

Then (11.35), (11.36), and (11.44) together imply the following lemma.

Lemma 11.1.4: *If the assumption of rotational invariance on η^i, $i = 1, 2, \ldots, 2s+1$, is satisfied, and if $\eta \in \mathcal{D}(P_0^{1/2})$, then, for all β, such that $\|\beta(\mathbf{p})\| \le 1$, $\forall \mathbf{p}$,*

$$\frac{(2\pi)^3}{m} \langle P_0 - \|\mathbf{P}\| \rangle_\eta \, \|\phi\| \, \|\psi\| \le |I_{\phi,\psi}| \le \frac{(2\pi)^3}{m} \langle P_0 + \|\mathbf{P}\| \rangle_\eta \, \|\phi\| \, \|\psi\|. \qquad (11.47)$$

As a consequence of this lemma, we see that both the operator A_σ in (11.23) and its inverse, A_σ^{-1}, are bounded, with

$$(A_\sigma^{-1} \phi)(k) = [\langle a_\sigma(k, P) \rangle_\eta]^{-1} \phi(k), \qquad \phi \in \mathfrak{H}_W^s. \qquad (11.48)$$

Indeed, collecting all of these results we obtain the following result.

Proposition 11.1.5: *Let η^i, $i = 1, 2, \ldots, 2s+1$, satisfy the condition of rotational invariance. Then, for every space-like affine section σ, the set of vectors $\mathfrak{S}_\sigma$ in (11.24) is a family of spin-s coherent states, forming a rank-$(2s+1)$ frame $\mathcal{F}\{\eta^i_{\sigma(\mathbf{q},\mathbf{p})}, A_\sigma, 2s+1\}$, if and only if $\eta \in \mathcal{D}(P_0^{1/2})$. The operator A_σ acts via multiplication by a bounded invertible function $\mathcal{A}_\sigma(k)$ given by (11.41) and A_σ^{-1} via multiplication by the function $\mathcal{A}_\sigma^{-1}(k)$. Moreover, the norm of A_σ is bounded by the estimate,*

$$\frac{(2\pi)^3}{m} \langle P_0 - \|\mathbf{P}\| \rangle_\eta \le \|A_\sigma\| \le \frac{(2\pi)^3}{m} \langle P_0 + \|\mathbf{P}\| \rangle_\eta, \qquad (11.49)$$

while its spectrum satisfies

$$\operatorname{Spec}(A_\sigma) \subset \frac{(2\pi)^3}{m} [\langle P_0 - \|\mathbf{P}\| \rangle_\eta, \langle P_0 + \|\mathbf{P}\| \rangle_\eta]. \qquad (11.50)$$

It is interesting to note that the estimates (11.49) and (11.50) are universal; i.e., the bounds are valid for *any* space-like affine section σ. Indeed,

$$\frac{(2\pi)^3}{m} \langle P_0 - \|\mathbf{P}\| \rangle_\eta = \inf_{\|\beta\| \le 1} \operatorname{Spec}(A_\sigma), \quad \sigma \equiv \sigma(\beta), \qquad (11.51)$$

$$\frac{(2\pi)^3}{m} \langle P_0 + \|\mathbf{P}\| \rangle_\eta = \sup_{\|\beta\| \le 1} \operatorname{Spec}(A_\sigma), \quad \sigma \equiv \sigma(\beta). \qquad (11.52)$$

In addition, the class of affine sections is stable under the action of $\mathcal{P}_+^\uparrow(1,3)$. If $\sigma : \Gamma \to \mathcal{P}_+^\uparrow(1,3)$ is any section, then, for arbitrary $(a, A) \in \mathcal{P}_+^\uparrow(1,3)$,

11.1. The Poincaré groups $\mathcal{P}_+^\uparrow(1,3)$ and $\mathcal{P}_+^\uparrow(1,1)$

$\sigma_{(a,A)}$ is again a section [see (7.4)], where

$$\sigma_{(a,A)}(\mathbf{q},\mathbf{p}) = (a,A)\sigma((a,A)^{-1}(\mathbf{q},\mathbf{p})) = \sigma(\mathbf{q},\mathbf{p})\,h((a,A),\,(a,A)^{-1}(\mathbf{q},\mathbf{p})), \tag{11.53}$$

$(a,A)^{-1}(\mathbf{q},\mathbf{p})$ being the translate of $(\mathbf{q},\mathbf{p})$ by $(a,A)^{-1}$ under the action (11.14) and

$$h((a,A),\,(a,A)^{-1}(\mathbf{q},\mathbf{p})) = \sigma(\mathbf{q},\mathbf{p})^{-1}(a,A)\sigma((a,A)^{-1}(\mathbf{q},\mathbf{p})) \in T \times SU(2).$$

Moreover, if the section σ defines the frame $\mathcal{F}\{\eta^i_{\sigma(\mathbf{q},\mathbf{p})}, A_\sigma, 2s+1\}$, then $\sigma_{(a,A)}$ defines the frame $\mathcal{F}\{\eta^i_{\sigma_{(a,A)}(\mathbf{q},\mathbf{p})}, A_{\sigma_{(a,A)}}, 2s+1\}$, where $A_{\sigma_{(a,A)}} = U(a,A)A_\sigma U(a,A)^*$.

Let now $\mathfrak{A}$ denote the class of all affine space-like sections, but with φ not necessarily assumed to be zero.

Proposition 11.1.6: If $\sigma \in \mathfrak{A}$, then $\sigma_{(a,A)} \in \mathfrak{A}$, for all $(a,A) \in \mathcal{P}_+^\uparrow(1,3)$.

In view of this result (see [20] for a proof), starting with any family of coherent states $\mathfrak{S}_\sigma$, we may generate an entire class of covariantly translated families $\mathfrak{S}_{\sigma_{(a,A)}}$ of other coherent states, using the natural action (11.53) of $\mathcal{P}_+^\uparrow(1,3)$ on the space of sections. If σ is characterized by β and φ, then the functions β' and φ' that characterize $\sigma_{(a,A)}$ may be computed explicitly without difficulty.

Some special cases of the affine admissible sections σ are of particular interest.

1. *The Galilean section σ_0*

This section is defined by

$$\beta(\mathbf{p}) = \beta_0(\mathbf{p}) = 0, \quad \vartheta(\mathbf{p}) = \vartheta_0(\mathbf{p}) = 0,$$
$$\beta_0^*(\mathbf{p}) = \frac{\mathbf{p}}{p_0}, \quad \vartheta_0^*(\mathbf{p}) = \frac{\mathbf{p}}{m}. \tag{11.54}$$

Thus, here $\|\beta_0(\mathbf{p})\| < 1$, $\|\beta_0^*(\mathbf{p})\| < 1$, $\forall \mathbf{p}$. From (11.54) and (11.37),

$$a_\sigma(k,p) = a_0(k,p) = \frac{(2\pi)^3}{m}\frac{k_0 p_0 - \mathbf{k}\cdot\mathbf{p}}{k_0}, \tag{11.55}$$

and using the rotational invariance of $|\eta(k)|^2$,

$$\mathcal{A}_\sigma(k) = \mathcal{A}_0(k) = \frac{(2\pi)^3}{m}\langle P_0\rangle_\eta\, \mathbb{I}_{2s+1}. \tag{11.56}$$

Hence,

$$A_\sigma = A_0 = \frac{(2\pi)^3}{m}\langle P_0\rangle_\eta\, I, \tag{11.57}$$

so that the frame is tight.

2. The Lorentz section σ_ℓ

This is the principal section $\sigma_\wp$ introduced in the previous chapter. This time

$$\boldsymbol{\beta}(\mathbf{p}) = \boldsymbol{\beta}_\ell(\mathbf{p}) = \boldsymbol{\beta}_0^*(\mathbf{p}), \qquad \boldsymbol{\vartheta}(\mathbf{p}) = \boldsymbol{\vartheta}_\ell(\mathbf{p}) = \boldsymbol{\vartheta}_0^*(\mathbf{p}); \tag{11.58}$$

in other words, the Galilean and Lorentz sections are duals to each other. Furthermore, for this section, one again gets a tight frame:

$$A_\sigma = A_\ell = (2\pi)^3 m \langle P_0^{-1} \rangle_\eta \, I. \tag{11.59}$$

Indeed, from (11.58) and (11.36), we see that $a_\sigma(k,p)$ does not depend on k:

$$a_\sigma(k,p) = a_\ell(k,p) = \frac{(2\pi)^3 m}{p_0}. \tag{11.60}$$

In fact, assuming rotational invariance, one sees from (11.36) that this property holds only if $\boldsymbol{\beta}^*(\mathbf{p}) \equiv \mathbf{0}$, that is, for the Lorentz section.

3. The symmetric section σ_s

This section is self-dual, being given by

$$\boldsymbol{\beta}(\mathbf{p}) = \boldsymbol{\beta}_s(\mathbf{p}) = \boldsymbol{\beta}_s^*(\mathbf{p}) = \frac{\mathbf{p}}{m + p_0}, \qquad \boldsymbol{\vartheta}(\mathbf{p}) = \boldsymbol{\vartheta}_s(\mathbf{p}) = \boldsymbol{\vartheta}_s^*(\mathbf{p}) = \frac{\mathbf{p}}{m + p_0}. \tag{11.61}$$

Again, $\|\boldsymbol{\beta}_s(\mathbf{p})\| < 1$, $\forall \mathbf{p}$. Proceeding in the same way as before, we find that the operator $A_\sigma = A_s$ is now given by

$$(A_s \phi)(k) = \mathcal{A}_s(k) \phi(k) = (2\pi)^3 \langle \frac{k_0 P_0 + m^2}{m(k_0 + P_0)} \rangle_\eta \phi(k), \qquad \phi \in \mathfrak{H}_W^s. \tag{11.62}$$

To determine the spectrum of A_s, note that the function

$$f(k_0) \equiv a_s(k,p) = (2\pi)^3 \frac{k_0 p_0 + m^2}{m(k_0 + p_0)} \tag{11.63}$$

satisfies the estimate $(2\pi)^3 \leq f(k_0) \leq (2\pi)^3 p_0/m$, $\forall k_0 \in [m, \infty)$. This implies that the spectrum of the operator $A_\sigma = A_s$ is the entire interval $[(2\pi)^3 \|\eta\|^2, (2\pi)^3 \langle P_0 \rangle_\eta / m]$, so that for this section the frame is never tight.

4. The limiting sections $\sigma_\pm$

In addition to these three space-like sections, it is worth mentioning two limiting cases, given by

$$\boldsymbol{\beta}(\mathbf{p}) = \boldsymbol{\beta}_+(\mathbf{p}) = \frac{\mathbf{p}}{\|\mathbf{p}\|}, \qquad \boldsymbol{\vartheta}(\mathbf{p}) = \boldsymbol{\vartheta}_+(\mathbf{p}) = \frac{m\mathbf{p}}{\|\mathbf{p}\|(p_0 - \|\mathbf{p}\|)}$$

$$\boldsymbol{\beta}_-(\mathbf{p}) = \boldsymbol{\beta}_+^*(\mathbf{p}) = -\frac{\mathbf{p}}{\|\mathbf{p}\|}, \qquad \boldsymbol{\vartheta}_-(\mathbf{p}) = \boldsymbol{\vartheta}_+^*(\mathbf{p}) = -\frac{m\mathbf{p}}{\|\mathbf{p}\|(p_0 + \|\mathbf{p}\|)} \tag{11.64}$$

In this limiting situation, $\|\beta_+(\mathbf{p})\| = \|\beta_-(\mathbf{p})\| = 1$, $\forall \mathbf{p}$. Thus, these two sections are light-like and duals of each other. Here again, the frame is never tight. In addition, each of these light-like sections saturates one of the bounds in (11.49), the lower bound for σ_+ and the upper bound for σ_-.

One can also consider time-like sections, with $\|\beta_+(\mathbf{p})\| > 1$, but then admissibility [i.e., convergence of the integral (11.30)] will require restrictions on the support of the function η. This situation seems to be generic. We have met it in the case of Euclidean CS in Section 10.3, and we will meet it again for the Galilei group in Section 11.2.

A final comment is in order here. The Galilean and the Lorentz sections are dual to each other, yet completely dissimilar. More generally, the nature of the operator A_σ depends strongly on the choice of the section. Every space-like affine section yields a multiplication operator, but the spectrum of the latter may be very different for different sections. The geometrical analysis made in [20] shows that the Lorentz section is in fact the only truly covariant section, as expected from the principal section, and thus should yield simpler results. We will see shortly that the same holds in one space dimension. In this sense, choosing a section σ is very similar to choosing a gauge in electrodynamics. There too, some choices are more convenient than other ones for a given problem.

11.1.2 The Poincaré group in 1+1 dimensions, $\mathcal{P}_+^\uparrow(1,1)$ (♣)

The (1+1)-dimensional case $\mathcal{P}_+^\uparrow(1,1)$ is entirely parallel to the previous one, for spin $s = 0$. Accordingly, we will be very brief, referring the reader to the original papers [12]–[16] for further details.

The group elements (a, Λ_p) of $\mathcal{P}_+^\uparrow(1,1) = \mathbb{R}^2 \rtimes SO_o(1,1)$ are parametrized by $a = (a_0, \mathbf{a}) \in \mathbb{R}^2$ and $\Lambda_p \in SO_o(1,1)$, the latter being a Lorentz boost indexed by $p = (p_0, \mathbf{p})$, where $p_0 = (\mathbf{p}^2 + m^2)^{1/2}$, $m > 0$.

The Wigner representation U_W of mass $m \geq 0$ reads as

$$(U_W(a, \Lambda_p)\phi)(k) = e^{ik \cdot a}\, \phi(\Lambda_p^{-1}k), \quad \phi \in \mathfrak{H}_m = L^2(V_m^+, dk/k_0). \quad (11.65)$$

Here, V_m^+ denotes the forward mass hyperbola of mass $m \geq 0$, $V_m^+ = \{k = (k_0, \mathbf{k}) \,|\, k^2 = k_0^2 - \mathbf{k}^2 = m^2, k_0 > 0\}$, $k \cdot a = k_0 a_0 - \mathbf{k} \cdot \mathbf{a}$, with invariant measure dk/k_0. Again, U_W is not square integrable over $\mathcal{P}_+^\uparrow(1,1)$, and we look for appropriate quotients. The natural phase spaces, i.e., the coadjoint orbits of $\mathcal{P}_+^\uparrow(1,1)$, are the following [18, 29]:

- for $m > 0$: $\Gamma = \mathcal{P}_+^\uparrow(1,1)/T$, where T is the subgroup of time translations

- for $m = 0$: $\Gamma_{l(r)} = \mathcal{P}_+^\uparrow(1,1)/T_{l(r)}$, where T_l, T_r are subgroups of lightlike translations.

Neither T nor $T_{l(r)}$ is the stability subgroup of any vector in $\mathcal{H}_m$, and, hence, we again have to use the general construction for building covariant coherent states.

Let us consider first the massive case, $m > 0$. The phase space $\Gamma = \mathcal{P}_+^\uparrow(1,1)/T$ carries canonical coordinates $(\mathbf{q},\mathbf{p})$ and the $\mathcal{P}_+^\uparrow(1,1)$-invariant measure $d\mathbf{q}\,d\mathbf{p}$. In the principal bundle $\mathcal{P}_+^\uparrow(1,1) \to \Gamma = \mathcal{P}_+^\uparrow(1,1)/T$, a section $\sigma : \Gamma \to \mathcal{P}_+^\uparrow(1,1)$ reads as

$$\sigma(\mathbf{q},\mathbf{p}) = \sigma_0(\mathbf{q},\mathbf{p})\left((f(\mathbf{q},\mathbf{p}),0),I\right) \equiv (\hat{q},\Lambda_p), \qquad (11.66)$$

where $\sigma_0(\mathbf{q},\mathbf{p}) = ((0,\mathbf{q}),\Lambda_p)$ is again called the Galilean section and the expression multiplying $\sigma_0(\mathbf{q},\mathbf{p})$ is an arbitrary element of T, indexed by the function f. As in the 1+3 case, we consider only *affine* sections, corresponding to $f(\mathbf{q},\mathbf{p}) = \mathbf{q}\cdot\vartheta(\mathbf{p})$. Thus, the section is characterized by the function ϑ. The class of affine sections is stable under the action $\sigma \mapsto \sigma_{(a,\Lambda_p)}$ of $\mathcal{P}_+^\uparrow(1,1)$, and it has the advantage that, for such sections, all calculations may be done explicitly.

As before, the section σ may be parametrized in terms of a momentum-dependent "speed" β:

$$\hat{q}_0 = \beta(\mathbf{p})\cdot\hat{\mathbf{q}}. \qquad (11.67)$$

Geometrically, (11.67) implies that, for fixed $\mathbf{p}$, the choice of β determines a particular reference frame in $\hat{q}$-space [13]. In particular, we see that the vector $\hat{q}$ is time-like, light-like, or space-like, whenever $|\beta| > 1$, $|\beta| = 1$ or $|\beta| < 1$, respectively, and this classification is also $\mathcal{P}_+^\uparrow(1,1)$-invariant (compare Proposition 11.1.1). Sections for which $|\beta| < 1$ will thus be called *space-like*.

We also introduce the dual speed [compare (11.21)]

$$\beta^*(\mathbf{p}) = \frac{\mathbf{p} - p_0\beta(\mathbf{p})}{p_0 - \mathbf{p}\cdot\beta(\mathbf{p})} = \frac{1}{p_0}(\mathbf{p} - m\vartheta(\mathbf{p})). \qquad (11.68)$$

One has, as before, $\beta^{**} = \beta$, which means that the sections come in dual pairs, and, moreover, the duality $\beta \leftrightarrow \beta^*$ preserves the space-like (respectively, time-like or light-like) character of the vector $\hat{q}$.

According to the general scheme, we now define the Poincaré CS

$$\eta_{\sigma(\mathbf{q},\mathbf{p})} = U_W(\sigma(\mathbf{q},\mathbf{p}))\eta, \ (\mathbf{q},\mathbf{p}) \in \Gamma, \qquad (11.69)$$

and the corresponding resolution operator

$$A_\sigma = \int_\Gamma |\eta_{\sigma(\mathbf{q},\mathbf{p})}\rangle\langle\eta_{\sigma(\mathbf{q},\mathbf{p})}|\,d\mathbf{q}\,d\mathbf{p}. \qquad (11.70)$$

Then, the results of [12]–[16] may be summarized as follows.

Theorem 11.1.7: *Let $\sigma : \Gamma = \mathcal{P}_+^\uparrow(1,1)/T \to \mathcal{P}_+^\uparrow(1,1)$ be any affine space-like section. Then, a vector η is admissible mod(T,σ) iff it is of finite energy, $\eta \in \mathcal{D}(P_0^{1/2})$, and every such vector generates a family of coherent*

11.1. The Poincaré groups $\mathcal{P}_+^\uparrow(1,3)$ and $\mathcal{P}_+^\uparrow(1,1)$

states indexed by points in the phase space Γ, which constitute a rank one frame:

$$\mathfrak{S}_\sigma = \{\eta_{\sigma(\mathbf{q},\mathbf{p})} = U_W(\sigma(\mathbf{q},\mathbf{p}))\eta, \ (\mathbf{q},\mathbf{p}) \in \Gamma\} \tag{11.71}$$

More precisely, A_σ is a positive, bounded multiplication operator with bounded inverse, $(A_\sigma\psi)(k) = \mathcal{A}_\sigma(k)\psi(k)$, where

$$\mathcal{A}_\sigma(k) = \int_{V_m^+} \mathcal{A}_\sigma(k,p) \, |\eta(p)|^2 \, \frac{d\mathbf{p}}{p_0}. \tag{11.72}$$

The kernel $\mathcal{A}_\sigma(k,p)$ is defined in terms of the function ϑ or the speed β defining the section σ as follows:

$$\mathcal{A}_\sigma(k,p) = \frac{2\pi}{m} \frac{k \cdot p}{k_0 - \mathbf{p} \cdot \vartheta(\Lambda_p^{-1}k/m)} \tag{11.73}$$

$$= \frac{2\pi m}{p_0 + \mathbf{p} \cdot \boldsymbol{\beta}^*(\Lambda_p^{-1}k/m)}. \tag{11.74}$$

The spectrum of the operator A_σ obeys the following universal bounds:

$$\frac{2\pi}{m}\langle P_0 - |\mathbf{P}|\rangle_\eta \leq \|A_\sigma\| \leq \frac{2\pi}{m}\langle P_0 + |\mathbf{P}|\rangle_\eta, \tag{11.75}$$

where $\langle \cdot \rangle_\eta \equiv \langle \eta | \cdot \eta \rangle$ denotes a mean value in the state η.

Since the resolution operator A_σ is a positive multiplication operator, so are its inverse and its square root. The latter fact allows us to compute explicitly weighted CS, as indicated in (7.55), Chapter 7, Section 7.3.

Proposition 11.1.8: *For any space-like affine section σ and admissible vector η, the frame $\mathfrak{S}_\sigma = \{\eta_{\sigma(\mathbf{q},\mathbf{p})}, \ (\mathbf{q},\mathbf{p}) \in \Gamma\}$ may be turned into a tight frame of weighted CS*

$$\eta_{\sigma(\mathbf{q},\mathbf{p})}^w = W_\sigma(\mathbf{q},\mathbf{p})\eta_{\sigma(\mathbf{q},\mathbf{p})}, \tag{11.76}$$

where $W_\sigma(\mathbf{q},\mathbf{p})$ is a measurable family of weighting operators, given explicitly by:

$$W_\sigma(\mathbf{q},\mathbf{p}) = \|\eta\|^{-2} C_\sigma^{-1/2}(p) |\eta_{\sigma(\mathbf{q},\mathbf{p})}\rangle \langle \eta_{\sigma(\mathbf{q},\mathbf{p})}|. \tag{11.77}$$

In this expression, $C_\sigma^{-1/2}(p)$ is the positive self-adjoint operator of multiplication by the function $\mathcal{A}_\sigma(k, \Lambda_k \bar{p})^{-1/2}$, where $\bar{p} = (p_0, -\mathbf{p})$.

We conclude with some concrete examples, which are essentially identical to those discussed in the 1+3 case. Among the class of admissible space-like sections, we recover the three remarkable cases discussed in the previous section, with almost identical properties. Once again, of course, the Lorentz section is the principal section.

1. *Galilean section* σ_0: $\vartheta(\mathbf{p}) = \mathbf{0} \Leftrightarrow \beta(\mathbf{p}) = \mathbf{0} \Leftrightarrow \beta^*(\mathbf{p}) = \dfrac{\mathbf{P}}{p_0}$
   $A_{\sigma_0}^\eta = 2\pi m^{-1}[\langle P_0\rangle_\eta\, I - \langle \mathbf{P}\rangle_\eta\, \mathbf{P}\, P_0^{-1}]$, and thus the frame can be made tight if one requires that $\langle \mathbf{P}\rangle_\eta = 0$. This is the one-dimensional remnant of the rotational invariance imposed in the 1+3 case.

2. *Lorentz section* σ_ℓ: $\vartheta(\mathbf{p}) = \dfrac{\mathbf{P}}{m} \Leftrightarrow \beta(\mathbf{p}) = \dfrac{\mathbf{P}}{p_0} \Leftrightarrow \beta^*(\mathbf{p}) = 0$
   $A_{\sigma_\ell}^\eta = 2\pi m \langle P_0^{-1}\rangle_\eta\, I$. Thus, the frame is tight for any admissible vector η, and this is the only section with this property [here, and only here, the kernel $\mathcal{A}_\sigma(k,p)$ does not depend on k].

3. *Symmetric section* σ_s: $\vartheta(\mathbf{p}) = \dfrac{\mathbf{P}}{m + p_0} \Leftrightarrow \beta(\mathbf{p}) = \beta^*(\mathbf{p}) = \dfrac{\mathbf{P}}{m + p_0}$
   The frame is not tight for any η.

Again, the first two sections are duals of each other and the third one is self-dual. Also, non-space-like sections can be considered, exactly as in the previous case.

As for admissible vectors, the following functions are convenient for explicit computations:

- Gaussian vector: $\eta_G(k) \sim e^{-k_0/U}$;
- binomial vector: $\eta_\alpha(k) \sim (1 + (k_0 - m)/U)^{-\alpha/2}$, $\alpha > 1/2$.

In both cases, U is a normalization constant, with the dimension of an energy. We will see in Section 11.3 that these two types of admissible vectors have attractive properties when comparing the Poincaré group $\mathcal{P}_+^\uparrow(1,1)$ with the anti-de Sitter group $SO_o(1,2)$ and the Galilei group $\widetilde{\mathcal{G}}(1,1)$ via the technique of group contraction.

Finally, a word can be said about orthogonality relations, in the spirit of Chapter 8, Section 8.2. In order to derive these relations, we have to compute the following matrix element and, in particular, to find sufficient conditions for its convergence:

$$I_\sigma^{\eta_1\eta_2}(\phi_2, \phi_1) = \langle \phi_2 | A_\sigma^{\eta_1\eta_2} | \phi_1 \rangle, \qquad (11.78)$$

where

$$A_\sigma^{\eta_1\eta_2} = \int_\Gamma |\eta_{1\,\sigma(\mathbf{q},\mathbf{p})}\rangle \langle \eta_{2\,\sigma(\mathbf{q},\mathbf{p})}|\, d\mathbf{q}\, d\mathbf{p}. \qquad (11.79)$$

In the case of a tight frame, we will get $A_\sigma^{\eta_1\eta_2} = \lambda_\sigma(\eta_1, \eta_2) I$, where $\lambda_\sigma(\cdot,\cdot)$ is a sesquilinear form. Then, one gets

$$I_\sigma^{\eta_1\eta_2}(\phi_2, \phi_1) = \lambda_\sigma(\eta_1, \eta_2)\, \langle \phi_2 | \phi_1 \rangle. \qquad (11.80)$$

If the form λ_σ is closed, we may write $\lambda_\sigma(\eta_1, \eta_2) = \langle C\eta_1 | C\eta_2\rangle$, as in the proof of Theorem 8.2.1, and ordinary orthogonality relations follow.

Coming back to our Poincaré CS, and proceeding exactly as before, one finds that the integral (11.79) converges if both η_1 and η_2 belong to the

11.1. The Poincaré groups $\mathcal{P}_+^\uparrow(1,3)$ and $\mathcal{P}_+^\uparrow(1,1)$

domain of $P_0^{1/2}$, and then $A_\sigma^{\eta_1 \eta_2}$ is the operator of multiplication by the function

$$A_\sigma^{\eta_1 \eta_2}(k) = \int_{V_m^+} \mathcal{A}_\sigma(k,p) \, \overline{\eta_2(p)} \, \eta_1(p) \, \frac{d\mathbf{p}}{p_0}, \tag{11.81}$$

with the kernel $\mathcal{A}_\sigma(k,p)$ given in (11.74). Of course, putting $\eta_1 = \eta_2 = \eta$, one recovers the operator $A_\sigma \equiv A_\sigma^{\eta\eta}$.

From (11.81), we can now explicitly compute the matrix element (11.78). Let $\eta_1, \eta_2 \in \mathcal{D}(P_0^{1/2})$ and $\phi_1, \phi_2 \in \mathfrak{H}$, and write $\rho_j = |\phi_j\rangle\langle\eta_j|$, $j = 1, 2$. Then, one has

$$\begin{aligned} I_\sigma^{\eta_1 \eta_2}(\phi_2, \phi_1) &= \int_\Gamma \overline{\langle U(\sigma(\mathbf{q},\mathbf{p}))\eta_2|\phi_2\rangle} \langle U(\sigma(\mathbf{q},\mathbf{p}))\eta_1|\phi_1\rangle d\mathbf{q}d\mathbf{p} \\ &= \int_\Gamma \overline{\mathrm{Tr}[U(\sigma(\mathbf{q},\mathbf{p}))^* \rho_2]} \, \mathrm{Tr}[U(\sigma(\mathbf{q},\mathbf{p}))^* \rho_1] \, d\mathbf{q}d\mathbf{p} \\ &= \langle \rho_2 | \mathcal{A}_\sigma \rho_1 \rangle_{\mathcal{B}_2(\mathfrak{H})}, \end{aligned} \tag{11.82}$$

where $\mathcal{A}_\sigma$ is a positive self-adjoint operator on $\mathcal{B}_2(\mathfrak{H})$:

$$(\mathcal{A}_\sigma |\phi\rangle\langle\eta|)(k,p) = \mathcal{A}_\sigma(k,p)\phi(k)\overline{\eta(p)}. \tag{11.83}$$

Consider first the case of a tight frame. For the Galilean section σ_0 with $\langle \mathbf{P} \rangle_\eta = 0$, we obtain

$$\lambda_{\sigma_0}(\eta_1, \eta_2) = \frac{2\pi}{m} \langle \eta_1 | P_0 \eta_2 \rangle, \tag{11.84}$$

which is manifestly a closed sesquilinear form. Hence, we have orthogonality relations, with the Duflo–Moore operator

$$C_0 = \sqrt{\frac{2\pi}{m}} P_0^{1/2} \quad \text{or} \quad K_0 = \frac{m}{2\pi} P_0^{-1}. \tag{11.85}$$

Similarly, for the Lorentz section σ_ℓ, we get

$$\lambda_{\sigma_\ell}(\eta_1, \eta_2) = 2\pi m \, \langle \eta_1 | P_0^{-1} \eta_2 \rangle, \tag{11.86}$$

and, thus,

$$C_\ell = \sqrt{2\pi m} P_0^{-1/2} \quad \text{or} \quad K_\ell = \frac{1}{2\pi m} P_0. \tag{11.87}$$

We see again how much the result depends on the choice of the section, even for tight frames. In addition, one gets in both cases a Wigner transform, as described in Chapter 8, Section 8.3,

$$(\mathcal{W}\rho)(\mathbf{q},\mathbf{p}) = \mathrm{Tr}[U(\sigma(\mathbf{q},\mathbf{p}))^* \mathcal{A}_\sigma^{-1} \rho], \tag{11.88}$$

which is an isometry from $\mathcal{B}_2(\mathfrak{H})$ into $L^2(\Gamma, d\mathbf{q}d\mathbf{p})$.

If the frame is not tight, however, we don't see any way of transforming (11.82) into a useful orthogonality relation.

11.1.3 Poincaré CS: The massless case

We turn now to the massless case. In 1+3 dimensions, no significant difference exists with respect to the massive case, since the limit $m \to 0$ is known to be nonsingular (but there is more freedom in choosing an appropriate phase space, as indicated in [Sch]). In 1+1 dimensions, however, the situation becomes more intricate. Indeed, because of the infrared divergences inherent to any massless quantum field theory in two dimensions, one is forced to work in an indefinite metric (Krein) space, with a nonunitary, indecomposable representation of $\mathcal{P}_+^\uparrow(1,1)$. Yet CS may be obtained by a suitable adaptation of the general formalism used so far, namely, the introduction of *quasi-sections*, as done in [29], which we follow.

Let us first recall that a *Krein space* $\mathfrak{K}$ is a vector space equipped with a nondegenerate sesquilinear form $\langle \cdot, \cdot \rangle$, such that $\mathfrak{K} = \mathfrak{K}_1 \oplus \mathfrak{K}_2$, where the sesquilinear form is positive definite (respectively, negative definite) on $\mathfrak{K}_1$ (respectively, $\mathfrak{K}_2$) and both $\mathfrak{K}_1$ and $\mathfrak{K}_2$ are complete (i.e., Hilbert spaces) with respect to the restriction of the sesquilinear form. It follows that any vector $f \in \mathfrak{K}$ may be decomposed as $f = f_1 + f_2$ ($f_j \in \mathfrak{K}_j$), with

$$\langle f, f \rangle = \|f_1\|^2 - \|f_2\|^2, \quad \|f_1\|^2 = \langle f_1, f_1 \rangle \geq 0, \quad \|f_2\|^2 = -\langle f_2, f_2 \rangle \geq 0.$$

The Krein space $\mathfrak{K}$ is called a *Pontrjagin space* if one of $\mathfrak{K}_1, \mathfrak{K}_2$ is finite dimensional. Finally, given a Krein space $\mathfrak{K}$ with indefinite inner product $\langle \cdot, \cdot \rangle$, a majorant Hilbert topology on $\mathfrak{K}$ is a Hilbert inner product (i.e., positive definite) $(\cdot, \cdot)$, such that

$$|\langle f, g \rangle| \leq (f, f)^{1/2} (g, g)^{1/2}, \; \forall f, g \in \mathfrak{K}.$$

A systematic study of Krein spaces and other indefinite inner product spaces may be found in [Bog].

Let us now come back to $\mathcal{P}_+^\uparrow(1,1)$. Starting from the relevant Wightman two-point function, which is not positive definite, one gets an indefinite sesquilinear form $\langle \cdot, \cdot \rangle$ on the Schwartz space $\mathcal{S}(\mathbb{R}^2)$. Upon completion with respect to an appropriate majorant Hilbert topology, one obtains the Krein space (actually, it is a Pontrjagin space with one negative dimension)

$$\mathfrak{K} = L^2(C_+, \frac{d\mathbf{k}}{|\mathbf{k}|}) \oplus V \oplus X, \tag{11.89}$$

where X and V are both one-dimensional subspaces, C_+ is the positive light cone, and the nonpositivity of $\langle \cdot, \cdot \rangle$ comes from the fact that the metric operator J is the identity on the first summand of $\mathfrak{K}$, but equals $\begin{pmatrix} 0 & 1 \\ 1 & 0 \end{pmatrix}$ on the complementary subspace $V \oplus X$. Then, the Wigner representation U_W may be defined on a dense subset of $\mathfrak{K}$. It is now only J-unitary, however, i.e., $\langle U_W(a, \Lambda)f, U_W(a, \Lambda)g \rangle = \langle f, g \rangle$, and not unitary for the associated Hilbert space inner product. Furthermore, U_W is neither irreducible nor completely reducible, but indecomposable.

11.1. The Poincaré groups $\mathcal{P}_+^\uparrow(1,3)$ and $\mathcal{P}_+^\uparrow(1,1)$

Since the positive light cone is disconnected, consisting of two half-lines, $C_+ = \mathbb{R}_*^+ \cup \mathbb{R}_*^-$, the Hilbert subspace $L^2(C_+ d\mathbf{k}/|\mathbf{k}|)$ may in turn be decomposed into a direct sum

$$L^2(C_+, \frac{d\mathbf{k}}{|\mathbf{k}|}) = \mathfrak{H}_l \oplus \mathfrak{H}_r, \tag{11.90}$$

where

$$\mathfrak{H}_l = L^2(\mathbb{R}^-, \frac{d\mathbf{k}}{|\mathbf{k}|}), \qquad \mathfrak{H}_r = L^2(\mathbb{R}^+, \frac{d\mathbf{k}}{|\mathbf{k}|}) \tag{11.91}$$

(left and right Hilbert spaces, respectively). Correspondingly, one may quotient the representation U_W and finally obtain two unitary irreducible representations $U_{l(r)}$ defined on $\mathfrak{H}_{l(r)}$. This amounts to considering the matrix elements

$$\langle \psi_1, U_W(a, \Lambda)\psi_2 \rangle_{\mathfrak{H}_{l(r)}}, \quad \psi_1, \psi_2 \in \mathfrak{H}_{l(r)}, \tag{11.92}$$

and associating with the so-defined sesquilinear forms the operators $U_{l(r)}(a, \Lambda)$. The final result is

$$(U_{l(r)}(a, \Lambda)\psi)(k) = e^{ika}\psi(\Lambda^{-1}k), \quad \psi \in \mathfrak{H}_{l(r)}. \tag{11.93}$$

These are the representations of $\mathcal{P}_+^\uparrow(1,1)$ that are used in the construction of systems of massless coherent states [but the resolution of the identity (11.98) that we will obtain in the end lives in the Krein space $\mathfrak{K}$].

We consider first the right coadjoint orbit $\Gamma_r = \mathcal{P}_+^\uparrow(1,1)/T_r$, corresponding to the representation U_r of $\mathcal{P}_+^\uparrow(1,1)$. A global parametrization of this orbit is given by (τ, p), $\tau \in \mathbb{R}, p > 0$, and it carries the $\mathcal{P}_+^\uparrow(1,1)$-invariant measure $d\mu(\tau, p) = d\tau dp/p$. Now, it turns out that the natural (Galilean) section $\sigma_n(\tau, p) = ((0, \tau), \Lambda_p)$ is not admissible, because the integral of the type (7.46),

$$c_r(\eta, \phi) = \int_{\Gamma_r} |\langle U_r(\sigma_n(\tau, p))\eta|\phi\rangle|^2 \, d\tau dp/p, \tag{11.94}$$

giving the admissibility condition, is infrared divergent (that is, at $p = 0$), for any $\eta, \phi \in \mathfrak{H}_r$. The following quasi-section, however, is admissible:

$$\sigma_r(\tau, p) = ((0, \frac{\tau}{p}), \Lambda_{\rho(p)}), \qquad \text{where} \qquad \rho(p) = \frac{p}{2} - \frac{1}{2p}. \tag{11.95}$$

A straightforward calculation then shows that a vector $\eta \in \mathcal{H}_r$ is admissible for the quasi-section (11.95) if and only if $\eta \in \mathcal{D}(H_r^{-1/2})$, where H_r is the usual momentum operator on $\mathfrak{H}_r$ (but now both H_r and H_r^{-1} are unbounded).

Using this quasi-section, we define *right coherent states* as the states:

$$\eta_{\sigma_r(\tau,p)}(k) = (U_r(\sigma_r(\tau,p))\eta)(k) = e^{i\frac{\tau}{p}k}\eta(\frac{k}{p}), \quad \eta \in \mathfrak{H}_r. \tag{11.96}$$

Then, an explicit calculation leads to the following (weak) identity:

$$\frac{1}{c_{\sigma_r}(\eta)} \int_{\Gamma_r} |U_r(\sigma_r(\tau,p))\eta\rangle\langle U_r(\sigma_r(\tau,p))\eta| \, d\mu(\tau,p) = I, \qquad (11.97)$$

where $c_{\sigma_r}(\eta) \equiv c_r(\eta,\eta)$ is given in (11.94). Thus, we get a genuine resolution of the identity and a tight frame. An intriguing fact, albeit not totally unexpected (since the phase spaces are the same in the two cases, namely, a half-plane), is that these right CS of $\mathcal{P}_+^\uparrow(1,1)$ are identical with the familiar wavelets associated to the $ax + b$ group, which we shall discuss in Chapter 12 [see also Chapter 8, Section 8.1, in particular (8.24)].

In the same way, we may construct a corresponding set of *left coherent states*. Collecting together all of these results, one obtains a resolution of the identity in the Krein space (11.89):

$$I = P_V + P_X + \frac{1}{c_{\sigma_l}(\phi)} \int_{\Gamma_l} |U_l(\sigma_l(\tau,p))\phi\rangle\langle U_l(\sigma_l(\tau,p))\phi| \, d\mu_l(\tau,p)$$
$$+ \frac{1}{c_{\sigma_r}(\psi)} \int_{\Gamma_r} |U_r(\sigma_r(\tau,p))\psi\rangle\langle U_r(\sigma_r(\tau,p))\psi| \, d\mu_r(\tau,p), \qquad (11.98)$$

where $\phi \in \mathfrak{H}_l$, $\psi \in \mathfrak{H}_r$, both admissible, and P_V, P_X denote the rank 1 projection operators on V, X, respectively. This result suggests that most of the CS formalism might possibly be extended to (simple) indefinite metric spaces.

11.2 The Galilei groups $\widetilde{\mathcal{G}}(1,1)$ and $\widetilde{\mathcal{G}} \equiv \widetilde{\mathcal{G}}(1,3)$ (♣)

In Chapter 9, Section 9.2.1, we have discussed at great length the structure of the extended Galilei group $\widetilde{\mathcal{G}} \equiv \widetilde{\mathcal{G}}(1,3)$, and we have constructed, by the Mackey method, a class of induced unitary irreducible representations, given in (9.72) and (9.74). Coming to CS, however, we have limited our analysis to the isochronous subgroup $\widetilde{\mathcal{G}}'$, because the restriction of these representations to $\widetilde{\mathcal{G}}'$ are square integrable mod $(\Theta \times SU(2))$, where Θ is the phase subgroup coming from the central extension. The corresponding parameter space is the phase space $\Gamma = \widetilde{\mathcal{G}}'/(\Theta \times SU(2)) \simeq \mathbb{R}^6$, and the associated Galilei CS constitute a tight frame, as seen in (9.87).

In the present section, we are going to extend the analysis to the whole extended Galilei group $\widetilde{\mathcal{G}}$. Exactly as for the Poincaré case, however, the rotation subgroup inherent to three space dimensions does not change things much, except, of course, for the possibility of having nonzero spin values. Indeed the essential features of the construction are already present in one space dimension. Hence, we shall mainly discuss here the corresponding group $\widetilde{\mathcal{G}}(1,1)$. Moreover, we shall be rather brief, since the procedure parallels that used above for the Poincaré group $\mathcal{P}_+^\uparrow(1,1)$. (Details may be found in [16].)

11.2. The Galilei groups $\tilde{\mathcal{G}}(1,1)$ and $\tilde{\mathcal{G}} \equiv \tilde{\mathcal{G}}(1,3)$ (♣)

Group elements of $\tilde{\mathcal{G}}(1,1)$ are parametrized as $g = (\theta, b, \mathbf{a}, \mathbf{v})$, where θ is the parameter corresponding to the central extension of mass m, $(b, \mathbf{a})$ are time and space translations, and $\mathbf{v}$ is the boost parameter. In momentum space, the unitary irreducible representation that we shall use reads as (compare (9.72), with $j = 0$ and $E_0 = 0$):

$$(\hat{U}(g)\psi)(\mathbf{k}) = \exp i(\theta + \frac{\mathbf{k}^2}{2m}b - \mathbf{k}\cdot\mathbf{a})\,\psi(\mathbf{k} - m\mathbf{v}), \quad \psi \in L^2(\mathbb{R}, d\mathbf{k}). \quad (11.99)$$

This representation is not square integrable over $\tilde{\mathcal{G}}(1,1)$ (the integral over b diverges); hence we have to take a quotient. Following the now standard procedure, we choose the associated coadjoint orbit of $\tilde{\mathcal{G}}(1,1)$, which is the quotient $\Gamma = \tilde{\mathcal{G}}(1,1)/(\Theta \times \mathcal{T})$, where $\Theta \times \mathcal{T}$ is the subgroup of $\tilde{\mathcal{G}}(1,1)$ consisting of phase changes and time translations. It carries global coordinates $(\mathbf{q}, \mathbf{p}) \equiv (\mathbf{a} - b\mathbf{v}, m\mathbf{v}) \in \mathbb{R}^2$ and the invariant measure is $d\mathbf{q}\,d\mathbf{p}$.

As in the Poincaré case, we first consider the Galilean section $\sigma_0 : \Gamma \to \tilde{\mathcal{G}}(1,1)$, which in this case is the principal section, and is given by

$$\sigma_0(\mathbf{q}, \mathbf{p}) = (0, 0, \mathbf{q}, \frac{\mathbf{p}}{m}). \quad (11.100)$$

Then, an arbitrary affine section is of the form (again, an irrelevant term $\varphi(\mathbf{p})$ has been set to zero):

$$\sigma(\mathbf{q}, \mathbf{p}) = \sigma_0(\mathbf{q}, \mathbf{p})\,(0, \mathbf{q}\cdot\vartheta(\mathbf{p}), \mathbf{0}, \mathbf{0})$$
$$= (\mathbf{q}\cdot\vartheta(\mathbf{p})\frac{\mathbf{p}^2}{2m}, \mathbf{q}\cdot\vartheta(\mathbf{p}), \mathbf{q} + \frac{\mathbf{p}}{m}\mathbf{q}\cdot\vartheta(\mathbf{p}), \frac{\mathbf{p}}{m}), \quad (11.101)$$

with ϑ a real continuous function that indexes the section.

Next, one looks for conditions to be imposed on admissible vectors η to ensure that the integral $I_\sigma^{\eta_1\eta_2}(\phi_2, \phi_1)$, the analogue of (11.78), converges and is explicitly calculable. As compared to the Poincaré case, the new feature here is that (except for the Galilean section) the Jacobian $\mathcal{J}_\mathbf{X}(\mathbf{k})$ corresponding to the change of variables $\mathbf{k} \mapsto \mathbf{X}(\mathbf{k})$ in the integral analogous to (11.30) is totally unwieldy, unless one imposes a restriction on the support of the admissible vectors η_j. Accordingly, one defines $\eta \in L^2(\mathbb{R}, d\mathbf{k})$ to be admissible $\mathrm{mod}(\Theta \times \mathcal{T}, \sigma)$ if it satisfies the following two conditions:

1. $\forall \mathbf{k} \in \mathbb{R}, \; \mathbf{p}\cdot\vartheta(\mathbf{k} - \mathbf{p}) \geq m$ implies $\eta(\mathbf{p}) = 0$;

2. The following integral converges:

$$\mathcal{A}_\sigma(\mathbf{k}) = 2\pi \int_{\mathbf{p}\cdot\vartheta(\mathbf{k}-\mathbf{p})<m} d\mathbf{p}\,\left(1 - \frac{\mathbf{p}}{m}\cdot\vartheta(\mathbf{k} - \mathbf{p})\right)^{-1}|\eta(\mathbf{p})|^2. \quad (11.102)$$

For the Galilean section, corresponding to $\vartheta \equiv 0$, every vector $\eta \in L^2(\mathbb{R}, d\mathbf{k})$ is admissible — again, this is the natural or fully covariant section. For any other affine section, the support of an admissible vector must be restricted, and, contrary to the Poincaré case, the admissibility of η depends now explicitly on the particular section σ. Notice that the integral in (11.102) defines a positive function, which is a constant iff ϑ is a constant. We summarize the results in the following proposition.

Proposition 11.2.1: Let $\sigma : \Gamma \to \widetilde{\mathcal{G}}(1,1)$ be an affine section. Then, any vector $\eta \in L^2(\mathbb{R}, d\mathbf{k})$ admissible mod $(\Theta \times \mathcal{T}, \sigma)$ generates a frame of CS indexed by points in phase space, given by the vectors

$$\eta_{\sigma(\mathbf{q},\mathbf{p})}(\mathbf{k}) = [\widehat{U}(\sigma(\mathbf{q},\mathbf{p}))\eta](\mathbf{k})$$
$$= \exp i \left(\frac{|\mathbf{k}-\mathbf{p}|^2}{2m} \mathbf{q} \cdot \vartheta(\mathbf{p}) - \mathbf{k} \cdot \mathbf{q} \right) \eta(\mathbf{k}-\mathbf{p}). \quad (11.103)$$

The corresponding resolution generator A_σ is a positive, bounded, invertible multiplication operator

$$(A_\sigma \psi)(\mathbf{k}) = \mathcal{A}_\sigma(\mathbf{k})\psi(\mathbf{k}), \ \psi \in L^2(\mathbb{R}, d\mathbf{k}), \quad (11.104)$$

with kernel $\mathcal{A}_\sigma(\mathbf{k})$ given in (11.102).

Here too, as in the Poincaré case (Proposition 11.1.8), each of the CS frames so obtained may be turned into a tight frame of weighted CS, using weighting operators similar to (11.77). We refer to [16] for details.

Among affine sections, we mention the following (rather trivial!) cases.

(1) The *Galilean section* σ_0, corresponding to $\vartheta(\mathbf{p}) = 0$: Every vector is admissible, and the frame is tight.

(2) *Constant sections*, with $\vartheta(\mathbf{p}) = \vartheta_c$, a positive constant: Here, one gets a tight frame for any vector η with support in $(-\infty, m\vartheta_c^{-1})$, such that $\langle (m - \mathbf{P} \cdot \vartheta_c)^{-1} \rangle_\eta < \infty$.

For these two sections, one may derive orthogonality relations and a Wigner transform, exactly as for $\mathcal{P}_+^\uparrow(1,1)$. the Duflo–Moore operator is

$$C_{\vartheta_c} = \sqrt{2\pi m} \, (m - \mathbf{P} \cdot \vartheta_c)^{-1/2} \quad (\vartheta_c \geq 0). \quad (11.105)$$

Other sections will in general require support restrictions on η and give nontight frames.

We conclude with two examples of admissible vectors, similar to those given in the previous section for $\mathcal{P}_+^\uparrow(1,1)$:

- Gaussian vector: $\eta_G(\mathbf{k}) \sim \exp(-\mathbf{k}^2/2mU)$;

- binomial vector: $\eta_\alpha(\mathbf{k}) \sim (1 + \mathbf{k}^2/2mU)^{-\alpha/2}$, $\alpha > 1/2$.

Here too U is a normalization factor, which may be interpreted as a kind of internal energy.

Exactly the same analysis may be made in the three-dimensional (3-D) case $\widetilde{\mathcal{G}} = \widetilde{\mathcal{G}}(1,3)$, with very similar results. In the simplest situation of spin $j = 0$, one may choose as parameter space the coadjoint orbit

$\Gamma = \widetilde{\mathcal{G}}/(\Theta \times \mathcal{T} \times SU(2)) \simeq \mathbb{R}^6$, with coordinates $(\mathbf{q}, \mathbf{p}) \equiv (\mathbf{a} - b\mathbf{v}, m\mathbf{v})$ and invariant measure $d\mathbf{q}\, d\mathbf{p}$.

11.3 The anti-de Sitter group $SO_o(1,2)$ and its contraction(s) (♣)

Among many other applications in physics [134], the group $SO_o(1,2)$ and its double covering $SU(1,1) \sim SL(2,\mathbb{R})$ play a central role in the kinematical symmetries of (1+1)-dimensional space–time. Different kinematical (or relativity) Lie groups are possible. Two of them have been considered in the previous sections, namely, the Galilei group $\widetilde{\mathcal{G}}(1,1)$ and the Poincaré group $\mathcal{P}_+^\uparrow(1,1)$. One can, however, distinguish (at least) seven different kinematics, if one adapts to (1+1)-dimensional space–times the classification of Bacry and Lévy-Leblond [58]. Restricting to 1+1 dimensions is just a matter of technical (but not conceptual!) simplification. These seven kinematics or relativities are well encapsulated in the following diagram concerning the relations between their respective Lie groups [133]:

$$
\begin{array}{ccccc}
SO_o(2,1) & \longrightarrow & \mathbb{R}^2 \rtimes SO_o(1,1) \equiv N_+ & & \\
\text{(de Sitter)} & & \text{(Newton)} & & \\
\searrow & & \downarrow & & \\
& \mathcal{P}_+^\uparrow(1,1) \longrightarrow & \widetilde{\mathcal{G}}(1,1) & \longrightarrow & \mathbb{R}^3 \\
& \text{(Poincaré)} & \text{(Galilei)} & & \text{(static)} \\
\nearrow & & \uparrow & & \\
SO_o(1,2) & \longrightarrow & \mathbb{R}^2 \rtimes SO(2) \equiv N_- & & \\
\text{(anti-de Sitter)} & & \text{(Newton)} & &
\end{array}
$$

Two of the relativities have "maximal" symmetry, i.e., their kinematical groups are the "de Sitterian" pseudo-orthogonal groups $SO_o(1,2)$ and $SO_o(2,1)$. The difference between them stems from the compact or noncompact nature of the time translation subgroup. Here, no physical unit is necessary to standardize their three (pseudo-) angular parameters. They are departure points for successive contractions (denoted by arrows on the diagram) until the ultimate "kinematics" is reached, where nothing moves (the static situation). At each step, some of the parameters acquire a physical dimension. Hence, they may be interpreted as a length, time, or momentum. Correspondingly, part of the original simple group structure breaks down into a (semi)direct product structure. Note that, on a quantum level, central extensions (trivial or not) of the respective groups have to be taken into account to get rid of infinite phases resulting from the contraction limits.

The anti-de Sitter group $SO_o(1,2)$ is of special interest, because it is the symmetry group for an elementary system that is simultaneously a deformation of the relativistic free particle (rest energy mc^2) and the harmonic oscillator (rest energy $\frac{1}{2}\hbar\omega$, where $\omega \equiv \kappa c$ and κ is the space–time curvature) [Ren, 137]. Indeed, writing the commutation rules for $\mathfrak{so}(1,2) \sim \mathfrak{su}(1,1)$,

$$[X_0, X_1] = mc^2\kappa^2 X_2, \quad [X_2, X_0] = \frac{1}{m}X_1, \quad [X_2, X_1] = \frac{1}{m^2 c}X_0, \qquad (11.106)$$

where X_0 is the compact generator, one can check that the limit $\kappa \to 0$ gives the Poincaré Lie algebra, whereas the limit $\kappa \to 0, c \to \infty, \kappa c = \omega$, gives the harmonic oscillator Lie algebra, renamed as the Newton (N_-) Lie algebra in the present context. The contraction toward the harmonic oscillator has been studied in [Ren], [133], [134], and [137]. Note that, in this context, the Gilmore–Perelomov CS for the discrete series of $SU(1,1)$ contract smoothly toward the Weyl-Heisenberg canonical CS [136]. On the other hand, the zero curvature limit toward $\mathcal{P}^\uparrow_+(1,1)$ and its representations was studied in [11] and [132], with the aim of analyzing the (singular) behavior of the $SU(1,1)$ Gilmore–Perelomov CS as $\kappa \to 0$. The problem was considered again in [89], [104], [135], and [255] to give a more rigorous treatment to the contraction procedure. The present section is strongly inspired by [255]. Henceforth, we put $m = 1, c = 1, \hbar = 1$ for convenience.

The definition of the contraction given in (4.155) at the level of the Lie algebras $\mathfrak{g}_1$ and $\mathfrak{g}_2$ is adapted to the present problem by defining the deformation map

$$\phi_\epsilon \equiv \Phi_\kappa : \mathfrak{g}_2 = \mathfrak{p}(1,1) \to \mathfrak{g}_1 = \mathfrak{su}(1,1), \quad \kappa \in \mathbb{R}^+_*, \qquad (11.107)$$

in the following way. First, we identify the Poincaré group $\mathcal{P}^\uparrow_+(1,1)$ with the group of (projective) 3×3 matrices

$$(a, \Lambda_p) \equiv \begin{pmatrix} \cosh\varphi & \sinh\varphi & a_0 \\ \sinh\varphi & \cosh\varphi & \mathbf{a} \\ 0 & 0 & 1 \end{pmatrix} = M(a, \varphi), \qquad (11.108)$$

with $\cosh\varphi = p_0, \sinh\varphi = \mathbf{p}$. The Lie algebra basis resulting from this parametrization is given by

$$f_0 = \frac{d}{da_0} M(a, \varphi)|_{a=0, \varphi=0} \qquad (11.109)$$

$$f_1 = \frac{d}{d\mathbf{a}} M(a, \varphi)|_{a=0, \varphi=0} \qquad (11.110)$$

$$f_2 = \frac{d}{d\varphi} M(a, \varphi)|_{a=0, \varphi=0}. \qquad (11.111)$$

We now define the invertible linear map $\Phi_\kappa : \mathfrak{p}(1,1) \to \mathfrak{su}(1,1)$ by

$$\Phi_\kappa : \begin{array}{rcl} f_0 & \mapsto & \kappa e_0 \\ f_1 & \mapsto & \kappa e_1 \\ f_2 & \mapsto & e_2, \end{array} \qquad (11.112)$$

11.3. The anti-de Sitter group $SO_o(1,2)$ and its contraction(s)

where $\{e_0, e_1, e_2\}$ is the basis of $\mathfrak{su}(1,1)$,

$$e_0 = \frac{1}{2}\begin{pmatrix} i & 0 \\ 0 & -i \end{pmatrix}, \quad e_1 = -\frac{1}{2}\begin{pmatrix} 0 & 1 \\ 1 & 0 \end{pmatrix}, \quad e_2 = \frac{1}{2}\begin{pmatrix} 0 & -i \\ i & 0 \end{pmatrix}, \tag{11.113}$$

with the commutation rules [compare to (11.106)]

$$[e_0, e_1] = e_2, \quad [e_2, e_0] = e_1, \quad [e_2, e_1] = e_0. \tag{11.114}$$

Then, one can see that

$$\lim_{\kappa \to 0} \Phi_\kappa^{-1}[\Phi_\kappa f_i, \Phi_\kappa f_j]_1 = [f_i, f_j]_2, \quad i, j = 0, 1, 2. \tag{11.115}$$

The corresponding map at the group level, $\Pi_\kappa : \mathcal{P}_+^\uparrow(1,1) \to SU(1,1)$, then reads as

$$\Pi_\kappa : M(a, \varphi) = e^{a_0 f_0 + \mathbf{a} f_1} e^{\varphi f_2} \mapsto e^{\kappa(a_0 e_0 + \mathbf{a} e_1)} e^{\varphi e_2} \tag{11.116}$$

Note the factorization formula

$$e^{\kappa(a_0 e_0 + \mathbf{a} e_1)} \equiv \begin{pmatrix} \alpha & \beta \\ \overline{\beta} & \overline{\alpha} \end{pmatrix} = e^{\theta' e_0} e^{\psi' e_1} e^{\varphi' e_2},$$

where the different parameters are defined or related by the following relations, where $\lambda \equiv \frac{1}{2}(a_0^2 - \mathbf{a}^2)^{1/2}$:

$$\alpha = \cos \kappa\lambda + ia_0 \frac{\sin \kappa\lambda}{2\lambda} \in \mathbb{C},$$

$$\beta = -\mathbf{a}\frac{\sin \kappa\lambda}{2\lambda} \in \mathbb{R},$$

$$\theta' = \arg\left(\cos 2\kappa\lambda + ia_0 \frac{\sin 2\kappa\lambda}{2\lambda}\right),$$

$$\psi' = \sinh^{-1}\left(\mathbf{a}\frac{\sin 2\kappa\lambda}{2\lambda}\right),$$

$$\tanh\frac{\varphi'}{2} = -i\frac{\beta - \overline{\alpha}\tanh\frac{\psi'}{2} e^{i\theta'}}{\alpha - \overline{\beta}\tanh\frac{\psi'}{2} e^{i\theta'}}.$$

Let us now recall the respective UIRs involved in the contraction scheme. For $SU(1,1)$, they are the representations U^j of the discrete series already described in Chapter 4, Section 4.2.2, and extended to the universal covering group of $SU(1,1)$. This means that the parameter $j = 1, 3/2, 2, 5/2, \ldots$ in (4.76) now runs over the continuous range

$$\frac{1}{2} < j = 1/\kappa < \infty. \tag{11.117}$$

We also recall the relevant Hilbert spaces and unitary actions. For typographical simplicity, we write $U_\kappa \equiv U^{1/\kappa}$, $\mathfrak{H}_\kappa \equiv \mathfrak{H}^{1/\kappa}$, $\mu_\kappa \equiv \mu_{1/\kappa}$,

$$\mathfrak{H}_\kappa = L^2(\mathcal{D}, (2/\kappa - 1)d\mu_\kappa), \tag{11.118}$$

$$d\mu_\kappa(z,\bar{z}) = (1-|z|^2)^{2/\kappa-2}\frac{dz\wedge d\bar{z}}{2\pi i}, \tag{11.119}$$

$$(U_\kappa(g)f)(z) = (\alpha-\bar{\beta}z)^{-2/\kappa}f\left(\frac{\bar{\alpha}z-\beta}{\alpha-\bar{\beta}z}\right), \tag{11.120}$$

$$\text{for } f\in\mathfrak{H}_\kappa \text{ and } g=\begin{pmatrix}\alpha & \beta\\ \bar{\beta} & \bar{\alpha}\end{pmatrix}\in SU(1,1).$$

For the Poincaré group, we shall choose another parametrization for the Wigner representation (11.65), better suited to the contraction procedure. To that effect, we introduce the carrier Hilbert space $\mathfrak{H} = L^2\left((-1,1); 2dx/(1-x^2)\right)$, obtained from $L^2(\mathcal{V}_m^+, d\mathbf{k}/k_0)$ by changing $\mathbf{k}$ into $x = \mathbf{k}/(k_0+1)$, i.e. $\mathbf{k} = 2x/(1-x^2)$, and putting $m = 1$. The (unit mass) Wigner representation $U_W(a,\Lambda_p) \equiv U(a,\varphi)$ is then given by

$$(U(a,\varphi)f)(x) = \exp -i\left(\frac{1+x^2}{1-x^2}a_0 + \frac{2x}{1-x^2}\mathbf{a}\right)f\left(\frac{x\cosh\frac{\theta}{2}+\sinh\frac{\theta}{2}}{\cosh\frac{\theta}{2}+x\sinh\frac{\theta}{2}}\right). \tag{11.121}$$

Let us consider an element f of the Fock-Bargmann space $\mathfrak{H}_\kappa$ on the disk $\mathcal{D}$ and its restriction, suitably weighted, to the purely imaginary diameter $\{z=ix,\ -1<x<1\}$:

$$I_\kappa(f)(x) \equiv (1-x^2)^{1/\kappa}f(ix). \tag{11.122}$$

The map I_κ from $\mathfrak{H}_\kappa$ to $\mathfrak{H}$ is called a *precontraction* (see Section 4.5.4). It has the following crucial property.

Theorem 11.3.1: *The precontraction I_κ is a continuous injection from $\mathfrak{H}_\kappa$ into $\mathfrak{H}$.*

Proof. Injectivity results from the holomorphy of f. Continuity results from the following majoration:

$$\|I_\kappa(f)\|_\mathfrak{H}^2 = 2\int_{-1}^1 |f(ix)|^2(1-x^2)^{2/\kappa-1}\,dx \leq \frac{\Gamma(2/\kappa+1)2^{2/\kappa}}{2/\kappa-1}\|f\|_{\mathfrak{H}_\kappa}^2, \tag{11.123}$$

that were proved in [255]. $\square$

Of course, the map I_κ is not surjective. For instance, the function $(1-x^2)^{1/\kappa}e^{1/(1-ix)}$ is in $\mathfrak{H}$ for all $\kappa < 2$, but $e^{1/(1-z)}$ is *not* in $\mathfrak{H}_\kappa$ for any $\kappa < 2$.

Let us put $\mathfrak{C}_\kappa = I_\kappa(\mathfrak{H}_\kappa)$. It is clear that, for $\kappa' < \kappa$, we have $\mathfrak{H}_{\kappa'} \supset \mathfrak{H}_\kappa$, and so $\mathfrak{C}_{\kappa'} \supset \mathfrak{C}_\kappa$. On the other hand, the functions of the form $(1-x^2)p(x)$, where p is a polynomial, form a dense subset of $\mathfrak{H}$, and they are also elements of $\mathfrak{C}_1$. Since $\mathfrak{C}_\kappa \supset \mathfrak{C}_1$ for all $\kappa \leq 1$, $\mathfrak{C}_\kappa$ is dense in $\mathfrak{H}$ for all $\kappa \leq 1$. (Actually the same holds true for any $\kappa < 2$.) From this discussion, the

11.3. The anti-de Sitter group $SO_o(1,2)$ and its contraction(s) (♣)

following commutative diagram results:

$$\ldots \mathfrak{H}_\kappa \longrightarrow \mathfrak{H}_{\kappa'} \ldots \qquad (\kappa' < \kappa), \qquad (11.124)$$
$$\searrow \quad \swarrow$$
$$\mathfrak{H}$$

where the arrows represent continuous injections. Hence the (Wigner) Hilbert space $\mathfrak{H}$ is the "universal arena" into which all spaces $\mathfrak{H}_\kappa$ are conveniently injected for the purpose of contraction. Accordingly, we propose the following definition.

Definition 11.3.2: *A family of states $\psi_\kappa \in \mathfrak{H}_\kappa$ is said to contract to $\psi \in \mathfrak{H}$ iff $I_\kappa(\psi_\kappa)$ tends to ψ in some (possibly weak) sense when $\kappa \to 0$.*

We are now in a position to see what happens to the Gilmore–Perelomov coherent states (4.99),

$$\frac{1}{\sqrt{2\pi}} \eta_{\sigma(z)}(z') = \left(\frac{2/\kappa - 1}{2\pi}\right)^{1/2} (1 - |z|^2)^{1/\kappa}(1 - z'\bar{z})^{-2/\kappa}, \qquad (11.125)$$

which are mapped to

$$\frac{1}{\sqrt{2\pi}} I_\kappa(\eta_{\sigma(z)})(x') = \left(\frac{2/\kappa - 1}{2\pi}\right)^{1/2} \left[\frac{(1 - x'^2)(1 - |z|^2)}{(1 - ix'\bar{z})^2}\right]^{1/\kappa}. \qquad (11.126)$$

When considered as a distribution, the action of this vector on $\phi \in C_0^\infty(-1,1)$ reads as

$$\frac{1}{\sqrt{2\pi}} \langle I_\kappa(\eta_{\sigma(z)})|\phi\rangle = 2\left(\frac{1-\kappa/2}{\pi\kappa}\right)^{1/2} \int_{-1}^{1} \left[\frac{(1-x'^2)(1-|z|^2)}{(1+ix'z)^2}\right]^{1/\kappa}$$
$$\times \phi(x') \frac{dx'}{1-x'^2}. \qquad (11.127)$$

This expression has in general no limit for $\kappa \to 0$, except for $z = ix$, $x \in (-1,1)$. In this case, we compute that (11.127) indeed converges to $2\phi(x)$.

So, when the CS label z is concentrated on the purely imaginary diameter of the disk, the limit of (11.126) is a Dirac distribution. This result could actually be expected from the fact that the Gilmore–Perelomov CS (11.125) *are* themselves a quantization of the open unit disk $\mathcal{D}$ considered as an $SU(1,1)$ coadjoint orbit or, better, as a classical phase space $\mathcal{D} \sim SU(1,1)/U(1)$ for a (massive) free particle in an anti-de Sitter spacetime. This phase space interpretation is well captured by the following $(\mathbf{a}, \mathbf{k})$ parametrization of $\mathcal{D}$:

$$\mathcal{D} \ni z = z(\mathbf{a}, \mathbf{k}; \kappa) = \frac{\sinh \kappa \mathbf{a} + i\mathbf{k} \cosh \kappa \mathbf{a}}{1 + k_0 \cosh \kappa \mathbf{a}}, \quad k_0 = \sqrt{1 + \mathbf{k}^2}, \; \mathbf{a}, \mathbf{k} \in \mathbb{R}.$$
$$(11.128)$$

The singular behavior of these CS is then obvious. They exactly reproduce the singular behavior of the parametrization (11.128) at zero curvature,

$$\lim_{\kappa \to 0} z(\mathbf{a}, \mathbf{k}; \kappa) = \frac{i\mathbf{k}}{k_0 + 1} = ix; \qquad (11.129)$$

i.e., the disk $\mathcal{D}$ contracts into the segment $(-i, i)$. Although the original complex structure is lost, this does not, however, imply a singular behavior at the level of the respective coadjoint orbits of $SU(1,1)$ and $\mathcal{P}_+^{\uparrow}(1,1)$. The limit (11.129) is just a (rather exotic) parametrization of the cylindric coadjoint orbit of the Poincaré group $\{(\mathbf{a}, k_0, \mathbf{k}) \mid k_0^2 - \mathbf{k}^2 = 1, k_0 > 0\}$, which is interpreted as the phase space for a massive free particle in Minkowski space–time.

The singular behavior of the Gilmore–Perelomov coherent states originates in the choice of the basis function $u_0(z) = 1$ in $\mathfrak{H}_\kappa$, the image of which, namely, $(1-x^2)^{1/\kappa}$, goes to zero as $\kappa \to 0$. The same holds for any element $u_n(z)$ of the orthonormal basis (4.93). It is thus necessary to choose some admissible vector η_κ in $\mathfrak{H}_\kappa$, such that the family

$$\eta_{\kappa g} = U_\kappa(g)\,\eta_\kappa, \ g \in SU(1,1), \qquad (11.130)$$

contracts to a Poincaré counterpart

$$\eta_{(a,\varphi)} = U(a,\varphi)\eta. \qquad (11.131)$$

Among them, it may be possible to select a subfamily of coherent states through a suitable choice of section $(a, \varphi) \equiv \sigma(\mathbf{q}, \mathbf{p})$, as in (11.66). Let us try, for instance, to recover in the contraction limit the Gaussian vector $\eta_G(k) \sim \exp(-k_0)$ mentioned in the previous section. The vector η_G has many interesting properties from a quantum as well as a statistical point of view [140]. One of them is related to the relativistic Heisenberg inequality

$$\Delta P_1 \, \Delta P_2 \geq \frac{1}{2}\langle P_0 \rangle, \qquad (11.132)$$

where the infinitesimal generator P_0 corresponds to the basis element f_0 in (11.111) in the representation $U(a, \varphi)$. Explicitly, we have (in the x-parametrization)

$$P_0 f(x) = \frac{1+x^2}{1-x^2} f(x) \quad \text{(energy)} \qquad (11.133)$$

$$P_1 f(x) = \frac{2x}{1-x^2} f(x) \quad \text{(momentum } P_1 \equiv \mathbf{P}) \qquad (11.134)$$

$$P_2 f(x) = i\frac{1-x^2}{2}\frac{d}{dx} f(x) \quad \text{(Lorentz boost } P_2 \equiv \mathbf{K}). \qquad (11.135)$$

The inequality (11.132) is saturated precisely when computed in the state η_G. This is related to the fact that the latter is annihilated by the operator $P_2 + iP_1$

$$(P_2 + iP_1)\eta_G = 0, \qquad (11.136)$$

11.3. The anti-de Sitter group $SO_o(1,2)$ and its contraction(s) (♣)

which should be compared with the Galilean (or Weyl–Heisenberg) analogue (2.15). The anti-de Sitterian counterpart of P_i is K_i:

$$K_i f = i \frac{d}{d\alpha} U_\kappa(e^{\alpha e_i})|_{\alpha=0}. \qquad (11.137)$$

So, it is natural to make a parallel with (11.136) and look for solutions to equations of the type

$$(AK_2 + iK_1)\psi_\alpha = \alpha\psi_\alpha, \qquad (11.138)$$

with $A > 0$, $\alpha \in \mathbb{C}$, and $\psi_\alpha \in \mathfrak{H}_\kappa$. These solutions saturate the following Heisenberg inequality [232]:

$$\Delta K_1 \Delta K_2 \geq \frac{\kappa}{2} \langle K_0 \rangle, \qquad (11.139)$$

if $A = \Delta K_2/\Delta K_1$. The set of solutions of (11.138) splits into three classes, according to the values taken by A.

1. $\underline{A > 1}$:

$$\psi_\alpha(z) = c_\alpha \left(z + i\sqrt{\frac{A+1}{A-1}}\right)^{-\kappa^{-1}(1-i\alpha/\sqrt{A^2-1})}$$

$$\times \left(z - i\sqrt{\frac{A+1}{A-1}}\right)^{-\kappa^{-1}(1+i\alpha/\sqrt{A^2-1})}, \qquad (11.140)$$

where c_α is a normalization constant.

2. $\underline{A = 1}$:

$$\psi_\alpha(z) = c_\alpha e^{\kappa^{-1}\alpha z}. \qquad (11.141)$$

These states, indexed by $\beta = \kappa^{-1}\alpha \in \mathbb{C}$, are the Barut–Girardello (BG) CS [65], in their Fock–Bargmann representation on the disk, i.e., as elements of $\mathfrak{H}_\kappa$. The alternative expression is

$$|\psi_\alpha\rangle \equiv |b_\beta\rangle = n_\beta^{-1/2} \sum_{n=0}^{\infty} \frac{\beta^n}{\sqrt{n!\Gamma(n+2/\kappa)}} |u_n\rangle, \qquad (11.142)$$

where the normalization factor n_β is given in terms of the modified Bessel function I_ν by

$$n_\beta = |\beta|^{1-2/\kappa} I_{2/\kappa-1}(2|\beta|). \qquad (11.143)$$

Note that the normalization factor c_α in (11.140) is given by

$$c_\alpha = [n_\beta \Gamma(2/\kappa)]^{-1/2}. \qquad (11.144)$$

The BG states yield the following resolution of the identity:

$$I = \int_\mathbb{C} |b_\beta\rangle \langle b_\beta| \, d\mu_{BG}(\beta), \qquad (11.145)$$

with the measure given by

$$\mu_{BG}(\beta) = \frac{1}{\pi} |\beta| I_{2/\kappa-1}(2|\beta|) K_{2/\kappa-1}(2|\beta|) d(|\beta|^2) d(\arg \beta). \quad (11.146)$$

The occurrence in this expression of the modified Bessel function of the third kind, K_ν, stems from the classical moment problem

$$\Gamma(n+1)\Gamma(n+2/\kappa) = \int_0^\infty u^n \rho(u)\, du, \quad (11.147)$$

with

$$\rho(u) = 2u^{1/\kappa} K_{2/\kappa-1}(2\sqrt{u}). \quad (11.148)$$

We emphasize that the BG states do not contract correctly for any $\alpha \in \mathbb{C}$, since the limit as $\kappa \to 0$ of the expression

$$I_\kappa \psi_\alpha(x) = [n_{\kappa^{-1}\alpha}\Gamma(2/\kappa)]^{-1/2} (1-x^2)^{1/\kappa} e^{i\alpha x/\kappa} \quad (11.149)$$

is not square integrable, although it may exist as a distribution supported on a set of measure zero.

3. $0 < A < 1$:

$$\psi_\alpha(z) = c_\alpha \left(z + \sqrt{\frac{1+A}{1-A}}\right)^{-\kappa^{-1}(1-\alpha/\sqrt{A^2-1})}$$

$$\times \left(z - \sqrt{\frac{1+A}{1-A}}\right)^{-\kappa^{-1}(1+\alpha/\sqrt{A^2-1})}, \quad (11.150)$$

It is interesting to note that similar distribution-valued CS appear in [240], in a different context, and in [277]. The states derived in the latter paper, called generalized intelligent states, are obtained as solutions of an equation similar to (11.138), and contain both Gilmore–Perelomov and Barut–Giradello CS.

In view of the effect of the contraction, only the states with nonzero values concentrated around the imaginary axis will be relevant in the present context. Now, the classical observables k_i corresponding to the K_i read [136] as

$$k_0 = \frac{1+z\bar{z}}{1-z\bar{z}}, \quad k_1 = i\frac{z-\bar{z}}{1-z\bar{z}}, \quad k_2 = \frac{z+\bar{z}}{1-z\bar{z}}. \quad (11.151)$$

Hence, k_2 must vanish in the limit $\kappa \to 0$, and this condition in turn entails $\Delta K_2 \to 0$ at the quantum level. In view of the equality

$$A^2(\Delta K_2)^2 + (\Delta K_1)^2 = A\kappa \langle K_0 \rangle, \quad (11.152)$$

which follows from (11.138), it seems natural to put $A = 1/\kappa$. Finally, by choosing the value 0 for the unessential parameter α, we are left with the

11.3. The anti-de Sitter group $SO_o(1,2)$ and its contraction(s) (♣)

"binomial" state
$$b_\kappa(z) = \left(z^2 + \frac{1+\kappa}{1-\kappa}\right)^{-1/\kappa}. \tag{11.153}$$

It is then easily checked that this state contracts to η_G (up to a factor):
$$\widetilde{b}_\kappa = I_\kappa b_\kappa, \quad \lim_{\kappa \to 0} \widetilde{b}_\kappa(x) = e^{-2/(1-x^2)} \equiv b(x) \sim \eta_G(k). \tag{11.154}$$

The square integrability of the representation then allows one to consider the family of $SU(1,1)$-transported vectors
$$b_{\kappa g} = U_\kappa(g) b_\kappa, \quad g \in SU(1,1), \tag{11.155}$$
as a (tight) frame, in the spirit of Theorem 8.1.3 (Section 8.1),
$$I = \frac{1}{c(b_\kappa)} \int_{SU(1,1)} |b_{\kappa g}\rangle \langle b_{\kappa g}| \, d\mu(g), \tag{11.156}$$
where
$$d\mu(g) = \frac{1}{2\pi^2} (1 - |z|^2)^{-2} d^2 z \, d\phi$$
is the Haar measure on $SU(1,1)$ expressed in terms of the factorization (4.66),
$$g = \sigma(z) k(\phi), \quad z \in \mathcal{D}, \; \phi \in [0, 2\pi). \tag{11.157}$$

The constant $c(b_\kappa)$ is the one defined in (8.8).

The important point concerning the CS $\psi_{\kappa g}$ is their behavior under contraction. A precise answer is given by the following results [255].

Theorem 11.3.3: *For fixed (a, φ), one has*
$$\lim_{\kappa \to 0} I_\kappa U_\kappa(\Pi_\kappa(a, \varphi)) b_\kappa = U(a, \varphi) b,$$
in the sense of the norm topology on $\mathfrak{H}$:
$$\lim_{\kappa \to 0} \| I_\kappa U_\kappa(\Pi_\kappa(a, \varphi)) b_\kappa - U(a, \varphi) b \|_\mathfrak{H} = 0, \quad \text{for } g = (a, \varphi) \in \mathcal{P}_+^\uparrow(1,1). \tag{11.158}$$

Note that (11.158) is a significative generalization of the contraction à la Dooley (see Section 4.5.4), in the sense that it involves states b_κ that depend on the deformation parameter κ, as it should be in the generic case.

The proof of Theorem 11.3.3 partially rests on the following limiting argument, which is interesting in itself.

Lemma 11.3.4: *Let f_κ be in $\mathfrak{H}$. Suppose that the family $\{f_\kappa\}$ satisfies the following two conditions:*
1. $\lim_{\kappa \to 0} f_\kappa = f$, almost everywhere on $(1,1)$;
2. $f_\kappa(x)/(1-x^2)$ is bounded on $(1,1)$, uniformly in κ.

Then,
$$\lim_{\kappa \to 0} \| f_\kappa - f \|_\mathfrak{H} = 0. \tag{11.159}$$

Theorem 11.3.3 in particular guarantees that the subfamily of $\{b_{\kappa g}, g \in SU(1,1)\}$, obtained by restricting $g \in SU(1,1)$ to (Cartan) sections (4.66),

$$b_{\kappa\sigma(z)} = U_\kappa(\sigma(z))b_\kappa, \quad g \in SU(1,1), \tag{11.160}$$

contracts in our sense to Poincaré coherent states of the form (11.69), defined precisely with the symmetric section (see Section 11.1.2)

$$(\mathbf{q}, \mathbf{p}) \mapsto \sigma_s(\mathbf{q}, \mathbf{p}) = ((\hat{q}_0, \hat{\mathbf{q}}), \Lambda_{\hat{\mathbf{p}}}), \tag{11.161}$$

where

$$\hat{q}_0 = p_0 \frac{\mathbf{q} \cdot \mathbf{p}}{p_0 + 1}, \quad \hat{\mathbf{q}} = p_0 \mathbf{q}, \quad \hat{\mathbf{p}} = \mathbf{p}. \tag{11.162}$$

More precisely, we have (see [11] for detailed calculations)

$$\Pi_\kappa \sigma(\mathbf{q}, \mathbf{p}) = \sigma(z(\hat{\mathbf{q}}, \hat{\mathbf{p}}, \kappa) + O(\kappa)), \tag{11.163}$$

with the notations (11.116) and (11.128).

Note that (11.162) implies the space–time relation

$$\hat{q}_0 = \frac{\hat{\mathbf{p}} \cdot \hat{\mathbf{q}}}{p_0 + 1}, \tag{11.164}$$

which has many interesting features from a kinematical point of view in Minkowskian geometry [11, 140].

More general Poincaré sections of the type (11.66) can be reached through contraction, that is, $\lim_{\kappa \to 0} \Pi_\kappa^{-1} \sigma_\lambda$, starting from sections

$$\sigma_\lambda(z) = \sigma(z_\lambda) k(\lambda), \tag{11.165}$$

where

$$\sigma_s(z) = \begin{pmatrix} \delta & \delta z \\ \delta \bar{z} & \delta \end{pmatrix}, \quad \delta = (1 - |z|^2)^{-1/2}, \tag{11.166}$$

$$k(\lambda) = \begin{pmatrix} e^{i\lambda/2} & 0 \\ 0 & e^{-i\lambda/2} \end{pmatrix} \in U(1), \tag{11.167}$$

and $\lambda : \mathcal{D} \to \mathbb{R}$ is a C^∞ function depending on κ in a smooth way. This section leads to a new parametrization of the open unit disk $\mathcal{D}$ analogous to a gauge transformation

$$z \mapsto z_\lambda = z_\lambda(\hat{\mathbf{q}}, \hat{\mathbf{p}}, \kappa) \equiv e^{i\lambda(z,\bar{z},\kappa)} z(\hat{\mathbf{q}}, \hat{\mathbf{p}}, \kappa), \quad \hat{\mathbf{q}} = \mathbf{q}. \tag{11.168}$$

If one makes a more precise assumption for the asymptotic behavior of λ,

$$\lambda(z, \bar{z}, \kappa) \equiv \lambda(\hat{\mathbf{q}}, \hat{\mathbf{p}}, \kappa) = \kappa \lambda_0(\hat{\mathbf{p}}) + \kappa \hat{\mathbf{q}} \cdot \boldsymbol{\lambda}_1(\mathbf{k}) + o(\kappa), \tag{11.169}$$

then the resulting Poincaré section is of affine type

$$((\hat{q}_0, \hat{\mathbf{q}}), \Lambda_{\hat{\mathbf{p}}}) = \lim_{\kappa \to 0} \Pi_\kappa^{-1} \sigma_\lambda(z)$$
$$\equiv \sigma_\lambda(\mathbf{q}, \mathbf{p}) \equiv ((0, \mathbf{q}), \Lambda_p)((\varphi(\mathbf{p}) + \mathbf{q} \cdot \boldsymbol{\vartheta}(\mathbf{p}), \mathbf{0}), I), \tag{11.170}$$

11.3. The anti-de Sitter group $SO_\circ(1,2)$ and its contraction(s) (♣)

where

$$\hat{q}_0 = \frac{\hat{\mathbf{q}} \cdot \hat{\mathbf{p}}}{p_0 + 1} + \hat{\mathbf{q}} \cdot \boldsymbol{\lambda}_1(\hat{\mathbf{p}}) + \lambda_0(\hat{\mathbf{p}}). \qquad (11.171)$$

Hence, comparing with (11.17) and (11.19), and taking into account the fact that φ is not equal to zero, we get

$$\boldsymbol{\lambda}_1(\hat{\mathbf{p}}) = \boldsymbol{\beta}(\mathbf{p}) - \frac{\hat{\mathbf{p}}}{p_0 + 1} \qquad (11.172)$$

$$\lambda_0(\hat{\mathbf{p}}) = \varphi(\mathbf{p})[p_0 - \mathbf{p} \cdot \boldsymbol{\beta}(\mathbf{p})]. \qquad (11.173)$$

A last point relative to the present discussion on sections in the two groups $SU(1,1)$ and $\mathcal{P}^\uparrow_+(1,1)$ is the following question. Let the coherent states $\eta_{\sigma(\mathbf{q},\mathbf{p})}$ be precisely those obtained by contraction from the CS $b_{\kappa\sigma(z)}$ given in (11.160). In that case, how does the resolution of the identity (11.156) in $SU(1,1)$ behave under precontraction and contraction, and similarly for the Poincaré resolution operator (11.70) under the action of I_κ^{-1}, when $\kappa \to 0$? This important question deserves attention, and no precise answer can be given at the moment.

Let us end this section by quoting another important result of [255] concerning contractions of (essentially) self-adjoint operators (called observables in a quantum context) on $\mathfrak{H}_\kappa$. Let B^κ be an observable on $\mathfrak{H}_\kappa$, such that $b_{\kappa g} \in \mathcal{D}(B^\kappa)$, and let $\widetilde{B}^\kappa$ be its precontracted partner:

$$\widetilde{B}^\kappa = I_\kappa B^\kappa I_\kappa^{-1}. \qquad (11.174)$$

Then, B^κ is said to contract to an observable B acting in $\mathfrak{H}_\kappa$ iff

$$\lim_{\kappa \to 0} \langle \widetilde{b}_{\kappa g'} | \widetilde{B}^\kappa | \widetilde{b}_{\kappa g} \rangle_\mathfrak{H} = \langle b_{g'} | B | b_g \rangle_\mathfrak{H}, \quad \forall g, g' \in \mathcal{P}^\uparrow_+(1,1). \qquad (11.175)$$

Let $p = p(\widetilde{K}_0^\kappa, \widetilde{K}_1^\kappa, \widetilde{K}_2^\kappa, \kappa)$ be a polynomial of the variables $\widetilde{K}_i^\kappa$.

Theorem 11.3.5: For any $g, g' \in \mathcal{P}^\uparrow_+(1,1)$, the following limit holds true:

$$\lim_{\kappa \to 0} \langle \widetilde{b}_{\kappa g'} | p(\widetilde{K}_0^\kappa, \widetilde{K}_1^\kappa, \widetilde{K}_2^\kappa, \kappa) | \widetilde{b}_{\kappa g} \rangle = \langle b_{g'} | p(P_0, P_1, P_2, 0) | b_g \rangle. \qquad (11.176)$$

Corollary 11.3.6: Let $f_\kappa \in \mathfrak{H}$ be of the form

$$f_\kappa(x) = P(x) \left(1 + \frac{\kappa d(\kappa)}{1 - x^2} \right)^{-a\kappa^{-1} + b + O(\kappa)} \frac{1}{(1 - x^2)^{n_o}}, \qquad (11.177)$$

where P is a polynomial, $a > 0$, $n_o \in \mathbb{N}_*$, and d is a nonnegative function defined for $\kappa > 0$, such that $0 < \lim_{\kappa \to 0} d(\kappa) < \infty$. Then, $p(\widetilde{K}_0^\kappa, \widetilde{K}_1^\kappa, \widetilde{K}_2^\kappa, \kappa) f_\kappa$ tends in the $\mathfrak{H}$-norm to $p(P_0, P_1, P_2, 0) f$, where f is the limit of f_κ.

In conclusion, anti-de Sitterian quantum observables, possibly with the exception of some pathological ones, contract to their Poincaré counterparts, and this contraction procedure involves covariant symbols on both sides.

We should note, however, that we make use in $\mathfrak{H}$ of a continuous family $\{b_g\}$ bigger than the set of Poincaré coherent states $\{b_{\sigma(\mathbf{q},\mathbf{p})}\}$.

Finally, similar contraction results can be found with quantum probes b_κ more general than (11.153), i.e., probes that do not necessarily saturate some Heisenberg inequality.

12
Wavelets

12.1 A word of motivation

Wavelet analysis is a particular time-scale or space-scale representation of signals that has become popular in physics, mathematics, and engineering in the last few years. The genesis of the method is interesting for the present book, so we will spend a paragraph outlining it. After the empirical discovery by Jean Morlet (who was analyzing microseismic data in the context of oil exploration [152]), it was recognized from the very beginning by Grossmann, Morlet, and Paul [156]–[160] that wavelets are simply coherent states associated to the affine group of the line (dilations and translations). Thus, immediately the stage was set for a far-reaching generalization, using the formalism developed in Chapter 8 (it is revealing to note that two out of those three authors are mathematical physicists). But then the wind changed. Meyer [217] and Mallat [212] made the crucial discovery that orthonormal bases of regular wavelets could be built, and even with compact support, as shown by Daubechies [99], by changing the perspective (of course, the orthonormal basis of the Haar wavelets was known since the beginning of the century, but these are piecewise constant, discontinuous functions). Group theory lost its priority in favor of the so-called multiresolution analysis (more about this in Section 13.1.1), which made contact with the world of signal processing and engineering. The theory then really caught the attention of practitioners, and it started to grow explosively.

In this chapter, we will nevertheless focus on the continuous wavelet transform or CWT, based on group representations, in one space (or time) dimension. This approach fits with the general trend of the book, but, more importantly, it is by far the most natural way to extend wavelets to higher dimensions and more general situations, which will be described in Chapter 14.

Let us begin with an intuitive discussion, to convey to the reader a feeling for the theory and to understand its success. As a matter of fact, most real life signals are nonstationary. They often contain transient components, sometimes physically very significant, and mostly cover a wide range of frequencies. In addition, frequently (but not always) a direct correlation exists between the characteristic frequency of a given segment of the signal and the time duration of that segment. Low-frequency pieces tend to last a long time, whereas high frequencies occur in general for a short moment only. Human speech signals are typical in this respect. Vowels have a relatively low mean frequency and last quite long, whereas consonants contain a wide spectrum, especially in the attack, and are often very short.

Clearly, the usual Fourier transform is inadequate for treating such signals. It gives a purely frequency domain representation and loses all information on time localization, which may be crucial. For this reason, signal analysts turn to *time-frequency* representations. The idea is that one needs *two* parameters: One, called a, refers to frequency; the other, b, indicates the position in the signal. Thus, assuming the transform to be linear on signals, one writes a general time-frequency transform of a signal s as

$$s(t) \leftrightarrow S(b, a) = \int_{\mathbb{R}} \overline{\psi_{ba}(t)}\, s(t)\, dt, \qquad (12.1)$$

where ψ_{ba} is the analyzing function and $\overline{\psi_{ba}}$ is its complex conjugate, $a > 0$, $b \in \mathbb{R}$. This concept of a time-frequency representation is in fact quite old and familiar: The most obvious example is simply a musical score! Clearly, it is important to know when to play a given note, and not only to which frequency it corresponds. Eloquent comments along this line by Ville and by de Broglie may be found in [Fla, p.9].

Among the possible linear time-frequency transforms, two stand out as particularly simple and efficient: The windowed or short-time Fourier transform (WFT) and the wavelet transform (WT). Both are particular cases of CS, only the group differs, and in fact they should be developed in parallel [Dau]. The essential difference between the two is in the way the frequency parameter a is introduced in the analyzing function. In both cases, b is simply a time translation. The kernels of the two transforms can be written as follows.

1. For the *WFT*, one uses

$$\psi_{ba}(t) = e^{it/a}\, \psi(t - b). \qquad (12.2)$$

Here, ψ is a window function and the a-dependence is a modulation ($1/a \sim$ frequency); the window has constant width, but the lower the value of a, the larger the number of oscillations in the window. Comparing with Section

2.2, we see that (12.2) yields exactly the canonical CS, associated to the Weyl–Heisenberg group.

2. For the *WT*, instead, one takes

$$\psi_{ba}(t) = \frac{1}{\sqrt{a}} \psi(\frac{t-b}{a}). \qquad (12.3)$$

The action of a on the function ψ is a dilation ($a > 1$) or a contraction ($a < 1$). The shape of the function is unchanged; it is simply spread out or squeezed. In particular, the effective support of ψ_{ba} varies as a function of a. As we will see below, (12.3) yields the CS associated to the (connected) affine group of the real line or $ax + b$ group.

Thus, already at this stage, we may infer that the wavelet transform will be better adapted to the analysis of the type of signals described above.

Notice that the requirement of linearity is nontrivial, for there exists a whole class of quadratic, or, more properly, sesquilinear, time-frequency representations. The prototype is the so-called Wigner–Ville transform, introduced originally by Wigner in quantum mechanics (in 1932!) [284] and extended by Ville to signal analysis [280]

$$W[s](b,a) = \int e^{-it/a} \, \overline{s(b - \frac{t}{2})} \, s(b + \frac{t}{2}) \, dt = 2 \langle \Pi s_{-b,-a} | s_{-b,-a} \rangle, \qquad (12.4)$$

where $s_{b,a} \equiv s_{ba}$ is defined in (12.2) and Π is the space reflection operator $(\Pi s)(x) = s(-x)$. A glance at (8.57) shows that the Wigner–Ville transform of $s(t)$ is just the Wigner distribution $W(q,p)$, for $q = b$, $p = 1/a$. Further information may be found in [Fla] or [91].

As it turns out, the examples discussed in this chapter are of a different character from all those described previously. On the mathematical side, they are distinguished by the fact that the underlying group is a group of space or space-time transformations that contains *dilations*. Also, the range of applications and the role is different. Whereas the previous cases were essentially concerned with quantum physics (atomic physics, optics, quantum mechanics, elementary particles, etc.), here we deal mostly with *classical* physics, even engineering: Signal and image processing, fluid mechanics, geophysics, astrophysics (notable exceptions are atomic and solid state physics [Be2]). In all of these cases, wavelets are used essentially as a tool for analyzing experimental data (e.g., time-frequency analysis of a signal). No explicit physical signification has been found for them so far. By contrast, the familiar Fourier transform is a mathematical tool, but it also has a physical support. For instance, the diffraction phenomenon in optics may be seen as the physical realization of the Fourier transform. In quantum mechanics also, the latter plays a central physical role, encoding the transition from position to momentum representation, and this is finally expressed in the form of Heisenberg uncertainty relations. Nothing similar is available for wavelets so far. Nevertheless, they have reached a remarkably wide range of applications.

Let us say a word here on dilations. Physical applications will be described briefly in the following sections. For every $n \geq 1$, we consider *global* dilations of $\mathbb{R}^n$ (zoom):

$$x \mapsto ax, \ a > 0, \ x \in \mathbb{R}^n. \tag{12.5}$$

These form a one-parameter, abelian group $\mathcal{D}^{(n)} \sim \mathbb{R}_*^+ \sim \mathbb{R}$, necessarily unimodular. The (left and right invariant) Haar measure is $d\mu(a) = a^{-1}\, da$. The group $\mathcal{D}^{(n)}$ has a natural unitary representation in the Hilbert space $L^2(\mathbb{R}^n, d^n x)$:

$$(U(a)f)(x) = a^{-n/2} f(a^{-1}x), \ a > 0. \tag{12.6}$$

Most of the groups we will consider below, the so-called *affine* or *similitude* groups, have the general structure

$$G_{\mathrm{aff}}^{(n)} = G^{(n)} \rtimes \mathcal{D}^{(n)}$$

of a semidirect product of $\mathcal{D}^{(n)}$ with an appropriate transformation group $G^{(n)}$ of $\mathbb{R}^n$, for instance, the Euclidean, Galilei or Poincaré groups — each of which is in fact the isometry group of a given geometry of space or space-time.

Now, each of these groups $G^{(n)}$ is itself a semidirect product $G^{(n)} = \mathbb{R}^n \rtimes S^{(n)}$, where $S^{(n)}$ is the corresponding homogeneous group, and $\mathcal{D}^{(n)}$ acts nontrivially on $\mathbb{R}^n$, according to (12.5). As a consequence, the affine group $G_{\mathrm{aff}}^{(n)}$ becomes *nonunimodular*. If we denote by db the Lebesgue measure on $\mathbb{R}^n$, and by $d\Sigma^{(n)}$ the Haar measure on $S^{(n)}$ (which is unimodular), then the left and right invariant Haar measures on $G_{\mathrm{aff}}^{(n)}$ read, respectively, as

$$d\mu(b, S, a) = \frac{db}{a^n}\, d\Sigma^{(n)}\, \frac{da}{a}, \quad d\mu_r(b, S, a) = db\, d\Sigma^{(n)}\, \frac{da}{a}. \tag{12.7}$$

Clearly, nonunimodularity does complicate the situation (the Duflo–Moore operator C of Theorem 8.2.1 is now nontrivial, and necessarily unbounded). The addition of dilations more than compensates, however, in that it restores *square integrability*. As we will see in Chapter 14, most of the affine groups of interest have a square integrable representation, possibly modulo the center. The only notable exception is the affine Weyl–Heisenberg group (Section 15.2), for which we will have to the use the general formalism developed in the previous chapters.

Remark: Our Fourier transform in $\mathbb{R}^n$ is defined as

$$\widehat{f}(\xi) = (2\pi)^{-n/2} \int_{\mathbb{R}^n} e^{-i\xi \cdot x}\, f(x)\, dx, \tag{12.8}$$

for $\xi, x \in \mathbb{R}^n$ and $\xi \cdot x = \xi_1 x_1 + \xi_2 x_2 + \ldots + \xi_n x_n$. Note that, for better homogeneity, we will use the variable x both for $n=1$, instead of the more familiar t of signal processing, and for $n > 1$, instead of **x**, as used in the preceding chapters.

12.2 Derivation and properties of the 1-D continuous wavelet transform (♣)

Take first the full affine group of the line $G_{\text{aff}} = \{(b,a) \,|\, b \in \mathbb{R},\, a \neq 0,\}$, with the natural action $x \mapsto ax + b$ and group law

$$(b,a)(b',a') = (b + ab', aa'). \tag{12.9}$$

Thus, G_{aff} is a semidirect product of the translation group $\mathbb{R}$ by the full dilation group $\mathbb{R}_*$: $G_{\text{aff}} = \mathbb{R} \rtimes \mathbb{R}_*$. The unit element is $(0,1)$, and the inverse of (b,a) is $(-a^{-1}b, a^{-1})$.

The group G_{aff} is nonunimodular, the left Haar measure is $d\mu(b,a) = |a|^{-2} da\, db$ and the right Haar measure is $d\mu_r(b,a) = |a|^{-1} da\, db$.

The whole theory of wavelets rests on the following central result [56].

Theorem 12.2.1: *Up to unitary equivalence, G_{aff} has a unique UIR, acting in $L^2(\mathbb{R}, dx)$, namely,*

$$(U(b,a)f)(x) = |a|^{-1/2} f\left(\frac{x-b}{a}\right) \equiv f_{ba}(x) \quad (a \neq 0, b \in \mathbb{R}), \tag{12.10}$$

or, on Fourier transforms,

$$\left(\widehat{U(b,a)f}\right)(\xi) = |a|^{1/2} \widehat{f}(a\xi) e^{-ib\xi} \quad (a \neq 0, b \in \mathbb{R}). \tag{12.11}$$

The representation U is square integrable, and a vector $\psi \in L^2(\mathbb{R}, dx)$ is admissible iff it satisfies the condition

$$c_\psi \equiv 2\pi \int_{-\infty}^{\infty} |\widehat{\psi}(\xi)|^2 \frac{d\xi}{|\xi|} < \infty. \tag{12.12}$$

Proof. Unitarity and irreducibility of U are immediate.

Uniqueness follows from the fact that the representation U is obtained by Mackey's standard method of induction (there is only one nontrivial orbit in ξ-space) [Bar, 56] (see Chapter 3).

Finally, the square integrability follows from a straightforward calculation:

$$\int_{G_{\text{aff}}} |\langle \widehat{U(b,a)\psi} | \widehat{\psi} \rangle|^2 \frac{da\, db}{a^2} =$$

$$= \iiiint e^{ib(\xi - \xi')} \overline{\widehat{\psi}(a\xi)}\, \widehat{\psi}(a\xi') \widehat{\psi}(\xi) \overline{\widehat{\psi}(\xi')}\, d\xi\, d\xi' \frac{da}{|a|} db$$

$$= 2\pi \iint |\widehat{\psi}(a\xi)|^2 |\widehat{\psi}(\xi)|^2 \frac{da}{|a|} d\xi$$

$$= 2\pi \|\psi\|^2 \int_{-\infty}^{+\infty} |\widehat{\psi}(\xi)|^2 \frac{d\xi}{|\xi|} < \infty$$

(the interchanging of integrals is justified by Fubini's theorem). □

In practice, the admissibility condition (12.12) (plus some regularity; $\psi \in L^1 \cap L^2$ suffices) is equivalent to a zero mean condition:

$$\psi \text{ admissible} \stackrel{(\Leftarrow)}{\Rightarrow} \widehat{\psi}(0) = 0 \Leftrightarrow \int_{-\infty}^{+\infty} \psi(x)\, dx = 0. \quad (12.13)$$

Indeed, if $\psi \in L^1(\mathbb{R})$, $\widehat{\psi}$ is continuous and (12.12) requires $\widehat{\psi}(0) = 0$. Conversely [98], if $\int \psi(x)\, dx = 0$ and $\int (1+|x|)^\alpha |\psi(x)|\, dx < \infty$ for some $\alpha > 0$ (a condition slightly stronger than $\psi \in L^1$), then $|\widehat{\psi}(\xi)| \leq |\xi|^\beta$, $\beta = \min\{\alpha, 1\}$, and (12.12) holds.

From now on, an admissible function will be called a *wavelet* [note that some authors [Woj] use more restrictive definitions in the context of multiresolution analysis (see Section 13.1.1)]. Thus, a wavelet ψ is by necessity an *oscillating* function, real or complex-valued (see the examples below), and this is in fact the origin of the term "wavelet" (note, however, that they were called "wavelets of constant shape" in the very first paper [156], because "wavelet" was already in use in the geophysics community, with quite a different meaning).

Let ψ be a wavelet and $s \in L^2(\mathbb{R})$ be a signal. Then, the continuous wavelet transform (CWT) of s with respect to ψ is the function $S \equiv T_\psi s$ given by the scalar product of s with the transformed wavelet ψ_{ba}:

$$\begin{aligned} S(b,a) &= \langle \psi_{ba} | s \rangle \\ &= |a|^{-1/2} \int_{-\infty}^{+\infty} \overline{\psi(a^{-1}(x-b))}\, s(x)\, dx \quad (12.14) \\ &= |a|^{1/2} \int_{-\infty}^{+\infty} \overline{\widehat{\psi}(a\xi)}\, \widehat{s}(\xi)\, e^{ib\xi}\, d\xi. \quad (12.15) \end{aligned}$$

Clearly, the map $T_\psi : s \mapsto S$ coincides, up to a constant, with the coherent state map W_K or W_η described in Sections 7.1.1 and 8.1, respectively. Thus, all properties listed below are immediate translations of the corresponding ones discussed there. Notice that a different normalization is often used in the examples, namely, replacing $|a|^{-1/2}$ by $|a|^{-1}$ in (12.10) and (12.14), which has the effect of enhancing small scales in transforms. This is called the L^1-normalization, because it ensures that the L^1 norm is conserved under dilation. Indeed, denoting $\widetilde{\psi}_{ba}(x) = |a|^{-1} f((x-b)/a)$, we have $\|\widetilde{\psi}_{ba}\|_1 = \|\psi\|_1$, $\forall b, a$. The L^2-normalization given above, however, is the only one that comes from a unitary representation of G_{aff}.

In practice, one often imposes on the analyzing wavelet ψ a number of additional properties, for instance, restrictions on the support of ψ and of $\widehat{\psi}$ (see Section 12.4 below). Or ψ may be required to have a certain number $N \geq 1$ of *vanishing moments* (by the admissibility condition (12.13), the moment of order 0 must always vanish):

$$\int_{-\infty}^{\infty} x^n \psi(x)\, dx = 0, \quad n = 0, 1, \ldots N \quad (12.16)$$

12.2. Derivation and properties of the 1-D continuous wavelet transform (♣)

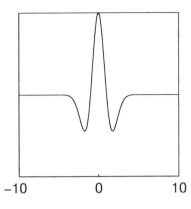
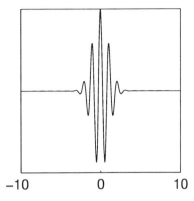

Figure 12.1. Two usual wavelets; (left) Mexican hat or Marr wavelet; (right) Real part of the Morlet wavelet ($\xi_o = 5.6$).

This property improves its efficiency at detecting singularities in the signal. Indeed, the transform (12.14) is then blind to the smoothest part of the signal, that which is polynomial of degree up to N — and less interesting, in general. Only the sharper part remains, including all singularities (jumps in the signal or one of its derivatives, for instance). For instance, if the first moment ($n = 1$) vanishes, the transform will erase any linear *trend* in the signal.

Before proceeding to the general properties of the continuous wavelet transform, let us give two examples, actually the most widely used wavelets, which are depicted in Figure 12.1.

1. *The Mexican hat or Marr wavelet*

This is simply the second derivative of a Gaussian:

$$\psi_H(x) = (1-x^2)\,e^{-x^2/2}, \quad \widehat{\psi_H}(\xi) = \xi^2\,e^{-\xi^2/2}. \tag{12.17}$$

It is a real wavelet, with two vanishing moments ($n = 0, 1$). Similar wavelets, with more vanishing moments, are obtained by taking higher derivatives of the Gaussian [158]:

$$\psi_H^{(m)}(x) = \left(\frac{1}{i}\frac{d}{dx}\right)^m e^{-x^2/2}, \quad \widehat{\psi_H^{(m)}}(\xi) = \xi^m\,e^{-\xi^2/2}. \tag{12.18}$$

2. *The Morlet wavelet*

This is just a modulated Gaussian:

$$\begin{aligned}\psi_M(x) &= \pi^{-1/4}\left(e^{i\xi_o x} - e^{-\xi_o^2/2}\right)e^{-x^2/2} \\ \widehat{\psi_M}(\xi) &= \pi^{-1/4}\left[e^{-(\xi-\xi_o)^2/2} - e^{-\xi^2/2}\,e^{-\xi_o^2/2}\right].\end{aligned} \tag{12.19}$$

In fact, the first term alone does *not* satisfy the admissibility condition, hence, the necessity of a correction. For ξ_o large enough (typically $\xi_o \geq 5.5$), however, this correction term is numerically negligible ($\leq 10^{-4}$). The Morlet wavelet is complex; hence, the corresponding transform $S(b,a)$ is also complex. This enables one to deal separately with the phase and the modulus of the transform, and the phase turns out to be a crucial ingredient in most algorithms used in applications such as feature detection.

Consider now the connected subgroup of G_{aff}, called G_+ or the $ax+b$ group:

$$G_+ = \{(b,a) : b \in \mathbb{R}, a > 0\} \equiv \mathbb{H}. \tag{12.20}$$

When restricted to G_+, the representation U splits into the direct sum of two unitarily inequivalent, square integrable, UIRs $U_\pm$, acting in the two Hardy spaces:

$$\begin{aligned} H_+(\mathbb{R}) &= \{f \in L^2(\mathbb{R}) \,|\, \widehat{f}(\xi) = 0 \text{ for } \xi < 0\}, \\ H_-(\mathbb{R}) &= \{f \in L^2(\mathbb{R}) \,|\, \widehat{f}(\xi) = 0 \text{ for } \xi > 0\}, \end{aligned} \tag{12.21}$$

and

$$L^2(\mathbb{R}) = H_+(\mathbb{R}) \oplus H_-(\mathbb{R}). \tag{12.22}$$

Elements of $H_+(\mathbb{R})$ [respectively, $H_-(\mathbb{R})$] extend to functions analytic in the upper [respectively, lower] complex half-plane, and accordingly they are called upper [respectively, lower] analytic signals [Lyn, Pap].

From now on, we shall mostly restrict ourselves to the representation U_+, acting in $H_+(\mathbb{R})$ [actually, Theorem 8.1.5 allows one to extend wavelet analysis from $H_+(\mathbb{R})$ to the full space $L^2(\mathbb{R})$]. Thus, a vector $\psi \in H_+(\mathbb{R})$ is admissible if it satisfies

$$c_\psi = 2\pi \int_0^\infty |\widehat{\psi}(\xi)|^2 \frac{d\xi}{\xi} < \infty. \tag{12.23}$$

If, in addition, $\widehat{\psi}$ is real, ψ is then called a *progressive* wavelet (notice that neither the Mexican hat nor the Morlet wavelet are progressive).

The general properties of the continuous wavelet transform may then by summarized into the following theorem, which is simply the particularization of Theorem 8.1.3 of Chapter 8.

Theorem 12.2.2: *Let $T_\psi : s \mapsto S$ be the wavelet transform with respect to the wavelet ψ. Then the following is true.*

1. *The map $W_\psi \equiv c_\psi^{-1/2} T_\psi$ is an isometry from the space $\mathfrak{H} = L^2(\mathbb{R})$ or $H_+(\mathbb{R})$ of finite energy signals (in the sense of signal processing) onto the space $\mathfrak{H}_\psi$ of wavelet transforms, which is a closed subspace of $L^2(\mathbb{H}, a^{-2} da\, db)$. In other words, W_ψ conserves energy:*

$$\iint_\mathbb{H} |S(b,a)|^2 \frac{da\, db}{a^2} = c_\psi \int_\mathbb{R} |s(x)|^2 \, dx. \tag{12.24}$$

12.2. Derivation and properties of the 1-D continuous wavelet transform (♣)

Equivalently, the wavelet ψ generates a resolution of the identity (tight frame):

$$c_\psi^{-1} \iint_{\mathbb{H}} |\psi_{ba}\rangle\langle\psi_{ba}| \frac{da\,db}{a^2} = I. \tag{12.25}$$

2. The projection $\mathbb{P}_\psi = W_\psi W_\psi^*$ from $L^2(\mathbb{H}, a^{-2}da\,db)$ onto $\mathfrak{H}_\psi$ is an integral operator, with (reproducing) kernel

$$K(b', a'; b, a) = c_\psi^{-1} \langle \psi_{b'a'} | \psi_{ba} \rangle. \tag{12.26}$$

Hence, a function $f \in L^2(\mathbb{H}, a^{-2}da\,db)$ is the wavelet transform of a signal if and only if it satisfies the reproducing property

$$f(b', a') = \iint_{\mathbb{H}} K(b', a'; b, a)\, f(b, a)\, \frac{da\,db}{a^2}. \tag{12.27}$$

3. By 1, the map W_ψ may be inverted on its range $\mathfrak{H}_\psi$ by its adjoint: $W_\psi^{-1} \restriction \mathfrak{H}_\psi = W_\psi^*$. As a consequence, the signal s may be recovered from its wavelet transform S with help of the reconstruction operator $R_\psi = T_\psi^*$:

$$s(x) = c_\psi^{-1}(R_\psi S)(x) = c_\psi^{-1} \iint_{\mathbb{H}} \psi_{ba}(x)\, S(b, a)\, \frac{da\,db}{a^2}. \tag{12.28}$$

4. Finally, the continuous wavelet transform is covariant with respect to the group G_+:

$$s(x - b_o) \mapsto S(b - b_o, a),\ b_o \in \mathbb{R}, \tag{12.29}$$
$$a_o^{-1/2} s(a_o^{-1} x) \mapsto S(a_o^{-1} b, a_o^{-1} a),\ a_o > 0.$$

The unitary operator $W_\psi : H_+(\mathbb{R}) \to \mathfrak{H}_\psi$ is the particularization to the $ax + b$ group of the coherent state map W_K or W_η described in Sections 7.1.1 and 8.1, respectively, and similarly in the case of the full affine group G_{aff}.

Clearly, the reconstruction formula (12.28) yields a decomposition of the signal into a linear superposition of the wavelets ψ_{ba} with coefficients $S(b, a)$. In the same way, one may reconstruct the signal by using a wavelet χ different from the analyzing wavelet ψ [Hol, 178],

$$s(x) = c_{\psi\chi}^{-1} \iint_{\mathbb{H}} \chi_{ba}(x)\, (T_\psi s)(b, a)\, \frac{da\,db}{a^2}, \tag{12.30}$$

provided χ and ψ satisfy the compatibility condition

$$0 < |c_{\psi\chi}| < \infty, \quad \text{where} \quad c_{\psi\chi} = \int_{-\infty}^{\infty} \overline{\hat{\psi}(\xi)}\, \hat{\chi}(\xi)\, \frac{d\xi}{|\xi|}. \tag{12.31}$$

When (12.31) holds, we say that χ is a *reconstruction wavelet* for ψ. Notice that such an extended reconstruction formula also follows from the general theory of square integrable representations discussed in Section 8.2, namely, the relation (8.52). Analogous reconstruction formulas may also be written for the WT based on the full affine group G_{aff} [Dau, 98].

We emphasize the covariance property (12.29), especially that with respect to translations. It is one of the reasons why the continuous wavelet transform is extremely useful for detecting particular features in signals. Indeed, translation covariance is lost in the discrete time WT discussed briefly in Section 13.1, and this is one of the drawbacks of the discrete wavelet schemes based on multiresolution.

Now, since the wavelet transform (12.14) is a scalar product, it makes sense to take for χ a singular function, provided the signal is correspondingly smooth [this is the basic idea of weak (Sobolev) derivatives or rigged Hilbert spaces [Gel]]. For instance, if the signal s is continuous, we may take for χ a Dirac measure (δ function), so that $\widehat{\chi} = 1$. Thus, if the analyzing wavelet ψ satisfies the condition resulting from (12.31),

$$\int_{-\infty}^{\infty} |\widehat{\psi}(\xi)| \frac{d\xi}{|\xi|} < \infty, \qquad (12.32)$$

then one gets the simplified reconstruction formula used by Morlet [157] (and originally due to Calderón):

$$s(x) = \int_0^{\infty} S(a,x) \frac{da}{a}. \qquad (12.33)$$

The extension of the continuous wavelet transform to distributions, in the context of rigged Hilbert spaces, is discussed in detail in the next section.

12.3 A mathematical aside: Extension to distributions

Before dealing with the practical problems of implementing the continuous wavelet transform and applying it to physical problems, we shall devote a few comments to a remarkable mathematical aspect, namely, its extension to distributions.

Let us consider for a while the standard rigged Hilbert space (RHS) [Gel]:

$$\mathcal{S}(\mathbb{R}) \subset L^2(\mathbb{R}) \subset \mathcal{S}^{\times}(\mathbb{R}), \qquad (12.34)$$

where $\mathcal{S}(\mathbb{R})$ is the Schwartz space of fast decreasing C^{∞} functions and $\mathcal{S}^{\times}(\mathbb{R})$ is the space of tempered distributions, that is, continuous *antilinear* functionals on $\mathcal{S}(\mathbb{R})$. Now, the sesquilinear form $< \cdot, \cdot >$ that puts $\mathcal{S}^{\times}$ and $\mathcal{S}$ in duality may be identified, up to a complex conjugation, with the inner product of L^2:

$$< f, s > = \langle s|f \rangle_{L^2(\mathbb{R})}, \quad f, s \in \mathcal{S}(\mathbb{R}). \qquad (12.35)$$

Note that the convention we have adopted here is the one used in the RHS theory. It allows both embeddings in (12.34) to be *linear*. Other advantages will be seen below. Note that the opposite convention is used in [Hol],

12.3. A mathematical aside: Extension to distributions

considering instead the space $\mathcal{S}(\mathbb{R})'$ of continuous *linear* functionals on $\mathcal{S}(\mathbb{R})$, and then an antilinear embedding of $\mathcal{S}$ into $\mathcal{S}'$.

Many operations from analysis, such as derivation, multiplication by a polynomial, convolution, map $\mathcal{S}(\mathbb{R})$ continuously into itself, with respect to its natural Fréchet topology. It follows that all of these operations extend by duality to $\mathcal{S}^\times(\mathbb{R})$. So does, in particular, the Fourier transform, which is an isomorphism of the triplet (12.34) onto itself.

Coming back to the continuous wavelet transform, observe from (12.14), that it *is* an L^2 inner product. Thus, the usual RHS reasoning applies: one of the two functions ψ or s may be taken as singular (a tempered distribution) provided the other one is correspondingly good. This justifies the use of plane waves or δ-functions as signals, or that of a δ-function as analyzing wavelet, as discussed in the previous section.

We may go further, however. Indeed, the continuous wavelet transform in many ways provides a generalization of the Fourier transform, so it is a natural question to ask whether it extends to distributions, and maps the triplet (12.34) onto a similar one. The answer to both questions is in fact yes, as shown by Holschneider [Hol, 173]. Of course, the situation is somewhat more complicated here, since the wavelet transform maps functions on $\mathbb{R}$, even $\mathbb{R}^+$ in the progressive case, into functions on the half-plane $\mathbb{H} \equiv \mathbb{R}^2_+$. This section is largely inspired by this work. Since our basic aim, however, is to give the flavor of the theory, we will mostly skip the proofs, which may be found in the original references.

The key tool in proving these properties is that of *localization*: Given the asymptotic behavior of a signal s as $x \to \pm\infty$, what is that of its wavelet transform $S(b, a)$? This aspect is particularly important in the present context. Indeed, we are talking here about localization in the half-plane $\mathbb{H}$, and the latter as a natural interpretation as phase space, as we will see in Section 12.4.1. Thus, what is involved here is localization in phase space, a key aspect of coherent state theory. (Notice that Holschneider [Hol, 173] always uses the L^1-normalization, so that the results reported here will in some cases differ slightly from his.)

For $\alpha, \beta \geq 0$, consider the functions

$$\psi_\alpha^+(x) = \begin{cases} (1+|x|^\alpha)^{-1}, & x \geq 0, \\ 0, & x < 0, \end{cases} \quad \psi_\alpha^-(x) = \begin{cases} 0, & x > 0, \\ (1+|x|^\alpha)^{-1}, & x \leq 0, \end{cases} \quad (12.36)$$

and

$$\psi_{\alpha\beta}(x) = \psi_\alpha^-(x) + \psi_\beta^+(x), \quad \psi_\alpha(x) \equiv \psi_{\alpha\alpha}(x). \quad (12.37)$$

A function s is called *polynomially localized* if $|s(x)| \leq c_\alpha \psi_\alpha(x)$ for some α. Define also

$$\phi_{\alpha\beta}(\xi) = \psi_\alpha^+\left(\frac{1}{\xi}\right)\psi_\beta^+(\xi), \quad \phi_\alpha(\xi) \equiv \phi_{\alpha\alpha}(\xi). \quad (12.38)$$

The function $\phi_{\alpha\beta}$ vanishes for $\xi < 0$ and tends to zero as ξ^α for $\xi \to 0$ and as ξ^β for $\xi \to \infty$. A progressive function s [that is, $\hat{s}(\xi) = 0$ for $\xi < 0$] is called *polynomially bandlimited* or *strip-localized* if its Fourier transform is majorized by some $\phi_{\alpha\beta}$; that is, $\hat{s}(\xi) \leq c\phi_{\alpha\beta}(\xi)$ a.e., for some constant $c > 0$.

Then, the basic localization result asserts that, if the wavelet ψ and the signal s are both polynomially localized (respectively, strip-localized), so is the wavelet transform $T_\psi s$:

Theorem 12.3.1: *(1) Suppose that*

$$|\psi(x)| \leq c|\psi_\alpha(x)| \quad \text{and} \quad |s(x)| \leq c|\psi_\alpha(x)|, \tag{12.39}$$

for some $\alpha > 1$. Then

$$(T_\psi s)(b, a) \sim c' \left(\frac{1}{1+a}\right)^{1/2} \psi_\alpha\left(\frac{b}{1+a}\right), \tag{12.40}$$

with a constant c' depending only on α.

(2) Suppose $\alpha, \beta, \alpha', \beta'$ are all nonnegative and $\alpha \neq \beta' - 1, \alpha' \neq \beta - 1$. Let the wavelet ψ and the signal s be both progressive and satisfy

$$|\hat{\psi}(\xi)| \leq \phi_{\alpha\beta}(\xi), \qquad |\hat{s}(\xi)| \leq \phi_{\alpha'\beta'}(\xi). \tag{12.41}$$

Then, the wavelet transform is also localized in scale,

$$|T_\psi s(b, a)| \leq c\sqrt{a}\,\phi_{\min(\alpha, \beta'-1)\min(\alpha'+1, \beta)}(a), \tag{12.42}$$

where c is independent of ψ and s.

In (1), the notation $r \sim s$ for two functions means as usual that $c^{-1} < r(t)\,s(t)^{-1} < c$ for some $c > 0$ and all t. Also, a similar result holds for general functions $\psi_{\alpha\beta}, \psi_{\alpha'\beta'}$. In (2), the restrictions $\alpha \neq \beta' - 1, \alpha \neq \beta - 1$ ensure the absence of logarithmic corrections. The proof is easy and may be found in [173].

Using this concept of localization, one may now build a RHS appropriate for the continuous wavelet transform. Denote by $\mathcal{S}_+(\mathbb{R})$ the set of functions in $\mathcal{S}(\mathbb{R})$ whose Fourier transform is arbitrarily well polynomially bandlocalized: For every $\alpha > 0$, there is a constant $c_\alpha > 0$, such that

$$|s(x)| \leq c_\alpha \psi_\alpha(x) \quad \text{and} \quad |\hat{s}(\xi)| \leq c_\alpha \phi_\alpha(\xi). \tag{12.43}$$

It turns out that $\mathcal{S}_+(\mathbb{R})$ is a *closed* subspace of $\mathcal{S}(\mathbb{R})$, consisting exactly of those functions whose Fourier transform vanishes for $\xi \leq 0$ (the progressive functions) — hence, the notation is consistent. The space $\mathcal{S}_+(\mathbb{R})$ carries a natural topology, given by the seminorms

$$\|s\|_\alpha^{(+)} = \sup_{\alpha' \in [0,\alpha]} \inf c_{\alpha'}, \tag{12.44}$$

where the infimum is taken over all constants $c_{\alpha'}$ that verify (12.43). It is easy to show that $\{\|\cdot\|_\alpha^{(+)}, \alpha > 0\}$ is an ordered set of seminorms:

$\|s\|_\alpha^{(+)} \leq \|s\|_\beta^{(+)}$ if $\alpha \leq \beta$. Thus, $\mathcal{S}_+(\mathbb{R})$ becomes a locally convex topological vector space. A sequence $s_n \in \mathcal{S}_+(\mathbb{R})$ tends to zero if $\lim_{n\to\infty} \|s_n\|_\alpha^{(+)} = 0$ for all $\alpha \geq 0$ (or, equivalently, for a directed set $\{\alpha_m\}$).

Now we turn to wavelet transforms. If both the wavelet ψ and the signal s are in $\mathcal{S}_+(\mathbb{R})$, Theorem 12.3.1 tells us that the wavelet transform $S(b,a)$ is well localized both in scale and position; namely, for every $\alpha > 0$, there is a constant $c_\alpha > 0$ such that

$$|S(b,a)| \leq c_\alpha \phi_\alpha(a)\, \psi_\alpha\left(\frac{b}{1+a}\right). \tag{12.45}$$

Denote the space of such functions by $\mathcal{S}(\mathbb{H})$. As above, this space becomes a locally convex topological vector space when equipped with the seminorms:

$$\|S\|_\alpha^{(2)} = \sup_{\alpha' \in [0,\alpha]} \inf c_{\alpha'}. \tag{12.46}$$

The two spaces $\mathcal{S}_+(\mathbb{R})$ and $\mathcal{S}(\mathbb{H})$ are for the continuous wavelet transform what Schwartz's space $\mathcal{S}(\mathbb{R})$ is for the Fourier transform, as results from the following theorem.

Theorem 12.3.2: *(1) Given a wavelet $\psi \in \mathcal{S}_+(\mathbb{R})$, the wavelet transform $T_\psi : \mathcal{S}_+(\mathbb{R}) \to \mathcal{S}(\mathbb{H})$ is continuous with respect to the respective topologies and the following estimate holds for $\alpha > 0$:*

$$\|T_\psi s\|_\alpha^{(2)} \leq c \|\psi\|_{2(\alpha+1)}^{(+)} \|s\|_{2(\alpha+1)}^{(+)}, \tag{12.47}$$

with some constant $c > 0$ depending on α only.

(2) Similarly, the reconstruction operator R_ψ is continuous from $\mathcal{S}(\mathbb{H})$ into $\mathcal{S}_+(\mathbb{R})$, and one has the estimate, for $\alpha > 1$,

$$\|R_\psi S\|_\alpha^{(+)} \leq c \|\psi\|_\alpha^{(+)} \|S\|_\alpha^{(2)}. \tag{12.48}$$

Given Theorem 12.3.2, the extension to distributions is easy. First we define the dual $\mathcal{S}_+^\times(\mathbb{R})$ [respectively, $\mathcal{S}^\times(\mathbb{H})$] as the space of continuous antilinear maps on $\mathcal{S}_+(\mathbb{R})$ [respectively, $\mathcal{S}(\mathbb{H})$]. [Holschneider [Hol] uses the space $\mathcal{S}'_+(\mathbb{R})$, defined as the dual of $\mathcal{S}_-(\mathbb{R})$.] Continuity of $F \in \mathcal{S}_+^\times(\mathbb{R})$ means, for instance, that there exists a constant $c > 0$ such that, for every $s \in \mathcal{S}_+(\mathbb{R})$,

$$|<F,s>| \leq c \|s\|_\alpha^{(+)} \tag{12.49}$$

for some $\alpha \geq 0$ (here, $<\cdot,\cdot>$ is the sesquilinear form defining the duality). Clearly, one has, exactly as for the Schwartz triplet (12.34),

$$\mathcal{S}_+(\mathbb{R}) \subset L^2(\mathbb{R}^+, d\xi) \subset \mathcal{S}_+^\times(\mathbb{R}), \tag{12.50}$$

and the duality form $<\cdot,\cdot>$ may be seen as an extension/restriction of the inner product of $L^2(\mathbb{R}^+)$:

$$<h,s> = \langle s|h\rangle_{L^2(\mathbb{R}^+)}, \quad h \in L^2(\mathbb{R}^+),\ s \in \mathcal{S}_+(\mathbb{R}). \tag{12.51}$$

Notice that, as before, this relation provides a *linear* embedding of $\mathcal{S}_+$ into $\mathcal{S}_+^\times$. The same triplet structure is present, of course, on the side of the transforms:

$$\mathcal{S}(\mathbb{H}) \subset L^2(\mathbb{H}, a^{-2} da db) \subset \mathcal{S}^\times(\mathbb{H}). \tag{12.52}$$

We equip both duals with the weak*-topology. This means, for instance, that a sequence $F_n \in \mathcal{S}_+^\times(\mathbb{R})$ tends to zero in $\mathcal{S}_+^\times(\mathbb{R})$ if

$$\lim_{n \to \infty} <F_n, s> = 0, \ \forall s \in \mathcal{S}_+(\mathbb{R}). \tag{12.53}$$

Now we are ready to extend the wavelet transform to distributions. Using the reconstruction operator R_ψ, we may write, by definition, for any $s \in \mathcal{S}_+(\mathbb{R})$, $\Phi \in \mathcal{S}(\mathbb{H})$:

$$\langle T_\psi s | \Phi \rangle_{L^2(\mathbb{H})} = \langle s | R_\psi \Phi \rangle_{L^2(\mathbb{R}_+)}. \tag{12.54}$$

Thus, using the identification (12.51), we may define the wavelet transform of $F \in \mathcal{S}_+^\times(\mathbb{R})$ by duality as the distribution $T_\psi F \in \mathcal{S}^\times(\mathbb{H})$ given by

$$<T_\psi F, \Phi> = <F, R_\psi \Phi>, \quad \forall \Phi \in \mathcal{S}(\mathbb{H}). \tag{12.55}$$

Similarly, the reconstruction operator R_ψ is defined on $\Psi \in \mathcal{S}^\times(\mathbb{H})$ by the relation:

$$<R_\psi \Psi, s> = <\Psi, T_\psi s>, \quad \forall s \in \mathcal{S}_+(\mathbb{R}). \tag{12.56}$$

Since all operations involved are continuous, we may conclude the following theorem.

Theorem 12.3.3: *(1) For any $\psi, \chi \in \mathcal{S}_+(\mathbb{R})$, the wavelet transform T_ψ and the reconstruction operator R_χ*

$$T_\psi : \mathcal{S}_+^\times(\mathbb{R}) \to \mathcal{S}^\times(\mathbb{H}), \quad R_\chi : \mathcal{S}^\times(\mathbb{H}) \to \mathcal{S}_+^\times(\mathbb{R})$$

are continuous for the respective weak-topologies.*

(2) If χ is a reconstruction wavelet for the analyzing wavelet ϕ, then

$$R_\chi T_\psi = c_{\psi \chi} I, \tag{12.57}$$

with $c_{\psi \chi} = \int_\mathbb{R} \overline{\hat{\psi}(\xi)} \, \hat{\chi}(\xi) \, |\xi|^{-1} d\xi$ and I is the identity operator on $\mathcal{S}_+^\times(\mathbb{R})$.

(3) Given $F \in \mathcal{S}_+^\times(\mathbb{R})$, its wavelet transform $\Phi = T_\psi F \in \mathcal{S}^\times(\mathbb{H})$ is the function

$$\Phi(b, a) \equiv (T_\psi F)(b, a) = <F, \psi_{ba}>.$$

The function Φ is C^∞ and polynomially bounded, with all its derivatives

$$\left| \Phi(b, a) \phi_\alpha(a) \psi_\alpha \left(\frac{b}{1+a} \right) \right| \leq c_\alpha, \quad \text{for some } \alpha \geq 0,$$

and it satisfies pointwise the reproducing equation

$$\Phi(b', a') = \iint_\mathbb{H} K_{\psi \chi}(b', a'; b, a) \Phi(b, a) \, \frac{da \, db}{a^2}, \tag{12.58}$$

with the reproducing kernel

$$K_{\psi\chi}(b', a'; b, a) = c_{\psi\chi}^{-1}\langle\psi_{b'a'}|\chi_{ba}\rangle. \tag{12.59}$$

(4) In addition, the reproducing equation (12.58) defines the projection of $\mathcal{S}^\times(\mathbb{H})$ onto the image $T_\psi \mathcal{S}_+^\times(\mathbb{R})$.

Sketch of the proof. (See [Hol] or [173] for details.)
Statement (1) follows from the continuity of all operations.
Statement (2) follows from the definitions of T_ψ and R_χ; given any $F \in \mathcal{S}_+^\times(\mathbb{R})$ and any function $s \in \mathcal{S}_+(\mathbb{R})$, one has:

$$< R_\chi T_\psi F, s > = < T_\psi F, T_\chi s > = < F, R_\psi T_\psi s > = c_{\psi\chi} < F, s >,$$

in virtue of the reconstruction formula on $\mathcal{S}_+(\mathbb{R})$.
As for (3), the relation $(T_\psi F)(b, a) = <F, \psi_{ba}>$ and the reproducing equation (12.58) follow from a direct computation (this involves exchanging some limits and some integrals, both being duly justified). Finally the C^∞ character of $T_\psi F$ and its polynomial boundedness follow from the following technical lemma, which has an independent interest. □

Lemma 12.3.4: *For every distribution $F \in \mathcal{S}_+^\times(\mathbb{R})$ and every function $s \in \mathcal{S}_+(\mathbb{R})$, one has*

$$\frac{\partial}{\partial b}\langle F, T^b s\rangle = \langle F, \frac{\partial}{\partial x} T^b s\rangle, \tag{12.60}$$

$$\frac{\partial}{\partial a}\langle F, D^a s\rangle = \langle F, x\frac{\partial}{\partial x} D^a s\rangle, \tag{12.61}$$

and, for some $\alpha \geq 0$ and $c > 0$,

$$\psi_\alpha(b) |\langle F, T^b s\rangle| \leq c, \tag{12.62}$$
$$\phi_\alpha(a) |\langle F, D^a s\rangle| \leq c, \tag{12.63}$$

where $T^b s = s(\cdot - b)$ and $D^a s = a^{-1} s(\cdot/a)$.

The conclusion is twofold. First, exactly as the Fourier transform, the continuous wavelet transform on progressive functions extends by duality to the appropriate spaces of distributions, including continuity. Second, since the continuous wavelet transform is after all a convolution, it has a regularizing effect: The wavelet transform of a tempered distribution is a C^∞ polynomially bounded function, exactly as in the usual theory of tempered distributions on $\mathbb{R}$ or $\mathbb{R}^2$. The only new fact is that here the wavelet transform maps functions on $\mathbb{R}^+$ into functions on the half-plane $\mathbb{H}$. Otherwise, all results follow the standard pattern.

12.4 Interpretation of the continuous wavelet transform

Besides the fact that the wavelet transform is a particular case of the CS machinery, it possesses other aspects that lead to a fruitful interpretation, both from the mathematical and the physical vantage points.

12.4.1 The CWT as phase space representation

As we have discussed in Section 12.2, the continuous wavelet transform derives from a UIR of the affine group G_{aff} or G_+, which is of the semidirect product type. Hence, the considerations developed in Section 10.2.2 apply, and the coadjoint orbits yield precious information concerning the representations. More precisely, according to the theory of Kirillov [Kir], each coadjoint orbit correspond to a unique (up to unitary equivalence) unitary irreducible representation. On the other hand, each coadjoint orbit is a *symplectic manifold*, precisely the kind of manifold that, in mechanics, is usually taken as the *phase space* of a classical system and starting point for the procedure of (geometric) quantization. Thus, we have to compute the coadjoint orbits of G_+ [Gui].

An easy calculation [29] shows that G_+ has three coadjoint orbits. The first one is $\mathcal{O}_o = \{0\} \times \mathbb{R}$ and corresponds to the identity representation. The other two are two-dimensional, namely, $\mathcal{O}_\pm = \mathbb{R}_*^\pm \times \mathbb{R} \equiv \mathbb{R}_\pm^2$, with Lebesgue measure, and correspond to the two representations U_+ and U_-. Thus, $\mathcal{O}_+$ is homeomorphic to G_+. Writing $\kappa = a^{-1}$, we see that the left Haar measure on $G_+ = \mathbb{H}$ reduces to the latter:

$$d\mu_l(b, a) = a^{-2} da\, db = d\kappa\, db. \tag{12.64}$$

Thus $(\mathbb{H}, d\mu_l)$ is a phase space associated to U_+, so that the CWT yields a *phase space realization* of signals [32, 119]. This fact is physically significant, in that it opens the way to the application of wavelets to quantum problems, in particular, in the quantization process. Similar considerations hold true in higher dimensions, as we shall see in the next chapter. Note also that the interpretation of a^{-1} as a momentum variable is implicit in the identification of the Wigner–Ville transform $W[s](b, a)$ with the Wigner distribution $W(q, p)$, for $q = b$, $p = 1/a$ [see the comments after (12.4)].

Note that the Poincaré half-plane we obtain here is also a coadjoint orbit of $SL(2, \mathbb{R}) \simeq SU(1, 1)$ (see Section 4.2.2). This is related to the fact that the representation U_+ extends to a discrete series representation of $SL(2, \mathbb{R})$, of which G_+ is a subgroup.

12.4.2 Localization properties and physical interpretation of the CWT

The efficiency of the CWT hinges on the interplay between two facts:
1. The admissibility condition for ψ, reduced to $\int \psi(x)\,dx = 0$;
2. The support properties of ψ.

Consider the latter first. We have seen above that it is essential to require that ψ be square integrable, possibly integrable, and satisfy the admissibility condition, but this is far from sufficient in practice.

On the contrary, one usually assumes ψ and $\widehat{\psi}$ to be as well localized as possible (compatible with the Fourier–Heisenberg uncertainty principle, of course). More specifically, assume that ψ has an "essential" support of width T, centered around zero, while $\widehat{\psi}$ has an essential support of width Ξ, centered around ξ_o. Then, the transformed wavelets ψ_{ba} and $\widehat{\psi}_{ba}$ have, respectively, an essential support of width aT around b and an essential support of width Ξ/a around ξ_o/a. Notice that the product of the two widths is constant (we know it has to be bounded below by a fixed constant, by Fourier's theorem). From this, we can characterize the behavior of the WT in the two extreme cases.

1. If $a \gg 1$, ψ_{ba} is a wide window, whereas $\widehat{\psi}_{ba}$ is very peaked around a small frequency ξ_o/a; this transform will be most sensitive to *low frequencies*.
2. If $a \ll 1$, ψ_{ba} is a narrow window and $\widehat{\psi}_{ba}$ is wide and centered around a high frequency ξ_o/a; this wavelet has a good localization capability in the time domain and is mostly sensitive to *high frequencies*.

Thus, we have obtained a tool that reproduces the correlation between duration and average frequency often encountered in real life signals, and announced in Section 12.1: Low frequency portions of the signal tend to be long, whereas high frequencies occur briefly in general.

Combining now these localization properties with the zero mean condition and the fact that ψ_{ba} acts like a filter (convolution in x-space),

$$S(b, a) = |a|^{-1/2} \int_{-\infty}^{+\infty} \overline{\psi(a^{-1}(x - b))}\, s(x)\, dx$$

$$= |a|^{1/2} \int_{-\infty}^{+\infty} \overline{\widehat{\psi}(a\xi)}\, \widehat{s}(\xi)\, e^{ib\xi}\, d\xi,$$

we see that the CWT performs a *local filtering*, both in position and in scale. The wavelet transform $S(b, a)$ is nonnegligible only when the wavelet ψ_{ba} matches the signal; that is, it selects the part of the signal, if any, that is concentrated around the time b and the scale a.

In order to get a physical interpretation of this feature, we notice that in signal analysis, as in classical electromagnetism, the L^2 norm is interpreted as the total energy of the signal. Therefore, the relation (12.24) suggests to interpret $|S(b, a)|^2$ as the energy density in the wavelet parameter space.

Therefore, if the wavelet is well localized, the local filtering effect means that the energy density of the transform will be concentrated on the significant parts of the signal. This is the key to all approximation schemes that make wavelets such an efficient tool.

Furthermore, combining the support properties of ψ with the covariance of the CWT under G_+, one sees that the CWT analysis has constant relative bandwidth: $\Delta \xi/\xi =$ constant, contrary to the WFT analysis, which has a constant bandwidth, $\Delta \xi =$ constant. This implies that it has a better resolution at high frequency, i.e., small scales: the continuous wavelet transform is a *singularity detector*.

In addition to its localization properties, the wavelet ψ is often required to have a certain number of vanishing moments, as we already mentioned in Section 12.2 [see (12.16]. This condition determines the capacity of the wavelet transform to detect and measure singularities. Indeed, if ψ has all its moments vanishing up to order $N \geq 1$, then it is blind to polynomials of degree up to N. Equivalently, it detects singularities down to the $(N + 1)$st derivative of the signal. This property is crucial for a whole class of applications, namely, the determination of local regularity of functions or measures, more generally, the characterization of singularities. Note that the latter aspect often requires more specialized wavelets, such as chirps [Jaf, Tor, 219].

All taken together, the CWT may be called a *mathematical microscope*, with optics ψ, position b, and global magnification $1/a$. In addition, by its very definition, the wavelet transform is an ideal tool for analyzing scale dependent features, in particular, *fractals* [Arn, 50]. A simple but striking example is that of the devil's staircase, that is, the function $f(x) = \int_0^x d\mu$, where μ is a uniform measure on the triadic Cantor set. The wavelet transform of f exhibits in a transparent way the fractal structure of the function, as shown in Figure 12.2.

Finally, as we shall see in Chapter 14, all considerations made in this section extend to higher dimensions, where all together they contribute to make the CWT into a remarkably efficient tool for image analysis (in two dimensions).

12.5 Discretization of the continuous WT: Discrete frames

The reproducing property (12.27) implies, as usual for CS, that the information content of the wavelet transform $S(b, a)$ is highly redundant. In fact, the signal has been unfolded from one to two dimensions, and this explains the practical efficiency of the CWT for disentangling parts of the signal

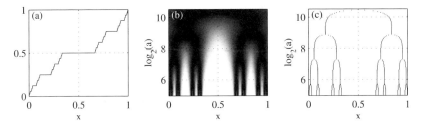

Figure 12.2. Unraveling a fractal function: (a) The devil's staircase; (b) its wavelet tranform (with the first derivative of a Gaussian); (c) the corresponding skeleton (see Section 12.6).

that live at the same time, but on different scales. This redundancy may be eliminated — this is the rationale behind the *discrete* wavelet transform that we will discuss below. In may also be exploited, in several ways. A first possibility is to obtain interesting interpolation properties, which in fact rely on the Lie group structure. We will skip the point here and refer the interested reader to [159]. What will concern us here is the observation that it must be possible to obtain the full information about the signal from a small subset of the values of the transform $S(b, a)$, for instance, a *discrete* subset. In fact, the actual (numerical) reconstruction of a signal from its wavelet transform requires the discretization of the integral in (12.28). What is the minimal sampling grid ensuring no loss of information?

The question may be answered at two levels. On the practical side, the minimal grid may be determined from the reproducing kernel K. The latter is the autocorrelation function of ψ and thus defines a correlation length in a and b. This approach certainly yields a qualitative answer, which may be made quantitative by defining a precise notion of *resolving power* [Chu, 32]. From a mathematical point of view, the answer lies in the theory of discrete *frames*, briefly described in Section 3.4 [Dau, 97].

Let us explain how frames enter naturally in the picture (the reasoning is valid for any linear transformation that yields a different representation of the signal, for instance, the windowed Fourier transform). Let $\Gamma = \{a_j, b_k, j, k \in \mathbb{Z}\}$ be a discrete lattice in the (a, b)-half-plane $\mathbb{H}$. We say that Γ yields a good discretization if an arbitrary signal $s(x)$ may be represented as a discrete superposition

$$s(x) = \sum_{j,k \in \mathbb{Z}} \langle \psi_{jk} | s \rangle \, \widetilde{\psi}_{jk}(x), \qquad (12.65)$$

where $\psi_{jk} \equiv \psi_{b_k a_j}$ and $\widetilde{\psi}_{jk}$ may be explicitly constructed from ψ_{jk}. We emphasize that (12.65) must be an *exact* representation; i.e., t no loss of information occurs as compared with the continuous reconstruction (12.28). Actually (12.65) means that the signal $s(x)$ may be replaced by the set

$\{\langle \psi_{jk}|s\rangle\}$ of its wavelet coefficients. Thus, we consider the frame operator

$$F : s \mapsto \{\langle \psi_{jk}|s\rangle\}.$$

Since $s \in L^2$, it is natural to require that the sequence of coefficients be also square integrable and that F be continuous from $L^2(\mathbb{R})$ to ℓ^2, i.e.,

$$\sum_{j,k \in \mathbb{Z}} |\langle \psi_{jk}|s\rangle|^2 \leq \mathsf{M}\|s\|^2, \quad \mathsf{M} > 0. \tag{12.66}$$

In addition, one wants the reconstruction of $s(x)$ from its coefficients to be numerically stable, that is, a small error in the coefficients implies a small error in the reconstructed signal. In particular, if the LHS of (12.66) is small, $\|s\|^2$ should also be small. Therefore, a constant $\mathsf{m} > 0$ must exist, such that

$$\mathsf{m}\|s\|^2 \leq \sum_{j,k \in \mathbb{Z}} |\langle \psi_{jk}|s\rangle|^2 \leq \mathsf{M}\|s\|^2 \tag{12.67}$$

(the lower bound indeed guarantees the numerical stability [Dau]). In other words, as indicated in Section 3.4, the set $\{\psi_{jk}\}$ constitutes a discrete *frame*, with *frame bounds* m and M.

This is precisely the point where that the basic difference arises between the *discretized* continuous wavelet transform and the discrete wavelet transform (DWT), which we will discuss briefly in Section 13.1. In the former case, the wavelet ψ is chosen *a priori* (with very few constraints, see above), and the question is whether one can find a lattice Γ such that $\{\psi_{jk}\}$ is a frame with decent frame bounds m, M. In the other approach, one usually imposes that the set $\{\psi_{jk}\}$ be an orthonormal basis and tries to construct a function ψ to that effect. The construction is rather indirect, and the resulting function is usually very complicated (often, it has a fractal behavior).

Of course, the practical question is: How does one build a good frame? In view of the discussion in Section 3.4, a "good" frame is a frame with a width as small as possible, to ensure that truncation of the expansion (3.41) yields a good approximation. Since this question is more general than the specific example of wavelets, we will postpone the discussion to Chapter 16, where we will treat in parallel various classes of CS, including wavelets.

12.6 Ridges and skeletons

Real signals are frequently very entangled and noisy, and their wavelet transform is difficult to understand. Yet, a clever exploitation of the intrinsic redundancy of the CWT is often able to bypass the difficulty. Now, reduction to a discrete subset is not the only way to exploit the redundancy. Another one, which in a sense is more intrinsic, is the restriction of the transform to its *ridges* [108], which we are going to describe now.

12.6. Ridges and skeletons

As a matter of fact, many signals are well approximated by a superposition of *spectral lines*

$$s(x) = \sum_l A_l(x) e^{i\xi_l x}, \qquad (12.68)$$

or, more generally, by a so-called *asymptotic* signal

$$s(x) = \sum_l A_l(x) e^{i\phi_l(x)}, \qquad (12.69)$$

where the amplitude $A_l(x)$ varies slowly with respect to the phase $\phi_l(x)$; i.e.,

$$\left| \frac{1}{A_l(x)} \frac{dA_l(x)}{dx} \right| \ll \left| \frac{d\phi_l(x)}{dx} \right|. \qquad (12.70)$$

Typical examples are spectra in NMR spectroscopy [108].

For a signal of this kind, the wavelet transform (12.14) in the position domain is a sum of rapidly oscillating integrals, and the essential contribution to each of them is given by the stationary points of the phase of the integrand. Assume for simplicity there is only one such point $x_s = x_s(a)$, which is defined as a solution of the equation

$$\frac{d\phi_l}{dx}(x_s) = \frac{\xi_o}{a}, \qquad (12.71)$$

where ξ_o is a basic frequency of the analyzing wavelet [typically, a Morlet wavelet (12.19)]. In fact, (12.71) means that the instantaneous frequency $d\phi_l(x)/dx$ coincides with the scaled frequency ξ_o/a of the wavelet. Then, the *ridge* of the wavelet transform is defined as the set of points (a, b) for which $x_s(a) = b$. These constitute a curve in the (a, b) half-plane, and a detailed analysis shows that, on this curve, the wavelet transform $S(b, a)$ coincides, up to a small correction, with the analytic signal $Z(b)$ associated to $s(x)$ [Lyn, 108, 161] (we recall that the analytic signal associated to the signal s is obtained by subtracting the negative frequency component of $\hat{s}$ [Lyn, Pap, Tor]; this is done by taking the Hilbert transform of s). In the general case, the wavelet transform of a signal has, of course, many ridges, each of them being essentially a line of local maxima. The set of all ridges is called the *skeleton* of the transform.

The conclusion is that the restriction of the wavelet transform $S(b, a)$ to its skeleton contains the whole information. In particular, the frequency modulation law $x^{-1} \arg\{s(x)\}$ of $s(x)$ is easily recovered from the skeleton. Thus, it is not necessary to compute the whole wavelet transform, but only its skeleton. This is of course much less costly computationally, because there are fast algorithms available. More importantly, the ridge concept is quite robust in the presence of noise, so that this technique will usually improve considerably the signal-to-noise ratio, thus, the efficiency of the wavelet analysis method. Spectacular examples may be found, for instance, in the analysis of (multi)fractal curves [Arn, 35, 50], or that of the fluctuations of the Earth's magnetic field [1].

12.7 Applications

The continuous wavelet transform has found a wide variety of applications in various branches of physics and signal processing. We will list here a representative selection; most of them, and many more, may be found in the proceedings volumes [Com, Me2, Me4], with the original references. In addition, a recent survey of physical applications is given in the volume [Be2]. In all cases, the CWT is primarily used for analyzing transient phenomena, detecting abrupt changes in a signal, or comparing it with a given pattern.

1. *Sound and acoustics*
   For historical reasons, the first applications of the CWT were in the field of acoustics. A few examples are musical synthesis, speech analysis [Mae], and modeling of the sonar system of bats and dolphins. Other examples include various problems in underwater acoustics, such as the disentangling of the different components of an underwater refracted wave and the identification of an obstacle (a submarine is a good example!).

2. *Geophysics*
   This is the origin of the method, which was designed in an empirical fashion by Morlet for analyzing the recordings of microseisms used in oil prospecting. More recently, the CWT has been applied to the analysis of various long time series of geophysical origin, e.g., in gravimetry (fluctuations of the local gravitational field), in geomagnetism (fluctuation of the Earth's magnetic field [1]), or in astronomy (fluctuations of the length of the day, variations of solar activity, measured by the sunspots, etc).

3. *Fractals, turbulence*
   As mentioned above, the CWT is an ideal tool for studying fractals, or, more generally, phenomena with particular properties under scale changes. Thus, it is quite natural that the CWT has found many applications in the analysis of (1-D and 2-D) fractals, artificial (diffusion-limited aggregates) or natural (arborescent growth phenomena). Related to these applicationsis the use of the CWT in the analysis of developed turbulence (identification of coherent structures, uncovering of hierarchical structure) [Arn, 50, 120, 121].

4. *Atomic physics*
   When an atom is hit by a short, intense laser pulse, it emits radiation

that covers a whole spectrum of harmonics of the laser frequency (experimentally, harmonics up to order 135 have been observed). This is a fast and complex physical process, which cannot be understood without a time-frequency analysis. This has been done, both with WFT and wavelets, yielding, for instance, the time profile of each individual harmonic [37, 45] and the effect of the polarization of the laser field on harmonic generation [46]. This technique leads to the controlled emission of ultrashort light pulses, in the 10^{-18}s (attosecond) range, a potentially very useful tool in many applications.

5. *Spectroscopy*
   This was one of the earliest and most successful applications, in particular, for NMR spectroscopy, for which the method proved extremely efficient in subtracting unwanted spectral lines or filtering out background noise. We will discuss this example in more detail below [42, 62].

6. *Analysis of local singularities*
   The strong point of the CWT is to detect singularities in a signal, but it yields also a fine characterization of their strengths, in particular, in the case of oscillating singularities [53, 54]

7. *Shape characterization*
   A particular case of analysis of local singularities is the determination of the shape of an object, a standard problem in image processing, for instance, in robotic vision. A novel approach [35] consists in treating the contour of the object as a complex curve in the plane and analyzing it with the 1-D CWT. The method benefits from all good properties of the wavelet transform, for instance, its robustness to noise, and looks promising for applications.

8. *Medical and biological applications*
   The CWT has been used for analyzing or monitoring various electrical or mechanical phenomena in the brain (EEG, VEP) or the heart (ECG) [Ald, Tho, 274]. It also yields good models for the auditory mechanism [100]. A recent success is the statistical analysis of correlations in DNA sequences, resolving a raging controversy among biologists [51].

9. *Industrial applications*
   Here again, the important aspect is monitoring, for instance, in detecting anomalies in the functioning of nuclear, electrical, or mechanical installations.

Before concluding this section, let us illustrate the use of the CWT by the example of NMR spectroscopy.

The physical phenomenon may be described as follows. When a sample is placed in a static magnetic field, nuclei with a magnetic moment align

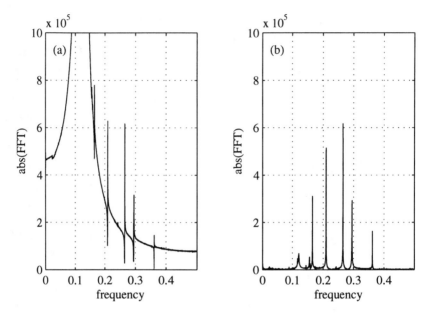

Figure 12.3. Application of the CWT in NMR spectroscopy: Subtraction of an unwanted peak. (a) The original spectrum; (b) the spectrum reconstructed after subtraction of the water peak.

along this applied field, resulting in a net magnetization. In its equilibrium state, the magnetization is static and does not induce a signal in the receiver antenna. In order to obtain information, one must first excite the nuclei with a radio frequency pulse. After such a pulse, the magnetizations precess around the static field at angular frequencies characteristic of their chemical environment and relax to their equilibrium state. This precession induces a signal in the receiver antenna. The signal to be analyzed is the Fourier transform (spectrum) of the damped response curve of the protons. It contains a large number of narrow peaks, the spectral lines, but many among them are useless, coming, for instance, from the protons of the solvent. These peaks, which may be quite big, must be subtracted, and the position of the relevant ones measured with precision. In addition, the spectra are often quite noisy and must be "cleaned" before any useful measurement can be performed.

The problem of suppression of the solvent peak is crucial in NMR spectroscopy, in particular, for the ^{1}H spectroscopy. As a consequence, various methods have been designed for achieving it efficiently, and one can distinguish two different lines. One approach is experimental, namely, one submits the sample to a particular sequence of RF pulses before the actual measurement (the so-called saturation recovery or selective inversion recovery methods [Ala, Gun]). The other one consists in processing the data

12.7. Applications   281

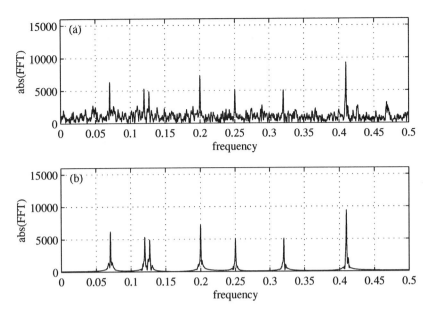

Figure 12.4. Noise filtering in spectroscopy: (a) The original spectrum; (b) The spectrum reconstructed after noise removal.

after the measurement, by various techniques of signal analysis (for instance, convolution in time-domain [201, 216] or statistical methods linked to singular value decomposition (SVD), see [80], [101], [249], and [278], and references therein). Compared with these standard methods, the wavelet technique (which resorts to the second approach) is both highly efficient and simple to implement.

A typical analysis of NMR spectra with help of wavelets is given in figures 12.3 and 12.4 [62, 108, 161]. The first one is an example of peak subtraction. The original spectrum (left) exhibits a huge parasite peak, due to the protons of the solvent (water), which masks to a large extent the interesting structures. The analysis consists in isolating this peak, on the CWT, subtracting it from the spectrum and reconstructing the remaining part. The result (right) is a spectrum in which all fine details are now clearly visible and have not been perturbed by the removal of the large peak. Indeed, the prominent structures appear at exactly the same place on the frequency (horizontal) axis in both pictures. The reason for the remarkable efficiency of the method in this case is that the huge line and the rest of the spectrum live at different scales; hence they are decoupled in the CWT ("unfolding") and can be readily separated with very little distortion.

Figure 12.4 is an example of noise filtering. The original signal (top) consists of a number of damped sinusoids enbedded in noise. The dominant peaks are localized with help of the CWT ridge algorithm, the remnant of the spectrum is subtracted, and the filtered spectrum is reconstructed, using one of the reconstruction formulas. The result (bottom) is quite spectacular and compares well with standard methods of noise suppression (signal filtering). This method has been applied successfully to real-life NMR spectra.

We may also remark that, in this problem as well as in the analysis of harmonic generation in laser–atom interaction, both the wavelet and the WFT (Gabor) analyses give very good, and in fact comparable, results. The lesson of those studies [37, 62] is that it is more important to choose the adequate range of parameters than to select a particular wavelet, or even a particular time-frequency method, wavelets, or WFT.

13
Discrete Wavelet Transforms

13.1 The discrete time or dyadic WT

As we have seen in Section 12.5, the discretization of the CWT leads, among other things, to the theory of frames. For many practical purposes of signal processing, a tight frame is almost as good as an orthonormal basis. Actually, if one stays with the standard wavelets, as we have done so far, one cannot do better, since these wavelets do not generate any orthonormal basis (like the usual coherent states). There are cases, however, in which an orthonormal basis is really required. A typical example is data compression, which is performed (in the simplest case) by removing all wavelet expansion coefficients below a fixed threshhold. In order to not introduce any bias in this operation, the coefficients have to be as decorrelated as possible, and, of course, an orthonormal basis is ideal in this respect.

Fortunately, it turns out that one can design wavelets that do generate orthonormal bases, and this leads to the *discrete time WT* (DWT). The key step was the discovery that almost all examples of orthonormal bases of wavelets can be associated to a multiresolution analysis [Me3, 212] and, furthermore, that the whole construction may be transcripted into the language of the so-called quadrature mirror filters (QMF). We describe briefly these two developments in turn. Further information may be found, for instance, in [Dau], [Rus], and [Vet].

13.1.1 Multiresolution analysis and orthonormal wavelet bases

A *multiresolution* or *multiscale analysis* of $L^2(\mathbb{R})$ is an increasing sequence of closed subspaces

$$\ldots \subset V_{-2} \subset V_{-1} \subset V_0 \subset V_1 \subset V_2 \subset \ldots, \tag{13.1}$$

with $\bigcup_{j\in\mathbb{Z}} V_j$ dense in $L^2(\mathbb{R})$ and $\bigcap_{j\in\mathbb{Z}} V_j = \{0\}$, such that

1. $f(x) \in V_j \Leftrightarrow f(2x) \in V_{j+1}$

2. There exists a function $\phi \in V_0$, called a *scaling* function, such that $\{\phi_{0,k} \mid k \in \mathbb{Z}\}$ is an orthonormal basis of V_0, where $\phi_{0,k}(x) = \phi(x-k)$.

Combining 1 and 2, one gets an orthonormal basis of V_j, $\{\phi_{j,k}(x) \equiv 2^{j/2}\phi(2^j x - k), k \in \mathbb{Z}\}$.

Each V_j can be interpreted as an approximation space: The approximation of $f \in L^2(\mathbb{R})$ at the resolution 2^j is defined by its projection onto V_j. The additional details needed for increasing the resolution from 2^j to 2^{j+1} are given by the projection of f onto the orthogonal complement W_j of V_j in V_{j+1}

$$V_j \oplus W_j = V_{j+1}, \tag{13.2}$$

and we have:

$$L^2(\mathbb{R}) = \bigoplus_{j\in\mathbb{Z}} W_j = V_{j_o} \oplus \left(\bigoplus_{j=j_o}^{\infty} W_j \right), \tag{13.3}$$

where j_o is an arbitrary lowest resolution level. Then, the theory asserts the existence of a function ψ, sometimes called the *mother* wavelet, explicitly computable from ϕ, such that $\{\psi_{j,k}(x) \equiv 2^{j/2}\psi(2^j x - k), j, k \in \mathbb{Z}\}$ constitutes an orthonormal basis of $L^2(\mathbb{R})$: These are the orthonormal *wavelets*. Like the $\{\phi_{j,k}\}$ above, they are indexed by points of the dyadic lattice, described in Section 12.5.

The construction of ψ proceeds as follows. First, the inclusion $V_0 \subset V_1$ yields the relation (called the scaling or refining equation, or a two-scale relation):

$$\phi(x) = \sqrt{2} \sum_{n\in\mathbb{Z}} h_n \phi(2x - n), \quad h_n = \langle \phi_{1,n} | \phi \rangle, \quad \sum_{n\in\mathbb{Z}} |h_n|^2 = 1. \tag{13.4}$$

Taking Fourier transforms, this gives

$$\hat{\phi}(2\xi) = m_0(\xi) \hat{\phi}(\xi), \quad \text{with } m_0(\xi) = \frac{1}{\sqrt{2}} \sum_{n\in\mathbb{Z}} h_n e^{-in\xi}. \tag{13.5}$$

Thus, m_0 is a 2π-periodic function, and it satisfies the relation

$$|m_0(\xi)|^2 + |m_0(\xi + \pi)|^2 = 1, \quad \text{a.e.} \tag{13.6}$$

Iterating (13.5), one gets the scaling function as the (convergent!) infinite product

$$\widehat{\phi}(\xi) = (2\pi)^{-1/2} \prod_{j=1}^{\infty} m_0(2^{-j}\xi). \tag{13.7}$$

By the same argument as in (13.4) and (13.5), a function $\psi \in W_0 \subset V_1$ may be defined by the relation

$$\widehat{\psi}(2\xi) = m_1(\xi)\,\widehat{\phi}(\xi), \tag{13.8}$$

where m_1 is another 2π-periodic function. As a consequence of the relation $V_0 \oplus W_0 = V_1$ and the orthonormality of the functions $\{\phi_{j,k}\}$, the functions m_0, m_1 must satisfy the identity

$$m_1(\xi)\,\overline{m_0(\xi)} + m_1(\xi + \pi)\,\overline{m_0(\xi + \pi)} = 0, \quad \text{a.e.} \tag{13.9}$$

The simplest solution is to put

$$m_1(\xi) = e^{i\xi}\,\overline{m_0(\xi + \pi)}, \tag{13.10}$$

which implies, in particular, $|m_0(\xi)|^2 + |m_1(\xi)|^2 = 1$, a.e. Then, one obtains

$$\psi(x) = \sqrt{2} \sum_{n \in \mathbb{Z}} (-1)^{n-1} h_{-n-1} \phi(2x - n), \tag{13.11}$$

or, equivalently, using the freedom in the choice of the solution to (13.9),

$$\psi(x) = \sqrt{2} \sum_{n \in \mathbb{Z}} (-1)^n h_{-n+1} \phi(2x - n). \tag{13.12}$$

Then, one proves that this function indeed generates an orthonormal basis with all required properties. Various additional conditions may then be imposed on the function ψ (hence, on the basis wavelets): arbitrary regularity, several vanishing moments (in any case, ψ has always mean zero, as in the CWT), symmetry, fast decrease at infinity, even compact support (see below).

13.1.2 Connection with filters and the subband coding scheme

Actually, the discussion above means that we have translated the multiresolution structure into the language of digital filters (by a filter, we mean a multiplication operator in frequency space or a linear convolution in the time variable). For instance, $m_0(\xi)$ is a filter, with Fourier coefficients h_n, $m_1(\xi)$ is another one, and $\{m_0, m_1\}$ are called quadrature mirror filters or QMF whenever (13.9) is satisfied. Then, the various restrictions imposed on ψ translate into suitable constraints on the filter coefficients h_n. For example, we will see below that ψ has compact support if only finitely many h_n differ from zero (one then speaks of a finite impulse response or FIR filter). The advantage of turning to that filter language is that efficient and well-understood algorithms are available, very similar to the so-called Laplacian pyramid familiar in signal processing [Vet].

Indeed, let $f \in V_0$. Then, using the decomposition $V_0 = V_{-1} \oplus W_{-1}$, one may write:

$$\begin{aligned} f &= \sum_{k \in \mathbb{Z}} c_{0,k}\, \phi_{0,k} \\ &= \sum_{k \in \mathbb{Z}} c_{1,k}\, \phi_{-1,k} + \sum_{k \in \mathbb{Z}} d_{1,k}\, \psi_{-1,k}. \end{aligned}$$

From the orthonormality of the bases, one gets immediately:

$$c_{1,k} = \sum_{n \in \mathbb{Z}} \overline{h_{n-2k}}\, c_{0,n}, \quad d_{1,k} = \sum_{n \in \mathbb{Z}} \overline{g_{n-2k}}\, c_{0,n}, \tag{13.13}$$

with $h_n = \langle \phi_{0,n} | \phi_{-1,0} \rangle = \langle \phi_{1,n} | \phi_{0,0} \rangle$ (of course, $\phi_{0,0} = \phi$), $g_n = \langle \phi_{0,n} | \psi_{-1,0} \rangle$. (In practice, the signals are real-valued functions, so that the coefficients h_n, g_n are real too.) Thus, the sequences $c_1 = (c_{1,k}), d_1 = (d_{1,k})$ are uniquely determined from c_0 by application of the filters h and g, acting by convolution and decimation. The filter h is a low-pass filter, g is a high-pass filter, and they allow perfect reconstruction, namely (simply by taking adjoints),

$$c_{0,n} = \sum_{k \in \mathbb{Z}} \left(h_{n-2k} c_{1,k} + g_{n-2k} d_{1,k} \right). \tag{13.14}$$

In addition, it turns out that h and g are conjugate quadrature filters (CQF), a special case of QMF characterized by the relation $g_n = (-1)^n h_{-n+1}$, that is, $h \equiv m_0$ and $g \equiv m_1$, as defined in (13.5) and (13.10), respectively. The interpretation of the relations (13.13) and (13.14) is the following. If c_0 describes the signal at a given resolution, then c_1 corresponds to its approximation at half the resolution and d_1 to the additional details needed for recovering the initial resolution.

Now, the key point is that this operation may be iterated. The next approximation c_2 and the corresponding details d_2 are obtained from the sequence c_1 by the *same* filters h and g, by relations analogous to (13.13). This procedure of splitting repeatedly a given signal c_n into its low-frequency component c_{n+1} and its high-frequency component d_{n+1} is a standard technique in signal processing, known as the *subband coding scheme* [Vet]. In the case in which the filter h (hence, also for g) has finite length (finite number of nonzero coefficients h_n), one may visualize the resemblance of this algorithm with the pyramidal one. Let L be the length of the filter h. Then, at level j, one has:

$$c_{j,k} = \sum_{n \in \mathbb{Z}} \overline{h_{n-2k}}\, c_{j-1,n}, \quad d_{j,k} = \sum_{n \in \mathbb{Z}} \overline{g_{n-2k}}\, c_{j-1,n}; \tag{13.15}$$

that is, every $c_{2,k}$ depends on L coefficients $c_{1,k}$, each of which in turn depends on L coefficients $c_{0,k}$, and so on, and similarly for the detail coefficients $d_{j,k}$. This indeed leads graphically to a pyramid of nonvanishing coefficients. In other words, at each resolution j, the wavelet coefficients may be obtained in terms of those at lower resolution by adding finer and finer details, and this is obtained by repeated application of the two filters h, corresponding to the scaling function ϕ (low pass), and g, corresponding to the wavelet ψ (high pass).

The construction made so far shows the validity of the implications: multiresolution analysis $\Leftrightarrow$ orthonormal wavelet basis $\Rightarrow$ conjugate quadrature filters. The crucial point is the opposite implication: Under what conditions does a pair of CQF filters h, g generate an orthonormal wavelet basis? This question was answered by Daubechies [Dau], in the form of a regularity condition (implying some vanishing moments). In addition, she showed that, if one takes for h a filter with a finite number of nonzero coefficients h_n, then one generates an orthonormal basis of wavelets with *compact support*.

This, however, is not sufficient to guarantee the efficiency of the method, because the rapidity of the algorithm depends crucially on the *length* of the filters involved, as in general for pyramidal schemes. This remark opens the way to various improvements, to which we will come back in Section 13.2.

On the other hand, it is clear from the definition of a multiresolution analysis that the general translation covariance is lost. The dyadic lattice is invariant only row by row, and the row $j = j_o$ is invariant under discrete translations by $2^{-j_o}k$, $k \in \mathbb{Z}$. This feature creates many difficulties in applications.

13.1.3 Generalizations

As we just saw, appropriate (CQF) filters generate orthonormal wavelet bases. This result turns out to be too rigid, however, and various generalizations have been proposed.

1. Biorthogonal wavelet bases
As we mentioned in Section 12.2, the wavelet used for reconstruction in the continuous wavelet tranform need not be the same as that used for decomposition, the two have only to satisfy a cross-compatibility condition. The same idea in the discrete case leads to biorthogonal bases; i.e., one has two hierarchies of approximation spaces, V_j and $\hat{V}_j$, with cross-orthogonality relations. This gives a better control, for instance, on the regularity or decrease properties of the wavelets [90].

2. Wavelet packets and best basis algorithm
The construction of orthonormal wavelet bases leads to a special subband coding scheme, rather asymmetrical: Each sequence c_j gets further decomposed into c_{j+1} and d_{j+1}, whereas the detail sequence d_j is left unmodified. Thus, more flexible subband schemes have been considered, called *wavelet packets* in which both subspaces V_{j-1} and W_{j-1} are decomposed at each

step [Me3, Wic, 92]. They provide rich libraries of orthonormal bases and strategies for determining the optimal basis in a given situation.

3. The lifting scheme: Second generation wavelets

One can go further and abandon the regular dyadic scheme and the Fourier transform altogether. Using the "lifting scheme" of Sweldens [273], one obtains the so-called *second-generation wavelets*, which are essentially custom-designed for any given problem. This approach uses the fact that, in the biorthogonal scheme, a given wavelet fixes its biorthogonal partner only up to a rather arbitrary function. Then, one starts from a very simple wavelet and gradually obtains the needed one by choosing successive, appropriate biorthogonal partners.

4. Integer wavelet transforms

In their standard numerical implementation, the classical (discrete) WT converts floating point numbers into floating point numbers. In many applications (data transmission from satellites, multimedia), however, the input data consist of integer values only and one cannot afford to lose information: Only lossless compression schemes are allowed. Recent developments have produced new methods that allow one to perform all calculations in integer arithmetic [84].

13.1.4 Applications

Finally, as far as applications are concerned, wavelet (bi)orthogonal bases and wavelet packets may be used for most problems previously treated with the continuous wavelet tranform (for precise references, see, for instance, the conference volumes [Me2] and [Me4]. Their main virtue is a remarkable efficiency in data compression. For achieving useful rates, one has to determine which information is really essential and which may be discarded with acceptable loss of signal quality. High compression rates have been achieved with the discrete wavelet transform, especially when wavelets are combined with vector coding. Significant results have also been obtained in speech analysis (signal segmentation, analysis–synthesis, recognition).

Another field in which wavelet bases have brought spectacular progress is numerical analysis. Very fast algorithms may be designed for matrix multiplication, and this has opened many doors.

Similarly, wavelet bases have been applied successfully to a whole class of hard problems in pure mathematics, such as the construction of universal unconditional bases for many function spaces (L^p, Sobolev, Besov, etc.), efficient resolution of partial differential equations, analysis of singular integral operators (Calderón–Zygmund operators), etc.

Finally, expansion into orthogonal or biorthogonal wavelet bases has brought a whole new perspective in *ab initio* structure calculations in atomic and in solid-state physics. For a review of these recent developments, see [Be2, Chap. 8] and [49].

13.2 Towards a fast CWT: Continuous wavelet packets

Besides the full discretization described in Section 12.5, and the discrete WT just discussed, an intermediate procedure exists, introduced in [116], under the name of infinitesimal multiresolution analysis. It consists in discretizing the scale variable alone, on an arbitrary sequence of values (not necessarily powers of a fixed ratio). This leads to fast algorithms that could put the continuous wavelet transform on the same footing as the discrete wavelet transform in terms of speed and efficiency, by extending the advantages of the latter to cases in which no exact QMF is available. Let us sketch the method. Further details may be found in [Tor].

Instead of the standard L^2-normalization used so far, it is more convenient to choose the L^1-normalization, namely, to use $\widetilde{\psi}_{b,a}(x) = a^{-1}\psi\left(a^{-1}(x-b)\right)$. Then, given a wavelet ψ, normalized to $c_\psi = 1$, one lumps together all low-frequency components in a scaling function

$$\Phi(x) = \int_1^\infty \psi\left(\frac{x}{a}\right) \frac{da}{a^2} = \frac{1}{x}\int_0^x \psi(s)\,ds, \qquad \widehat{\Phi}(\xi) = \int_1^\infty \widehat{\psi}(a\xi)\,\frac{da}{a}, \tag{13.16}$$

and introduces the integrated wavelet

$$\Psi(x) = \int_{1/2}^1 \psi\left(\frac{x}{a}\right) \frac{da}{a^2} = \frac{1}{x}\int_x^{2x} \psi(s)\,ds, \qquad \widehat{\Psi}(\xi) = \int_{1/2}^1 \widehat{\psi}(a\xi)\,\frac{da}{a}. \tag{13.17}$$

These functions satisfy two-scale relations:

$$\Psi(x) = 2\Phi(2x) - \Phi(x), \qquad \widehat{\Psi}(\xi) = \widehat{\Phi}(\xi/2) - \widehat{\Phi}(\xi). \tag{13.18}$$

Next, one chooses a regular grid, as opposed to the dyadic one used in the discrete case, namely,

$$\Phi_x^j \equiv \widetilde{\Phi}_{x,2^{-j}} = 2^j \Phi(2^j(\cdot - x)), \qquad \Psi_x^j = 2^j \Psi(2^j(\cdot - x)). \tag{13.19}$$

Although the resulting transform yields a redundant signal representation, it has the great advantage over the conventional DWT of maintaining (integer) translation covariance. Then, exactly as in (13.3), one gets a discrete reconstruction formula:

$$s(x) = \langle \Phi_x^{j_o}|s\rangle + \sum_{j=j_o}^\infty \langle \Psi_x^j|s\rangle. \tag{13.20}$$

Then, assume there exist two functions μ_0, μ_1 satisfying the following relations, analogous to (13.5), (13.8):

$$\widehat{\Phi}(2\xi) = \mu_0(\xi)\widehat{\Phi}(\xi), \qquad \widehat{\Psi}(2\xi) = \mu_1(\xi)\widehat{\Phi}(\xi), \quad \text{a.e.} \tag{13.21}$$

These functions are *not* necessarily 2π-periodic. Since using the regular grid means sampling $\Phi(x)$ at unit rate, however, we have to assume that the function $\widehat{\Phi}$ is essentially supported in $[-\pi,\pi]$. Therefore, since the functions μ_0, μ_1 always appear in a product with $\widehat{\Phi}$, according to the relations (13.21), it is reasonable to approximate the functions μ_0, μ_1 in a neighborhood of zero by 2π-periodic functions m_0, m_1. In fact [Tor], there exists a unique pair m_0, m_1 that minimizes the quantities

$$\nu(\mu_i, m_i) = \left[\int_{\mathbb{R}} |(\mu_i(\xi) - m_i(\xi))\widehat{\Phi}(\xi)|^2\right]^{1/2}, \quad i = 0, 1,$$

namely,

$$m_0(\xi) = \frac{\sum_{k \in \mathbb{Z}} \overline{\widehat{\Phi}(\xi + 2k\pi)}\, \widehat{\Phi}(2\xi + 4k\pi)}{\sum_{k \in \mathbb{Z}} |\widehat{\Phi}(\xi + 2k\pi)|^2}, \quad (13.22)$$

$$m_1(\xi) = \frac{\sum_{k \in \mathbb{Z}} \overline{\widehat{\Phi}(\xi + 2k\pi)}\, \widehat{\Psi}(2\xi + 4k\pi)}{\sum_{k \in \mathbb{Z}} |\widehat{\Phi}(\xi + 2k\pi)|^2}. \quad (13.23)$$

These approximate filters m_0, m_1, which are called *pseudo-QMF*, satisfy the identity $m_0(\xi) + m_1(\xi) = 1$, a.e.

More flexibility is obtained if one subdivides the scale interval $[1/2, 1]$ into n subbands, by $a_o = 1/2 < a_1 < \ldots < a_n = 1$. In that case, one ends up with one scaling function $\Phi(t)$ and n integrated wavelets $\Psi_i(x)$, $i = 0, \ldots n - 1$, corresponding to integration from a_{i-1} to a_i. An additional improvement consists in periodizing the signal and computing filters m_0, m_1 of the same length as the signal. The resulting pyramidal algorithm has a complexity equal to one half of the traditional fast Fourier transform value $\mathcal{O}(N \log_2^2 N)$. Thus, one obtains a very fast implementation of the continuous wavelet transform, truly competitive with the discrete wavelet transform. The preliminary applications of this algorithm look very promising, both in 1-D and in 2-D [Van, 147].

13.3 Wavelets on the finite field $\mathbb{Z}_p$ (♣)

The construction of the discrete WT described in Section 13.1, as well as its biorthogonal generalization [Dau] are incompatible with the group-theoretical structure. The set of points $\{(k2^j, 2^j), j, k \in \mathbb{Z}\} \subset G_+$ defining the above basis is *not* a subgroup of G_+. (But it is a subgroup of the dyadic affine subgroup $\{(k2^j, 2^l), k, j, l \in \mathbb{Z}\} \subset G_+$.) A recent construction, however, goes a long way towards a group-theoretical understanding of the DWT, namely, wavelets on the finite field $\mathbb{Z}_p$ [125] (note that the field $\mathbb{Z}_p$ is sometimes denoted by $\mathbb{F}_p$ in the mathematical literature [Lan]). Since this is an immediate and interesting application of the CS formalism, we sketch it here.

Instead of $\mathbb{R}$, one considers the set $\mathbb{Z}_p = \mathbb{Z}/p\mathbb{Z}$ of remainders modulo p. When p is a prime number, $\mathbb{Z}_p$ is a field, with addition and multiplication

13.3. Wavelets on the finite field $\mathbb{Z}_p$ (♣)

modulo p; the elements of $\mathbb{Z}_p$ are then simply $\{0, 1, \ldots, p-1\}$. The affine transformations of $\mathbb{Z}_p$, viz. $n \mapsto an + b, a \in \mathbb{Z}_p^*, b \in \mathbb{Z}_p$, are well defined and form a group called the affine group G_p, with the usual group law of G_{aff} (taken modulo p). The group G_p has a natural unitary representation on $\ell^2(\mathbb{Z}_p)$:

$$(U(b,a)f)(n) = f_{ba}(n) \equiv f(\frac{n-b}{a}). \tag{13.24}$$

Let E denote the closed subspace of $\ell^2(\mathbb{Z}_p)$ defined by

$$E = \{f \in \ell^2(\mathbb{Z}_p) : \sum_{n=0}^{p-1} f(n) = 0\}, \tag{13.25}$$

so that $\ell^2(\mathbb{Z}_p) = \mathbb{C} \oplus E$. This decomposition reduces the representation U, and the restriction of U to E is irreducible (square integrability is automatic here, since G_p is finite). Accordingly, a wavelet transform may be defined, as usual, as the linear map $T_\psi : \ell^2(\mathbb{Z}_p) \to \ell^2(G_p)$ given by

$$(T_\psi f)(b,a) = \langle \psi_{ba} | f \rangle = \sum_{n=0}^{p-1} \overline{\psi_{ba}(n)} f(n). \tag{13.26}$$

Noting that $\psi_{ba}(n) = (\Delta_a f)(n-b)$, where $(\Delta_a f)(n) = f(a^{-1}n)$ is the discrete dilation operator acting on $\ell^2(\mathbb{Z}_p)$, we may also write this wavelet transform as

$$(T_\psi f)(b,a) = (\Delta_a \tilde{\psi} * f)(b), \tag{13.27}$$

where $\tilde{\psi}(n) \equiv \overline{\psi(-n)}$.

The map T_ψ is an isometry from E to $\ell^2(G_p)$, up to the constant $c_\psi = p\|\psi\|^2$, and, thus, it yields an inversion formula:

$$f(n) = \frac{1}{c_\psi} \sum_{(b,a) \in G_p} (T_\psi f)(b,a) \psi_{ba}(n), \quad f \in E. \tag{13.28}$$

A similar result exists for the full space $\ell^2(\mathbb{Z}_p)$.

When applied to explicit signals, this discrete WT leads to efficient decompositions and reconstructions. More interesting — and surprising — is the result of [125], according to which a discrete signal over $\mathbb{Z}_p$ may be interpreted as a sampled version of a continuous one. This interpretation requires the introduction of deformed dilations ("pseudodilations"), and these lead automatically to an algorithmic structure very similar to the multiresolution structure (the so-called "algorithme à trous" [Com]). Similar constructions have been performed for wavelets on an arbitrary finite field $\mathbb{F}$ [183], and on $\mathbb{Z}$ as well [40].

The key word here is "sampling", which transforms a continuous signal into a discrete one. This brings us to the general discretization problem.

13.4 Algebraic wavelets

A different way of generalizing the discrete WT consists in replacing the usual natural numbers, and the dyadic numeration underlying the multiresolution approach, by another system of numeration. A simple, though unusual, example is based on the golden mean $\tau = \frac{1}{2}(1+\sqrt{5})$, and we shall describe it in some detail in Section 13.4.1, following [138]. Two interesting aspects emerge. One is a construction that is a genuine generalization of the standard multiresolution and yields a corresponding orthonormal wavelet basis, the so-called τ-Haar basis. The other is the occurrence of a quasiperiodic structure at each multiresolution level, instead of the usual lattice structure. Finally, more general cases will be indicated in Section 13.4.2, namely, wavelet systems based on particular Pisot numbers, first studied in [139]. Three justifications can be given for such a lengthy treatment. First, it is an instructive pedagogical exercise. Next, we are going to construct the τ-equivalent of the Haar basis, the simplest and oldest orthonormal wavelet basis. Finally, the perspective of using Pisot numbers for generalizing multiresolution analysis is new.

13.4.1 τ-wavelets of Haar on the line (♣)

In this section, we will construct the Haar basis obtained by using as a scaling factor the irrational number $\tau = \frac{1}{2}(1+\sqrt{5})$, instead of the usual factor of two [138]. The algebraic nature of τ, based on the equation $\tau^2 = \tau + 1$, entails the τ-adic property

$$\frac{1}{\tau^j} = \frac{1}{\tau^{j+1}} + \frac{1}{\tau^{j+2}}, \; j \in \mathbb{Z}. \tag{13.29}$$

This equation provides a subdivision of the unit interval into two parts

$$A = [0,1] = [0, \frac{1}{\tau}] \cup [\frac{1}{\tau}, 1]. \tag{13.30}$$

Eq.(13.30) is the starting point of an iterative sequence of subdivisions of A into intervals of the type

$$A_{j,b} = \left[\frac{b}{\tau^j}, \frac{b+1}{\tau^j}\right], \; j \in \mathbb{N}, A_{0,0} = A, \tag{13.31}$$

$$A_{j,b} = A_{j+1,\tau b} \cup A_{j+2,\tau^2 b + \tau}, \tag{13.32}$$

where b is a τ-*integer* satisfying some boundedness condition [138]. Let us first explain the concept of a τ-integer. It is derived from the system of numeration based on the irrational number τ [73, 256]. Each nonnegative real number x can be expanded in powers of τ with power coefficients equal to 0 or 1:

$$x = \xi_j \tau^j + \xi_{j-1} \tau^{j-1} + \ldots + \xi_l \tau^l + \ldots, \qquad (13.33)$$

with

$$\xi_l \in \{0, 1\}, \; \xi_l \xi_{l-1} = 0. \qquad (13.34)$$

The second condition in (13.34) means that no partial sum like $\tau^l + \tau^{l-1}$ occurs in (13.33), since it equals τ^{l+1}. The coefficients ξ_l are computed with the aid of the so-called Rényi algorithm. First, the exponent j in (13.33) is the highest integer, such that

$$\tau^j \leqslant x < \tau^{j+1}, \qquad (13.35)$$

and so $\xi_j = [x/\tau^j] =$ integer part of $x/\tau^j = 1$ Then, we put $r_j = \{x/\tau^j\}$ = fractional part of x/τ^j, and the other digits ξ_l, $l < j$, are determined recursively from (ξ_j, r_j):

$$\xi_l = [\tau r_{l+1}], \; r_l = \{\tau r_{l+1}\}. \qquad (13.36)$$

Equivalently, we have

$$\xi_{j-l-1} = [\tau T_\tau^l (r_j)], \qquad (13.37)$$

where, for any $y \in [0, 1]$,

$$T_\tau y \equiv (\tau y) \bmod 1. \qquad (13.38)$$

In this system of numeration based on τ, the set of nonnegative τ-integers is the (lexicographically ordered) strictly increasing sequence of real numbers having only nonnegative powers of τ in their τ-expansion:

$$\mathbb{Z}_\tau^+ = \{0, 1, \tau, \tau^2, \tau^2 + 1, \tau^3, \tau^3 + 1, \tau^3 + \tau, \tau^4, \ldots\}. \qquad (13.39)$$

The set $\mathbb{Z}_\tau$ of τ-integers is then defined by

$$\mathbb{Z}_\tau = \mathbb{Z}_\tau^+ \cup (-\mathbb{Z}_\tau^+). \qquad (13.40)$$

The numbers in $\mathbb{Z}_\tau^+$ are the nodes of a quasiperiodic chain, called the Fibonacci tiling of the positive real line with two types of tiles, the long one with length 1 and the short one with length $1/\tau$.

$\mathbb{Z}_\tau^+$ can also be obtained by an inductive procedure. Let us introduce the increasing sequence of finite sets

$$B_N = \{x \in \mathbb{Z}_\tau^+ \mid 0 \leqslant x < \tau^N\}, \; N \in \mathbb{N}. \qquad (13.41)$$

We then have the recurrence formulas:

$$B_N = \tau B_{N-1} \cup (1 + \tau^2 B_{N-2}), \qquad (13.42)$$
$$B_N = B_{N-1} \cup (\tau^{N-1} + B_{N-2}). \qquad (13.43)$$

We can now express the boundedness condition on b in (13.31) in a precise form as follows.

The τ-adic interval $A_{j,b} = \left[\frac{b}{\tau^j}, \frac{b+1}{\tau^j}\right]$ appears at a certain step of the subdivision process (13.32), starting with (13.30), if b satisfies the condition

$$b \in \tau B_{j-1}, \tag{13.44}$$

which means the following
1. b is an "even" positive τ-integer, i.e. $b \in \tau \mathbb{Z}_\tau^+$;
2. $0 \leq b \leq \tau^j - 1$. $\tag{13.45}$

We show in Figure 13.1 (following [138]) the first steps of this τ-adic subdivision of the unit interval. Note the occurrence of the sequence of Fibonacci numbers,

$$F_j = F_{j-1} + F_{j-2}, \ F_0 = 1, \ F_1 = 1, \tag{13.46}$$

in the counting of intervals at given j:

$$F_j = \#\{\tau B_{j-1}\} = \#\{A_{j,b}, \ b \in \tau B_{j-1}\}. \tag{13.47}$$

Also, note that the jth subdivision step gives rise to the following a.e. partition of the interval A:

$$A = \bigcup_{j \leq k \leq 2j} \bigcup_{b \in E_{j,k}} A_{k,b}. \tag{13.48}$$

Here, the sets $E_{j,k} \subset \tau B_{k-1}$ are defined recursively by

$$E_{j,k} = E_{j-1,k-1} \cup (\tau^{k-1} + E_{j-1,k-2}). \tag{13.49}$$

Another approach to this subdivision procedure consists in using affine semigroup actions. It will give us the opportunity to bridge the gap between the set formalism examplified by (13.48) and the wavelet construction. We denote again by (b, a) the generic element of the affine group G_{aff} of the line described in Section 12.2, with action on $\mathbb{R}$ given by $(b, a)x = ax + b$. It is then clear that the generic interval $A_{j,b}$ in (13.31) is the result of an affine action on the original A:

$$A_{j,b} = (b\tau^{-j}, \tau^{-j}) A. \tag{13.50}$$

Now, the subdivision (13.30) of the original A into two subintervals involves two basic affine elements,

$$A = \left(0, \frac{1}{\tau}\right) A \cup \left(\frac{1}{\tau}, \frac{1}{\tau^2}\right) A. \tag{13.51}$$

The first one, $\delta_0 \equiv (0, 1/\tau)$, is a contraction with ratio $1/\tau$ from the origin, whereas the second one, $\delta_1 \equiv (1/\tau, 1/\tau^2)$, is a $1/\tau^2$ contraction from the upper bound 1. Therefore, the generic element $(b\tau^{-j}, \tau^{-j})$ in (13.50) can

13.4. Algebraic wavelets 295

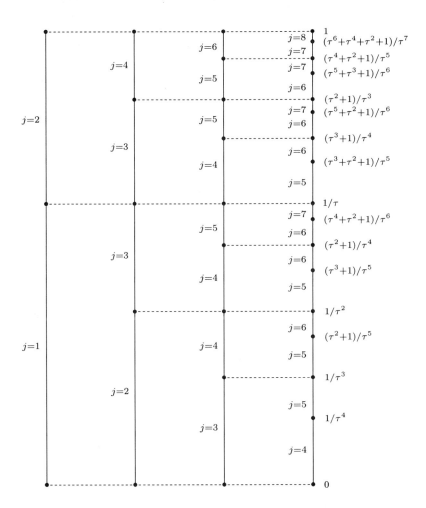

Figure 13.1. The first four steps of the τ-subdivision of the interval $[0, 1]$.

be viewed as a finite product of factors picked in the set $\{\delta_0, \delta_1\}$:

$$\left(\frac{b}{\tau^j}, \frac{1}{\tau^j}\right) = \delta_{\epsilon_n} \delta_{\epsilon_{n-1}} \ldots \delta_{\epsilon_1}, \quad \text{with } \epsilon_l \in \{0, 1\}. \tag{13.52}$$

Here, we see how a semigroup structure emerges from our construction. We shall call Γ the semigroup freely generated by the identity and the two contractions:

$$\Gamma = \{\delta_{\epsilon_n} \delta_{\epsilon_{n-1}} \ldots \delta_{\epsilon_1}, \ \epsilon_l \in \{0, 1\}, \ \delta_0 = \left(0, \frac{1}{\tau}\right), \ \delta_1 = \left(\frac{1}{\tau}, \frac{1}{\tau^2}\right)\}. \tag{13.53}$$

The generic element of Γ is easily computed:

$$\delta_{\epsilon_n} \delta_{\epsilon_{n-1}} \ldots \delta_{\epsilon_1} = \left(\frac{1}{\tau} \sum_{l=1}^{n} \epsilon_l \left(\frac{1}{\tau}\right)^{\sum_{k=l+1}^{n} 2^{\epsilon_k}}, \left(\frac{1}{\tau}\right)^{\sum_{l=1}^{n} 2^{\epsilon_l}}\right). \tag{13.54}$$

Comparing with (13.52), we obtain

$$j = \sum_{l=1}^{n} 2^{\epsilon_l}, \quad b = \sum_{l=1}^{n} \epsilon_l \, \tau^{(\sum_{k=1}^{l} 2^{\epsilon_k} - 1)}, \tag{13.55}$$

and we see how any element $b \in \tau B_{j-1}$ can also be written in a form that is not its τ-expansion of the type (13.33), but nevertheless involves only 0 and 1 as coefficients of powers of τ.

We now come to the explicit construction of the τ-Haar basis in $L^2[0, 1]$. For that purpose, we need a "mother" wavelet, called here a τ-Haar wavelet.

Definition 13.4.1: *The τ-Haar wavelet is the function on the real line $\mathbb{R}$ defined as*

$$h^\tau(x) = \begin{cases} \tau^{-1/2}, & \text{for } 0 \leq x \leq 1/\tau, \\ -\tau^{1/2}, & \text{for } 1/\tau < x \leq 1, \\ 0, & \text{otherwise.} \end{cases} \tag{13.56}$$

One can verify directly that

$$\int_0^1 (h^\tau(x))^2 \, dx = \frac{1}{\tau^2} + \tau \left(1 - \frac{1}{\tau}\right) = 1 \tag{13.57}$$

and that

$$\int_0^1 h^\tau(x) \, dx = \tau^{-1/2} \frac{1}{\tau} - \tau^{1/2} \frac{1}{\tau^2} = 0. \tag{13.58}$$

Thus, h^τ is orthogonal to $\chi_{[0,1]} \equiv \chi_A$, the characteristic function of the unit interval. Now, Definition 13.4.1 can be advantageously recast into a (semi)group representation language. As in Section 12.2, we consider the natural unitary representation of G_{aff} by operators acting on $L^2(\mathbb{R})$:

$$(U(g)f)(x) = a^{-1/2} f(g^{-1}x), \quad g = (b, a), \ a > 0, b \in \mathbb{R},$$

$$= a^{-1/2} f\left(\frac{x - b}{a}\right). \tag{13.59}$$

Then, we have (almost everywhere)
$$\begin{aligned} h^\tau(x) &= \tau^{-1/2}\chi_A(\tau x) - \tau^{1/2}\chi_A(\tau^2 x - \tau) \\ &= \left(\frac{1}{\tau}U(\delta_0) - \frac{1}{\tau^{1/2}}U(\delta_1)\right)\chi_A(x), \end{aligned} \qquad (13.60)$$

with the notations of (13.53) and (13.59).

We can now assert the following theorem.

Theorem 13.4.2: *The system*
$$\{\chi_A(x),\, \tau^{j/2}h^\tau(\tau^j x - b) \equiv h^\tau_{j,b}(x);\ j \in \mathbb{N},\ b \in \tau B_{j-1}\} =$$
$$= \{\chi_A,\, U(g)h^\tau;\ g \in \Gamma\} \qquad (13.61)$$
is an orthonormal basis in $L^2[0,1]$.

Proof. The proof is based on the properties of the τ-adic intervals $A_{j,b} = [b\tau^{-j}, (b+1)\tau^{-j}]$, as follows.

1. Either two τ-adic intervals do not overlap or one is contained in the other.

2. If one τ-adic interval is strictly contained in the other, then it is contained in one of the two subintervals obtained at the next τ-subdivision, either the left one or the right one.

Properties 1 and 2 entail the orthogonality of the set (13.60). The normalization is straightforward.

In order to be more explicit and prove completeness, let us now adopt a subdivision procedure, encoded by (13.48), which is different from the one shown in Figure 13.1. We want to have two different interval lengths only, a large one (L) and a small one (S), at each subdivision step. The next step consists in subdividing large intervals only, according to the rule
$$L^{(n)} \to L^{(n+1)} \cup S^{(n+1)},\ S^{(n)} \to L^{(n+1)}, \qquad (13.62)$$

which reminds us of the substitution rule for building the Fibonacci chain [138] or Penrose tilings. Therefore, at the jth subdivision step, one gets the following partition:
$$A = \left(\bigcup_{b \in \lambda^{(j)}} A^{(j)}_{j,b}\right) \cup \left(\bigcup_{b \in \sigma^{(j)}} A^{(j)}_{j+1,b}\right), \qquad (13.63)$$

where the sets $\lambda^{(j)} \subset \tau B_{j-1}$ (for large) and $\sigma^{(j)} \subset \tau B_j$ (for small) are recursively defined by
$$\begin{aligned} \lambda^{(j)} &= \tau \lambda^{(j-1)} \cup \sigma^{(j-1)} \\ \sigma^{(j)} &= \tau^2 \lambda^{(j-1)} + \tau. \end{aligned} \qquad (13.64)$$

The first steps of such a "minimal" subdivision procedure are shown in Figure 13.2. Note that the number of subintervals $A^{(j)}$ appearing at the

298   13. Discrete Wavelet Transforms

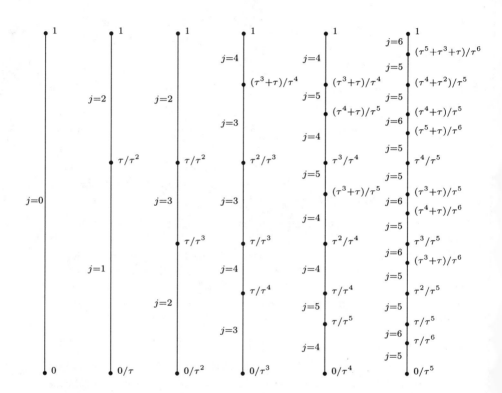

Figure 13.2. The first six steps of the alternative τ-subdivision of the interval $[0,1]$.

13.4. Algebraic wavelets

jth step is now the Fibonacci F_{j+1}, with $F_0 = 1$, $F_1 = 1$. Next, we consider two families of closed subspaces of $L^2[0,1]$ defined for $n \geq 0$ as follows.

- $t_n = \text{span}\{\chi_A, h^\tau_{j,b}; 0 \leq j < n, b \in {}_{\prime} B_{j-1}\}$,
  $t_0 = \text{span}\{\chi_A\}$, $t_1 = \text{span}\{\chi_A, h^\tau\}$. (13.65)

- u_n is the linear span of all functions in $L^2[0,1]$ that are constant on all intervals $A^{(n)}$ appearing at the nth subdivision step (13.63); that is

$$u_n = \text{span}\{\chi_{A^{(n)}}\}. \quad (13.66)$$

Clearly, we have, for all $n \in \mathbb{N}$,

$$t_n \subset t_{n+1}, \quad u_n \subset u_{n+1}, \quad (13.67)$$

and, since $\tau^{-1} A^{(n)}$ is one of the $A^{(n+1)}$,

$$f(x) \in X_n \Rightarrow f(\tau x) \in X_{n+1}, \; X = t \text{ or } u. \quad (13.68)$$

Then, the completeness of the system (13.61) results from the fact that $\cup_n u_n$ is dense in $L^2[0,1]$, together with the following lemma. □

Lemma 13.4.3: *For every $n \in \mathbb{N}$, one has $t_n = u_n$.*

Proof. The relation is proved by induction. Trivially $t_0 = u_0$. Next, we have from (13.64):

$$u_{n+1} = u_n \cup \text{span}\left\{\{\chi_{A^{(n+1)}_{n+1,b}}; b \in \tau \lambda^{(n)}\} \cup \{\chi_{A^{(n+1)}_{n+2,b}}; b \in \sigma^{(n+1)}\}\right\}, \quad (13.69)$$

and we suppose that $u_n \subset t_n$, which in particular implies that

$$\chi_{A_{n,b}} \in t_n \text{ for any } b \in \tau B_{n-1}. \quad (13.70)$$

Let us show that the new characteristic functions added to u_n in (13.69) and resulting from the subdivision of $A^{(n)}_{n,b}$ belong to t_{n+1}. Indeed, we have, at the first subdivision step,

$$\chi_{A_{1,0}} = \frac{1}{\tau} \chi_A + \frac{1}{\tau^{3/2}} h^\tau,$$
$$\chi_{A_{2,\tau}} = \frac{1}{\tau^2} \chi_A - \frac{1}{\tau^{3/2}} h^\tau, \quad (13.71)$$

and then by applying the affine operator $U((\tau^{-n} b, \tau^{-n}))$ on both sides of these equations, we obtain

$$\chi_{A^{(n+1)}_{n+1,\tau b}} = \frac{1}{\tau} \chi_{A^{(n)}_{n,b}} + \frac{1}{\tau^{(n+3)/2}} h^\tau_{n,b},$$
$$\chi_{A^{(n+1)}_{n+2,\tau^2 b+\tau}} = \frac{1}{\tau^2} \chi_{A^{(n)}_{n,b}} - \frac{1}{\tau^{(n+3)/2}} h^\tau_{n,b}. \quad (13.72)$$

Conversely, we have

$$t_{n+1} = t_n \cup \{h_{n,b}^\tau; b \in \tau B_{n-1}\} \tag{13.73}$$

and we suppose that $t_n \subset u_n$. Then, we note that $h_{n,b}^\tau$ is equal to $\tau^{\frac{n-1}{2}}$ (respectively, $-\tau^{\frac{n+1}{2}}$) on $A_{n+1,\tau b}^{(n+1)}$ (respectively, $A_{n+2,\tau^2 b+\tau}^{(n+1)}$) and is zero everywhere else. □

Now, let us see how one can extend the previous construction to the whole real line, or rather to $L^2(\mathbb{R})$. The procedure is based on relaxing the constraints $j, n \in \mathbb{N}$ and $b \in \tau B_{j-1}$ in (13.61) and (13.65). First, we extend the definition (13.66) to arbitrary $n \in \mathbb{Z}$ and $b \in \tau \mathbb{Z}_\tau^+$ in the following way:

U_n is the set of all functions in $L^2(\mathbb{R})$ that are constant on all (large) intervals $A_{n,b}^{(n)} = [\frac{b}{\tau^n}, \frac{b+1}{\tau^n}]$, $A_{n,-b-1}^{(n)} = [-\frac{b+1}{\tau^n}, -\frac{b}{\tau^n}] = -A_{n,b}^{(n)}$, with $b \in \tau \mathbb{Z}_\tau^+$, and on all (small) intervals $A_{n+1,b}^{(n)}$, $A_{n+1,-b-1}^{(n)}$ with $b \in \tau \mathbb{Z}_\tau^{+\text{odd}}$, where $\mathbb{Z}_\tau^{\text{odd}} = \mathbb{Z}_\tau \backslash \tau \mathbb{Z}_\tau$.

Note that U_0 corresponds to the subdivision of the line into intervals of length 1 or $1/\tau$ separating two successive τ-integers. Actually, the subdivision procedure that underlies the definition of the sequence $\{U_n\}$ exactly parallels (13.62): At the nth step, the whole line is divided into large intervals (of length τ^{-n}) and small ones (of length τ^{-n-1}). The next $(n+1)$st step consists in dividing the large intervals and leaving unchanged the small ones, which then become large.

The sets U_n have the following properties, which hold for all $n \in \mathbb{Z}$:

$$U_n \subset U_{n+1} \tag{13.74}$$

$$f(x) \in U_n \Leftrightarrow f(\tau x) \in U_{n+1}. \tag{13.75}$$

Similarly, we extend definition (13.65) as follows:

$$T_n = \text{span}\{h_{j,b}^\tau, h_{j,-b-1}^\tau; j < n, b \in \tau \mathbb{Z}_\tau^+\}, \tag{13.76}$$

with $n \in \mathbb{Z}$. It is clear that we also have

$$T_n \subset T_{n+1} \tag{13.77}$$

$$f(x) \in T_n \Leftrightarrow f(\tau x) \in T_{n+1}. \tag{13.78}$$

Then, Lemma 13.4.3 extends to these U_n and T_n, as follows.

Lemma 13.4.4: *For every $n \in \mathbb{N}$, one has $U_n = T_n$.*

Proof. Because of (13.75) and (13.78), it is enough to prove that $U_0 = T_0$.
Each $h_{j,b}^\tau$ (respectively, $h_{j,-b-1}^\tau$) for $j < 0$ is constant on any large interval $[b, b+1]$ (respectively, $[-b-1, -b]$) for $b \in \tau \mathbb{Z}_\tau^+$ and on any small

interval $[b/\tau, (b+1)/\tau]$ (respectively, $[-(b+1)/\tau, -b/\tau]$) for $b \in \tau \mathbb{Z}_\tau^{+\mathrm{odd}}$. For instance,

$$h_{-1,0}^\tau(x) = \tau^{-1} \chi_A(x) - \chi_{A_{1,\tau}}(x). \tag{13.79}$$

Thus, $T_0 \subset U_0$. Now, the characteristic function χ_A belongs to T_0. This stems from the expansion formula

$$\tau^{-1/2} \sum_{j<0} \tau^{j/2} h_{j,0}^\tau(x) = \chi_A(x). \tag{13.80}$$

This series is absolutely convergent in $L^2(\mathbb{R})$ since $\|\tau^{j/2} h_{j,0}^\tau\|_2 = \tau^{j/2}$ and $j < 0$, and one checks that the left-hand side equals

- 0, for $x < 0$, since $\operatorname{supp} h_{j,0}^\tau(x) = [0, \tau^{-j}]$;

- 1, for $0 \leqslant x \leqslant 1$, since, for $j < 0$ and $x \in [0, \tau^{-j-1}] \supset [0,1]$, $\tau^{j/2} h_{j,0}^\tau(x) = \tau^j h^\tau(\tau^j x) = \tau^{j-1/2}$, and

$$\tau^{-1/2} \sum_{j<0} \tau^{j-1/2} = \frac{1}{\tau} \left(\frac{1}{1-1/\tau} - 1 \right)$$
$$= 1;$$

- 0, for $x > 1$, since on $\tau^{r-1} \leqslant x \leqslant \tau^r$, $r \geqslant 1$, we have

$$\tau^{-1/2} \sum_{j<0} \tau^j h^\tau(\tau^j x) = \tau^{-1/2} \sum_{s=1}^\infty \frac{1}{\tau^s} h^\tau\left(\frac{x}{\tau^s}\right)$$
$$= \tau^{-1/2} \left[-\tau^{-1/2} \frac{1}{\tau^r} + \tau^{-1/2} \sum_{s=r+1}^\infty \frac{1}{\tau^s} \right]$$
$$= 0.$$

Now, if $\chi_A \in T_0$, the same holds for $\chi_{A_{0,b}}(x) = \chi_A(x-b)$ and for $\chi_{A_{0,-b-1}}(x) = \chi_A(x+b+1)$, just by translating (13.80). On the other hand, $\chi_{A_{1,\tau}}$ is in T_0, as results from (13.79) and (13.80). Now, translating (13.79) by $\frac{b-\tau}{\tau}$, which belongs to $\tau^2 \mathbb{Z}_\tau^+$ if $b \in \tau \mathbb{Z}_\tau^{+\mathrm{odd}}$, we obtain

$$\chi_{A_{1,b}}(x) = \chi_{A_{0,\frac{b-\tau}{\tau}}}(x) - h_{-1,\frac{b-\tau}{\tau^2}}^\tau(x). \tag{13.81}$$

This proves that characteristic functions of small intervals in $\mathbb{R}^+$ belong to T_0. The same holds for negative small intervals by replacing b by $-b-1$ in (13.81). Thus, we have shown that $U_0 = T_0$ and therefore that $U_n = T_n$ for all $n \in \mathbb{Z}$. $\square$

Since $\cup_{n \in \mathbb{Z}} U_n$ is dense in $L^2(\mathbb{R})$ and the system $\{h_{j,b}^\tau, h_{j,-b-1}^\tau; j \in \mathbb{Z}, b \in \tau \mathbb{Z}_\tau^+\}$ is orthonormal, we get

Theorem 13.4.5: *The system*

$$\{\tau^{j/2} h^\tau(\tau^j x - b), \tau^{j/2} h^\tau(\tau^j x + b + 1); j \in \mathbb{Z}, b \in \tau\mathbb{Z}_\tau^+\} \quad (13.82)$$

is an orthonormal basis of $L^2(\mathbb{R})$, called the τ-Haar basis of $L^2(\mathbb{R})$.

At this point, we note that, strictly speaking, the sequence $\{U_n = T_n\,;\,n \in \mathbb{Z}\}$ of subspaces of $L^2(\mathbb{R})$ is not a multiresolution analysis as defined in Section 13.1.1 [Woj]. We give here a more restrictive definition adapted to the present framework.

Definition 13.4.6: *A τ-multiresolution analysis is a sequence $(V_j)_{j\in\mathbb{Z}}$ of subspaces of $L^2(\mathbb{R})$ with the following properties:*

1. $\ldots \subset V_{-1} \subset V_0 \subset V_1 \ldots$;
2. span $\cup_{j\in\mathbb{Z}} V_j = L^2(\mathbb{R})$;
3. $\cap_{j\in\mathbb{Z}} V_j = \{0\}$;
4. $f(x) \in V_j$ if and only if $f(\tau^{-j} x) \in V_0$;
5. *There exists a function $\Phi \in V_0$, called a scaling function, such that the system $\{\Phi(x-b),\ \Phi(x+b+1);\ b \in \tau\mathbb{Z}_\tau^+\} \cup \{\Phi(\tau x - b),\ \Phi(\tau x + b + 1);\ b \in \tau\mathbb{Z}_\tau^{+\,\text{odd}}\}$ is an orthonormal basis in V_0.*

What is missing here with respect to the standard dyadic multiresolution of Section 13.1.1 is the counterpart of the invariance of V_0 under translations by integers. Indeed, we cannot assert that $f(x) \in V_0$ if and only if $f(x-b)$ and $f(x+b+1) \in V_0$ for all $b \in \tau\mathbb{Z}_\tau^+$, since the latter is not closed under addition. As a matter of fact, we have the Meyer property

$$\mathbb{Z}_\tau + \mathbb{Z}_\tau \subset \mathbb{Z}_\tau + \{0, \pm\frac{1}{\tau}, \pm\frac{1}{\tau^2}\}. \quad (13.83)$$

(A set Λ has the Meyer property if $\Lambda - \Lambda \subset \Lambda + F$, where F is a finite set [Mel, 218].) In the τ-Haar case, the scaling function is obviously the characteristic function χ_A of the unit interval.

Finally, let us end this section by proposing the following definition for τ-wavelets.

Definition 13.4.7: *A τ-wavelet is a function $\psi(x) \in L^2(\mathbb{R})$ such that the following family of functions is an orthonormal basis in the Hilbert space $L^2(\mathbb{R})$:*

$$\{\psi_{j,b}(x) \equiv \tau^{j/2} \psi(\tau^j x - b),\ \psi_{j,-b-1}(x) \equiv \tau^{j/2} \psi(\tau^j x + b + 1)\},$$

where j is an arbitrary integer and b is an "even" positive τ-integer; i.e., $b \in \tau\mathbb{Z}_\tau^+$.

We have already given an example of such a τ-wavelet, namely, the τ-Haar wavelet. Less trivial examples are not known at the moment. Perhaps a more tractable requirement would be to replace the existence of an orthonormal basis in Definitions 13.4.6 and 13.4.7 by a Riesz basis.

13.4.2 Pisot wavelets, etc. (♣)

Of course, the procedure that was followed in the previous section could tentatively be extended to *any* real number $\beta > 1$. Indeed, for such a β, there exists a numeration system based on the so-called Rényi β-expansion of real numbers, analogous to (13.33)

$$\mathbb{R}_+ \ni x = \sum_{l=-\infty}^{j} \xi_l \beta^l \equiv \xi_j \xi_{j-1} \ldots \xi_0 \xi_{-1} \ldots, \qquad (13.84)$$

where j is the largest integer, such that $\beta^j \leqslant x < \beta^{j+1}$. The positive integers ξ_l take their values in the alphabet (for β noninteger) $\{0, 1, 2, \ldots, [\beta]\}$ and are computed with the so-called greedy algorithm. One defines recursively

$$\xi_j = [x/\beta^j], \quad r_j = \{x/\beta^j\}, \text{ and for } l < j, \; \xi_l = [\beta r_{l+1}], \quad r_l = \{\beta r_{l+1}\}. \qquad (13.85)$$

A β-integer is a real number that has a vanishing fractional part in its β-expansion. The set of β-integers is denoted by

$$\mathbb{Z}_\beta = \{\pm(\xi_j \beta^j + \xi_{j-1}\beta^{j-1} + \ldots + \xi_1\beta + \xi_0)\} = \mathbb{Z}_\beta^+ \cup (-\mathbb{Z}_\beta^+). \qquad (13.86)$$

Some configurations $\xi_j \xi_{j-1} \ldots \xi_0 \xi_{-1} \ldots$ in this definition are not possible. What is allowed and what is forbidden in the set of β-expansions is completely determined by the so-called Rényi β-expansion of 1 [256]:

$$\begin{aligned} d(1, \beta) &= t_1 \beta^{-1} + t_2 \beta^{-2} + \ldots \\ &= 0.t_1 t_2 \ldots t_l \ldots, \quad t_l \in \{0, 1, 2, \ldots [\beta]\}. \end{aligned} \qquad (13.87)$$

This expansion is reminiscent of the identity $1 = 0.999\ldots9\ldots$ in the decimal system. Viewed in an iterative way, it corresponds to an infinite sequence of subdivisions of the interval $[0, 1]$, like the one in (13.30), starting from

$$1 = \underbrace{\frac{1}{\beta} + \ldots + \frac{1}{\beta}}_{t_1 \text{ times}} + \frac{\{\beta\}}{\beta}.$$

Eq.(13.87) is obtained by a process analogous to (13.85):

$$t_1 = [\beta], \; r_1 = \{\beta\}, \ldots, t_l = [\beta r_{l-1}], \quad r_l = \{\beta r_{l-1}\}, \ldots,$$

or, equivalently,

$$t_l = [\beta T_\beta^{l-1}(1)], \qquad (13.88)$$

where

$$T_\beta(x) = \beta x \,(\text{mod } 1). \qquad (13.89)$$

Then, we have the following β-expansion rule [241].

Proposition 13.4.8: *No infinite sequence of positive integers is present in any β-expansion if the sequence itself and all its (one-sided) translates are lexicographically larger than or equal to*

$$t_1 t_2 \ldots \quad \text{if the latter is infinite,}$$

and to:

$$(t_1 t_2 \ldots t_{m-1}(t_m - 1))^\omega, \quad \text{if } d(1, \beta) = 0.t_1 t_2 \ldots t_{m-1} t_m \text{ is finite.}$$

[here $(\cdot)^\omega$ means that the word in parentheses is indefinitely repeated.]

Thus, once $d(1, \beta)$ is known, it becomes possible in principle (but the process may turn out to be unpracticable!) to build $\mathbb{Z}_\beta$ by following the lexicographic order of the allowed sequence.

The countable set $\mathbb{Z}_\beta$ is naturally self-similar and symmetric with respect to the origin:

$$\beta \mathbb{Z}_\beta \subset \mathbb{Z}_\beta, \quad \mathbb{Z}_\beta = -\mathbb{Z}_\beta. \tag{13.90}$$

It tiles the line with intervals separating two nearest neighbors $x_i < x_{i+1}$. The length of the intervals, $l_i = x_{i+1} - x_i$, may then take its value in a finite or countably infinite set, depending on the nature of the number β. Let us assume that the latter is a Pisot–Vijayaraghavan (for short, PV or Pisot) number, i.e., an algebraic integer that is a solution of the equation

$$X^m = a_{m-1} X^{m-1} + \ldots + a_1 X + a_0, \quad a_i \in \mathbb{Z}, \tag{13.91}$$

such that all other solutions $\beta^{(i)}$ of (13.91) (the Galois conjugates of β) have a modulus strictly smaller than 1,

$$\beta^{(0)} = \beta > 1, \quad |\beta^{(i)}| < 1, \; i = 1, \ldots, m - 1. \tag{13.92}$$

Then, we have the following crucial result [73].

Theorem 13.4.9: *If β is a Pisot number, then the Rényi β-expansion of 1 is eventually periodic:*

$$d(1, \beta) = 0.t_1 t_2 \ldots t_m (t_{m+1} \ldots t_{m+p})^\omega. \tag{13.93}$$

It follows that $\mathbb{Z}_\beta$ is a self-similar tiling of the line with a *finite* set of different tiles. The lengths of the tiles are $\{T_\beta^i(1), \; 0 \leqslant i \leqslant m+p-1\}$. More precisely, the lengths take their values in the set

$$1, \beta - t_1, \beta^2 - t_1, \beta - t_2, \ldots, \beta^{m+p-1} - t_1 \beta^{m+p-2} - \cdots - t_{m+p-1}. \tag{13.94}$$

This result suggests that it is conceivable to extend to Pisot numbers the algebraic approach described in the previous section and construct "Pisot wavelets." As a matter of fact, we can propose the following definition as the exact counterpart of Definitions 13.4.6 and 13.4.7.

Definition 13.4.10: *A β-multiresolution analysis is a sequence $(V_j)_{j \in \mathbb{Z}}$ of subspaces of $L^2(\mathbb{R})$ with the following properties.*

1. $\ldots \subset V_{-1} \subset V_0 \subset V_1 \ldots$;
2. span $\cup_{j\in\mathbb{Z}} V_j = L^2(\mathbb{R})$;
3. $\cap_{j\in\mathbb{Z}} V_j = \{0\}$;
4. $f(x) \in V_j$ if and only if $f(\beta^{-j}x) \in V_0$;
5. There exists a function $\Phi \in V_0$, called a scaling function, such that the collection of all its suitably "translated-dilated" copies on the set of β-integers is an orthonormal basis in V_0.

Definition 13.4.11: Given a function $\psi \in L^2(\mathbb{R})$, define the family of functions

$$\mathcal{I} = \{\psi_{j,b}(x) \equiv \beta^{j/2}\,\psi(\beta^j x - b),\ \psi_{j,-b-l_i}(x) \equiv \beta^{j/2}\,\psi(\beta^j x + b + l_i)\},$$

where j is an arbitrary integer, b is an appropriate element of $\mathbb{Z}_\beta^+$, and l_i runs over all possible $(m+p)$ interval lengths (13.94). Then, ψ is called a β-wavelet if the family $\mathcal{I}$ is an orthonormal basis in the Hilbert space $L^2(\mathbb{R})$.

14
Multidimensional Wavelets

14.1 Going to higher dimensions

Exactly as in one dimension, multidimensional wavelets may be derived from the similitude group of $\mathbb{R}^n$ ($n > 1$), consisting of dilations, rotations, and translations. Of course, the most interesting case for applications is $n = 2$, where wavelets have become a standard tool in image processing, including radar imaging [Me2, Me4]. Also, $n = 3$ may have a practical importance, since some important physical phenomena are intrinsically multiscale and 3-D. Typical examples may be found in fluid dynamics, for instance, the appearance of coherent structures in turbulent flows, or the disentangling of a wave train in acoustics. In such cases, a 3-D wavelet analysis is likely to yield a deeper understanding [57]. The same comment is valid for the applications of the WT in quantum physics (quantum mechanics, atomic physics, solid-state physics, etc.). A good reference for the latter is the survey volume [Be2].

More importantly, the derivation of the CWT in dimensions larger than two requires in certain cases (axisymmetric wavelets) the general CS construction developed at length in the previous chapters, because the natural parameter space is no longer the similitude group itself, but rather a coset space (although here we are still in the Gilmore–Perelomov setting, the dividing subgroup is the isotropy subgroup of the wavelet). This opens the door to a host of other generalizations, namely, wavelets on manifolds on which a group action is defined. We will discuss below several instances, namely, wavelets on spheres and various kinds of time-dependent wavelets.

The latter could prove to be useful tools for motion tracking, including in relativistic situations (here, the relevant group would be the Galilei or the Poincaré group augmented by dilations, that is, the corresponding affine groups; we will cover these examples in Section 15.3). Remarkably enough, most of these cases, including wavelets on spheres, require the full-fledged extension of the CS method, beyond Gilmore–Perelomov.

14.2 Mathematical analysis (♣)

An n-dimensional signal of finite energy is represented by a complex-valued function $s \in L^2(\mathbb{R}^n, d^n x)$. The natural operations, usually applied to a signal s, are obtained by combining three elementary transformations: (1) translation $s(x) \mapsto s(x-b)$, (2) dilation $s(x) \mapsto a^{-n/2} s(a^{-1}x)$, and (3) rotation $s(x) \mapsto s(R^{-1}x)$, where $b \in \mathbb{R}^n$ is the displacement parameter, $a > 0$ is the dilation parameter and $R \in SO(n)$ is an $n \times n$ rotation (orthogonal) matrix. In particular

- for $n = 2$, $R \equiv R(\theta)$, a 2×2 rotation matrix, parametrized by an angle $\theta \in [0, 2\pi)$;
- for $n = 3$, $R \equiv R(\alpha, \beta, \gamma)$, a 3×3 rotation matrix, parametrized for instance by the three Euler angles α, β, γ.

These three operations generate the n-dimensional Euclidean group with dilations [Mur, 225], also known as the similitude group of $\mathbb{R}^n$, $SIM(n) = \mathbb{R}^n \rtimes (\mathbb{R}_*^+ \times SO(n))$. This group $SIM(n)$ has the following natural action on a signal s, in position and Fourier space, respectively:

$$s_{b,a,R}(x) = [U(b,a,R)s](x) = a^{-n/2} s(a^{-1} R^{-1}(x-b)) \quad (14.1)$$
$$\widehat{s_{b,a,R}}(k) = [\widehat{U}(b,a,R)\hat{s}](k) = a^{n/2} \hat{s}(a R^{-1} k) e^{-ik \cdot b}. \quad (14.2)$$

We note in passing that one could replace the group $SO(n)$ by $O(n)$ in the definition of rotations, that is, include also mirror symmetries (or more general discrete operations), but this has never been considered.

The basic structure of the n-dimensional WT is the same as in the 1-D case, as results from the following theorem.

Theorem 14.2.1: *The operator family $U(\cdot)$ defined in (14.1) is a unitary irreducible representation of $SIM(n)$ in $L^2(\mathbb{R}^n, d^n x)$, and it is unique up to unitary equivalence. This representation is square integrable: A vector $\psi \in L^2(\mathbb{R}^n, d^n x)$ is admissible iff it satisfies the condition*

$$c_\psi = (2\pi)^n A_{n-1} \int_{\mathbb{R}^n} |\hat{\psi}(k)|^2 \frac{d^n k}{|k|^n} < \infty, \quad (14.3)$$

where $|k|^2 = k \cdot k$ and $A_{n-1} = \prod_{k=2}^{n-1} \frac{2\pi^{k/2}}{\Gamma(k/2)}$ is the volume of $SO(n-1)$.

Proof. The proof is elementary. The unitarity of U is obvious. Irreducibility follows from the fact that, for every nonzero $s \in L^2(\mathbb{R}^n, d^n x)$, the linear span of the set

$$\mathcal{D} = \{s_{b,a,R} = U(b, a, R)s, \ (b, a, R) \in SIM(n)\} \qquad (14.4)$$

is dense in $L^2(\mathbb{R}^n, d^n x)$. Indeed, let $f \in L^2(\mathbb{R}^n, d^n x)$ be orthogonal to $\mathcal{D}$; that is, $\langle s_{b,a,R} | f \rangle = 0, \ \forall \, (b, a, R) \in SIM(n)$. This means

$$\langle \widehat{s_{b,a,R}} | \widehat{f} \rangle = a^{n/2} \int_{\mathbb{R}^n} e^{ik \cdot b} \, \overline{\widehat{s}(a R^{-1}k)} \, \widehat{f}(k) \, d^n k = 0, \quad \forall \, (b, a, R) \in SIM(n),$$

which implies that $\overline{\widehat{s}(a R^{-1} k)} \, \widehat{f}(k) = 0$ a.e., for all $(a, R) \in \mathbb{R}_*^+ \times SO(n)$. Since the action of the latter on $\mathbb{R}^n$ is transitive, this implies that $\widehat{f}(k) = 0$ a.e., thus, $f = 0$.

Uniqueness follows from the fact that the representation U is obtained by Mackey's standard method of induction, since there is only one nontrivial orbit in k-space [Bar, Mur, 225] (see Section 10.2.4).

Finally, the square integrability follows again from the explicit calculation of the L^2 norm of the matrix element of $U(b, a, R)$ with respect to the left Haar measure $dg \equiv a^{-(n+1)} da \, dR \, d^n b$, where dR is the Haar measure on $SO(n)$,

$$\iiint_{SIM(n)} |\langle U(b, a, R)\psi | \psi \rangle|^2 \, \frac{da}{a^{n+1}} \, dR \, d^n b = c_\psi \|\psi\|^2, \qquad (14.5)$$

where c_ψ is the constant defined in (14.3). $\qquad \square$

Actually the two results stated in this theorem, namely, uniqueness and square integrability of the representation U, already follow from the analysis of semidirect products made in Section 9.1. Nevertheless, we found it instructive to give the present direct proof. Interestingly, the presence of dilations is crucial both for the uniqueness and for the square integrability of U. Now the restriction of U to the rotation group $SO(n)$ is the quasiregular representation, which decomposes into the direct sum of all irreducible vector representations. In other words, the inclusion of dilations forbids the appearance of spinorial representations of $SO(n)$.

A nonzero signal ψ that satisfies the admissibility condition (14.3) is called an *n-dimensional wavelet*. Again, if ψ is regular enough ($\psi \in L^1(\mathbb{R}^n, d^n x) \cap L^2(\mathbb{R}^n, d^n x)$ suffices), the admissibility condition (14.3) simply means that the wavelet has zero mean:

$$\widehat{\psi}(0) = 0 \ \Leftrightarrow \ \int_{\mathbb{R}^n} \psi(x) \, d^n x = 0. \qquad (14.6)$$

In addition, both ψ and $\widehat{\psi}$ are supposed to be well localized. Clearly, $\psi_{b,a,R}$ satisfies all of these conditions whenever ψ does.

Given a finite energy signal $s \in L^2(\mathbb{R}^n, d^n x)$, its continuous wavelet transform (with respect to the fixed wavelet ψ), $S \equiv W_\psi s$ is given by:

$$S(b, a, R) = \langle \psi_{b,a,R} | s \rangle \tag{14.7}$$

$$= a^{-n/2} \int_{\mathbb{R}^n} \overline{\psi(a^{-1} R^{-1}(x - b))} s(x) \, d^n x \tag{14.8}$$

$$= a^{n/2} \int_{\mathbb{R}^n} e^{ib \cdot k} \overline{\hat{\psi}(aR^{-1}k)} \, \hat{s}(k) \, d^n k. \tag{14.9}$$

The most interesting case arises when the wavelet ψ is *axially symmetric*, i.e. $SO(n-1)$-invariant. Then, one can replace everywhere $SO(n)$ by $SO(n)/SO(n-1) \simeq S^{n-1}$, the unit sphere in $\mathbb{R}^n$. The rotation R becomes $R \equiv R(\varpi)$, $\varpi \in S^{n-1}$, and the CWT is $S \equiv S(a, \varpi, b) \in L^2(X, d\nu)$, where the parameter space is

$$\begin{aligned} X &= SIM(n)/SO(n-1) \\ &= \mathbb{R}^n \rtimes (\mathbb{R}_*^+ \times SO(n)/SO(n-1)) \\ &= \mathbb{R}^n \rtimes (\mathbb{R}_*^+ \times S^{n-1}) \simeq \mathbb{R}^n \times \mathbb{R}_*^n, \end{aligned} \tag{14.10}$$

with $SIM(n)$-invariant measure

$$d\nu(x) = \frac{da}{a^{n+1}} \, d\varpi \, d^n b. \tag{14.11}$$

Furthermore, while $b \in \mathbb{R}^n$ corresponds to the position variables, the pair (a^{-1}, ϖ) may be interpreted as a spatial frequency in spherical polar coordinates. To be sure, writing $\kappa = a^{-1}$, the volume element on X becomes

$$d\nu(x) = \kappa^{n-1} d\kappa \, d\varpi \, d^n b \simeq d^n k \, d^n b, \tag{14.12}$$

the (Lebesgue) volume element of $\mathbb{R}^{2n}$. As for $n = 1$ (see Section 12.4.1), one may easily compute the coadjoint orbits of $SIM(n)$. It turns out that there are only two: A trivial one, $\mathcal{O}_o = \{0\} \times \mathbb{R}^n$, associated to the identity representation, and one of dimension $2n$, $\mathcal{O}_U = \mathbb{R}_*^n \times \mathbb{R}^n$, with canonical variables $k \in \mathbb{R}_*^n$, $x \in \mathbb{R}^n$, which corresponds to the representation U. Thus, $\mathcal{O}_U \simeq (X, d^n k \, d^n b)$, so that here too the CWT is a phase space representation [119].

From now on, we will restrict ourselves to the case of an axially symmetric wavelet ψ. Of course, this covers also the fully isotropic case ($SO(n)$-invariant wavelet), where S does not depend on ϖ at all.

The main properties of the wavelet transform may be summarized as follows (compare with Theorem 12.2.2).

Theorem 14.2.2: *Let the map $W_\psi : s \mapsto c_\psi^{-1/2} S$ be defined by*

$$(W_\psi s)(b, a, \varpi) = c_\psi^{-1/2} \langle \psi_{b,a,\varpi} | s \rangle, \; s \in L^2(\mathbb{R}^n, d^n x). \tag{14.13}$$

Then:

1. W_ψ conserves the norm of the signal:

$$\iiint_X |S(b,a,\varpi)|^2 \, \frac{da}{a^{n+1}} \, d\varpi \, d^n b \;=\; c_\psi \int_{\mathbb{R}^n} |s(x)|^2 \, d^n x, \qquad (14.14)$$

i.e. it is an isometry from the space of signals into the space of transforms, which is a closed subspace $\mathfrak{H}_\psi$ of $L^2(X, d\nu)$. Equivalently, the family of wavelets $\{\psi_{b,a,\varpi}\}$, with $a > 0, \varpi \in S^{n-1}, b \in \mathbb{R}^n$, generates a resolution of the identity:

$$c_\psi^{-1} \iiint_X |\psi_{b,a,\varpi}\rangle \langle \psi_{b,a,\varpi}| \, \frac{da}{a^{n+1}} \, d\varpi \, d^n b = I. \qquad (14.15)$$

2. The adjoint of W_ψ, restricted to $\mathfrak{H}_\psi$, yields the reconstruction formula:

$$s(x) = c_\psi^{-1} \iiint_X \psi_{b,a,\varpi}(x) \, S(b,a,\varpi) \, \frac{da}{a^{n+1}} \, d\varpi \, d^n b \; . \qquad (14.16)$$

3. The projection from $L^2(X, d\nu)$ onto $\mathfrak{H}_\psi$ is an integral operator, with (reproducing) kernel

$$K(b',a',\varpi'|b,a,\varpi) = c_\psi^{-1} \langle \psi_{b',a',\varpi'}|\psi_{b,a,\varpi}\rangle, \qquad (14.17)$$

that is, the autocorrelation function of ψ.

4. The CWT is covariant with respect to the group $SIM(n)$:

$$\begin{aligned} s(x - b_o) &\mapsto S(b - b_o, a, \varpi), \; b_o \in \mathbb{R}^n, \\ s(R_o^{-1}x) &\mapsto S(R_o^{-1}b, a, R_o^{-1}\varpi), \; R_o \in SO(n), \\ a_o^{-n/2} s(a_o^{-1}x) &\mapsto S(a_o^{-1}b, a_o^{-1}a, \varpi), \; a_o > 0. \end{aligned} \qquad (14.18)$$

No proof is needed, since this theorem is the exact adaptation to the case at hand of Theorem 12.2.2.

As we can see, all of the formulas are the exact analogues of those given in the previous section for $n = 1$. It follows that the interpretation of the n-dimensional CWT is entirely similar to that given previously. The only new element is the presence of the rotation degree of freedom, indexed by the variable $R \in SO(n)$, or $\varpi \in S^{n-1}$ in the axisymmetric case. Here also, the covariance is crucial for applications and it is lost in the common version of the 2-D discrete WT (see Section 14.3.4). In particular, the presence of the rotation parameter makes the multidimensional CWT sensitive to *directions*, provided one uses an oriented wavelet. As we will see in Section 14.3.3, many applications are based on this property.

This remark brings us precisely to the question of the choice of the analyzing wavelet ψ. Two cases are possible, depending on the problem at hand. (1) If one wants to perform a pointwise analysis, that is, when no oriented features are present or relevant in the signal, one may choose an analyzing wavelet ψ that is invariant under rotation [full $SO(n)$ invariance]. (2) When the aim is to detect directional features in a signal, for instance, to perform directional filtering, one has to use a wavelet that is *not* rotation invariant. The best angular selectivity will be obtained if ψ is *directional*, which means that its (essential) support in spatial frequency space is contained in a convex cone with apex at the origin. Since it may sound counterintuitive, this definition requires a word of justification. According to (14.9), the wavelet acts as a filter in k-space (multiplication by the function $\widehat{\psi}$). Suppose the signal $s(x)$ is strongly oriented, for instance, a long segment along the x-axis. Then, its Fourier transform $\widehat{s}(k)$ is essentially supported in the plane $k_x = 0$. In order to detect such a signal, with a good directional selectivity, one needs a wavelet ψ supported in a narrow cone in k-space. Then the WT is negligible unless $\widehat{\psi}(k)$ is essentially concentrated on the support of $\widehat{s}(k)$: Directional selectivity demands a restriction of the support of $\widehat{\psi}$, not ψ.

Let us examine in more detail some examples of n-dimensional wavelets of each kind. In most cases, they are the obvious generalizations of those given in Chapter 12 for $n = 1$.

(1) Isotropic wavelets
• *The Mexican hat or Marr wavelet*
This is a real, rotation-invariant wavelet, given by the Laplacian of a Gaussian:

$$\psi_H(x) = -\Delta \exp(-\frac{1}{2}|x|^2), \quad \Delta = \partial^2_{x_1} + \partial^2_{x_2} + \ldots + \partial^2_{x_n}$$
$$= (n - |x|^2) \exp(-\frac{1}{2}|x|^2). \qquad (14.19)$$

One also uses higher order Laplacians of the Gaussian:

$$\psi_H^{(m)}(x) = (-\Delta)^m \exp(-\frac{1}{2}|x|^2). \qquad (14.20)$$

For increasing m, these wavelets have more and more vanishing moments and are thus sensitive to increasingly sharper details. An interesting technique [51] is to analyze the same signal with several wavelets $\psi_H^{(m)}$, for different m. The features common to all transforms surely belong to the signal and are not artifacts of the analysis.

• *Difference wavelets*
An interesting class consists of wavelets obtained as the difference between a function h and a contracted version of the latter. If h is a smooth nonnegative function, integrable and square integrable, with all moments of order 1 vanishing, then the function ψ given by the relation

$$\psi(x) = h(x) - \alpha^{-2} h(\alpha^{-1} x) \quad (\alpha > 1) \qquad (14.21)$$

is easily seen to be a wavelet satisfying the admissibility condition (14.6). Such difference wavelets have an additional advantage in that they lead to interesting and fast algorithms [116]. We will come back to this point.

A typical example is the "difference-of-Gaussians" or DOG wavelet, obtained by taking for h a Gaussian. The DOG filter is a good substitute

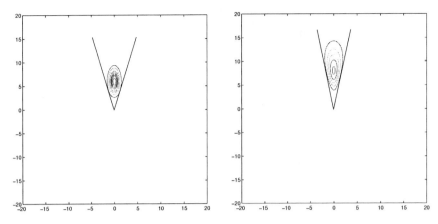

Figure 14.1. Two directional 2-D wavelets, in k space, seen in level curves: (left) The Morlet wavelet $\widehat{\psi}_M$ ($k_o = (0,6), \epsilon = 5$); (right) the Cauchy wavelet $\widehat{\psi}_{44}^{(C_{10})}$ with supporting cone $C_{10} = C(-10°, 10°)$, rotated by 90° for the sake of comparison.

for the Mexican hat (for $\alpha = 1.6$, their shapes are extremely similar), frequently used in psychophysics works [DeV]. We will encounter a similar wavelet in the CWT on the two-sphere, in Section 15.1.1.

(2) Directional wavelets
• *The Morlet wavelet*
This one is the prototype of an oriented wavelet:

$$\psi_M(x) = \exp(ik_o \cdot x) \exp(-\frac{1}{2}|Ax|^2) + \text{corr. term.} \tag{14.22}$$

The parameter $k_o \in \mathbb{R}^n$ is the wave vector, and $A = \text{diag}[\epsilon_1^{-1/2}, \epsilon_2^{-1/2}, \ldots]$, $\epsilon_j \geq 1$, is an $n \times n$ anisotropy matrix. The correction term enforces the admissibility condition $\widehat{\psi}_M(0) = 0$, but it is numerically negligible for $|k_o| \geq 5.6$ and will usually be dropped. The most useful choice is the axially symmetric Morlet wavelet, with wave vector k_o pointing in the x_n direction and rotation invariant around that axis. For $n = 3$, this yields the following wavelet:

$$\psi_{ax}(x) = \exp(ik_o z) \exp -\frac{1}{2}[\epsilon^{-1}(x^2 + y^2) + z^2]. \tag{14.23}$$

The 2-D Morlet wavelet is shown, in k-space, in Figure 14.1 (left).
• *The conical wavelets*
In order to achieve a genuinely oriented wavelet, it suffices to consider a smooth function $\widehat{\psi}^{(C)}(k)$ with support in a strictly convex cone C in spatial frequency space and behaving inside C as $P(k_1, \ldots, k_n)e^{-\zeta \cdot k}$, with $\zeta \in C$ and $P(.)$ denoting a polynomial in n variables. Alternatively, one may replace the exponential by a Gaussian, which gives a better localization in

spatial frequency. Thus one obtains the class of *conical* wavelets. Let us study in some detail the former case, called the *Cauchy* wavelets.

Take first $n = 2$. Let $\mathcal{C} \equiv \mathcal{C}(\alpha,\beta) = \{k \in \mathbb{R}^2 \,|\, \alpha \leq \arg k \leq \beta\}$ be the convex cone given by the directions α and β, with corresponding unit vectors e_α, e_β. The dual cone, also convex, is

$$\widetilde{\mathcal{C}} \equiv \mathcal{C}(\tilde{\alpha},\tilde{\beta}) = \{k \in \mathbb{R}^2, k \cdot k' > 0, \;\forall k' \in \mathcal{C}(\alpha,\beta)\}, \qquad (14.24)$$

where $\tilde{\beta} = \alpha + \pi/2$, $\tilde{\alpha} = \beta - \pi/2$, and therefore $e_{\tilde{\alpha}} \cdot e_\beta = e_{\tilde{\beta}} \cdot e_\alpha = 0$, whereas $e_{\tilde{\alpha}} \cdot e_\alpha = e_{\tilde{\beta}} \cdot e_\beta = \sin(\beta - \alpha)$. Given a fixed vector $\eta \in \widetilde{\mathcal{C}}(\tilde{\alpha},\tilde{\beta})$, we define the 2-D Cauchy wavelet in spatial frequency variables [32, 33, 38]:

$$\widehat{\psi}^{(\mathcal{C})}_{lm}(k) = \begin{cases} (k \cdot e_{\tilde{\alpha}})^l \, (k \cdot e_{\tilde{\beta}})^m \, e^{-k \cdot \eta}, & k \in \mathcal{C}(\alpha,\beta); \\ 0, & \text{otherwise}. \end{cases} \qquad (14.25)$$

The Cauchy wavelet $\widehat{\psi}^{(\mathcal{C})}_{ml}(k)$ is strictly supported in the cone $\mathcal{C}(\alpha,\beta)$ and the parameters $l, m \in \mathbb{N}^*$ give the number of vanishing moments on the edges of the cone. An explicit calculation yields the following result:

$$\psi^{(\mathcal{C})}_{lm}(x) = \text{const.} \, (z \cdot e_\alpha)^{-l-1} \, (z \cdot e_\beta)^{-m-1}, \qquad (14.26)$$

where we have introduced the complex variable $z = x + i\eta \in \mathbb{R}^2 + i\widetilde{\mathcal{C}}$. We show in Figure 14.1 (right) the wavelet $\widehat{\psi}^{(\mathcal{C})}_{44}(k)$ for $\mathcal{C} = \mathcal{C}(-10°, 10°)$. This is manifestly a highly directional filter.

Actually, the origin of the name "Cauchy" is the following example. For $\alpha = 0, \beta = \pi/2, \eta = e_{\pi/4}$, and $l = m = 1$, one gets

$$\psi^{(\mathcal{C})}_{11}(x) = \frac{1}{2\pi} \, (1 - ix)^{-2}(1 - iy)^{-2}, \qquad (14.27)$$

i.e., the product of two 1-D Cauchy wavelets [Hol], that is, derivatives of the Cauchy kernel $(z-t)^{-1}$. Of course, this example is of little use in practice. The main interest of Cauchy wavelets is their good angular selectivity, which requires a narrow cone. For applications, it turns out that the wavelet $\psi^{(\mathcal{C}_{10})}_{44}$, with support in the cone $\mathcal{C}_{10} = \mathcal{C}(-10°, 10°)$ has properties very similar to those of the Morlet wavelet (14.22) with $|k_o| = 5.6$, except that here the opening angle of the cone is totally controllable (see Section 14.3.2 for the calibration problem). For a Morlet wavelet, on the contrary, the cone gets narrower for increasing $|k_o|$, but then the amplitude decreases as $\exp(-|k_o|^2)$. In that sense, Cauchy wavelets are better adapted. This may be related to the fact that they have minimal uncertainty, as discussed in the next section.

An alternative possibility is to replace in (14.25) the exponential by a Gaussian centered on the axis $\zeta_{\alpha\beta} = e_{\frac{1}{2}(\alpha+\beta)}$ of the cone, $\exp(-|k - a_o \zeta_{\alpha\beta}|^2)\; (a_o > 0)$. The resulting conical wavelet is very similar to the previous one, except that it is more concentrated in spatial frequency space, since it is also localized in scale, around the central scale a_o. Although the pure Gaussian is well peaked, however, the addition of a large number of vanishing moments tends to spread it. Thus, one can achieve an even better scale localization by using an appropriate width for the Gaussian. We have found empirically that the following function, strongly peaked along the x-axis, has an almost ideal behavior:

$$\widehat{\psi}_{lm\sigma}^{(C)}(k) = \begin{cases} (k \cdot e_{-\tilde{\alpha}})^l \, (k \cdot e_{\tilde{\alpha}})^m \, e^{-\frac{1}{2}\sigma(k_x - k_o)^2}, & k \in C(-\alpha, \alpha), \\ 0, & \text{otherwise,} \end{cases} \quad (14.28)$$

with $k_o = \frac{(\sigma-1)}{\sigma}(l+m)^{1/2}$. This shows how wavelets may be tailored for specific applications. For instance, the conical wavelet (14.28), with supporting cone $C_{10} = C(-10°, 10°)$, has been used in the analysis of the Penrose tiling depicted in Figure 14.4.

Let us take now $n = 3$ and consider the convex simplicial (or pyramidal) cone $C(\alpha, \beta, \gamma)$ defined by the three unit vectors $e_\alpha, e_\beta, e_\gamma$, the angle between any two of them being smaller than π. The dual cone is also simplicial, namely, $\widetilde{C} = C(\tilde{\alpha}, \tilde{\beta}, \tilde{\gamma})$, where $e_{\tilde{\alpha}} = e_\beta \wedge e_\gamma$ is orthogonal to the β-γ face, and similarly for $e_{\tilde{\beta}} = e_\gamma \wedge e_\alpha$ and $e_{\tilde{\gamma}} = e_\alpha \wedge e_\beta$. With these notations, given a vector $\eta \in \widetilde{C}$ and $l, m, n \in \mathbb{N}^*$, we define a 3-D Cauchy wavelet in spatial frequency space as:

$$\widehat{\psi}_{lmn}^{(C,\eta)}(k) = \begin{cases} (k \cdot e_{\tilde{\alpha}})^l \, (k \cdot e_{\tilde{\beta}})^m \, (k \cdot e_{\tilde{\gamma}})^n \, e^{-k \cdot \eta}, & k \in C(\alpha, \beta, \gamma), \\ 0, & \text{otherwise.} \end{cases}$$
$$(14.29)$$

Here too, the expression for the 3-D wavelet in position space may be obtained explicitly as

$$\psi_{lmn}^{(C,\eta)}(x) = \frac{i^{l+m+n+3}}{2\pi} \, l! \, m! \, n! \cdot \det A \cdot \frac{(e_{\tilde{\alpha}} \cdot e_\alpha)^l \, (e_{\tilde{\beta}} \cdot e_\beta)^m \, (e_{\tilde{\gamma}} \cdot e_\gamma)^n}{(z \cdot e_\alpha)^{l+1} \, (z \cdot e_\beta)^{m+1} \, (z \cdot e_\gamma)^{n+1}}, \quad (14.30)$$

where A is the matrix that transforms the unit vectors e_1, e_2, e_3 into the triple $e_\alpha, e_\beta, e_\gamma$ and we have written $z = x + i\eta$. From the expressions (14.29) and (14.30), one may then obtain other 3-D Cauchy wavelets, for instance, one supported in a circular cone. Take a circular convex cone, aligned on the positive k_z-axis, with total opening angle $2\theta_o$ ($0 < \theta_o < \pi/2$). In spherical polar coordinates $k = (|k|, \theta, \phi)$, the interior of the cone is simply $C(\theta_o) = \{k \in \mathbb{R}^3 \mid \theta \leq \theta_o\}$. Then, an axisymmetric 3-D Cauchy wavelet supported in this cone may be defined, for instance, by

$$\widehat{\psi}_m^{(\theta_o)}(k) = \begin{cases} |k|^l (\tan^2 \theta_o - \tan^2 \theta)^m \, e^{-|k|\cos \theta}, & 0 \leq \theta \leq \theta_o; \\ 0, & \text{otherwise.} \end{cases} \quad (14.31)$$

Again, $m \in \mathbb{N}$ defines the number of vanishing moments on the surface of the cone, that is, the regularity of the wavelet. For very small θ_o, this wavelet lives inside a narrow pencil: It clearly evokes the beam of a searchlight — a vivid illustration of the wavelet as a directional probe!

If we note that the expression on the right-hand side of (14.31) may be written as $|k|^{l'} (\tan^2 \theta_o \, k_z^2 - k_x^2 + k_y^2)^m \, e^{-k_z}$, we see that all of these wavelets are built on the same model, namely, $F(k)^m e^{-k_z}$, where $F(k) = 0$ is the equation of the cone. Clearly, the whole construction extends to any number of dimensions $n \geq 2$, and the last remark gives a hint on how to design Cauchy or conical wavelets adapted to a general cone.

An interesting property of these conical wavelets, in any dimension, is their analyticity. Indeed, since the function $\widehat{\psi}_{lm}^{(C)}(k)$ has support in the convex cone C and is of fast decrease at infinity, it follows from general theorems [StW] that its Fourier transform $\psi^{(C)}(x)$ is the boundary value of a function $\psi_{lm}^{(C)}(z)$, holomorphic in the tube $\mathbb{R}^n + i\widetilde{C}$. This is seen from the expressions (14.26) and (14.30) of the Cauchy wavelets, for which the inverse Fourier transform can be computed explicitly.

Another interpretation is that we have here in fact a construction of an n-D progressive wavelet. Indeed, the definitions (14.25) and (14.29) show that the Cauchy wavelet is obtained by taking the directional derivative of the exponential $\exp(-k \cdot \eta)$ and then taking an analytic signal associated to it, by putting to zero the part that lives outside of the convex cone C. The same is true for more general conical wavelets of the type (14.28). In that sense, the conical wavelets yield a genuine multidimensional generalization of the 1-D Hardy functions obtained as Fourier transforms of progressive wavelets [156], much more so than the so-called 2-D Hardy functions defined in [95].

14.3 The 2-D case

In this section, we examine some additional aspects of 2-D wavelets, since this is the most important case for applications (image processing mainly, but not only).

14.3.1 Minimality properties

As we have seen in Chapter 2, the canonical coherent states have the characteristic property of *minimal uncertainty*, which means that they saturate the inequality in the Heisenberg uncertainty relations (2.2), and this is interpreted by saying that canonical CS are quantum states whose behavior is closest to classical. What about wavelets, which are the coherent states associated to the similitude groups?

According to the standard discussion in quantum mechanics textbooks [Coh, Got], two observables of a quantum system, represented by self-adjoint operators A and B, obey the uncertainty relation

$$\Delta A . \Delta B \geq \frac{1}{2} |\langle [A, B] \rangle|, \tag{14.32}$$

where
$$\Delta A \equiv \Delta_\phi A = \sqrt{\langle A^2 \rangle - \langle A \rangle^2} \tag{14.33}$$
denotes the variance of A in the state ϕ and $\langle C \rangle = \langle \phi | C \phi \rangle$ is the average of the operator C in the state ϕ. The state ϕ is said to have *minimal uncertainty* if equality holds in (14.32), which happens iff
$$(A - \langle A \rangle)\phi = -i\lambda_o (B - \langle B \rangle)\phi, \tag{14.34}$$
for some $\lambda_o > 0$. (The notion of minimal uncertainty has also been emphasized, and the result (14.34) proven, by Gabor in his pioneering paper on communication [129].)

In the case of canonical CS, the noncommuting operators are Q and P, that is, the infinitesimal generators of translations in phase space. In order to apply this concept to 2-D wavelets, we return to the discussion following Theorem 14.2.1, which showed that the 2-D CWT is also a phase space representation. Accordingly, we consider the infinitesimal generators of the transformation (14.1) or its equivalent (14.2) in k-space and denote them by P_1 and P_2 for translations, D for dilations, and J for rotations. Among these are four nonzero commutators, namely
$$[D, P_1] = iP_1, \quad [J, P_2] = -iP_1, \quad [D, P_2] = iP_2, \quad [J, P_1] = iP_2, \tag{14.35}$$
but the first two transform into the last two under a rotation by $\pi/2$. Thus, it is enough to consider the uncertainty relations for the first pair:
$$\Delta D . \Delta P_1 \geq \frac{1}{2} |\langle P_1 \rangle|, \qquad \Delta J . \Delta P_2 \geq \frac{1}{2} |\langle P_1 \rangle|. \tag{14.36}$$
Then, according to (14.34), a vector $\widehat{\psi}$ saturates these inequalities iff it satisfies the following system of equations:
$$\begin{array}{l} (D + i\lambda_1 P_1)\widehat{\psi}(k) = (\langle D \rangle + i\lambda_1 \langle P_1 \rangle)\widehat{\psi}(k) \\ (J + i\lambda_2 P_2)\widehat{\psi}(k) = (\langle J \rangle + i\lambda_2 \langle P_1 \rangle)\widehat{\psi}(k) \end{array} \quad (\lambda_1, \lambda_2 > 0). \tag{14.37}$$
Solving this system of partial differential equations in polar coordinates, one finally obtains that a real wavelet $\widehat{\psi}$ is minimal with respect to the first pair of commutation relations (14.35) iff it vanishes outside some convex cone $\mathcal{C}$ in the half-plane $k_x > 0$ and is exponentially decreasing inside:
$$\widehat{\psi}(k) = \begin{cases} c|k|^\kappa \, e^{-\lambda \, k \cdot e_1} & (\kappa > 0, \, \lambda > 0), \quad k \in \mathcal{C}, \\ 0, & \text{otherwise.} \end{cases} \tag{14.38}$$
Now, if we had chosen the second pair in (14.35) instead, we would have obtained a convex cone in the lower half-plane $k \cdot e_2 < 0$. Combining the two results, we see that the wavelet $\widehat{\psi}$ is minimal with respect to all commutation relations (14.35) simultaneously iff its support is contained in the lower right quarter plane, $k \cdot e_1 > 0, \, k \cdot e_2 < 0$. Since the whole construction is rotation invariant, this in turn means that the opening angle of the supporting cone must be strictly smaller than $\pi/2$.

318   14. Multidimensional Wavelets

Proposition 14.3.1:
A real wavelet ψ has minimal uncertainty iff it vanishes outside some convex cone $\mathcal{C}$, with apex at the origin and opening angle $\Phi < \pi/2$, and is exponentially decreasing inside:

$$\widehat{\psi}(k) = \begin{cases} c\,|k|^{\kappa}\, e^{-\lambda\, k\cdot\eta} & (\kappa > 0,\ \lambda > 0,\ \eta \in \widetilde{\mathcal{C}}),\quad k \in \mathcal{C} \\ 0, & \text{otherwise.} \end{cases} \tag{14.39}$$

In other words, $\widehat{\psi}$ must be of the form

$$\widehat{\psi}(k) = c\,\chi_{\mathcal{C}}(k)\,|k|^{\kappa}\, e^{-\lambda\, k\cdot\eta} \quad (\kappa > 0,\ \lambda > 0), \tag{14.40}$$

where $\chi_{\mathcal{C}}$ is the characteristic function of $\mathcal{C}$, or a smoothened version thereof.

We may now impose some degree of regularity (vanishing moments) at the boundary of the cone by taking an appropriate linear superposition of such minimal wavelets ψ. Thus, we obtain finally

$$\widehat{\psi}^{\mathcal{C}}(k) = c\,\chi_{\mathcal{C}}(k)\, F(k)\, e^{-\lambda\, e_1 \cdot k}, \quad (\lambda > 0) \tag{14.41}$$

where $F(k)$ is a polynomial in k_x, k_y, vanishing at the boundaries of the cone $\mathcal{C}$, including the origin. Clearly, a Cauchy wavelet with $\eta = e_1$ is of this type.

Other minimal wavelets may be obtained if one includes commutators with elements of the enveloping algebra, i.e., polynomials in the generators. For instance, taking the commutator between D and the Laplacian $-\Delta = P_1^2 + P_2^2$, one finds a whole family of minimal isotropic wavelets, among them all powers of the Laplacian, Δ^n, acting on a Gaussian [34]. For $n = 2$, this gives the 2-D isotropic Mexican hat [94]. By comparison, the same problem in 1-D was posed and solved by Klauder [190], with the result that the minimal wavelets are the equivalent of the Cauchy wavelets popularized in [Pau, 244], namely, $\widehat{\psi}_m(\xi) = \xi^m\, e^{-\xi}$, for $\xi \geq 0$ ($m > 0$), and 0 for $\xi < 0$.

14.3.2 Interpretation, visualization problems, and calibration

As we have said above, all formulas for the 2-D continuous wavelet transform are the exact analogues of the corresponding ones in 1-D, which we discussed in Chapter 12. Hence, the interpretation of the CWT as a singularity scanner is also the same. In practice, indeed, one assumes the wavelet ψ to be well localized both in position space (x) and in spatial frequency space (k). From this, one deduces again that wavelet analysis operates at constant *relative bandwidth*, $\Delta|k|/|k| = $ const. Therefore, the analysis is most efficient at high frequencies or small scales, and so it is particularly apt at detecting *discontinuities* in images, either point singularities (contours, corners) or directional features (edges, segments).

In addition, the wavelet ψ is often required to have a number of *vanishing moments* (as, for instance, the mth order Mexican hat or the

Cauchy wavelets), which increases its capacity at detecting the singular part of images, in particular, the discontinuities (where the most significant information lies).

In conclusion, as in the 1-D case, the 2-D wavelet transform may be interpreted as a mathematical, direction selective, microscope, with optics ψ, magnification $1/a$, and orientation-tuning parameter θ [50]. Notice that the magnification is global, independent of the direction, but one has the additional property of directivity, given by the rotation angle θ.

Now comes a striking difference between 1-D and 2-D analyses. In 1-D, the wavelet parameters (a, b) span a half-plane, so that a lot of information may be obtained visually (one may say that the wavelet transform unfolds the signal from 1-D to 2-D, thus decoupling parts that live at different scales). Some sophisticated methods of feature detection exploit this, for instance, the WTMM method (wavelet tranform modulus maxima) [50, 108], which is based on the ridges of the transform. But in 2-D, one faces a problem of visualization, since the transform $S(b, a, \theta)$ is a function of four variables, two position variables $b \in \mathbb{R}^2$ and the pair $(a, \theta) \in \mathbb{R}_*^+ \times [0, 2\pi) \simeq \mathbb{R}_*^2$.

Since it is impossible to compute and visualize the full transform, it is necessary to fix certain of the variables a, θ, b_x, b_y. Six choices of 2-D sections are possible, but the phase space interpretation discussed above indicates that two of them are more natural. Either (a, θ) or (b_x, b_y) are fixed, and the WT is treated as a function of the two remaining variables. Thus, we obtain the two standard representations [28, 32, 33] of image analysis.

(1) The *position or aspect-angle representation:* a and θ are fixed and the CWT is considered as a function of position b alone. Alternatively, one may use polar coordinates, in which case the variables are interpreted as *range* $|b|$ and *perception angle* α, a familiar parametrization of images.

(2) The *scale-angle representation:* For fixed b, the CWT is considered as a function of scale a and anisotropy angle θ, i.e., of spatial frequency. In other words, one looks at the full transform as through a keyhole located at b and observes all scales and all directions at once.

The position representation is the most familiar one, and it is useful for the general purposes of image processing: Detection of position, shape, and contours of objects; pattern recognition; and image filtering by resynthesis after elimination of unwanted features (for instance, noise). The scale-angle representation will be particularly interesting whenever scaling behavior (as in fractals) or angular selection is important, in particular, when directional wavelets are used. In fact, both representations are needed for a full understanding of the properties of the continuous wavelet transform in all four variables.

320   14. Multidimensional Wavelets

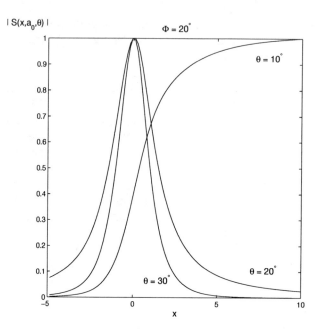

Figure 14.2. Testing the angular selectivity of the Cauchy wavelet $\psi_{44}^{(C_{10})}$ with the semi-infinite rod signal. The figure shows the modulus of the CWT as a function of x, for various values of the misorientation angle θ.

In addition to these two familiar representations, there are four other 2-D sections, obtained by fixing two of the four variables $(a, \theta, |b|, \alpha)$ and analyzing the CWT as a function of the remaining two. Among these, the *angle-angle representation* might be useful for applications [31]. Here, one fixes the range $|b|$ and the scale a and considers the CWT at all perception angles α and all anisotropy angles θ. This case is particularly interesting, because the parameter space is now compact (it is a torus).

The last point we want to address in this more practical section is that of *calibration*. We have compared the continuous wavelet transform to a (mathematical) microscope, so one should determine its *resolving power*. This is particularly true in 2-D: How well does the wavelet tool separate scales or directions? Interestingly enough (from our point of view), it turns out that a large part of the answer resides in the reproducing kernel K. This was to be expected, since K is in fact the autocorrelation function of the wavelet. Of course, K has to be studied in all four variables (a, θ, b_x, b_y), that is, both in the position and in the scale-angle representations. The latter, in particular, yields the angular information, and that is where directional wavelets in fact originate from.

In addition, one also uses benchmark signals for testing definite properties. In order to give the flavor of the technique evoked here, let us give an example of wavelet calibration. For testing the angular selectivity of the

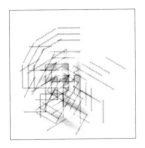  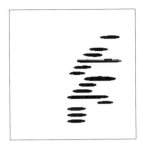

Figure 14.3. Directional filtering with the CWT: (a) The pattern; (b) CWT with a Cauchy wavelet supported in $\mathcal{C}(-10°, 10°)$; and (c) the same after thresholding.

wavelet $\psi_{44}^{(\mathcal{C}_{10})}$, one uses as benchmark signal a semi-infinite rod, sitting along the positive x-axis, and modeled with a δ-function,

$$s(\vec{x}) = \vartheta(x)\,\delta(y), \qquad (14.42)$$

where $\vartheta(x)$ is the step function. Then, one computes the CWT of s as a function of x, for the fixed scale $a = 1$. Thus, θ is the "misorientation" of the wavelet with respect to the signal (the rod). The result, shown in Figure 14.2, is that $\psi_{44}^{(\mathcal{C}_{10})}$ detects the orientation of the rod with a precision of the order of $10°$. Indeed, for $\theta = 0$, the WT is a "wall," increasing smoothly from zero, for $x \leq -5$, to its asymptotic value (normalized to 1) for $x \geq 5$. Then, for increasing misorientation θ, the wall gradually collapses and essentially disappears for $\theta > 10°$. Only the tip of the rod remains visible, and, for large θ ($\theta > 45°$), it gives a sharp peak. In the same way, one can also test the angular selectivity as a function of the opening angle of the support cone. We refer the reader to the original articles for more details [28, 32, 38].

The outcome of the calibration process is that a wavelet $\widehat{\psi}(k)$ with support in a narrow cone in spatial frequency space is highly selective in direction. In that case, the CWT offers the possibility of directional filtering, and this is crucial in many applications. In order to see what this means and to show the efficiency of a directional wavelet for that purpose, we consider in Figure 14.3(a) a pattern made of rods in many different directions. Applying the CWT with a Cauchy wavelet, oriented horizontally [that is, $\widehat{\psi}(k)$ is concentrated along the k_y-axis], selects all rods that have roughly the same direction (b), whereas the other ones, which are misaligned, yield only a faint signal corresponding to their tips. Since this is in fact noise, one performs a thresholding for removing it, thus getting an improved picture (c). The same two operations are then repeated with various successive orientations of the wavelet. In this way, one can count the number of objects that lie in any particular direction (similar results may be obtained with a Morlet wavelet [32]).

14.3.3 Practical applications of the CWT in two dimensions

Among the multidimensional wavelets, the case $n = 2$ is certainly the most interesting for applications, in particular, in image processing. The 2-D continuous wavelet transform has been used by a number of authors, in a wide variety of physical or engineering problems [Com, Me2, Me4]. In all cases, its main use is for the *analysis* of images, that is, the detection of specific features, for instance, a hierarchical structure or particular discontinuities, such as edges, filaments, contours, and boundaries between areas of different luminosity. Of course, the type of wavelet chosen, oriented or not, depends on the precise aim.

Among the most successful applications, we may quote the following.

1. *Contour detection, character recognition:*
   The contour or the edges of an object are discontinuities in luminosity. Hence, the CWT will detect them with efficiency [Mur, 28]. Possible applications are segmentation of images (e.g., medical images) or character recognition [31].

2. *Symmetry detection*
   The 2-D CWT with a highly directional wavelet (e.g., conical) allows one to detect discrete rotational or combined dilation–rotation symmetries, even locally, in objects like geometrical figures, Penrose tilings, or diffraction patterns of quasicrystals [38]. More details are given below.

3. *Analysis of astronomical images*
   The CWT has been used for several purposes in astronomy: Noise filtering (background sky), with a technique known as "unsharp masking," unraveling of the hierarchical structure of a galactic nebula, or that of the universe itself (galaxy counts, detection of galaxy clusters, or voids) [270]. Here, one usually couples an isotropic CWT analysis with statistical methods.

4. *Analysis of 2-D fractals*
   The CWT is an ideal tool for analyzing 2-D (multi)fractals, either artificial (numerical snowflakes, diffusion-limited aggregates) or natural (electrodeposition clusters, various arborescent phenomena), since the scaling behavior is the crucial aspect. Particular applications include the measurement of the fractal dimensions and the unraveling of universal laws (mean angle between branches, azimuthal Cantor structures, etc.), with help of the WTMM method [48, 50, 171]. In addition, a sophisticated statistical treatment is necessary (fractals are never exact), which leads to the so-called thermodynamical formalism of multifractal analysis [52]. Here too, wavelet maxima are the appropriate tool for characterizing the multifractals [179].

5. *Analysis of 2-D turbulence in fluids*

   Two-dimensional developed turbulence in fluids is a field in which the CWT gives new insights, in particular, concerning the localization of small scales in the distribution of energy or enstrophy and the evolution of coherent structures [120, 121]. Other applications in fluid dynamics include the visualization and measurement of a velocity field with the help of an oriented wavelet (see below).

6. *Texture analysis*

   The determination and classification of textures in images is an old and difficult problem, with many potential applications. Because most textures are oriented, it is natural to try and use the 2-D CWT with directional wavelets for attacking the problem. Some progress on the texture problem has been achieved recently along these lines [150, 180]. One of the key steps is the generalization to 2-D of the algorithm for measuring the instantaneous frequency of the signal (which becomes here the local wave vector) and the systematic use of the ridge or skeleton of the CWT, both familiar in 1-D (see Section 12.6).

7. *Medical physics and psychophysics*

   Medical imaging, in particular, 2-D NMR imaging and tomography, is an important application, and much work is in progress in this field. Another one is modeling of human vision, for instance, the definition of local contrast in images. This notion is based on a nonlinear extension of the CWT, which leads to interesting mathematical developments, under the name of infinitesimal multiresolution analysis [Duv, 27, 116], but also to faster algorithms, even competitive with the discrete WT.

As an illustration of the physical applications of 2-D wavelets, we give three examples, all of them of a directional nature. The first two concern fluid mechanics, and both rely on the possibility of directional filtering, described in the previous section.

The first example is a straightforward application of the method described above [Wis, 285]. The aim is to measure the velocity field of a 2-D turbulent flow around an obstacle. Velocity vectors are materialized by small segments (tiny plastic balls are injected into the stream and photographed with a fast CCD camera). Then, the WT with a Morlet wavelet is computed twice. First, the WT selects those vectors that are closely aligned with the wavelet; then, the analysis is repeated with a wavelet oriented in the orthogonal direction, thus, completely misoriented with respect to the selected vectors. Now the WT sees only the tips of the vectors and their length may be easily measured. Using appropriate thresholdings, the complete velocity field may thus be obtained, in a totally automated fashion, with an efficiency at least comparable to that of more traditional methods.

A second example concerns the disentangling of a wave train, represented by a linear superposition of damped plane waves. The problem originates from underwater acoustics: When a point source emits a sound wave above the surface of water, the wave hitting the surface splits into several components of very different characteristics (called, respectively, direct, lateral, and transient), and the goal is to measure the parameters of all components. This phenomenon has been analyzed successfully with the WT both in 1-D [265] and in 2-D [32]. In the latter case, the signal representing the underwater wave train is taken as a linear superposition of damped plane waves,

$$f(x) = \sum_{n=1}^{N} c_n \, e^{i k_n \cdot x} \, e^{-l_n \cdot x}, \tag{14.43}$$

where, for each component, k_n is the wave vector, l_n is the damping vector, and c_n is a complex amplitude. Then, using successively the scale-angle and the position representations described above, one is able to measure all of the $6N$ parameters of this signal with surprising ease and precision. As for the 3-D version, it is not more difficult, except for the numerical cost.

Our third example shows the CWT as a symmetry scanner. Many objects are invariant, at least locally, under specific rotations, sometimes combined with dilation with a given factor (inflation), such as, for instance, regular geometrical figures (a square, a hexagon), Penrose tilings, or the diffraction pattern of quasicrystals. For such objects, the 2-D CWT with a highly directional wavelet (e.g., conical) is an efficient symmetry detector. The relevant tool is the so-called *scale-angle measure* of the object (sometimes called the *wavelet spectrum*). For a signal $s(x)$, this is the function

$$\mu_s(a, \theta) = \int_{\mathbb{R}^2} |S(b, a, \theta)|^2 \, d^2 b, \tag{14.44}$$

which may also be interpreted as the partial energy density in the scale-angle variables. If only the rotational behavior is required, one may further integrate μ_s over a. When the object is invariant under rotation by $2\pi/n$, this function is $2\pi/n$-periodic in θ. If the object is invariant under dilation by a factor a_o, the same behavior is visible in the function $\mu_s(a, \theta)$. Thus, in the case of an inflation invariance, $\mu_s(a, \theta)$ is a doubly periodic function.

We give in Figure 14.4 an application of this analysis on a Penrose tiling (top). The figure on the bottom left shows the scale-angle measure $\mu_s(a, \theta)$ itself, and the one on the right its local maxima. On this picture, it is clear that the object is invariant under dilation by τ, the golden mean, under rotation by $\pi/5$, but also under a combined rotation–dilation symmetry, namely a rotation by $\pi/10$, together with a dilation by a factor $\lambda = 1.36$. In this particular case, to achieve a better scale localization, we have used the conical wavelet (14.28), with parameters $m = n = 4$, $\sigma = 16$.

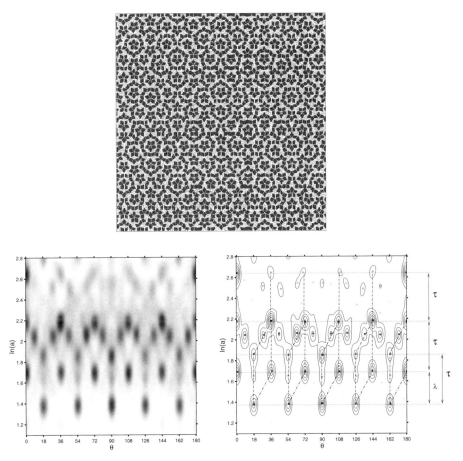

Figure 14.4. Symmetry detection with the CWT: (top) A Penrose tiling; (bottom) the corresponding scale-angle measure $\mu_S(a,\theta)$, obtained with a conical wavelet of the type (14.28) supported in $\mathcal{C}(-10°, 10°)$ (left); the same in contour levels (right); this pattern has a rotation symmetry by $\pi/5$, a dilation symmetry by τ, and a mixed symmetry, consisting of a rotation by $\pi/10$, combined with a dilation by $\lambda = 1.36$. Homologous maxima are linked by a line segment.

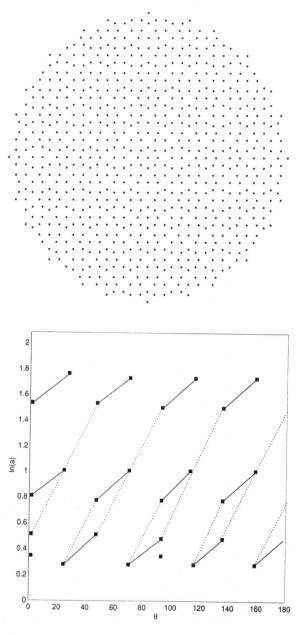

Figure 14.5. Similar analysis of an octagonal pattern: (a) The pattern; (b) the local maxima of its scale-angle measure $\mu_S(a,\theta)$; this pattern has a rotation symmetry by $\pi/4$ and two distinct mixed symmetries, consisting of a rotation by $\pi/8$, combined with a dilation by $\delta_1 = \sqrt{2}\cos(\pi/8)$, respectively, $\delta_2 = 2\cos(\pi/8)$. Homologous maxima are linked by a line segment, continuous for δ_1 and dashed for δ_2.

Vatious types of quasiperiodic lattices may be treated by the same method and they all give similar results. An example is given in Figure 14.5. As can be read from the graph of the scale-angle measure, this pattern has, in addition to the obvious invariance under rotation by $\pi/4$, *two* distinct combined rotation–dilation symmetries, namely, rotation by $\pi/8$ with a dilation by a factor $\delta_1 = \sqrt{2}\cos(\pi/8)$, respectively, $\delta_2 = 2\cos(\pi/8)$ (the first one is only approximate). Other examples include a tiling with fivefold symmetry built from the root diagram of a certain infinite-dimensional Lie algebra [43]; or a pattern of symmetry 12 generated by the Faraday instability in fluid dynamics [44]. In each case, the pattern possesses combined rotation–dilation symmetries, in addition to the obvious rotation symmetry. Remarkably enough, in all cases analyzed so far, these symmetries, previously unkown, were discovered on the graph of the scale-angle measure, *not* on the pattern itself. This illustrates the power of the method. Furthermore, a technique exists guaranteeing that no other hidden symmetry is present (see [38] and [179] for further details). An obvious application is the analysis of the diffraction patterns of physical quasicrystals or nanotubes, and there the hope is to uncover some physical properties of the object. At this stage, however, the question is open.

An interesting new feature appears in this type of situation: We have here (approximate) invariance under a *semigroup* (dilation + rotation). This opens the way to a whole new class of problems in wavelet (or CS) theory: What happens if groups are replaced by semigroups? The question is fascinating, but very little is known so far in this direction, except in the discrete abelian case, which we will discuss in detail in Chapter 16, Section 16.5.

14.3.4 The discrete WT in two dimensions

As we have seen in Section 13.1, a key step in the success of the 1-D *discrete* WT was the discovery that almost all examples of orthonormal bases of wavelets may be derived from a multiresolution analysis and, furthermore, that the whole construction may be transcripted into the language of QMF filters. In the 2-D case, the situation is exactly the same, as we shall sketch in this section. Further information may be found in [Me3] or [98].

The simplest approach consists in building a 2-D multiresolution analysis simply by taking the direct (tensor) product of two such structures in 1-D, one for the x direction and one for the y direction.

If $\{V_j, j \in \mathbb{Z}\}$ is a multiresolution analysis of $L^2(\mathbb{R})$, then $\{\widetilde{V}_j = V_j \otimes V_j, j \in \mathbb{Z}\}$ is a multiresolution analysis of $L^2(\mathbb{R}^2)$. Writing again $\widetilde{V}_j \oplus \widetilde{W}_j = \widetilde{V}_{j+1}$, it is easy to see that this 2-D analysis requires one scaling function, $\Phi(x,y) = \phi(x)\,\phi(y)$, but three wavelets:

$$\Psi^h(x,y) = \phi(x)\,\psi(y), \quad \Psi^v(x,y) = \psi(x)\,\phi(y), \quad \Psi^d(x,y) = \psi(x)\,\psi(y). \tag{14.45}$$

As the notation suggests, Ψ^h detects preferentially horizontal edges, that is, discontinuities in the vertical direction, Ψ^v and Ψ^d, vertical and oblique edges, respectively.

From these three wavelets, one gets an orthonormal basis of $\widetilde{V}_j$ as $\{\Phi^j_{kl}(x,y) = \phi_{j,k}(x)\,\phi_{j,l}(y),\ k,l \in \mathbb{Z}\}$, and one for $\widetilde{W}_j$ in the same way, namely, $\{\Psi^{\alpha,j}_{kl}(x,y), \alpha = h,v,d \text{ and } k,l \in \mathbb{Z}\}$.

As in the 1-D case, the implementation of this construction rests on the pyramidal algorithm of Mallat [211, 212]. The technique consists in translating the multiresolution structure into the language of QMF filters and putting suitable constraints on the filter coefficients. For instance, ψ has compact support if only finitely many of them differ from zero.

Clearly, this construction enforces a Cartesian geometry, with the horizontal and the vertical directions playing a preferential role. This is natural for certain types of images, such as in television, but is poorly adapted for detecting edges in arbitrary directions. Other solutions are possible, however. More isotropic wavelets may be obtained, either [Mar] by superposition of wavelets with specific orientation tuning or by choosing a different way of dilating, using a nondiagonal 2-D dilation matrix (which amounts to dilating by a noninteger factor) [98]. Consider, for instance, the following dilation matrices:

$$D_0 = \begin{pmatrix} 2 & 0 \\ 0 & 2 \end{pmatrix}, \quad D_1 = \begin{pmatrix} 1 & 1 \\ 1 & -1 \end{pmatrix}, \quad D_2 = \begin{pmatrix} 1 & 1 \\ -1 & 1 \end{pmatrix}. \quad (14.46)$$

The matrix D_0 corresponds to the usual dilation scheme by powers of 2, whereas D_1 and D_2 lead to the so-called "quincunx" scheme [Fea]. Actually, only one wavelet is needed in this scheme, instead of three. This is consistent with the theorem according to which the number of independent wavelets needed in a given multiresolution scheme equals $|\det D| - 1$, where D is the dilation matrix used.

The scheme based on orthonormal wavelet bases is, however, too rigid for most applications and various generalizations have been proposed, such as biorthogonal wavelet bases or wavelet packets or second-generation wavelets, exactly as in 1-D. In particular, the lack of translation covariance is a serious defect, for which various remedies have been proposed, for instance, the use of a Cartesian grid instead of the dyadic one — which makes a step toward the continuous transform, as in the construction of pseudo-QMFs, discussed in Section 14.3.5.

Another generalization of the standard DWT is the extension to other numeration systems, as we have discussed in one dimension in Section 13.4. Precisely, the case of τ-wavelets in the plane has already been explored in [141]. The aim is to construct wavelet bases that would be beter adapted to the intrinsic geometry of certain patterns, such as the various aperiodic tilings developed for studying quasicrystals [72]. This new orientation reveals the need for a more general algebraic structure than the one commonly used in multiresolution. Significant progress in this direction has been achieved recently [40], and we shall devote the final section of the book, Section 16.5 to a thorough discussion of these results.

Applications of the DWT

As with other methods, wavelet bases may be applied to all standard problems of image processing. The main problem of course is data compression, and, for achieving useful rates, one has to determine which information is really essential and which may be discarded with acceptable loss of image quality. Significant results have been obtained in the following directions:

1. representation of images in terms of *wavelet maxima* [215], as a substitute for the familiar zero-crossing schemes [Mar];

2. in particular, application of this maxima representation to the detection of edges and, more generally, detection and analysis of local singularities [213];

3. image compression, combining the previous wavelet maxima method for contours and biorthogonal wavelet bases for texture description [126];

4. denoising, by combining the DWT with statistical techniques [110].

Some applications are less conventional. For instance, a technique based on biorthogonal wavelet bases is nowadays used by the FBI for the identification of fingerprints [81]. The advantages over more conventional tools are the ease of pattern identification and the superior compression rates, which allow one to store and transmit a much bigger amount of information in real time. Another striking application is the deconvolution of noisy images from the Hubble Space Telescope, by a technique combining the DWT with a statistical analysis of the data [Bo1, 78, 257]. The results compare favorably in quality with those obtained by conventional methods, but the new method is much faster. Yet another field of applications (although it was done before the wavelet techniques were born) is constructive quantum field theory: Various perturbation expansions (the so-called "cluster expansion") used in the analysis of field theory models [67] are in fact discrete wavelet expansions (the summation over scales, indexed by j, was originally motivated by renormalization group arguments). Finally, one should also quote a large amount of work under development in the field of high definition television (HDTV), in which wavelet techniques are being actively exploited; here again, the huge compression rates make them specially interesting.

14.3.5 Continuous wavelet packets in two dimensions

Continuous wavelet packets may be defined in 2-D by a straightforward extension of the 1-D construction, described in Section 13.2. The full benefit of this approach, however, shows if one works in polar coordinates, for then the directional variable may be controlled. Starting from an isotropic wavelet ψ, one gets an isotropic scaling function $\widehat{\Phi}(|k|)$ and a family of isotropic integrated wavelets $\widehat{\Psi}_i^{\text{iso}}(|k|)$. Now, one can do better and design directional pseudo-QMFs as follows. Let $\{\eta_l(\theta), l = 1, \ldots d\}$ be a resolution of the identity consisting of C^∞, 2π-periodic, functions of compact support, so that $\sum_{l=1}^{d} \eta_l(\theta) = 1$. Then, one obtains a family of directional integrated wavelets, in polar coordinates, as

$$\widehat{\Psi}_{i,l}(|k|,\theta) = \widehat{\Psi}_i^{\text{iso}}(|k|)\, \eta_l(\theta), \tag{14.47}$$

and, indeed, one has

$$\sum_{l=1}^{d} \widehat{\Psi}_{i,l}(|k|,\theta) = \widehat{\Psi}_i^{\text{iso}}(|k|). \tag{14.48}$$

The net result is a discrete, fast, implementation of the 2-D CWT, including the directional degree of freedom. Here, the technique of periodizing the signal and adapting the length of the pseudo-QMFs to that of the signal becomes crucial for obtaining a fast algorithm. When this is done, the result looks very promising, as shown in preliminary applications, for instance, in directional filtering [Van, 147].

15
Wavelets Related to Other Groups

15.1 Wavelets on the sphere and similar manifolds

Several applications exist in which data to be analyzed are defined on a sphere, in geophysics or astronomy, of course, but also in statistics and other instances (then spheres of dimension higher than two might occur). If one is interested only in very local features, one may ignore the curvature and work on the tangent plane, but when global aspects become important (description of plate tectonics on the Earth, for instance), one needs a genuine generalization of wavelet analysis to the sphere. Several authors have studied this problem, with various techniques, mostly discrete (see, for instance [269] for an efficient solution, based on second-generation wavelets). To preserve the rotational invariance of the sphere, however, a continuous approach is clearly necessary. A solution has been proposed in [176], with several ad hoc assumptions. It turns out that the general formalism developed in this book yields an elegant solution to the problem [39] and, in particular, allows one to derive all assumptions of [176].

15.1.1 The two-sphere (♣)

We start with the two-sphere S^2 and consider signals in $L^2(S^2, d\varpi)$, where $d\varpi = \sin\theta \, d\theta \, d\phi$ denotes the usual Lebesgue measure on the sphere. The natural operations on such signals are motions on the sphere (translations) and local dilations. The former are given by rotations from $SO(3)$. Dilations around the North Pole are obtained by considering ordinary dilations in the tangent plane and lifting them to S^2 by stereographic projection from the South Pole (which corresponds to the point at infinity in the plane); thus, a dilation by a becomes a nonlinear map $\theta \mapsto \theta_a$ acting on the azimuthal angle:

$$\tan \frac{\theta_a}{2} = a \tan \frac{\theta}{2}. \tag{15.1}$$

As for dilations around any other point $\varpi \in S^2$, it suffices to bring ϖ to the North Pole by a rotation $\rho \in SO(3)$, perform the dilation and go back by the inverse rotation ρ^{-1}. Obviously translations and dilations do not commute. The only group combining only $SO(3)$ and the dilation group $A \equiv \mathbb{R}_*^+$, however, is their direct product. Indeed, $SO(3)$ has no outer automorphisms, which prevents the construction of a nontrivial semidirect product with $\mathbb{R}_*^+ \sim \mathbb{R}$.

A solution to this difficulty is to embed the two groups into the Lorentz group $SO_o(3,1)$, using the Iwasawa decomposition $SO_o(3,1) = SO(3) \cdot A \cdot N$, where $A \sim \mathbb{R}$ and N is a one-dimensional abelian group, isomorphic to the complex plane $\mathbb{C}$ (see Section 4.5.2). The justification of this is that the Lorentz group is the conformal group of the sphere S^2, and both rotations and dilations are conformal transformations [39]. Thus, the parameter space of the CWT is $X = SO_o(3,1)/N$, and a natural section is $\sigma(\rho, a) = \rho \cdot a \cdot 1$. Furthermore, $S^2 = SO_o(3,1)/P$, where $P = SO(2) \cdot A \cdot N$, the minimal parabolic subgroup [Kna, Lip], is the isotropy subgroup of the South Pole and $SO(2)$ consists of rotations around the z-axis. Hence, $SO_o(3,1)$ acts transitively on S^2.

In order to compute this action explicitly, two methods are available. The first one, which extends in a straightforward way to the same problem in higher dimensions [36], is to exploit the Iwasawa decomposition. The other one, specific to the dimension 2, is to go to $SL(2,\mathbb{C})$, the double covering of $SO_o(3,1)$, and to use the Gauss decomposition $SL(2,\mathbb{C}) = Z_+ B_-$, where $Z_+ \simeq \mathbb{C}$ (see Section 4.5.2). Thus, S^2 is now identified with the Riemann sphere and $\mathbb{C} = SL(2,\mathbb{C})/B_-$. The restriction to S^2 of the natural projection is the stereographic projection Φ (hence, it is bijective). In this way, the action $z \mapsto (\alpha z + \beta)(\gamma z + \delta)^{-1}$ of $SL(2,\mathbb{C})$ on $\mathbb{C}$ is lifted to S^2 by Φ^{-1}, or, more properly, that of $SL(2,\mathbb{C})/Z_2 \sim SO_o(3,1)$, which is simply transitive. This shows that the natural group to use is indeed the Lorentz group. For dilations, in particular, one recovers (15.1), by both methods. Clearly, we are in the general situation described in Chapter 7, Section 7.3. Since N corresponds to translations in the plane, P cannot leave any function invariant, unless it is constant. Thus, the CS we are going to define are not of the Gilmore–Perelomov type.

The next step is to find an appropriate UIR of $SO_o(3,1)$ in the space of signals $L^2(S^2, d\varpi)$. A possible choice is the following principal series representation [Kna, Lip], which is induced by the trivial character of $A \cdot N$:

$$[U(g)f](\varpi) = \lambda(a, \varpi)^{1/2} f(g^{-1}\varpi), \quad g = \rho \cdot a \cdot n. \tag{15.2}$$

15.1. Wavelets on the sphere and similar manifolds

In this relation, $\lambda(a, \varpi)$ is the correcting factor (Radon–Nikodym derivative) that takes into account the fact that the measure $d\varpi$ on S^2 is not invariant under dilations. For $\varpi = (\theta, \phi)$, one gets simply

$$\lambda(a, \varpi) = 4a^2 \left[(a^2 - 1)\cos\theta + (a^2 + 1)\right]^{-2}.$$

This representation is infinite dimensional. Its restriction to $SO(3)$ is the quasi-regular representation

$$[U_{qr}(\rho)f](\varpi) = f(\rho^{-1}\varpi), \; \rho \in SO(3), \tag{15.3}$$

which decomposes into the direct sum of all the familiar $(2l+1)$-dimensional representations, $l = 0, 1, \ldots$. The key point is that U is square integrable mod (N, σ).

Proposition 15.1.1: *The representation U, given in (15.2), is square integrable modulo the subgroup N and the section σ. A nonzero vector $\eta \in L^2(S^2, d\varpi)$ is admissible mod(N, σ) iff there exists $c > 0$, independent of l, such that*

$$\frac{1}{2l+1} \sum_{m=-l}^{l} \int_0^\infty \frac{da}{a^3} |\langle Y_l^m | \eta_a \rangle|^2 < c, \tag{15.4}$$

where Y_l^m denotes the usual spherical harmonic and $\eta_a = U(\sigma(e, a))\eta$ corresponds to a pure dilation.

The proof, as usual, consists in an explicit calculation, using the properties of Fourier analysis on the sphere ($\langle Y_l^m | \psi_a \rangle$ is a Fourier coefficient).

Thus, when ψ is admissible, the family $\{\psi_{\sigma(x)}, x \in X\}$ is a continuous family of CS, but in fact, we have more [39].

Proposition 15.1.2: *For any admissible vector η such that $\int_0^{2\pi} d\varphi\, \eta(\theta, \varphi) \neq 0$ (for instance, if η is axisymmetric), the family $\{\eta_{\sigma(x)}, x \in X\}$ is a continuous frame; that is, there exist constants $\mathsf{m} > 0$ and $\mathsf{M} < \infty$ such that*

$$\mathsf{m} \|\phi\|^2 \leq \int_X d\nu(x) |\langle \eta_{\sigma(x)} | \phi \rangle|^2 \leq \mathsf{M} \|\phi\|^2, \; \forall \phi \in L^2(S^2, d\varpi). \tag{15.5}$$

These two propositions yield the basic ingredient for writing the CWT on S^2. The wavelets on the sphere are the functions $\psi_{\rho,a} = U(\sigma(\rho, a))\psi$, with ψ admissible. Then, the CWT reads, with $U_S(\rho, a) = U(\sigma(\rho, a))$,

$$\begin{aligned} S(\rho, a) &= \langle U(\sigma(\rho, a))\psi | s \rangle \\ &= \int_{S^2} d\varpi\, \overline{[U_S(\rho, a)\psi](\varpi)}\, s(\varpi) \\ &= \int_{S^2} d\varpi\, \overline{\psi_a(\rho^{-1}\varpi)}\, s(\varpi). \end{aligned} \tag{15.6}$$

This spherical CWT has all of the desired properties, except covariance. Indeed, it is easily shown by a direct calculation [39] that:

1. The spherical CWT (15.6) is covariant under motions on S^2: For any $\rho_o \in SO(3)$, the transform of the rotated signal $s(\rho_o^{-1}\omega)$ is the function $S(\rho_o^{-1}\rho, a)$.

2. It is *not* covariant under dilations. Indeed, the wavelet transform of the dilated signal $\lambda(a_o, \omega)^{1/2} s(a_o^{-1}\omega)$ is $\langle U(g)\psi|s\rangle$, with $g = a_o^{-1}\rho a$, and the latter, while a well-defined element of $SO_o(3,1)$, is *not* of the form $\sigma(\rho', a')$.

This negative result of course reflects the fact that the parameter space X of the spherical CWT is not a group.

Now the full admissibility condition (15.4) of a wavelet ψ is somewhat complicated to use in practice, since it requires the evaluation of nontrivial Fourier coefficients. There is, however, a simpler, although only necessary, condition, namely,

$$\int d\varpi \, \frac{\psi(\theta, \varphi)}{1 + \cos\theta} = 0, \quad \varpi = (\theta, \varphi). \tag{15.7}$$

(Hence, ψ must vanish sufficiently fast when one approaches the South Pole.) This is a zero mean condition, so that we have the filtering effect, as usual. Thus, a genuine CWT on the sphere has been obtained. One should notice that the poles do not play any particular role in this CWT, since the sphere S^2 is a homogeneous space under $SO(3)$: All points of S^2 are really equivalent.

Typical wavelets satisfying the necessary condition (15.7) are difference wavelets, as in the flat case (14.21):

$$\psi_\phi = \phi - \frac{1}{\alpha} D^\alpha \phi, \quad \alpha > 1, \tag{15.8}$$

where ϕ is a smoothing function and $D^\alpha = U_s(e, \alpha)$ is again a pure dilation. Such a wavelet is fully admissible if ϕ is sufficiently regular at the poles. The simplest one, which is the spherical equivalent of the DOG, corresponds to $\phi_G(\theta, \varphi) = \exp(-\tan^2 \frac{\theta}{2})$. This is an axisymmetric wavelet, which yields an efficient detection of discontinuities on the sphere [39]. It is shown in Figure 15.1.

An additional bonus is that this CWT on the sphere has the expected Euclidean limit. By this, we mean the following: Consider instead of the unit sphere S^2 a sphere S_R^2 of radius R, and let $R \to \infty$. In technical terms, one performs a group contraction, keeping fixed the subgroup $SO(2)$ of rotations around the z axis, using the technique described in Section 4.5.4, with $\epsilon = R^{-1}$. Then, S_R^2 becomes the plane $\mathbb{R}^2$, the group $SO(3)$ becomes the Euclidean group of $\mathbb{R}^2$, and $\sigma(SO(3) \cdot \mathbb{R})$ becomes $SIM(2)$, with corresponding transitive action. Furthermore, the representation $U_R \equiv U_{S_R}$ becomes the natural representation U of (14.1) of $SIM(2)$, in the following sense. The representation U_R is realized in the Hilbert space $\mathfrak{H}_R = L^2(S_R^2, R^2 d\varpi)$, and U is realized in $\mathfrak{H} = L^2(\mathbb{R}^2, d^2 x)$. For each R, we choose $\mathcal{D}_R = \mathcal{D} = \mathcal{C}_0(\mathbb{R}^2)$,

15.1. Wavelets on the sphere and similar manifolds   335

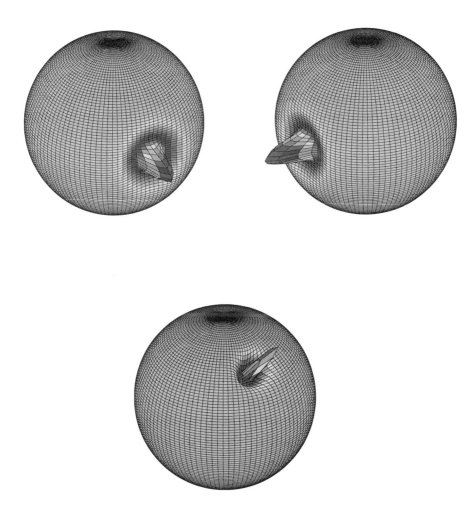

Figure 15.1. The spherical DOG wavelet. (Top line) The wavelet at scale $a = 0.25$, positioned on the equator $\theta = 90°$, with longitude $\varphi = -10°$ (left), respectively, $\varphi = -90°$ (right). (Bottom) The same at scale $a = 0.125$ and position $\theta = 45°$, $\varphi = -10°$.

the space of continuous functions of compact support, which is dense in $\mathfrak{H}$. The family of injections is defined as

$$(I_R f)(r,\varphi) = \frac{4R^2}{4R^2 + r^2} f\left(2\arctan\frac{r}{2R}, \varphi\right), \qquad (15.9)$$

where we have used polar coordinates (r, φ) in the plane. It is easily shown that I_R is in fact unitary from $\mathfrak{H}_R$ onto $\mathfrak{H}$. Then, one shows [39] that, for every $\phi \in \mathcal{D}$ and $g \in SIM(2)$, one has

$$\lim_{R \to \infty} \|I_R U_R(\Pi_R(g)) I_R^{-1}\phi - U(g)\phi\|_{\mathfrak{H}} = 0, \qquad (15.10)$$

which means precisely that the representation U of $SIM(2)$ is a contraction of the family of representations U_R of $SO_o(3,1)$ as $R \to \infty$. Finally, admissible vectors tend exactly to admissible vectors [also, the necessary admissibility condition (15.7) goes into the corresponding one (14.6) in the plane]. For instance, the spherical DOG wavelet tends to the usual flat space DOG. Therefore, the wavelet analysis on S^2 goes into the usual wavelet analysis in the plane.

In conclusion, the formulas (15.6) yield a genuine CWT on the sphere, entirely derived from group theory, following the formalism of general coherent states developed in this book. In addition, the Euclidean limit is valid, with a precise group-theoretical formulation. Preliminary tests with the spherical DOG wavelet show that it has the expected capability of detecting discontinuities, whether they lie at one of the poles of the sphere or not. The only remaining problem is of a computational nature. Indeed the formula (15.6) requires a pointwise convolution on the sphere, which is very time-consuming. This is not specific to wavelet analysis, however, it simply reflects the lack of an efficient convolution algorithm on the sphere and, in particular, the difficulty of finding an appropriate discretization of the latter.

Speaking about the sphere, it may be useful to emphasize that the wavelets described here have nothing to do with the coherent states associated to $SU(2)$, the spin CS, whose parameter space is the Bloch sphere S^2, which we have described in Section 7.2.1. Here the sphere is the manifold on which wavelets and signals live, whereas the parameter space is the noncompact manifold $X = SO_o(3,1)/N \sim SO(3) \cdot \mathbb{R}$.

15.1.2 Generalization to other manifolds

The whole scheme may be generalized to higher dimensions, essentially verbatim [36]. It can also be extended to other setups, for instance, a CWT on a two-sheeted hyperboloid [39]. In $\mathbb{R}^3$, this means $H^2 = SO_o(2,1)/SO(2)$, and the stereographic projection from either "pole" is available, mapping one sheet onto the interior of the unit disk in the plane tangent to the other pole, and the other sheet onto the exterior. Now this suggests a further generalization. In both cases, S^2 and H^2, the unit disk, image of one sheet or one hemisphere, is a classical domain. Also, the stereographic projection has a group-theoretical origin, in terms of a Gauss decomposition, as seen above. This paves the way for the generalization of the CWT to a whole class of homogeneous spaces (Riemannian symmetric spaces) [Hel]. The only problem is to find a suitable larger group containing dilations. The Lorentz group may be used in the case of the sphere or the hyperboloid, because it is the corresponding conformal group. For more general Riemannian symmetric spaces, however, the latter is not larger than the motion group; hence, another solution is needed. The case $n = 2$ gives a hint: The group $SL(2, \mathbb{C})$ is not only the double covering of $SO_o(3,1)$, it is also its complexification. Thus, a possible candidate could be the complexified group of the motion group of the manifold. At this stage, however, this is only a speculation.

Another direction that looks promising is to consider the wavelet problem in the general framework of representation theory of semisimple groups. Let G be a noncompact, connected, semisimple Lie group containing subgroups that can be interpreted as motions and dilations of a certain manifold Y. In mathematical terms, the existence of wavelets associated to G amounts to find an isometry from $L^2(X)$ to $L^2(Y)$, where X is either a subgroup G_o of G or a quotient $X = G/H$. In the first case, the central tool is a square integrable or discrete series representation of G_o; in the second one, a representation of the relative discrete series of X. At this stage, no general theory is available, but results have been obtained in a number of cases.

1. Let, again, $G = KAN$ be the Iwasawa decomposition of G (Section 4.5.2), with minimal parabolic subgroup $P = MAN$, and let $\overline{N} = \theta(N)$, where θ is the Cartan involution of G [Hel, Kna] In this case, under specific conditions that involve only $\overline{N}$, a wavelet transform exists for any measurable subset S of MA, with $X = \overline{N} \times S$ and U a principal series representation of G (induced by P), acting in $L^2(\overline{N})$ [187].

A particular case is that of $G = SL(n+2, \mathbb{R})$, with $X = \overline{N} \times A_1 \simeq H_n \times \mathbb{R}$, where $H_n \equiv G_{WH}$ is the Weyl–Heisenberg group in n dimensions and $A_1 \subset A$. Here also $Y = \overline{N} \simeq H_n$ [188].

2. Other wavelets on H_n have been obtained using closed subgroups of $SU(n+1, 1)$, either the affine Weyl–Heisenberg group [208] or the minimal parabolic subgroup P [164]. More precisely, let $U^{n+1} = \{(z, z_{n+1}) \in \mathbb{C}^{n+1} :$ Im $z_{n+1} > |z|^2\}$ be the Siegel upper half-plane. Then H_n is the boundary of U^{n+1} and P may be considered as the affine automorphism group of U^{n+1}, in the sense that it consists of operations that may be interpreted as translations, dilations, and rotations on U^{n+1}. Here again, P has a natural unitary representation in $L^2(H_n)$, which decomposes into a direct sum of square integrable representations, each of which generates a wavelet transform on H_n (one recovers the case of the affine Weyl–Heisenberg group simply by ignoring the rotations; we will treat this case in detail in Section 15.2).

3. The previous result extends to the group of real symmetric matrices [165]. Let $\mathcal{H}_n = \{z = x + iy \in \mathbb{C}^{n \times n},\, z^t = z, y \text{ real and positive definite}\}$ be the tube domain over the cone of real positive $n \times n$ symmetric matrices, and B be its Shilov boundary, that is, the space of $n \times n$ real symmetric matrices. The symplectic group $Sp(n, \mathbb{R})$ acts on $\mathcal{H}_n$, and it contains two automorphism groups of B, called P_o and P. The elements of P_o may be identified with translations and dilations, while P contains, in addition, rotations from $SO(n)$. Both P_o and P have UIRs on $L^2(B)$, which are square integrable, and B essentially reduces to the nilpotent component in the Iwasawa decomposition of $Sp(n, \mathbb{R})$. Thus, we are in a situation very similar to the previous one.

In all of those examples, one gets a wavelet transform, either on the Weyl–Heisenberg group H_n or on the space of real symmetric matrices, starting from various transitive group actions.

15.2 The affine Weyl–Heisenberg group (♣)

Combining dilations with the Weyl–Heisenberg group gives a very natural transformation group for signal processing, the affine Weyl–Heisenberg group G_{aWH}. It contains as subgroups both the affine group, which leads to the wavelet transform, and the Weyl–Heisenberg group, which gives the windowed Fourier transform (see Section 12.1), so the hope is that G_{aWH} might yield a tool that combines the advantages of the two techniques. CS associated with G_{aWH} have been constructed in [275] and [276] (see also [170]).

A generic group element $g = (b, \omega, a) \in G_{aWH}$ is the product of a translation b, a frequency shift ω, and a dilation $a \neq 0$. The group G_{aWH} is unimodular, and its invariant Haar measure is $|a|^{-1} db\, d\omega\, da$. G_{aWH} has a unique unitary irreducible representation, acting in $L^2(\mathbb{R}, dx)$, namely,

$$[U(b, \omega, a)f](x) = |a|^{-1/2} e^{i\omega x} f(\frac{x-b}{a}). \qquad (15.11)$$

(Actually, this is a projective representation, since we have dropped the phase factor coming from the central extension in G_{WH} throughout.) In any case, this representation is not square integrable. One way to obtain (weighted) CS is to multiply the Haar measure by some density function $\varrho(\omega)$ (at the price of losing the covariance with respect to frequency translations) and then indeed one gets a resolution of the identity [275].

Another possibility is to restrict oneself to suitable subgroups of G_{aWH}. Along this line, one may proceed in two essentially equivalent ways, as follows.

1. Find a suitable subgroup of G_{aWH} such that the restriction of (15.11) to it becomes square integrable. A physically natural choice is to

15.2. The affine Weyl–Heisenberg group (♣)

consider frequency translations as functions of the dilation parameter; that is, consider elements of G_{aWH} of the form $(b, \omega(a), a)$. Enforcing the subgroup property then gives (this amounts to solve a cohomological problem, i.e., to find a cocycle)

$$\omega_\lambda(a) = \lambda(\frac{1}{a} - 1), \quad \lambda \in \mathbb{R}, \tag{15.12}$$

and the subgroup $G_\lambda = \{(b, \omega_\lambda(a), a), a \neq 0, b \in \mathbb{R}\}$. The restriction of U to G_λ is indeed square integrable and yields ordinary CS.

One can also drop the requirement that G_λ be a subgroup and show by direct calculation that the most general function ω that yields a tight frame is

$$\omega(a) = \frac{\lambda}{a} + \mu, \quad \lambda, \mu \in \mathbb{R}. \tag{15.13}$$

In both cases, however, the result is only a trivial modification of the usual wavelet analysis, which may nevertheless have some practical interest, because it leads to a substantial gain in computing time.

2. Apply the general theory of Section 7.3. Define the one-parameter subgroup $H_\lambda = \{(0, \omega_\lambda(a), a) \in G_{aWH}\} \simeq \mathbb{R}$ and the corresponding homogeneous space $X_\lambda = G_{aWH}/H_\lambda$. With coordinates $(b, \omega) \in \mathbb{R}^2$, X_λ has the G_{aWH}-invariant measure $d\nu(b, \omega) = db\, d\omega$. Define a section $\sigma : X_\lambda \to G_{aWH}$ as $\sigma(b, \omega) = (b, \omega, \beta(b, \omega))$, with $\beta : X_\lambda \to \mathbb{R}_*^+$, a piecewise differentiable function. Then, a straightforward calculation shows that the most general section yielding a resolution of the identity (a tight frame) is given by a function β of the form

$$\beta(b, \omega) \equiv \beta(\omega) = (c\omega + d)^{-1}, \quad c, d \in \mathbb{R}, \omega \neq -d/c. \tag{15.14}$$

For $c = 0$, one gets a constant section, which leads to a standard Gabor analysis, whereas $c \neq 0$ gives wavelet analysis.

Thus, in order to get something new, one has to consider nontight frames. A natural solution has been found by Torrésani [276], and, essentially, the same result has been recovered in [170]. Given a vector $\eta \in L^2(\mathbb{R}, dx)$, define the positive function

$$\widehat{\chi}(\xi) = \int_\mathbb{R} |\widehat{\eta}((\xi - \omega)\beta(\omega))|^2 |\beta(\omega)|\, d\omega. \tag{15.15}$$

Call the function $\beta : \mathbb{R} \to \mathbb{R}_*^+$ admissible for η if the function $\widehat{\chi}$ is bounded and bounded away from zero, i.e., there exists two positive constants K_1, K_2 such that

$$0 < K_1 \leq \widehat{\chi}(\xi) < K_2 < \infty, \quad \text{a.e.}$$

Proposition 15.2.1: Let $G_{aWH} = \{(b, \omega, a)\}$ be the affine Weyl–Heisenberg group and U be its natural unitary representation (15.11) in $L^2(\mathbb{R}, dx)$. Let $\beta : \mathbb{R} \to \mathbb{R}_*^+$ be a piecewise differentiable function and

$\sigma_\beta : (b,\omega) \mapsto (b,\omega,\beta(\omega))$ be the corresponding section in the bundle $G_{aWH} \to X_\lambda = G_{aWH}/H_\lambda \simeq \mathbb{R}^2$, where H_λ is the subgroup of G_{aWH} of all elements of the form $(0, \lambda(a^{-1} - 1), a)$. Let $\eta \in L^2(\mathbb{R}, dx)$ be such that the function β is admissible for η. Then, the corresponding section σ_β is admissible for (U, η) and the vector η generates a frame of CS:

$$\begin{aligned}\eta_{\sigma_\beta(b,\omega)}(x) &= [U(\sigma_\beta(b,\omega))\eta](x), \\ &= |\beta(\omega)|^{-1/2} e^{i\omega x} \eta\left(\beta(\omega)^{-1}(x-b)\right), \\ & \quad (b,\omega) \in X_\lambda \simeq \mathbb{R}^2. \end{aligned} \quad (15.16)$$

The resolution generator A_{σ_β} is the operator of convolution by a function χ, which is the inverse Fourier transform of $\widehat{\chi}$ given in (15.15). Finally, a tight frame of weighted CS $\widetilde{\eta}_{b\omega}$ is obtained explicitly, in Fourier space, as

$$\widehat{\widetilde{\eta}_{b\omega}}(\xi) = \widehat{\chi}(\xi)^{-1/2} \widehat{\eta_{\sigma_\beta(b,\omega)}}(\xi). \quad (15.17)$$

The proof of this proposition reduces to a straightforward computation of the matrix element

$$\langle f | A_{\sigma_\beta} f \rangle = \iint_{\mathbb{R}^2} |\langle U(\sigma_\beta(b,\omega))\eta | f \rangle|^2 \, db \, d\omega = \int_{\mathbb{R}} \widehat{\chi}(\xi) |\widehat{f}(\xi)|^2 \, d\xi.$$

The rest is immediate.

In order to be admissible, a function β must satisfy a rather mild technical condition given in [276], which means, essentially, that it does not oscillate too wildly. Several explicit examples may be found in that paper, including of course the affine function $\beta(\omega) = (c\omega + d)^{-1}$ discussed previously. Another interesting example [170, 276] is the function

$$\beta_{comp}(\omega) = \begin{cases} 1, & \text{for } |\omega| \leq \omega_o, \\ \frac{\lambda}{|\omega| - \omega_o + \lambda}, & \text{for } |\omega| > \omega_o, \end{cases} \quad (15.18)$$

where λ and ω_o are two positive constants. The associated CS, called composite wavelet packets, behave as Gabor CS for low frequencies ($|\omega| < \omega_o$) and as wavelets for high frequencies. In this way, one truly obtains the interpolation between the two standard methods of time-frequency analysis. It may be noted that such a mixed behavior is quite common in physiological sensors, like the ear or the eye [DeV, Mar].

The CS constructed in Proposition 15.2.1 offer interesting perspectives in signal processing, for they constitute a continuous version of the widely used wavelet packets [Me2, 92] (see Section 13.1.3). (The objects described here have nothing to do with what we have called continuous wavelet packets in Section 13.2, following [116].)

An alternative approach, which has become popular in the signal processing community, is to ignore the problem of square integrability and

15.2. The affine Weyl–Heisenberg group (♣)

work directly with functions

$$h_{h,\omega,a}(x) \equiv (U(b,\omega,a)h)(x) = |a|^{-1/2} e^{i\omega x} h(\frac{x-b}{a}), \quad (b,\omega,a) \in G_{aWH}, \tag{15.19}$$

where the window h is a real C^∞ function in $L^2(\mathbb{R}, dx)$, with nonvanishing integral, $\int h(x)\, dx \neq 0$. A typical example is a Gaussian window, and it is of course optimal in the Gabor sense (i.e., with respect to uncertainty relations or phase space localization). Such functions $h_{b,\omega,a}$ are called *time-frequency atoms* (one usually takes $a > 0$, since it is meant as a scale variable). For a given function h, any family of such atoms is called a *dictionary*. The goal is then to find a suitable dictionary $\{h_\gamma, \gamma \in \Gamma \subset G_{aWH}\}$, usually countable, that allows us to expand into it an arbitrary function $f \in L^2(\mathbb{R}, dx)$, in such a way that the expansion is well adapted to the particular class of signals to be analyzed. An efficient algorithm to that effect has been introduced by Mallat and Zhang, under the name of *matching pursuit* [214], and it offers a good alternative to the best basis algorithm [92] mentioned in Section 13.1.3, case 2. Since the dictionary is discrete in general, however, this is really a discretization problem, so we will postpone the discussion to Chapter 16, Section 16.3.1.

The whole construction extends to n dimensions [185]. The n-dimensional affine Weyl–Heisenberg group, also denoted G_{aWH}, is a semidirect product of the n-dimensional Weyl–Heisenberg group (translations in position and in spatial frequency) by the group $\mathbb{R}_*^+ \times SO(n)$ of dilations and rotations. The unitary irreducible representations of G_{aWH} may be constructed by an extension of Kirillov's method of coadjoint orbits (G_{aWH} is not solvable for $n > 2$). Among them is the (projective) Stone–von Neumann representation in $L^2(\mathbb{R}^n, dx)$:

$$[U(q,p,a,R)f](x) = a^{-n/2} e^{ip\cdot(x-q)} f\left(R^{-1}(\frac{x-q}{a})\right), \tag{15.20}$$

with $R \in SO(n)$ and the other variables as before. This representation is not square integrable; hence, one has to find appropriate homogeneous spaces.

As shown in [185], many tight frames may be constructed for the following three spaces, by the same method as in the 1-D case:

1. $X_1 = G_{aWH}/H_1$, where H_1 is the subgroup of dilations.

2. $X_2 = H_2 \backslash G_{aWH}$, where H_2 is the subgroup of spatial frequency translations. Notice that here G_{aWH} acts on X_2 on the right. It follows that the resulting CS do *not* coincide with the multidimensional wavelets mentioned in Section 6.2.

3. $X_3 = G_{aWH}/H_3$, where $H_3 = \mathbb{R}_*^+ \times SO(n)$ is the subgroup of dilations and rotations.

In addition, nontight frames and the associated weighted CS may be obtained explicitly in all three cases, by the same method as before.

15.3 The affine or similitude groups of space–time

As mentioned at the end of Section 12.1, the similitude group associated to a given geometry is obtained by adding dilations to the corresponding isometry group. Both the $ax + b$ group and $SIM(n)$ are of this type, the latter being just the affine Euclidean group, and both lead to interesting CS, namely, wavelets. Thus, it is natural to look for CS associated to other similitude or affine groups.

The wavelets presented so far, however, are designed for analyzing static signals or images. When it comes to time-varying phenomena, such as moving objects or successive images (such as on a TV screen or in a movie), an additional time dependence must be included; that is, we have to look for affine groups in space–time. Here too, the general CS scheme is applicable, once one has identified the appropriate group of space–time transformations and its representation. We will first sketch the simplest solution [115]. More sophisticated ones, based on the affine Galilei group or the affine Poincaré group, will be discussed afterwards.

15.3.1 Kinematical wavelets (♣)

In n space dimensions, a time-dependent signal of finite energy may be represented by a function $s \in L^2(\mathbb{R}^{n+1}, d^n x dt)$. On such signals, the required operations are space translations ($\mathbb{R}^n$), time translations ($\mathbb{R}$), space rotations ($SO(n)$), and two independent dilations for space and time variables, respectively. The appropriate form of the latter becomes evident if one notices that our visual system introduces a correlation between the size and the speed of a moving object: In order to be visible, fast-moving objects have to be wide and narrow objects have to be slow. The following transformations (which are mathematically equivalent to two independent dilations) reproduce that pattern:

$$x \mapsto a\, c^{1/n+1} x, \qquad t \mapsto a\, c^{-n/n+1} t. \qquad (15.21)$$

The corresponding unitary operators in $L^2(\mathbb{R}^{n+1}, d^n x dt)$ read as

- global dilation : $(D^a s)(x,t) = a^{-(n+1)/2} s(a^{-1}x, a^{-1}t),$
$$a > 0,\ x \in \mathbb{R}^n,\ t \in \mathbb{R}; \qquad (15.22)$$

- speed tuning : $(A^c s)(x,t) = s(c^{1/n+1}x, c^{-n/n+1}t),$
$$c > 0,\ x \in \mathbb{R}^n,\ t \in \mathbb{R}. \qquad (15.23)$$

15.3. The affine or similitude groups of space–time

The two operations commute, so that scale analysis and speed analysis are independent, as they should be.

Putting all this together, one arrives at the following group:

$$G_{n,1} = \mathbb{R}^{n+1} \rtimes (\mathbb{R}_*^+ \times \mathbb{R}_*^+ \times SO(n)) \times \mathbb{Z}_2, \tag{15.24}$$

with elements $(b, \tau; a, c; R, \epsilon)$, where $b \in \mathbb{R}^n$ and $\tau \in \mathbb{R}$ denote space and time translations, respectively, $a > 0$ is a global dilation, $c > 0$ is the speed tuning, $R \in SO(n)$ is a rotation, and $\epsilon = \pm 1$ is time inversion (this additional parameter ensures the irreducibility of the representation that will be used). This group $G_{n,1}$ is a semidirect product of the type we have studied in Chapter 10.

The group law is

$$(b, \tau; a, c; R, \epsilon)(b', \tau'; a', c'; R', \epsilon') =$$
$$= (b + ac^{1/n+1} Rb', \tau + ac^{-n/n+1} \tau'; aa', cc'; RR', \epsilon\epsilon'). \tag{15.25}$$

The group $G_{n,1}$ is nonunimodular, with left and right Haar measures given, respectively, by

$$d\mu_l = \frac{da}{a} \frac{dc}{c} \frac{d^n b \, d\tau \, dR}{a^{n+1}}, \quad d\mu_r = \frac{da}{a} \frac{dc}{c} d^n b \, d\tau \, dR. \tag{15.26}$$

As in the previous cases, the group $G_{n,1}$ has a natural unitary representation U in the space $L^2(\mathbb{R}^{n+1}, d^n x dt)$ of finite energy signals:

$$[U(g)s](x, t) \equiv s_g(x, t) = a^{-(n+1)/2} s\left(\frac{c^{1/n+1}}{a} R(x - b), \epsilon \frac{t - \tau}{ac^{n/n+1}}\right),$$
$$g = (b, \tau; a, c; R, \epsilon) \in G_{n,1}. \tag{15.27}$$

This representation is irreducible and square integrable, and the admissibility condition reads as

$$c_\psi \equiv c_n \iint_{\mathbb{R}^{n+1}} \frac{|\widehat{\psi}(k, \omega)|^2}{|k|^n |\omega|} d^n k \, d\omega < \infty, \tag{15.28}$$

where $\widehat{\psi}(k, \omega)$ is the usual Fourier transform of $\psi(x, t)$ and c_n is a dimensional constant.

From there on, everything follows the traditional pattern. The wavelet transform is defined as $(T_\psi s)(g) = \langle \psi_g | s \rangle$, and it has all of the properties we have encountered previously. In particular, T_ψ is a multiple of an isometry (energy conservation):

$$\sum_{\epsilon = \pm 1} \int_{G_{n,1}} |(T_\psi s)(g)|^2 \, d\mu_l(g) = \frac{c_\psi}{2} \iint_{\mathbb{R}^{n+1}} |s(x, t)|^2 \, d^n x \, dt,$$
$$g = (b, \tau; a, c; R, \epsilon), \tag{15.29}$$

and there is an inverse transform (reconstruction formula):

$$s(x, t) = 2c_\psi^{-1} \int_{G_{n,1}} (T_\psi s)(g) \psi_g(x, t) \, d\mu_l(g). \tag{15.30}$$

In practice, one should take an axisymmetric wavelet, as in the purely spatial case (Section 14.2). Then, the rotation R is replaced again by a point ϖ on the unit sphere, the CWT lives on the space $G_{n,1}/SO(n-1) = \mathbb{R}^{n+1} \times \mathbb{R}_*^+ \times \mathbb{R}_*^n \times \mathbb{Z}_2$, and the wavelets so obtained are of the Gilmore–Perelomov type. The rest is as before.

Thanks to the filtering property in a and c, this CWT (called *kinematical*) is efficient in detecting moving objects: The angular parameter ϖ detects the direction of the target, the dilation parameter a catches its size, and the new parameter c adjusts the speed of the wavelet to that of the target. In more than one space dimension, the parameter c identifies the *speed* of the target, that is, the modulus of its velocity, but the angular variables will detect the direction of the movement, thus, the full velocity vector. Therefore the spatiotemporal CWT is a tool for motion tracking or target detection. Clearly, plenty of applications exist in which such a technique might be used [202].

Actually, one may go slightly further. In the case of kinematical wavelets, the motion (or relativity) group consists only of space–time translations and rotations. One can replace it by a more sophisticated one, either the Galilei or the Poincaré group, depending on which physical context one is working. If one then combines either of these with space–time dilations, one obtains the corresponding *affine* group. Wavelets for these groups have been constructed also, and we shall discuss them in the next section. Note that, in the nonrelativistic framework, it is not clear which time-dependent wavelets are the most efficient, the Galilean ones discussed in Section 15.3.2 or the purely kinematical ones described here (which are simpler).

15.3.2 The affine Galilei group (♣)

Let us begin by the most general case, that of the affine Galilei group in n space dimensions, obtained by combining the Galilei group with independent space and time dilations. Yet, a choice must be made here.

Indeed, in quantum mechanics, the natural approach would be to consider the semidirect product $\mathcal{G}_o \rtimes \mathcal{D}_2$ of the pure Galilei group $\mathcal{G}_o$ with a 2-D dilation group $\mathcal{D}_2 \equiv \mathcal{D}_2^{(n)} \equiv (\mathbb{R}_*^+)^2 \simeq \mathbb{R}^2$ and then to construct projective representations of the resulting group, by the standard method of central extensions [204], discussed in Section 4.5.3. A straightforward computation [41], however, shows that the only central extensions of this group by $\mathbb{R}$ are of the form $\mathcal{G}_o \rtimes G_{WH}$, where G_{WH} is the Weyl–Heisenberg group, itself a central extension of $\mathcal{D}_2$. In particular, the central extension procedure fails to generate mass, as it does in the usual situation, without dilations. The alternative is to take *first* a central extension of $\mathcal{G}_o$ and then the semidirect product with $\mathcal{D}_2$. The first step produces the extended Galilei group $\widetilde{\mathcal{G}}^{(m)}$, corresponding to the extension parameter $m \geq 0$ [204, 281]. As we shall see below, however, m should *not* be interpreted as mass, but rather as a

15.3. The affine or similitude groups of space–time

mass unit or mass scale (the physical mass varies under dilations, whereas m is fixed).

Thus, for physical reasons, we will consider as affine Galilei group the semidirect product $\mathcal{G}_{\text{aff}}^{(m)} = \widetilde{\mathcal{G}}^{(m)} \rtimes \mathcal{D}_2$, for $m > 0$, (actually, this group is an extension of $\mathcal{G}_o \rtimes \mathcal{D}_2$ by $\mathbb{R}$, but a noncentral one). We shall denote a generic element of $\mathcal{G}_{\text{aff}}^{(m)}$ by

$$g = (\theta, b, \mathbf{a}, \mathbf{v}, R, \lambda_0, \lambda), \tag{15.31}$$

where $\theta \in \mathbb{R}$ is the extension parameter in $\widetilde{\mathcal{G}}^{(m)}$, $b \in \mathbb{R}$ and $\mathbf{a} \in \mathbb{R}^n$ are the time and space translations, respectively, $\mathbf{v} \in \mathbb{R}^n$ is the boost parameter, $R \in SO(n)$ is a rotation, and $\lambda_0 \in \mathbb{R}_*^+, \lambda \in \mathbb{R}_*^+$, are time and space dilations, respectively. The action of g on space–time is then

$$\begin{cases} \mathbf{x} & \mapsto \lambda R \mathbf{x} + \lambda_0 \mathbf{v} t + \mathbf{a}, \\ t & \mapsto \lambda_0 t + b. \end{cases} \tag{15.32}$$

The group law reads as

$$gg' = (\theta + \frac{\lambda^2}{\lambda_0}\theta' + m(\lambda \mathbf{v} \cdot R\mathbf{a}' + \frac{1}{2}\lambda_0 \mathbf{v}^2 b'), b + \lambda_0 b', \mathbf{a} + \lambda R \mathbf{a}' + \lambda_0 \mathbf{v} b',$$
$$\mathbf{v} + \frac{\lambda}{\lambda_0} R \mathbf{v}', RR', \lambda_0 \lambda_0', \lambda \lambda'). \tag{15.33}$$

Moreover, the center of the group is trivial [that is, $\mathcal{G}_{\text{aff}}^{(m)}$ is a noncentral extension of $\widetilde{\mathcal{G}}^{(m)}$; it is also the automorphism group of $\widetilde{\mathcal{G}}^{(m)}$].

In order to build space–time wavelets associated to $\mathcal{G}_{\text{aff}}^{(m)}$, we consider a unitary representation of spin zero [41]. It may be obtained by direct unitary implementation of the action (15.32) on space–time, or by the Mackey method of induced representations, discussed in Section 10.2.4. The Hilbert space is $\mathfrak{H} = L^2(\mathbb{R}^n \times \mathbb{R}^2, d\mathbf{k}\, dE\, d\gamma)$, and the representation reads as

$$[U(g)\psi](\mathbf{k}, E, \gamma) = \lambda^{n+2/2}\, e^{i(\gamma\theta + Eb - \mathbf{k}\cdot\mathbf{a})}\, \psi(\mathbf{k}', E', \gamma'), \tag{15.34}$$

with

$$\mathbf{k}' = \lambda R^{-1}(\mathbf{k} - m\gamma \mathbf{v}), \quad E' = \lambda_0(E - \mathbf{k}\cdot\mathbf{v} + \frac{1}{2}m\gamma \mathbf{v}^2), \quad \gamma' = \frac{\lambda^2}{\lambda_0}\gamma. \tag{15.35}$$

These relations, which are in fact a part of the coadjoint action of $\mathcal{G}_{\text{aff}}^{(m)}$, show that the real mass parameter is the combination $m\gamma$, and it varies under dilations, as it should.

Now, since the sign of $(E - \frac{\mathbf{k}^2}{2m\gamma})$ and that of γ are both invariant under the transformation (15.35), the representation U splits into the direct sum of four irreducible subrepresentations U_j ($j = \pm\pm$), corresponding to the decomposition of $\mathbb{R}^n \times \mathbb{R}^2$ into four disjoint subsets according to the signs of these two invariants. In addition, the full representation U, and thus each subrepresentation U_j, is square integrable over $\mathcal{G}_{\text{aff}}^{(m)}$. For definiteness we choose U_{++}, corresponding to $\mathfrak{H}_{++} = L^2(\mathfrak{D}_{++}, d\mathbf{k}\, dE\, d\gamma)$, where

$$\mathfrak{D}_{++} = \{(\mathbf{k}, E, \gamma) \in \mathbb{R}^n \times \mathbb{R}^2 \mid \gamma > 0, E - \frac{\mathbf{k}^2}{2m\gamma} > 0\}. \tag{15.36}$$

Then, indeed, an explicit calculation shows that

$$\int_{\mathcal{G}_{\text{aff}}^{(m)}} |\langle U_{++}(g)\eta|\theta\rangle|^2\, dg = c\|\theta\|_{++}^2 \iiint_{\mathfrak{D}_{++}} d\mathbf{k}\, dE\, d\gamma \frac{|\eta(\mathbf{k}, E, \gamma)|^2}{\gamma^{n+1}(E - \frac{\mathbf{k}^2}{2m\gamma})}, \tag{15.37}$$

so that a vector $\eta \in \mathcal{H}_{++}$ is admissible iff the last integral converges. There is obviously a dense set of such admissible vectors. For any of them, η, one gets a tight frame of Galilean wavelets, indexed by $\mathcal{G}_{\text{aff}}^{(m)}$, as $\eta_g = U_{++}(g)\eta$ ($g \in \mathcal{G}_{\text{aff}}^{(m)}$), with the expression (15.34) for the representation. These are of course CS of the Gilmore–Perelomov type.

In addition, it is possible to construct sets of wavelets indexed by fewer parameters by taking quotients $X = \mathcal{G}_{\text{aff}}^{(m)}/H$ and appropriate sections $\sigma : X \to \mathcal{G}_m^{\text{aff}}$. The simplest example is obtained if we take for H the subgroup $\mathcal{D}_0 \equiv \mathbb{R}_*^+$ of time dilations λ_0. The corresponding coset space $X = \mathcal{G}_{\text{aff}}^{(m)}/\mathcal{D}_0$ is parametrized by points $x = (\theta, b, \mathbf{a}, \mathbf{v}, R, \lambda)$. We consider first the basic section $\sigma_1 : X \to \mathcal{G}_m^{\text{aff}}$,

$$\sigma_1(x) = (\theta, b, \mathbf{a}, \mathbf{v}, R, 1, \lambda). \tag{15.38}$$

A straightforward calculation shows that the admissibility condition reduces to setting $E - \frac{\mathbf{k}^2}{2m\gamma} = 1$ (or a constant) in the integral (15.37); that is, a vector $\eta \in \mathfrak{H}_{++}$ is admissible $\text{mod}(\mathcal{D}_0, \sigma_1)$ iff the following integral converges:

$$\iint_{\gamma > 0} d\mathbf{k}\, d\gamma\, \gamma^{-(n+1)} |\eta(\mathbf{k}, 1 + \frac{\mathbf{k}^2}{2m\gamma}, \gamma)|^2 < \infty. \tag{15.39}$$

We consider now a general section:

$$\sigma_\beta(x) = \sigma_1(x)\, (0, 0, \mathbf{0}, \mathbf{0}, I, \beta(x), 1) = (\theta, b, \mathbf{a}, \mathbf{v}, R, \beta(x), \lambda), \tag{15.40}$$

where $\beta : X \to \mathcal{D}_0 \equiv \mathbb{R}_*^+$ is a Borel function and $\beta(x) \in \mathcal{D}_0$ represents a time dilation. Again, this corresponds to a relation between the variables $(\mathbf{k}, E, \gamma)$, which may conveniently be written as $f_\beta(\mathbf{k}, E, \gamma) = 0$. Thus, the admissibility condition $\text{mod}(\mathcal{D}_0, \sigma_\beta)$ reads as

$$\iiint_{\mathfrak{D}_{++}} d\mathbf{k}\, dE\, d\gamma \frac{|\eta(\mathbf{k}, E, \gamma)|^2}{\gamma^{n+1}(E - \frac{\mathbf{k}^2}{2m\gamma})} \delta(f_\beta(\mathbf{k}, E, \gamma)) < \infty. \tag{15.41}$$

For any such admissible vector η, one obtains a dense set of CS, indexed by $X = \mathcal{G}_{\text{aff}}^{(m)}/\mathcal{D}_0$ and given by $\eta_{\sigma_\beta(x)} = U_{++}(\sigma_\beta(x))\eta$, where $U_{++}(\sigma_\beta(x))$ is the representation (15.34), with λ_0 replaced by $\beta(x)$ in (15.35). Of course, for $\beta(x) \equiv 1$, one recovers the basic section σ_1.

An interesting example is given by $\beta(x) = \lambda^2$, i.e., the Schrödinger case. The corresponding constraint relation is simply $\gamma = $ const, which we normalize to $\gamma = 1$, thus getting the admissibility condition:

$$\iint_{E-\frac{\mathbf{k}^2}{2m}>0} d\mathbf{k}\,dE\, \frac{|\eta(\mathbf{k},E,1)|^2}{E-\frac{\mathbf{k}^2}{2m}} < \infty. \tag{15.42}$$

The γ-integration has disappeared in (15.42), because inserting the factor $\delta(\gamma-1)$ in (15.41) is equivalent to quotienting out the subgroup Θ of phase factors, in addition to the time dilation subgroup $\mathcal{D}_0$.

15.3.3 The (restricted) Schrödinger group (♣)

Another possibility is to impose from the beginning the relation $\lambda_0 = \lambda^2$, which is equivalent to restricting oneself to the Galilei–Schrödinger group $\mathcal{G}_S^{(m)} = \widetilde{\mathcal{G}}^{(m)} \rtimes \mathcal{D}_S$, where $\mathcal{D}_S$ is the corresponding one-dimensional subgroup of $\mathcal{D}_2$. The rationale of this restriction is that $\mathcal{G}_S^{(m)}$ is the maximal kinematical invariance group of the free Schrödinger and heat equations, that is, the largest group of space–time transformations that map solutions of these equations into other solutions [231, 247] (in Niederer's definition, the full invariance group, called the Schrödinger group, contains in addition the so-called *expansions*, which are more or less the Galilean equivalent of pure conformal transformations). This restriction leads to several drastic modifications.

1. The group law changes; in particular, the addition of phase factors becomes

$$\theta'' = \theta + \theta' + m(\lambda \mathbf{v} \cdot R\mathbf{a}' + \frac{1}{2}\lambda^2 \mathbf{v}^2 b'). \tag{15.43}$$

As a consequence, the center of $\mathcal{G}_S$ is the one-dimensional subgroup Θ of phase factors, isomorphic to $\mathbb{R}$.

2. Therefore, $\mathcal{G}_S^{(m)}$ is a central extension of $\mathcal{G}_o \rtimes \mathcal{D}_S$ by $\Theta \equiv \mathbb{R}$, and we have [$\triangleleft_m$ denotes a central extension with (mass) parameter m]:

$$\mathcal{G}_S^{(m)} = \widetilde{\mathcal{G}}^{(m)} \rtimes \mathcal{D}_S \equiv (\mathbb{R} \triangleleft_m \mathcal{G}_o) \rtimes \mathcal{D}_S = \mathbb{R} \triangleleft_m (\mathcal{G}_o \rtimes \mathcal{D}_S) \tag{15.44}$$

Thus, in the Schrödinger case, the two operations $\triangleleft_m$ and $\rtimes$ commute. We emphasize that $\mathcal{D}_S$ is the *only* one-dimensional subgroup of $\mathcal{D}_2$ that admits a central extension by $\mathbb{R}$ [41].

3. The parameter γ is now invariant, by (15.35), and the dependence on θ in the representation reduces to a trivial phase. Accordingly, it may be factored out, exactly as for the Weyl–Heisenberg group. Thus we may fix $\gamma = 1$ as above, so that m becomes the mass, as it appears in the Schrödinger equation [281].

4. The restriction to $\mathcal{G}_S^{(m)}$ of the representation (15.34) cannot be square integrable, since $\mathcal{G}_S^{(m)}$ has a noncompact center $\Theta \simeq \mathbb{R}$. After quotient-

ing out this center, the representation space reduces to $\mathfrak{H}_S = L^2(\mathbb{R}^n \times \mathbb{R}, d\mathbf{k}\, dE)$. In technical terms, the original space is a direct integral over γ, and we are taking the restriction to a single component, corresponding to $\gamma = 1$. The corresponding reduced representation, U_S, splits into the direct sum of two irreducible ones, $U_S = U_+^S \oplus U_-^S$, corresponding to the decomposition $\mathfrak{H}_S = \mathfrak{H}_+^S \oplus \mathfrak{H}_-^S$, where $\mathfrak{H}_\pm^S = L^2(\mathfrak{D}_\pm^S, d\mathbf{k}\, dE)$, with $\mathfrak{D}_\pm^S = \{(\mathbf{k}, E) \in \mathbb{R}^n \times \mathbb{R} \mid E - \frac{\mathbf{k}^2}{2m} \gtrless 0\}$ (the two subspaces $\mathfrak{H}_\pm^S$ might be called Schrödinger–Hardy spaces; they are the analogues of the usual Hardy spaces on $\mathbb{R}$, i.e., the subspaces of progressive, respectively, anti-progressive, wavelets).

5. The restriction to the Galilei group $\widetilde{\mathcal{G}}^{(m)}$ of the representation U_S decomposes into a direct integral of irreducible representations

$$L^2(\mathbb{R}^n \times \mathbb{R}, d\mathbf{k}\, dE) \simeq \int_\mathbb{R}^\oplus \mathfrak{H}_E\, dE, \quad \mathfrak{H}_E \simeq L^2(\mathbb{R}^n, d\mathbf{k})$$

$$U_S(\theta, b, \mathbf{a}, \mathbf{v}, R, 1, 1) = \int_\mathbb{R}^\oplus U_E(\theta, b, \mathbf{a}, \mathbf{v}, R)\, dE,$$

where U_E is equivalent to the usual unitary irreducible representation of $\widetilde{\mathcal{G}}^{(m)}$.

6. As expected from the discussion above, the two representations $U_\pm^S$ are square integrable modulo the center Θ, with admissibility condition [compare (15.42)]

$$\iint_{\mathfrak{D}_\pm^S} d\mathbf{k}\, dE\, \frac{|\widehat{\eta}(\mathbf{k}, E)|^2}{E - \frac{\mathbf{k}^2}{2m}} < \infty, \quad \widehat{\eta} \in \mathfrak{H}_\pm^S, \tag{15.45}$$

where $\widehat{\eta}(\mathbf{k}, E)$ denotes the $(n+1)$-dimensional Fourier transform of $\eta(\mathbf{x}, t)$. In other words, η is admissible iff it belongs to the domain of $H_S^{-1/2}$, where $H_S = i\partial_t + \frac{\Delta}{2m}$ is the free Schrödinger operator. There is a dense set of admissible vectors η and each of them generates a tight frame of CS, of the Gilmore–Perelomov type.

An interesting class of Schrödinger wavelets consists of functions of the form

$$\eta(\mathbf{x}, t) = H_S\, \chi(\mathbf{x}, t), \quad H_S = i\partial_t + \frac{\Delta}{2m}, \tag{15.46}$$

where χ is a suitable element of $\widehat{\mathfrak{H}}_\pm^S$, such as, for instance:

- the Schrödinger–Marr wavelet:

$$\eta_{SM}(\mathbf{x}, t) = \left(i\partial_t + \frac{\Delta}{2m}\right) e^{-(x^2 + t^2)/2};$$

- the Schrödinger–Cauchy wavelet:

$$\psi_{SC}(\mathbf{x}, t) = \left(i\partial_t + \frac{\Delta}{2m}\right)(t+i)^{-1} \prod_{j=1}^n (x_j + i)^{-1}.$$

The choice (15.46) is based on the following property. Let ψ be a wave function, solution of the Schrödinger equation $[H_S - V(\mathbf{x})]\psi(\mathbf{x},t) = 0$, with a potential $V(\mathbf{x})$. Then, the wavelet transform of ψ with respect to the wavelet (15.46) is given by

$$\langle U_+^S(g)\eta \,|\, \psi \rangle = \langle U_+^S(g)\chi \,|\, V\psi \rangle, \quad g = (b, \mathbf{a}, \mathbf{v}, R, \lambda) \in \mathcal{G}_S^{(m)}/\Theta. \quad (15.47)$$

In particular, if ψ is a solution of the *free* Schrödinger equation, its wavelet transform with respect to η is identically zero! The proof of (15.47) is immediate if one notices the identity

$$U_+^S(g)\, H_S\, U_+^S(g)^{-1} = \lambda^2\, H_S,$$

which expresses that $\mathcal{G}_S^{(m)}$ is the invariance group of the free Schrödinger equation. It remains to be seen, however, whether Schrödinger wavelets of the type (15.46) will be useful in solving actual poblems in quantum mechanics. A typical instance in which the answer might be positive is the description of quasiclassical (Rydberg) states of atoms.

15.3.4 The affine Poincaré group (♣)

In the relativistic case, the situation changes completely. First, by covariance, only one scale parameter can exist, common to space and time variables. Hence, the $(1+n)$-dimensional affine Poincaré group, also called the Weyl–Poincaré or the similitude group, takes the form of a semidirect product: $SIM(1,n) = \mathbb{R}^{n+1} \rtimes (\mathbb{R}_*^+ \times SO_o(1,n))$. Next, the natural representation, obtained once again by the Mackey method, is of the Wigner type, but acts in $\mathfrak{H} = L^2(C_n, dm(k))$, where $C_n = \mathbb{R}_*^+ \times (SO_o(1,n)/SO(n))$ is the *solid* forward light cone in $\mathbb{R}^{n+1}$ and $dm(k) = (k^2)^{-(n+1)/2}\, dk_0\, d\mathbf{k}$, with $k^2 = k_0^2 - \mathbf{k}^2$. In the spin 0 case, this representation reads as

$$[U(b,a,\Lambda)f](k) = e^{ik \cdot b} f\left((a\Lambda)^{-1}k\right), \quad (b,a,\Lambda) \in SIM(1,n),$$
$$f \in L^2(C_n, dm(k)). \quad (15.48)$$

The new fact is that this unitary irreducible representation *is* square integrable, so that standard CS may be constructed. Explicit examples have been obtained in [Unt] and [77]. Moreover, as shown in the latter paper, one can construct *discrete* subsets of that family of CS that still constitute a frame. We will discuss these results and some of their generalizations in Chapter 16.

In addition, this work has an interesting geometrical content, which warrants a detailed discussion. The starting point is to embed $SIM(1,n)$ into $SO_o(2, n+1)$, the conformal group of Minkowski space–time (note the analogy with the method described in Section 15.1 for deriving wavelets on the sphere). Indeed, the representation (15.48) is simply the restriction to $SIM(1,n)$ of a projective representation of $SO_o(2, n+1)$ [exactly as the familiar massless representations of $\mathcal{P}_+^\uparrow(1,3)$]. Then, consider the complex

tube
$$\mathbb{T} = \mathbb{R}^{n+1} - iC_n \equiv \{z = k - ip,\ k \in \mathbb{R}^{n+1},\ p \in C_n\}. \tag{15.49}$$

The tube $\mathbb{T}$ is a homogeneous space for $SO_o(2, n+1)$, since $\mathbb{T} \sim SO_o(2, n+1)/SO(2) \times SO(n+1)$, and a Riemannian globally symmetric space, realizing Cartan's classical domain BD I [Hel]. Hence, $SIM(1,n)$ acts on $\mathbb{T}$, by $(b, a, \Lambda)z = a^{-1}\Lambda z - b$. Fix $\omega = 0 - i(1, 0, \ldots, 0)$ as base point in $\mathbb{T}$, and let $z = g\omega$, $g = (b, a, \Lambda) \in SIM(1, n)$. Let ψ_ω now be an element of $\mathfrak{H}$ and assume it is an admissible vector for the representation U of $SIM(1, n)$ given in (15.48). Then, following the general pattern, ψ_ω generates a family of CS of $SIM(1, n)$, indexed by the points $z \in \mathbb{T}$:
$$\psi_z \equiv \psi_{g\omega} = U(g)\psi_\omega, \quad g = (b, a, \Lambda) \in SIM(1, n). \tag{15.50}$$

Moreover, these CS constitute a tight frame, for one has
$$\int_\mathbb{T} |\langle \psi_z | f \rangle|^2 d\mu(z) = c\|f\|_\mathfrak{H}^2,\ f \in \mathfrak{H} = L^2(C_n, dm(k)), \tag{15.51}$$

where, for $z = k - ip$, $d\mu(z) = (p^2)^{n+1} dp\, dk$ is the $SO_o(2, n+1)$-invariant measure on $\mathbb{T}$. In the case $n = 1$, the admissibility condition reads [77] as
$$\int_{C_1} \frac{|\psi_\omega(k)|^2}{k^2} dm(k) < \infty, \tag{15.52}$$

and it is manifestly satisfied for a dense set of elements of $\mathfrak{H}$, which proves the square integrability of the representation U. A similar admissibility condition holds for $n > 1$.

A particularly interesting example of admissible vector, given in [Unt], is the (unnormalized) function
$$\psi_\omega^G(k) = (k^2)^{-(n+1)/2} e^{-k_0}. \tag{15.53}$$

This function and the CS generated from it have been used by Unterberger as a basic tool for his comprehensive relativistic generalization of Weyl's operational calculus (under the name of Fuchs and Klein–Gordon calculus). One may notice that the admissible vector ψ_ω^G coincides, up to the scaling factor $(k^2)^{-(n+1)/2}$, with the Poincaré Gaussian vector η_G defined in Section 11.1.2, and originally introduced in [184] (see the discussion in [13] and [16, Section 7]). Also, the connection between CS of the affine Poincaré group, the conformal group, and the tube $\mathbb{T}$ is mentioned in [93] and [209] (in conjunction with the latter reference, we recall that the conformal group of Minkowski space–time is isomorphic to $SO_o(2, 4)/\mathbb{Z}_2 \simeq SU(2, 2)/\mathbb{Z}_4$).

Thus, Unterberger's approach suggests a class of spaces X for which our general formalism applies, namely, the Riemannian globally symmetric spaces of noncompact type [Hel, Per]. Indeed, if $X = G/H$ is such a space, in other words, a noncompact classical domain, then there exists a smooth global section $\sigma : X \to G$ [Hel, Theorem VI.1.1], and, hence, the principal bundle $G \to X = G/H$ is trivializable. Typical examples, besides $\mathbb{T}$ itself, are $SU(1,1)/U(1)$, discussed at length in Chapter 4, Section 4.2.2, and in Chapter 11, Section 11.3, or $SU(2,2)/S(U(2) \times U(2))$. In the general case, G is a connected, noncompact, semisimple Lie group and H is the maximal compact subgroup of G. Considerable literature exists about the realization of the discrete series (i.e., square integrable) representations of G on the corresponding symmetric space G/H (see, for instance [124], [237], and [259], and the references contained in those papers). On the other hand, in the case of a Riemannian globally symmetric space of compact type, where $X = G/H$ is compact, no smooth global section $X \to G$ exists (although a global Borel section always does), but then any unitary irreducible representation of G is square integrable $\mod(H, \sigma)$, for any measurable section σ.

To conclude, let us mention that the 1+1 dimensional case has also been studied in [74] in the context of the classification of three-parameter extensions of the affine group, leading to suitable generalized Wigner functions.

16
The Discretization Problem: Frames, Sampling, and All That

In the preceding chapters, we have encountered both continuous and discrete frames (the latter are more traditional), but in fact the two are necessarily linked. Suppose one is dealing with a continuous frame, of rank one for simplicity, $\int_X |\eta_x\rangle\langle\eta_x|\, d\nu(x) = A$. When it comes to numerical calculation, the integral has to be discretized, so that in effect one always restricts oneself to a *discrete* subset of X. The question is whether this restriction will imply a loss of information and, hence, the *frame discretization* problem arises: Given a rank one frame $\{\mathfrak{H}, F, A\}$, can one find a discrete subset $\Lambda \subset X$ such that the vectors $\{\eta_{x_j}, x_j \in \Lambda\}$ constitute a discrete frame:

$$\sum_{x_j \in \Lambda} |\eta_{x_j}\rangle\langle\eta_{x_j}| = A' \ ? \tag{16.1}$$

Here, the resolution operator A' need *not* coincide with the original operator A. In fact, often it does not, since a continuous tight frame may very well contain discrete nontight ones. Examples will appear soon.

When the frame comes from some group representation U, $\eta_x = U(\sigma(x))\eta$, $x \in X$, the existence of such a discrete subframe depends on two factors, the fiducial vector η and the subset $\Lambda \subset X$. Some results in this direction have been obtained for specific groups, and we will survey them in the next few sections.

In fact, the impetus for constructing good discrete frames comes from signal processing, more precisely, from wavelets [97], as explained in Chapter 12, Section 12.5. These results revived interest in the question of discretization, and soon it was extended to the Gabor or windowed Fourier transform, in other words, the canonical CS. Thus, once again, we are back into old problems of quantum mechanics!

On the other hand, the same problem may be addressed from the point of view of signal processing. The question then is reformulated as follows: How does one sample a continuous signal without losing information? This is again an old problem, which has been revitalized in the context of Gabor or wavelet frames, and new developments are under way, to which we shall devote Section 16.5, before concluding.

16.1 The Weyl–Heisenberg group or canonical CS

We will begin our discussion with the case of canonical CS, equivalently, Gabor frames. This is a world in itself with a vast literature. Fortunately, the recent volume [Fei] gives a detailed and up-to-date survey of the field, to which we refer for further information and original references.

In this case, the geometry of the parameter space, i.e., phase space, is Euclidean. Typically, the discrete parameter set $\Lambda = \{(q_n, p_m), m, n \in \mathbb{Z}\}$ is a square (or rectangular) lattice in the (q, p)-plane and the frame theorem simply says (in the notation of Section 2.2) that a frame $\{\psi_{mn} \equiv \psi_{\sigma(q_n, p_m)}\}$ is obtained if the density of Λ is larger than a critical value. This is the well-known result of von Neumann concerning the density of canonical coherent states [Per], which is closely related to standard theorems known in different circles under the names of Fourier, Heisenberg (uncertainty relations), Nyquist, or Shannon. More precisely, let $\Lambda = q_o \mathbb{Z} \times p_o \mathbb{Z}$, and consider the corresponding family $\{\psi_{mn}\}$, where

$$\psi_{mn}(x) = e^{imp_o x} \psi(x - nq_o), \quad m, n \in \mathbb{Z}. \qquad (16.2)$$

[Compared to (2.33), we have dropped the constant phases.] Thus, the area of the unit cell of the lattice Λ is $S = q_o p_o$. Then [Fei, Per, 63, 245], the following is true.

(1) If $q_o p_o > 2\pi$ (undersampling), the family $\{\psi_{mn}\}$ is not complete.

(2) If $q_o p_o < 2\pi$ (oversampling), the family $\{\psi_{mn}\}$ is overcomplete and remains so if one removes a finite number of points from Λ.

(3) If $q_o p_o = 2\pi$ (critical sampling), the family $\{\psi_{mn}\}$ is complete and remains so if one removes any single point from Λ, but becomes noncomplete if one removes two or more points.

This result is purely mathematical, but it has an immediate translation in physical terms, if one remembers that the (q, p)-plane is the phase space of a quantum system, as discussed in Chapter 2. Restoring the conversion factors, we see that a cell of area 2π in the (q, p)-plane corresponds to a cell of area $2\pi\hbar$ in the phase plane. Thus, the result above means that the system $\{\psi_{mn}\}$ is complete if there is, on average, not less than one CS

16.1. The Weyl–Heisenberg group or canonical CS

in a cell of area $2\pi\hbar$ in the phase plane. A corresponding statement is less obvious in the case of wavelets, for lack of an explicit physical interpretation of the latter.

Coming back to the language of frames, this result may be reformulated as follows [Dau]: Frames, even tight frames, with good time-frequency localization exist for $q_o p_o < 2\pi$; no frame exists for $q_o p_o > 2\pi$; in the critical case $q_o p_o = 2\pi$, frames do exist, but with bad localization properties, as a consequence of the celebrated Balian–Low theorem (BLT) given below. In particular, orthonormal bases exist *only* in the critical case, and thus are necessarily poorly localized. This results from the following inequalities, where m and M are the frame bounds of the Gabor frame generated by $\psi \in L^2(\mathbb{R})$:

$$\mathrm{m} \leq \frac{2\pi}{q_o p_o}\|\psi\|^2 \leq \mathrm{M}. \tag{16.3}$$

Theorem 16.1.1 (Balian–Low theorem): *Let* $\psi \in L^2(\mathbb{R})$ *and let* $q_o, p_o > 0$ *satisfy* $q_o p_o = 2\pi$. *If* $\{\psi_{mn}\}$ *is a Gabor frame, then*

$$either \quad \int_{-\infty}^{\infty} |x\psi(x)|^2\, dx = \infty \quad or \quad \int_{-\infty}^{\infty} |k\widehat{\psi}(k)|^2\, dk = \infty. \tag{16.4}$$

The statement (16.4) means that ψ and $\widehat{\psi}$ cannot both have fast decrease simultaneously (remember fast decrease of $\widehat{\psi}$ means smoothness of ψ); i.e., the Gabor frame $\{\psi_{mn}\}$ cannot be well localized in phase space. Actually this theorem may be extended in various ways, including the generalization to irregular sets $\{q_n, p_m, m, n \in \mathbb{Z}\}$. A thorough discussion may be found in [Dau, Chapter 4] or in [Fei, Chapter 2].

In his original paper [129], Gabor used a Gaussian as the basis function ψ, on the grounds that it minimizes the joint uncertainty in phase space (in other words, it saturates the uncertainty relations, as we have seen in Chapter 2, Section 2.1. Although this yields a frame in the oversampling case, however, it does *not* in the critical case [59]. In addition, the corresponding expansion

$$\phi = \sum_{m,n \in \mathbb{Z}} \langle \psi_{mn} | \phi \rangle \, \widetilde{\psi}_{mn} \tag{16.5}$$

does not converge in L^2, but only in the sense of distributions.

An interesting extension of the Balian–Low theorem has been given by Zak [288]. The setup is a (quasi)-planar gas of electrons in a strong magnetic field perpendicular to the plane. As is well known, the energy levels of an electron in such a system form a discrete ladder, called the Landau levels, each of which is infinitely degenerate. Thus, the problem arises of finding a good orthonormal basis for each of those levels, preferably made of well-localized states. Now, the extended BLT of Zak asserts that a good localization of these states, which are in fact Gabor states, is impossible if they are taken orthogonal: An electron in a magnetic field (in the z-direction) cannot be well localized both in the x and y directions. As a matter of fact, exactly the same result was obtained in [30] with orthonormal wavelets, which shows that we have here a genuine physical effect.

For many aspects of the theory of Gabor frames, including the proof of the Balian–Low theorem, an essential tool is the *Zak transform*, also called the kq-representation [Dau, Fei, 59]. This transform is defined as follows:

$$(Zf)(s,t) = \sum_{l \in \mathbb{Z}} e^{2\pi i t l} f(s-l), \tag{16.6}$$

and it can be shown to be unitary from $L^2(\mathbb{R})$ to $L^2([0,1] \times [0,1])$. In the present context, its main virtue lies in the relation

$$(Zg_{mn})(s,t) = e^{2\pi i m s} e^{-2\pi i n t} (Zg)(s,t), \quad m, n \in \mathbb{Z}, \tag{16.7}$$

where, as above,

$$g_{mn}(x) = e^{2\pi i m x} g(x-n), \quad m, n \in \mathbb{Z}.$$

Thus, in the kq representation, the *commuting* displacement operators are simultaneously diagonalized [precisely, this representation was introduced originally for the sake of defining symmetric coordinates $k = \frac{2\pi}{a} t$ (quasi-momentum) and $q = as$ (quasicoordinate) in the dynamics of electrons in solids (a is the lattice spacing)]. In other words, we are using here for constructing a basis the canonical CS corresponding to an *abelian* subgroup of the Weyl–Heisenberg group.

A closely related approach has been developed in [Bo2] and [79]. The idea is to start from a *discrete* subgroup of the Weyl–Heisenberg group, $G_{dWH}^{(N)}$, made of commuting translations in position and momentum, exactly as above. This subgroup is obtained by restricting the action of G_{WH} to the torus $\mathbb{T}^2 = \mathbb{R}^2/\Lambda$, with the quantization condition $q_o p_o = 2\pi\hbar N$, $N \in \mathbb{N}$, and consists of the elements $\{(\theta, \frac{nq_o}{N}, \frac{mp_o}{N}), \theta \in \mathbb{R}, m, n \in \mathbb{Z}\}$. This "quantized" torus may be taken as the phase space of a quantum particle, and the aim of the work is to study the corresponding dynamical system, both in the classical setup (where it can be chaotic) and in the quantum context. To that effect, one constructs unitary representations of the discrete Weyl–Heisenberg group $G_{dWH}^{(N)}$ and the corresponding CS. These CS precisely provide the (quantization) link between the two frameworks, and the wavelet or CS transform essentially reduces to the Zak transform. The restriction of the usual (Schrödinger) UIR of G_{WH} to its discrete subgroup $G_{dWH}^{(N)}$ is highly reducible and decomposes into a direct integral of N-dimensional UIRs. Correspondingly, the restriction of the CS transform to any of these subspaces of dimension N almost coincides with the familiar FFT.

As a last remark, we ought to mention again that the Zak transform is the key tool for the construction of CS on the circle S^1, or, more generally, on a torus, as developed in [151].

16.2 Wavelet frames

Clearly, the question of the existence of a frame obtained by discretizing a given CS transform must take into account the geometry of the parameter space, that is, the lattice Λ must be invariant under discrete operations from the invariance group (which in general do not form a discrete subgroup). In the Weyl–Heisenberg case, this requirement leads to choosing a rectangular lattice in the phase plane, as we have just seen. Similarly, for the 'ax + b' group, corresponding to 1-D wavelets, the geometry of the time-scale half-plane is non-Euclidean, and the lattice Λ must be invariant under discrete dilations and translations:

- for the scales, one chooses naturally $a_j = a_o^j, j \in \mathbb{Z}$, for some $a_o > 1$;
- for the times, one takes $b_k = k\, b_o\, a_o^j, j, k \in \mathbb{Z}$.

Thus,
$$\psi_{jk}(x) = a_o^{-j/2}\, \psi(a_o^{-j}x - kb_o), \quad j, k \in \mathbb{Z}. \tag{16.8}$$

The most common choice is $a_o = 2$ (octaves!) and $b_o = 1$, which results in
$$\psi_{jk}(x) = 2^{-j/2}\, \psi(2^{-j}x - k), \quad j, k \in \mathbb{Z}. \tag{16.9}$$

We emphasize that this so-called *dyadic* lattice $\{(k2^j, 2^j),\ j, k \in \mathbb{Z}\}$ is exactly the same as the lattice that indexes the DWT, although the two approaches are totally different (see Section 13.1). Actually, other lattices may be used here too, for instance, based on the golden mean τ, as described (for the discrete WT) in Section 13.4.

For a given choice of ψ, a_o, one finds a range of values of b_o such that $\{\psi_{jk}\}$, as given in (16.8), is a frame. Detailed results may be found in [Dau] and [98]. Here, we will restrict ourselves to the following simplified version.

Theorem 16.2.1: Let ψ and a_o be such that:

1. $\displaystyle\inf_{1 \leq \xi \leq a_o} \sum_{j=-\infty}^{\infty} |\widehat{\psi}(a_o^j \xi)|^2 > 0;$

2. $|\widehat{\psi}(\xi)| \leq C\, |\xi|^\alpha\, (1 + |\xi|)^{-\gamma},\ \alpha > 0,\ \gamma > \alpha + 1.$

Then, there exists a b_{oo} such that $\{\psi_{jk}\}$ constitutes a frame for all choices $b_o < b_{oo}$.

Both the Mexican hat and the Morlet wavelet satisfy the conditions of the proposition for the dyadic case, $a_o = 2, b_o = 1$; thus, they both generate discrete frames on the dyadic lattice (and, in fact, on a more general class of hyperbolic lattices [98, 99]). Explicit values for the corresponding frame bounds m, M may be found in [Dau]. For the Mexican hat, one finds m = 3.223, M = 3.596, so that M/m = 1.116. This is not terribly good, and the Morlet wavelet is even worse. This defect is remedied in practice by "densifying" the lattice Λ, typically, replacing in (16.9) the exponent j by $\frac{\nu}{N}j$, $\nu = 0, 1, \ldots, N-1$ (N is called the number of voices). Taking, for instance, $N = 4$, one gets M/m = 1.007 for the Mexican hat. This of course improves the situation considerably. For further details, we refer the reader to [Dau]. If the speed is the determining criterion, however, one can do better by using the pseudo-QMF algorithm described in Chapter 13, Section 13.2. The latter is in fact close in spirit to the procedure explained here, but more efficient.

Interestingly, in the wavelet case, the relation analogous to (16.3) reads, for the wavelet ψ, as

$$\mathsf{m} \leq \frac{2\pi}{b_o} \sum_{j \in \mathbb{Z}} |\widehat{\psi}(a_o^j \xi)|^2 \leq \mathsf{M}, \tag{16.10}$$

for all $\xi \neq 0$. This implies, in particular, that the sum should be almost constant, a strong requirement indeed, which is rarely satisfied. By contrast, the inequalities (16.3) do not impose any restriction on ψ. The difference simply reflects the different properties of the two groups underlying the construction. The Weyl–Heisenberg group is unimodular, so that every vector is admissible, whereas the $ax+b$ group is not, hence, the nontrivial constraints on ψ for it to be being admissible.

As a final remark, we note that the same results hold true for n-dimensional wavelets associated with $SIM(n)$, for the same wavelets [Mur, 225].

Alternative approaches to the $ax+b$ group

Of course, the simplest situation occurs when the lattice Λ is the orbit of a discrete subgroup $\Lambda \subset G$. Unfortunately, this is seldom possible: already, in the 1-D wavelet case, the dyadic lattice is not of this type. However, in that particular case the difficulty may be circumvented [196] by embedding the $ax+b$ group G_+ into $SL(2, \mathbb{R})$ and then restricting it to the discrete subgroup $SL(2, \mathbb{Z})$: The function $\widehat{\psi}_o(k) = e^{-k}k^s$ generates a frame over $SL(2, \mathbb{Z})$ whenever $s = \frac{1}{2}, 1, \frac{3}{2}, 2, \frac{5}{2}$, corresponding to the lowest discrete series representations of $SL(2, \mathbb{R})$. These results, and the technique used for proving them, strongly suggest that arithmetic ideas are likely to play a role in his context.

Another possibility that goes in the same direction is to consider the dyadic lattice $\Lambda = \{(k2^j, 2^j), j, k \in \mathbb{Z}\}$ as a subgroup of the dyadic affine subgroup $\{(k2^j, 2^l), j, k, l \in \mathbb{Z}\}$ of G_+, as mentioned in Chapter 13, Section

13.3. This remains, however, unexplored up to now. Another possibility still is to introduce *dilation semigroups*. Indeed, as is clear from (13.3), the actual parameter space of a multiresolution analysis is not the full dyadic lattice, but only half of it, namely, (we put $j_o = 0$ for simplicity) $\Lambda_+ = \{(k2^j, 2^j), k \in \mathbb{Z}, j = 0, 1, 2, \ldots\}$. Then, Λ_+ is the semidirect product $\Lambda_+ = \mathbb{Z} \rtimes \mathcal{A}$ of $\mathbb{Z}$ by the abelian multiplicative semigroup $\mathcal{A} = \{2^j, j = 0, 1, 2, \ldots\}$. This point of view turns out to be extremely fruitful, and we shall develop it in detail in Section 16.5.

16.3 Frames for affine semidirect products

16.3.1 The affine Weyl–Heisenberg group

As mentioned in Section 15.2, the *matching pursuit* algorithm [214] is based on a suitable discretization of the natural UIR (15.11) of the affine Weyl–Heisenberg group G_{aWH}, leading to a particular dictionary of time-frequency atoms. The rationale of this approach is the following. An efficient representation of signals is obtained by expanding them into a basis that reflects their essential characteristics. For instance, in 1-D, a sharp impulse will be well decomposed into functions well concentrated in time (or space), whereas spectral lines are better represented by waveforms that have a narrow frequency support (the same argument is used in selecting a wavelet for a given type of signals). When the signal contains both of those elements, no single representation (frame, basis) will suffice completely, be it Fourier, Gabor, or wavelets. They are simply not flexible enough. Thus, one can use instead [214] time-frequency atoms $\{h_\gamma, \gamma \in \Gamma\}$, where $\gamma \equiv (b_j, \omega_k, a_m)$ runs over a countable subset of G_{aWH}. In other words, one needs a suitable discrete family of CS of G_{aWH}.

For particular choices of Γ, the family reduces to Gabor or wavelet frames. For instance, if we fix the scale parameter at a_o and set $\gamma = \gamma_{mn} = (nq_o, m\omega_o, a_o)$, $m, n \in \mathbb{Z}$, we obtain a Gabor frame as in (16.2). On the other hand, choosing $\gamma = \gamma_{jk} = (kb_o a_o^j, a_o^{-j}\omega_o, a_o^j)$, $j, k \in \mathbb{Z}$, one gets the wavelet frame or basis (16.8). In that sense, time-frequency atoms truly interpolate between Gabor and wavelet frames, as already noted in [Tor] and [276].

This being said, the key idea of the matching pursuit algorithm is to choose the successive vectors in the optimal family in a recursive and adaptive way by projecting the function to be analyzed on the linear span of the subfamily obtained at the previous step and minimizing, in the sense of the norm, the remaining difference. This algorithm converges and yields a very flexible and economical method of representing signals and images that contain a large spectrum of conflicting characteristics simultaneously.

16.3.2 The affine Poincaré groups (♣)

We have seen in Section 15.3.4 that the (spin 0) representation U of the affine Poincaré group in $1+n$ dimensions, $SIM(1,n)$, acting in $\mathfrak{H} = L^2(C_n, dm(k))$ and given in (15.48), is square integrable, with admissibility condition (15.52). Starting from this condition, a family of discrete tight frames for $SIM(1,n)$ can be obtained [77], which we now describe. As in the case of the affine Weyl–Heisenberg group, the corresponding lattice $\Lambda \subset C_n$ is generated by dyadic translations and Lorentz transformations, that is, operations of the underlying group.

Take first $n=1$. Let $u = u(k_0, \mathbf{k})$ be a real continuous function, supported in a square of side b_o^{-1} contained in the cone $C_1 = \{k \in \mathbb{R}^2 | k^2 > 0, k_0 > 0\}$. In order to discretize the action of $SIM(1,n)$, fix a scale $a_o > 1$ and choose a sequence of real numbers $\{t_l, l \in \mathbb{Z}\}$, such that

$$\sum_{l,j \in \mathbb{Z}} \frac{[u(a_o^j \Lambda_{-l} k)]^2}{a_o^{2j} k^2} = 1, \ \forall k \in C_1, \qquad (16.11)$$

where

$$\Lambda_l = \begin{pmatrix} \cosh t_l & \sinh t_l \\ \sinh t_l & \cosh t_l \end{pmatrix} \in SO_o(1,1), \quad \Lambda_{-l} = \Lambda_l^{-1}.$$

Using these ingredients, define the following countable set of wavelets:

$$u_{ljmn}(k) = e^{i(a_o^j \Lambda_{-l} k) \cdot \theta} u(a_o^j \Lambda_{-l} k), \ l,j \in \mathbb{Z}, \ \theta = (m,n) b_o \in b_o \mathbb{Z}^2. \quad (16.12)$$

Then, an explicit calculation shows that the family $\{u_{ljmn}, l,j,m,n \in \mathbb{Z}\}$ is a tight frame whenever the function u satisfies the condition (16.11). Moreover, one has

$$\sum_{l,j,m,n \in \mathbb{Z}} |\langle u_{ljmn} | f \rangle|^2 = b_o^{-2} \|f\|^2. \qquad (16.13)$$

Here, the factor b_o^{-2} is the area of the support of u.

In order to obtain explicit functions satisfying the condition (16.11), it suffices to introduce hyperbolic coordinates

$$k_0 = \mu \cosh t, \quad \mathbf{k} = \mu \sinh t, \quad \mu > 0, \ t \in \mathbb{R},$$

and to take a separable function $u(k) \equiv \tilde{u}(\mu, t) = u_1(\mu) u_2(t)$. Then,

$$u(a_o^j \Lambda_{-l} k) = \tilde{u}(a_o^j \mu, t - t_l) = u_1(a_o^j \mu) u_2(t - t_l),$$

and it is enough to choose C^∞ functions of compact support

$$v_1(\mu) = \mu^{-2} [u_1(\mu)]^2, \quad v_2(t) = [u_2(t)]^2,$$

such that

$$\sum_{j \in \mathbb{Z}} v_1(a_o^j \mu) = 1, \quad \sum_{l \in \mathbb{Z}} v_2(t - t_l) = 1.$$

16.3. Frames for affine semidirect products

It should be noted that these discrete wavelets are not exactly the discretized version of the CS associated with the unitary representation (15.48) of the affine Poincaré group; they correspond rather to the discretization of the action

$$[\widetilde{U}(b,a,\Lambda)f](k) = e^{i[(a\Lambda)^{-1}k]\cdot b} f\left((a\Lambda)^{-1}k\right),$$
$$(b,a,\Lambda) \in SIM(1,1), \; f \in L^2(C_1, dm(k)); \quad (16.14)$$

that is,

$$\widetilde{U}(b,a,\Lambda) = U(0,a,\Lambda)\, U(b,1,I).$$

This action, however, differs from the true representation (15.48),

$$U(b,a,\Lambda) = U(b,1,I)\, U(0,a,\Lambda),$$

only by a phase:

$$[\widetilde{U}(b,a,\Lambda)f](k) = e^{ik\cdot(a^{-1}\Lambda b - b)}[U(b,a,\Lambda)f](k).$$

Then, an explicit calculation, similar to the one made in Chapter 11, Section 11.1.1, shows that this action is square integrable as well; only the admissibility condition is slightly modified. Actually, the states $\widetilde{U}(b,a,\Lambda)\eta$ generated by the action of $\widetilde{U}$ on an admissible vector η form a tight frame, and the result just discussed means this remains true upon discretization. Strictly speaking, of course, these states are *not* coherent states, but weighted CS (described in Chapter 7, Section 7.3), though they play the same role.

Exactly the same construction may be performed in dimension $n > 1$. One starts again from a real-valued continuous function u supported in an $(n+1)$-dimensional hypercube of side b_o^{-1}, contained in the forward cone C_n. One fixes a scale $a_o > 1$ and chooses a countable family of Lorentz matrices, $\{\Lambda_l \in SO_o(1,n), l \in L\}$, with L a countable index set, such that

$$\sum_{j\in\mathbb{Z}, l\in L} \frac{[u(a_o^j \Lambda_l^{-1} k)]^2}{a_o^{(n+1)j}(k^2)^{(n+1)/2}} = 1, \quad \forall k \in C_n. \quad (16.15)$$

Next, one introduces in the cone C_n hyperbolic coordinates (μ, ω), $\mu > 0$, $\omega \in V_1^+ = \{k = (k_0, \mathbf{k}) \in \mathbb{R}^{n+1} | k^2 = k_0^2 - \mathbf{k}^2 = 1, k_0 > 0\}$, and writes $u(k) = u_1(\mu)\, u_2(\omega)$. If the functions

$$v_1(\mu) = \mu^{-(n+1)}[u_1(\mu)]^2, \quad v_2(\omega) = [u_2(\omega)]^2,$$

verify the identities

$$\sum_{j\in\mathbb{Z}} v_1(a_o^j \mu) = 1, \quad \sum_{l\in L} v_2(\Lambda_l^{-1}\omega) = 1, \quad (16.16)$$

then the function u satisfies the condition (16.15).

In that case, the family of wavelets

$$u_{lj\theta}(k) = e^{ia_o^j(\Lambda_l^{-1}k\cdot\theta)}\, u(a_o^j \Lambda_l^{-1} k), \; l \in L, j \in \mathbb{Z}, \theta \in b_o\mathbb{Z}^{n+1}, \quad (16.17)$$

is again a tight frame and one has:

$$\sum_{l,j,\theta} |\langle u_{lj\theta}|f\rangle|^2 = b_o^{-(n+1)} \|f\|^2. \tag{16.18}$$

Here too, the factor on the right-hand side is the volume of the support of u. The only difficulty is to find a suitable function v_2, and the solution given in [77] is based on the existence of a discrete subgroup Γ of $SO_o(1,n)$ such that the quotient $SO_o(1,n)/\Gamma$ is compact. Then, $\mathcal{V}_1^+/\Gamma$ is compact as well, so that it is possible to find a compact subset (a cube or a ball) $\mathcal{X}$ of $\mathcal{V}_1^+$ such that

$$\mathcal{V}_1^+ = \bigcup_{\gamma \in \Gamma} \gamma(\mathcal{X}).$$

Hence, if w is a continuous function with support in $\mathcal{X}$, then the function

$$v_2(\omega) = \frac{w(\omega)}{\sum_{\gamma \in \Gamma} w(\gamma^{-1}\omega)}$$

satisfies the required identity (16.16).

This result is in fact general, however, and the technique developed in [77] extends immediately to a whole class of semidirect product groups, as we now show.

16.3.3 Discrete frames for general semidirect products

The affine Poincaré group $SIM(1,n) = \mathbb{R}^n \rtimes (\mathbb{R}_*^+ \times SO_o(1,n))$ is a particular case of a semidirect product of the type $G = V \rtimes S$, with $V = \mathbb{R}^n$ and $S = \mathbb{R}_*^+ \times K$, where $\mathbb{R}_*^+$ corresponds to dilations and K is a connected, semisimple Lie group. This class was treated, in increasing generality, in [102], [71], [127] and [23], and we have discussed it at length in Chapters 9 and 10. As we have seen, the Duflo–Moore operator C is, in all cases, a multiplication operator by a positive function $C(k)$, which may be bounded or unbounded, and similarly for the inverse $C(k)^{-1}$. Thus, in general, we get CS systems, but not necessarily frames. As shown in [23], however, these CS systems always contain a discrete frame, which may be tight or nontight. The proof is a direct generalization of that used in [77] for the affine Poincaré group, and we sketch it.

The key step is the existence of a suitable discrete subgroup Γ of S. As in Chapter 9, Section 9.1.2, we consider the square integrable unitary representation U of G induced by $G_o = \mathbb{R}^n \rtimes S_o$, where $S_o \subset K \subset S$ is the compact stabilizer of a fixed point $k_o \in \widehat{\mathbb{R}}^n$. Let $\widehat{\mathcal{O}}$ be the orbit of k_o, which necessarily has positive Lebesgue measure in $\widehat{\mathbb{R}}^n$. Then, the basic topological facts underlying the construction of frames are summarized in the following lemma.

16.3. Frames for affine semidirect products

Lemma 16.3.1 : *(1) There exists a discrete subgroup $S_\Gamma = \{s_\gamma, \gamma \in \Gamma\}$ of S and a compact subset $\mathcal{K}$ of $\hat{\mathcal{O}}$ such that*

$$\hat{\mathcal{O}} = \bigcup_{\gamma \in \Gamma} s_\gamma(\mathcal{K});$$

(2) For any nonempty subset $\mathcal{X} \subset \hat{\mathcal{O}}$, there exists in S a finite set $S_I = \{s_\nu, \nu \in I\}$, with I a finite index set, such that:

$$\hat{\mathcal{O}} = \bigcup_{\gamma \in \Gamma, \nu \in I} s_\gamma s_\nu(\mathcal{X}).$$

The connection between the statements (1) and (2) of the lemma is that $\{s_\nu(\mathcal{X}), \nu \in I\}$ is a finite covering of $\mathcal{K}$, which always exists, since the latter is compact.

In the most general case, the representation $U \equiv U(b,s), b \in \mathbb{R}^n, s \in S$ is induced by a finite-dimensional cyclic representation L of the compact group $S_o \subset K$, of dimension N. We consider first the case $N = 1$, that is, the equivalent of the spin 0 case for the Poincaré group. In other words, L is the trivial representation of S_o. Then, one proceeds exactly as in the Poincaré case.

Let $\mathcal{X}$ be an open subset of $\mathcal{O}$ containing k_o and $\psi \in L^2(\hat{\mathcal{O}})$ be a function in the representation space of L, with support in $\mathcal{X}$. Let $S_I = \{s_\nu, \nu \in I\}$ be the finite subset of S determined by Lemma 16.3.1, and let b_o be the side of the smallest hypercube in $\hat{\mathcal{O}}$ containing $\mathcal{X}$. Then, a straightforward calculation shows that the vectors

$$\psi_{\gamma\nu\theta} = U(0, s_\gamma s_\nu)\, U(\theta, I)\, \psi, \quad \gamma \in \Gamma,\ \nu \in I,\ \theta \in b_o \mathbb{Z}^n, \tag{16.19}$$

form a discrete frame (which may be taken as a tight frame by properly adjusting the constants). Explicitly, the CS (or wavelets) (16.19) are given by the following functions:

$$\psi_{\gamma\nu\theta}(k) = e^{i[(s_\gamma s_\nu)^{-1}k]\cdot\theta}\, u((s_\gamma s_\nu)^{-1}k), \quad \gamma \in \Gamma,\ \nu \in I,\ \theta \in b_o \mathbb{Z}^n. \tag{16.20}$$

Clearly, these wavelets reduce to those of the affine Poincaré group, given in (16.17), if we take $K = SO_o(1, n)$ and $s = a\Lambda$ in the definition of the group G.

In the general case, the inducing representation L of S_o is a finite-dimensional cyclic representation, of dimension N, acting in a Hilbert space $\mathfrak{K}$, and the induced representation U of G is given generically by the form [see (10.80)]

$$(U(b,s)\phi)(k) = e^{ik\cdot b}\, L([h_0(s^{-1}, k)]^{-1})\phi(h^{-1}k),$$

where $h_0(s, k)$ denotes the element of S_o (the cocycle) associated with s in the inducing construction (see Section 10.2.4). In that case, the construction of a frame goes through in exactly the same way. The only difference is that one has to take care of the action of S_o on the support of the generating function ψ. More precisely, given a cyclic vector $v \in \mathfrak{K}$ for V, one can find N elements $\{t_i \in S_o, i = 1, \ldots, N\}$, such that $\{V(t_i)v, i = 1, \ldots, N\}$ is an algebraic basis of $\mathfrak{K}$. Then, one shows that, for any $w \in \mathfrak{K}$, there are two positive constants m, M, such that, for all k in a neighborhood of k_o,

$$\text{m} \|w\|_{\mathfrak{K}} \leq \sum_{i=1}^{N} |\langle L([h_0(t_i^{-1}, k)]^{-1})v | w\rangle|^2 \leq \text{M} \|w\|_{\mathfrak{K}}. \tag{16.21}$$

From this relation, one then deduces, by the same explicit calculation as before, that the family of CS

$$\psi_{\gamma\nu i\theta} = U(0, s_\gamma s_\nu t_i)\, U(\theta, I)\, \psi, \quad \gamma \in \Gamma,\ \nu \in I,\ i = 1, \ldots, N,\ \theta \in b_o \mathbb{Z}^n, \tag{16.22}$$

is a frame. Whether this frame can be made tight has to be decided in a specific case, in terms of the constants m, M, appearing in (16.21).

The remarkable fact about this result is that all groups of the type considered, namely, semidirect products of the form $G = \mathbb{R}^n \rtimes (\mathbb{R}_*^+ \times K)$ are amenable to the same conclusion. In all cases, discrete frames may be constructed by the same discretization method, and this point is obviously important for applications.

16.4 Groups without dilations: The Poincaré groups (♣)

For the affine groups discussed in the previous sections, the presence of dilations was the crucial factor that made the relevant representation square integrable. If we drop them, we are forced back to CS of the general type. The most important of these are the CS of the relativity groups. Explicit results have been obtained for the Poincaré group $\mathcal{P}_+^\uparrow(1,1)$ in one time and one space dimension, and we will survey them in this section. As will be clear from the sequel, the technique for discretizing the CS is essentially the same. The new feature here, however, is the need of a section σ : $\mathcal{P}_+^\uparrow(1,1)/T \to \mathcal{P}_+^\uparrow(1,1)$. As we have seen in Section 11.1, the nature of the continuous frames depends on the choice of the section. Naturally, we expect the same behavior for the discrete frames, and this is indeed the case.

As for the Poincaré group $\mathcal{P}_+^\uparrow(1,3)$, in particular, for the representation of spin $s = 0$, we have seen in Section 11.1 that results concerning continuous CS frames are essentially the same as for $\mathcal{P}_+^\uparrow(1,1)$, for all three standard sections. One should expect to have the same situation in the discrete case, but no proof has been given so far.

Thus, we concentrate on the Poincaré group $\mathcal{P}_+^\uparrow(1,1)$. Discrete frames for this group have been obtained in [186], which we now describe. From the Wigner representation (11.65), the CS of $\mathcal{P}_+^\uparrow(1,1)$ for an arbitrary

16.4. Groups without dilations: The Poincaré groups (♣)

section σ, indexed by the function ϑ, take the form

$$\eta_{\sigma(\mathbf{q},\mathbf{p})}(k) = (U(\sigma(\mathbf{q},\mathbf{p})\eta))(k) = e^{ik.\hat{q}}\eta(\Lambda_p^{-1}k)$$
$$= e^{-i\mathbf{X}_\mathbf{p}(k)\mathbf{q}}\eta(\Lambda_p^{-1}k), \quad (16.23)$$

where

$$\mathbf{X}_\mathbf{p}(\mathbf{k}) = \mathbf{k} - (\Lambda_p^{-1}k)_0\,\vartheta(\mathbf{p}), \quad (16.24)$$

and η is an admissible vector in $\mathfrak{H} = L^2(V_m^+, d\mathbf{k}/k_0)$; that is,

$$\int_\mathbb{R} |\eta(k)|^2\, d\mathbf{k} < \infty. \quad (16.25)$$

We proceed now to discretize these CS. For $l \in \mathbb{Z}$, let $p_l = (p_{l0}, \mathbf{p}_l)$ be a discretization of p. Then, for an arbitrary section $\sigma(\mathbf{q},\mathbf{p})$ and $n, l \in \mathbb{Z}$, we write the discretized version of the coherent states in (16.23) as

$$\eta_{n,l}(k) = e^{-i\mathbf{X}_l(\mathbf{k})\,\mathbf{q}_{n,l}}\,\eta(\Lambda_l^{-1}k), \qquad \mathbf{q}_{n,l} = n\,\Delta\mathbf{q}_l, \quad (16.26)$$

where $\Delta\mathbf{q}_l > 0$ is to be fixed later and $\mathbf{X}_l(\mathbf{k})$ and Λ_l are, respectively, the discretized forms of $\mathbf{X}_\mathbf{p}(\mathbf{k})$ and Λ_p. Let

$$a = (a_0, \mathbf{a}) \in V_m^+ \quad \text{and} \quad b = (b_0, \mathbf{b}) \in V_m^+, \quad (16.27)$$

and suppose that $\eta(k) = 0$ if $\mathbf{k} \notin [\mathbf{a}, \mathbf{b}]$; i.e., the length of the support of $\eta(k)$ is $\mathbf{b} - \mathbf{a}$. Then, $\eta(\Lambda_l^{-1}k) = 0$ if

$$\mathbf{k} \notin [\Lambda_l a, \Lambda_l b] = \left[\frac{a_0\mathbf{p}_l + a\mathbf{p}_{l0}}{m}, \frac{b_0\mathbf{p}_l + b\mathbf{p}_{l0}}{m}\right]$$

and the length of the support of $\eta(\Lambda_l^{-1}k)$, as a function of $\mathbf{k}$, is

$$[(b_0 - a_0)\mathbf{p}_l + (\mathbf{b} - \mathbf{a})p_{l0}]/m.$$

Let $\eta(\Lambda_l^{-1}k) = \widetilde{\eta}(\mathbf{X}_l)$, where $\mathbf{X}_l = \mathbf{X}_l(\mathbf{k})$ is a function of $\mathbf{k}$ and

$$\widetilde{b}_l = \Lambda_l b = \frac{1}{m}(b_0 p_{l0} + \mathbf{b}\mathbf{p}_l, b_0\mathbf{p}_l + \mathbf{b}p_{l0}) \quad (16.28)$$

$$\widetilde{a}_l = \Lambda_l a = \frac{1}{m}(a_0 p_{l0} + \mathbf{a}\mathbf{p}_l, a_0\mathbf{p}_l + \mathbf{a}p_{l0}). \quad (16.29)$$

Thus, the length, L_l, of the support of $\widetilde{\eta}(\mathbf{X}_l)$ is given by

$$L_l = \mathbf{X}_l(\widetilde{b}_l) - \mathbf{X}_l(\widetilde{a}_l) = \frac{1}{m}[(b_0 - a_0)\mathbf{p}_l + (\mathbf{b} - \mathbf{a})p_{l0}] - (b_0 - a_0)\vartheta(\mathbf{p}_l), \quad (16.30)$$

where $\vartheta(\mathbf{p}_l)$ is the discretized form of $\vartheta(\mathbf{p})$.

Let us now formally define the discretized frame operator,

$$T = \sum_{n,l=-\infty}^{\infty} |\eta_{n,l}\rangle\langle\eta_{n,l}|. \quad (16.31)$$

16. The Discretization Problem: Frames, Sampling, and All That

To examine the convergence of the operator sum in (16.31), we proceed as in Section 11.1.1 and consider, for arbitrary $\phi, \psi \in \mathfrak{H}$, the sum

$$I_{\phi,\psi} = \langle \phi | T \psi \rangle = \sum_{n,l=-\infty}^{\infty} \langle \phi | \eta_{n,l} \rangle \langle \eta_{n,l} | \psi \rangle$$

$$= \sum_{n,l=-\infty}^{\infty} \int_{V_m^+} \int_{V_m^+} e^{[i\{\mathbf{X}_l(\mathbf{k}) - \mathbf{X}_l(\mathbf{k}')\}\mathbf{q}_{n,l}]}$$

$$\times \overline{\phi(k)}\, \eta(\Lambda_l^{-1} k)\overline{\eta(\Lambda_l^{-1} k')}\, \psi(k') \frac{d\mathbf{k}}{k_0} \frac{d\mathbf{k}'}{k_0'}. \quad (16.32)$$

Writing $\Delta q_l = \frac{2\pi}{L_l}$, where L_l is the length of the support of $\widetilde{\eta}(\mathbf{X}_l)$ obtained in (16.30), we can write

$$I_{\phi,\psi} = \sum_{n=-\infty}^{\infty} \sum_{l=-\infty}^{\infty} \int_{V_m^+} \int_{V_m^+} e^{[i\{\mathbf{X}_l(\mathbf{k}) - \mathbf{X}_l(\mathbf{k}')\}\frac{2\pi}{L_l} n]}$$

$$\times \overline{\phi(k)}\, \widetilde{\eta}(\mathbf{X}_l(\mathbf{k}))\, \overline{\widetilde{\eta}(\mathbf{X}_l(\mathbf{k}'))}\, \psi(k') \frac{d\mathbf{k}}{k_0} \frac{d\mathbf{k}'}{k_0'}. \quad (16.33)$$

Since

$$d\mathbf{X}_l(\mathbf{k}') = \frac{1}{k_0'} \left[k_0' - \underline{(\Lambda_l^{-1} k')}\, \vartheta(\mathbf{p}_l) \right] d\mathbf{k}', \quad (16.34)$$

changing variables $\mathbf{k}' \mapsto \mathbf{X}_l(\mathbf{k}')$ yields

$$I_{\phi,\psi} = \sum_{l=-\infty}^{\infty} \int_{-\infty}^{\infty} d\mathbf{X}_l \int_{V_m^+} \frac{d\mathbf{k}'}{k_0'} \sum_{n=-\infty}^{\infty} e^{[i\{\mathbf{X}_l(\mathbf{k}) - \mathbf{X}_l(\mathbf{k}')\}\frac{2\pi}{L_l} n]}$$

$$\times \frac{1}{k_0 - \underline{(\Lambda_l^{-1} k)}\vartheta(\mathbf{p}_l)}\, \widetilde{\eta}(\mathbf{X}_l(\mathbf{k}))\, \overline{\widetilde{\eta}(\mathbf{X}_l(\mathbf{k}'))}\, \overline{\phi(k)}\, \psi(k'). \quad (16.35)$$

Using the relation

$$\sum_{n=-\infty}^{\infty} e^{i\frac{2\pi n}{L}(x-x')} = L\, \delta(x-x'), \quad (16.36)$$

we get

$$I_{\phi,\psi} = \sum_{l=-\infty}^{\infty} \int_{-\infty}^{\infty} d\mathbf{X}_l \int_{V_m^+} \frac{d\mathbf{k}'}{k_0'} \delta(\mathbf{X}_l(\mathbf{k}) - \mathbf{X}_l(\mathbf{k}'))$$

$$\times \frac{L_l}{k_0 - \underline{(\Lambda_l^{-1} k)}\vartheta(\mathbf{p}_l)}\, \widetilde{\eta}(\mathbf{X}_l(\mathbf{k}))\, \overline{\widetilde{\eta}(\mathbf{X}_l(\mathbf{k}'))}\, \overline{\phi(k)}\, \psi(k'). \quad (16.37)$$

16.4. Groups without dilations: The Poincaré groups (♣)

Assuming that the δ-integration can be performed, the above expression may be rewritten as

$$I_{\phi,\psi} = \langle \phi | T \psi \rangle = \int_{V_m^+} \frac{d\mathbf{k}}{k_0} \overline{\phi(k)} \, \psi(k) \sum_{l=-\infty}^{\infty} \frac{L_l}{k_0 - (\Lambda_l^{-1}k)\,\vartheta(\mathbf{p}_l)} |\eta(\Lambda_l^{-1}k)|^2. \quad (16.38)$$

Thus, like its continuous counterpart, T is a multiplication operator

$$(T\phi)(k) = T(k)\phi(k), \quad (16.39)$$

with

$$T(k) = \sum_{l=-\infty}^{\infty} \frac{L_l}{k_0 - (\Lambda_l^{-1}k)\vartheta(\mathbf{p}_l)} |\eta(\Lambda_l^{-1}k)|^2. \quad (16.40)$$

Hence the boundedness and invertibility of T depend on the boundedness and strict positivity of the function $T(k)$.

For $t, \varphi_l, \varphi_a, \varphi_b \in \mathbb{R}$, let

$$\begin{aligned}
k_0 &= m \cosh t, & a_0 &= m \cosh \varphi_a, & p_{l0} &= m \cosh \varphi_l, & b_0 &= m \cosh \varphi_b, \\
\mathbf{k} &= m \sinh t, & \mathbf{a} &= m \sinh \varphi_a, & \mathbf{p}_l &= m \sinh \varphi_l, & \mathbf{b} &= m \sinh \varphi_b.
\end{aligned} \quad (16.41)$$

Now, substituting (16.30) and (16.41) in (16.40), and writing $T(k) \equiv T(t)$, we obtain,

$$T(t) = 2 \sinh\left(\frac{\varphi_b - \varphi_a}{2}\right) \sum_{l=-\infty}^{\infty} \frac{\cosh((\frac{\varphi_b + \varphi_a}{2}) + \varpi(\varphi_l))}{\cosh(t - \varphi_l + \varpi(\varphi_l))} |\eta(t - \varphi_l)|^2, \quad (16.42)$$

where we have written

$$\beta^*(\mathbf{p}_l) \equiv \beta^*(\varphi_l) = \tanh \varpi(\varphi_l), \quad (16.43)$$

with β^* the dual speed (11.68) characterizing the section σ and ϖ is an arbitrary (continuous) real function.

Let $\varphi_l = l\varphi_0$, where $\varphi_0 > 0$ (fixed). Then, T takes the form:

$$T(t) = 2 \sinh\left(\frac{\varphi_b - \varphi_a}{2}\right) \sum_{l=-\infty}^{\infty} \frac{\cosh((\frac{\varphi_b + \varphi_a}{2}) + \varpi(l\varphi_0))}{\cosh(t - l\varphi_0 + \varpi(l\varphi_0))} |\eta(t - l\varphi_0)|^2. \quad (16.44)$$

For any compactly supported function η, with support length L, the sum in (16.44) contains at best $L/\varphi_0 + 1$ terms; i.e., it is a finite sum. We observe that each term in the sum is positive and bounded for any $t \in \mathbb{R}$, and, consequently, the function $T(t)$ is strictly positive and bounded; hence, the operator T is bounded with bounded inverse. Thus, the discretized version $\eta_{n,l}$ of the coherent states in (16.23) form a *discrete frame*. Depending on the choice of Ψ^* and η, this frame could be tight or nontight. In fact, the frame is tight iff $T(t)$ is a constant function.

We analyze the situation for the three specific choices of the section σ, described in Section 11.1.2.

1. *The Galilean section σ_0*

   For this section, $\vartheta(\varphi_l) = 0$, which implies that $\varpi(l\varphi_0) = l\varphi_0$; hence, we can write

   $$T(t) = \frac{2\sinh\left(\frac{\varphi_b - \varphi_a}{2}\right)}{\cosh t} \sum_{l=-\infty}^{\infty} \cosh\left(\frac{\varphi_b + \varphi_a}{2} + l\varphi_0\right) |\eta(t - l\varphi_0)|^2. \tag{16.45}$$

   For different values of t, $T(t)$ is different, so, in this case, the frame is never tight. By contrast, in the corresponding continuous case, the frame is tight for certain functions η (see Section 11.1.2).

2. *The Lorentz section σ_ℓ*

   For this section $\vartheta(\varphi_l) = \sinh\varphi_l$, so that $\varpi(l\varphi_0) = 0$ and

   $$T(t) = 2\sinh\left(\frac{\varphi_b - \varphi_a}{2}\right) \cosh\left(\frac{\varphi_b + \varphi_a}{2}\right) \sum_{l=-\infty}^{\infty} \frac{|\eta(t - l\varphi_0)|^2}{\cosh(t - l\varphi_0)}. \tag{16.46}$$

   If $|\eta(t - l\varphi_0)|^2 = \cosh(t - l\varphi_0)$ on the support of η, that is, for $\theta_a \leq t - l\theta_0 \leq \theta_b$, and zero outside, then

   $$T(t) = 2\sinh\left(\frac{\varphi_b - \varphi_a}{2}\right) \cosh\left(\frac{\varphi_b + \varphi_a}{2}\right) N(a, b) \tag{16.47}$$

   where $N(a, b) = 1 + \lfloor \frac{t - \varphi_a}{\varphi_0} \rfloor - \lceil \frac{t - \varphi_b}{\varphi_0} \rceil$ is the number of terms contributing to the sum (16.44). Here, we have used

   $$\begin{aligned} \lfloor n \rfloor &= n_o & \text{if} \quad n_o \leq n < n_o + 1, \\ \lceil n \rceil &= n_o + 1 & \text{if} \quad n_o < n \leq n_o + 1, \end{aligned} \quad \text{for some } n_o \in \mathbb{N}. \tag{16.48}$$

   Thus, $T(t)$ is a constant function, and, consequently, the frame is tight.

3. *The symmetric section σ_s*

   For this section, $\vartheta(\varphi_l) = \dfrac{\sinh\varphi_l}{1 + \cosh\varphi_l}$; hence, $\varpi(l\varphi_0) = \dfrac{l\varphi_0}{2}$ and

   $$T(t) = 2\sinh\left(\frac{\varphi_b - \varphi_a}{2}\right) \sum_{l=-\infty}^{\infty} \frac{\cosh(\frac{\varphi_b + \varphi_a + l\varphi_0}{2})}{\cosh(t - \frac{l\varphi_0}{2})} |\eta(t - l\varphi_0)|^2. \tag{16.49}$$

   In this case, $T(t)$ is not a constant function, and so the frame is never tight.

Now, if ϖ is a constant function, say, $\varpi(l\varphi_0) = \alpha$, then

$$T(t) = 2\sinh\left(\frac{\varphi_b - \varphi_a}{2}\right)\cosh\left(\frac{\varphi_b + \varphi_a}{2} + \alpha\right)\sum_{l=-\infty}^{\infty}\frac{|\eta(t - l\varphi_0)|^2}{\cosh(t - l\varphi_0 + \alpha)}. \tag{16.50}$$

Here again, if we have $|\eta(t - l\varphi_0)|^2 = \cosh(t - l\varphi_0 + \alpha)$ on the support of $\eta(t - l\varphi_0)$, $T(t)$ becomes a constant operator and, hence, the frame is tight.

For a general section β (characterized by the function ϖ), the discretized coherent states [see (16.23)] are

$$\eta_{n,l}(t) = \exp\left[-i\pi n\frac{\sinh(t - l\varphi_0 + \varpi(l\varphi_0))}{\sinh(\frac{\varphi_b - \varphi_a}{2})\cosh(\frac{\varphi_b + \varphi_a}{2} + \varpi(l\varphi_0))}\right]\eta(t - l\varphi_0). \tag{16.51}$$

In that case, we can write the reconstruction formula, for any $\phi \in \mathfrak{H}$, as

$$\phi(t) = \sum_{n,l=-\infty}^{\infty}\langle\eta_{n,l}(t)\,|\,\phi\rangle\, T(t)^{-1}\,|\,\eta_{n,l}(t)\rangle. \tag{16.52}$$

16.5 A group-theoretical approach to discrete wavelet transforms

16.5.1 Generalities on sampling

As already mentioned, the signal processing approach to the discretization problem consists in the *sampling* process, that is, evaluating a continuous signal on a discrete set of points. The goal is, of course, not to lose information in the process, and criteria to that effect are given by various forms of Shannon's sampling theorem. Basically, only one answer exists, namely, the density of sampling points must be high enough.

A priori, one has two ways of sampling a signal. The most natural one is to do it at *regular* intervals, $t_n = nT$, $n \in \mathbb{Z}$, where T^{-1} is the sampling rate. As a result, the sampled signal may retain some form of discrete translation invariance, at least in a statistical sense (stationary signals). In practice, however, the sampling is often *irregular*. Then, provided the sampling intervals $i_n = t_{n+1} - t_n$, are bounded below, many results are known, and this is a very active field of research [122, 153], in which number theory plays a nonnegligible role. Since the group-theoretical approach is the backbone of this volume, however, we will not discuss this topic of irregular sampling any further and simply refer the reader to the literature [Fei, 122].

Instead, we will concentrate on the regular case, starting from the issues raised in the treatment of wavelets on $\mathbb{Z}_p = \mathbb{Z}/p\mathbb{Z}$ introduced in Section 13.3. Indeed, these wavelets on $\mathbb{Z}_p$ give a fresh look to the problem, precisely because they import some ideas from number theory.

16.5.2 Wavelets on $\mathbb{Z}_p$ revisited (♣)

Because $\mathbb{Z}_p$ is a finite field, the dilation by $a \mod p$, $\Delta_a : f(n) \mapsto f(a^{-1}n)$, preserves the number of nonzero values of the wavelet: The set of nonzero values $\{f(a^{-1}n), n \in \mathbb{Z}_p\}$ is just a permutation of the set $\{f(n), n \in \mathbb{Z}_p\}$. Therefore, the discrete dilation operator Δ_a preserves the length of the support of f, unlike the continuous dilation operator used in Chapter 12 (this remark is the starting point of the "algorithme à trous" introduced in [177]). Thus the discrete case described in Section 13.3 cannot be the sampled version of the continuous case.

In order to correct this situation, one introduces *pseudodilation* operators [125]

$$\mathcal{D}_a = K_a \Delta_a, \tag{16.53}$$

where K_a is a bounded operator on $\ell^2(\mathbb{Z}_p)$. Then, the crucial lemma reads as follows.

Lemma 16.5.1: *Let T_b and Δ_a denote, respectively, the translation and dilation operators on $\ell^2(\mathbb{Z}_p)$:*

$$\begin{aligned}(T_b f)(n) &= f(n-b), \\ (\Delta_a f)(n) &= f(a^{-1}n),\end{aligned} \quad b, n \in \mathbb{Z}_p, \, a \in \mathbb{Z}_p^*.$$

Then, the operators $\pi(b,a) = T_b \mathcal{D}_a$, where $\mathcal{D}_a$ is the pseudodilation (16.53), form a representation of the affine group G_p iff K_a is a convolution operator,

$$\mathcal{D}_a = F_a * \Delta_a, \tag{16.54}$$

where the filter F_a satisfies the compatibility relations

$$F_{aa'} = F_a * \Delta_a F_{a'} \tag{16.55}$$

or, equivalently, in Fourier space,

$$\widehat{F}_{aa'}(k) = \widehat{F}_a(k) \, \widehat{F}_{a'}(ak). \tag{16.56}$$

The corresponding representation of G_p is unitary iff

$$|\widehat{F}_a(k)| = 1, \quad \text{for all } a \in \mathbb{Z}_p^*, \, k \in \mathbb{Z}_p.$$

Functions F_a that satisfy (16.55) and (16.56) are called *compatible filters*, in analogy with the notion of filter sketched in Chapter 13, Section 13.1.2.

Given a compatible filter F_a, the corresponding wavelet transform is defined as in (13.27):

$$(T_\psi f)(b,a) = (\mathcal{D}_a \widetilde{\psi} * f)(b) = (\Delta_a \widetilde{\psi} * (\widetilde{F}_a * f))(b), \tag{16.57}$$

where $\widetilde{\psi}(x) = \overline{\psi(-x)}$.

As a result of Lemma 16.5.1, one gets for this new WT the same algorithmic structure as in the usual multiresolution analysis on $\mathbb{R}$, described

16.5. A group-theoretical approach to discrete wavelet transforms

in Section 13.1.1:

$$\begin{aligned}(T_\psi f)(b, aa') &= \Delta_{aa'}\tilde{\psi} * (\tilde{F}_{aa'} * f)(b) \\ &= \Delta_{aa'}\tilde{\psi} * (\Delta_a \tilde{F}_{a'} * (\tilde{F}_a * f))(b).\end{aligned} \qquad (16.58)$$

Indeed, writing the same relation on $\ell^2(\mathbb{Z})$ with $a = 2^{j-1}$, $a' = 2$ (the structure is the same), one gets

$$(T_\psi f)(b, 2^j) = \Delta_{2^j}\tilde{\psi} * \left(\Delta_{2^{j-1}}\tilde{F}_2 * (\tilde{F}_{2^{j-1}} * f)\right)(b), \qquad (16.59)$$

in which one recognizes the standard pyramidal algorithm of the discrete wavelet transform (see Section 13.1.2). Indeed, these relations mean that the computation of the wavelet coefficients at scale 2^j can be performed through a pyramidal algorithm involving only dilated copies of two filters, F_2 (which stands for the low-pass filter) and ψ (band-pass filter). Since only the true dilation Δ_a is involved, all filters have constant length, and one has a fast algorithm.

In conclusion, every compatible filter F_a gives rise to an efficient wavelet transform on $\mathbb{Z}_p$. It remains to characterize these compatible filters, and the relations (16.55) and (16.56) show that this is a group cohomology problem; namely, F_a is a one-cocycle over the multiplicative group $\mathbb{Z}_p^*$, with values in the p-dimensional torus $\mathbb{T}^p = \{(c_1, c_2, \ldots, c_p), |c_j| = 1\}$, in the unitary case, in $\mathbb{C}^p$ in general.

Actually, unitary equivalent representations of G_p yield the same transform, so one needs only to characterize the unitary dual $\widehat{G}_p$ or, correspondingly, equivalence classes of cocycles F_a. In other words, one has to identify the first cohomology group $H^1(\mathbb{Z}_p^*, \mathbb{T}^p)$, or $H^1(\mathbb{Z}_p^*, \mathbb{C}^p)$ in the nonunitary case.

Surprisingly, it turns out that the general solution of this problem may be obtained explicitly [183], once one notes that the multiplicative group $\mathbb{Z}_p^* = \mathbb{Z}_p \setminus \{0\}$ is isomorphic to the cyclic group Z_{p-1} of order $p-1$. Let us give a concrete example for $p = 5$. One has, successively,

$\mathbb{Z}_5 = \{0, 1, 2, 3, 4\}$ with addition mod 5;

$\mathbb{Z}_5^* = \{1, 2, 3, 4\}$ with multiplication mod 5;

$Z_4 = \{1, i, -1, -i\} = \{1, \omega, \omega^2, \omega^3\}$ with $\omega = i = e^{i2\pi/4}$.

Then, the correspondence

$$\begin{array}{ll} 1 \leftrightarrow 1, & 3 \leftrightarrow \omega^3 \\ 2 \leftrightarrow \omega, & 4 \leftrightarrow \omega^2 \end{array}$$

gives the required group isomorphism between $\mathbb{Z}_5^*$ and Z_4.

In these terms, the result of [183] is that the group $H^1(Z_{p-1}, \mathbb{T}^p)$, which labels the nonequivalent unitary cocycles, is itself isomorphic to $\mathbb{Z}_p^* \simeq Z_{p-1}$. Indeed, the most general cocycle may be described as follows. Let v be a function on $\mathbb{Z}_p$ such that $|v(n)| = 1$, i.e., $v : \mathbb{Z}_p \to \mathbb{T}$. Then, $\mathbb{Z}_p^*$ acts on such functions as

$$(a \cdot v)(n) = v(a^{-1}n), \ a \in \mathbb{Z}_p^*, \ n \in \mathbb{Z}_p.$$

Lemma 16.5.2: *Let τ be a generator of the cyclic group $\mathbb{Z}_{p-1} \simeq \mathbb{Z}_p^*$. Then, the functions $F_a^{(\tau)}$ defined as*

$$\widehat{F_a^{(\tau)}}(k) = \prod_{j=0}^{r-1}(\tau^j \cdot v)(k), \quad \text{where } a = \tau^r \in \mathbb{Z}_p^*, \ k \in \mathbb{Z}_p,$$

are compatible filters giving rise to unitary pseudodilation representations of G_p. The elements $\tau^j \cdot v$ are to be viewed as functions $\mathbb{Z}_p \to \mathbb{T}$, and v must satisfy the relations

$$\prod_{k \in \mathbb{Z}_p^*} v(k) = 1 \ \text{ and } \ v(0)^{p-1} = 1.$$

Moreover, these functions $F_a^{(\tau)}$, $\tau \in \mathbb{Z}_{p-1}$, are all unitary compatible filters.

Exactly the same result holds for an arbitrary finite field $\mathbb{F}$, as shown in [183], but we refer the reader to the original paper for the details.

16.5.3 Wavelets on a discrete abelian group

The case of $\mathbb{Z}_p$ (or, more generally, $\mathbb{F}$) is still particular, in the sense that the dilations Δ_a form a multiplicative abelian group $\mathbb{Z}_p^*$ that acts on $\ell^2(\mathbb{Z}_p)$. Thus, as stated in Section 13.3, the affine group G_p is the semidirect product $\mathbb{Z}_p \rtimes \mathbb{Z}_p^*$, and the general theory of group-related CS applies.

This is not the case in the usual multiresolution context: No dilation group acts on $\ell^2(\mathbb{Z})$ (the space of discrete signals), only an abelian multiplicative *semigroup* $\mathcal{A} = \{2^j, j = 0, 1, 2, \ldots\}$. Thus, a purely group-theoretical formulation of the discrete WT is not available. Recent results [40], however, go a long way towards this goal and effectively provide the missing link between the DWT and the CWT. We will conclude this section with a rather detailed survey of these results, because we feel they are both important and promising.

Actually, it is more instructive to treat a general abstract case, the multiresolution situation being just a (particularly simple) example. Therefore, we consider a locally compact abelian (LCA) group G, a lattice Γ in G, and an abelian semigroup $\mathcal{A}$ with unit (called a *commutative monoid* in the mathematical literature [Lan]) acting on Γ (in the multiresolution case, $G = \mathbb{R}, \Gamma = \mathbb{Z}$, and $\mathcal{A} = \{2^j, j = 0, 1, 2, \ldots\}$). We use the notions and the terminology introduced in Chapter 4, Section 4.4. Notice that transforms on LCA groups have been considered before, in [153] for the Gabor case, and in [Hol, 175] for the wavelet case. The setup of the latter is a hierarchy

16.5. A group-theoretical approach to discrete wavelet transforms

of three LCA groups

$$J \subset H \subset G,$$

with H and J discrete, G/H compact (that is, H is a lattice), and H/J finite. A typical relevant example is $G = \mathbb{R}$, $H = \mathbb{Z}$, $J = 2\mathbb{Z}$. The aim of [Hol] and [175], however, is to generalize the notion of quadrature mirror filters (QMFs), and particular attention is paid to the sampling problem. In the paper discussed here, on the contrary, the goal is to provide an abstract, group-theoretical setting for the pyramidal algorithms of the DWT, using the cohomological language introduced in [125, 183].

Compatible filters: The general case

In order to make things precise, we start by defining the action of a semigroup $\mathcal{A}$ on the lattice Γ contained in the LCA group G.

Definition 16.5.3: *(1) Let $\mathcal{A}$ be a semigroup with unit element 1 and Γ a lattice in G. An action of $\mathcal{A}$ on Γ is a mapping $(a, x) \in \mathcal{A} \times \Gamma \mapsto ax \in \Gamma$, such that: (i) $a(a'\gamma) = (aa')\gamma$, $\forall a, a' \in \mathcal{A}$, $\gamma \in \Gamma$; and (ii) $1\gamma = \gamma$, $\forall \gamma \in \Gamma$. The action of $\mathcal{A}$ is one-to-one if for any $a \in \mathcal{A}$, the mapping $\gamma \mapsto a\gamma$ is one-to-one.*

(2) We say that the action of $\mathcal{A}$ on Γ is compatible (with the group law) if

$$a(\gamma\gamma') = (a\gamma)(a\gamma'), \quad \forall a \in \mathcal{A}, \, \gamma, \gamma' \in \Gamma. \tag{16.60}$$

(3) Assume the action of $\mathcal{A}$ on Γ is compatible and one-to-one. We say that a divides $\gamma \in \Gamma$ if there exists a (unique) $\gamma' \in \Gamma$ such that $\gamma = a\gamma'$. In such a case we write $a|\gamma$, and $\gamma' = a^{-1}\gamma$.

Note that $a^{-1}\gamma$ only denotes an element γ' of Γ such that $a\gamma' = \gamma$; this does *not* mean that a is invertible in $\mathcal{A}$.

From now on, we shall assume that the action of $\mathcal{A}$ on Γ is compatible and one-to-one. In addition, we suppose that, for each $a \in \mathcal{A}$, $a \neq 1$, $a\Gamma$ is a lattice of G and the quotient $Q_a = \Gamma/a\Gamma$ is a nontrivial finite group of order $d(a)$. The action of $\mathcal{A}$ on Γ induces an action on $\widehat{\Gamma}$, *the dual action*, in the natural way: For any $\gamma \in \Gamma$, $\chi \in \widehat{\Gamma}$, and $a \in \mathcal{A}$, $\langle a\chi, \gamma \rangle = \langle \chi, a\gamma \rangle$. The dual action is compatible, but not necessarily one-to-one. For convenience, we shall call *sequences* the functions on Γ.

As discussed in the previous section, no natural group of dilations acts on $\ell^2(\mathbb{Z})$, only semigroups $\mathcal{A} \subset \mathbb{Z}_+^*$, and these have a compatible and one-to-one action. As in the case of $\mathbb{Z}_p$, we first consider the natural dilation on $\ell^2(\mathbb{Z})$.

Definition 16.5.4: *The natural dilation by $a \in \mathcal{A}$ is the bounded operator Δ_a, defined on $\ell^2(\Gamma)$ as follows: For any sequence $\{u_\gamma, \gamma \in \Gamma\} \in \ell^2(\Gamma)$,*

$$(\Delta_a u)_\gamma = \begin{cases} u_{a^{-1}\gamma}, & \text{if } a | \gamma, \\ 0, & \text{otherwise.} \end{cases} \tag{16.61}$$

For each $a \in \mathcal{A}$, we thus obtain a map Δ_a that dilates an arbitrary sequence $u = (u_\gamma)_{\gamma \in \Gamma}$ by inserting zeros.

The action of Δ_a in the Fourier domain is as follows: For all $u \in \ell^2(\Gamma)$,

$$\widehat{\Delta_a u}(\chi) = \widehat{u}(a\chi), \quad \forall \chi \in \widehat{\Gamma}. \tag{16.62}$$

These maps Δ_a ($a \in \mathcal{A}$) satisfy a number of obvious properties. For instance,

$$\Delta_a \Delta_{a'} = \Delta_{aa'}, \quad \forall a, a' \in \mathcal{A}, \tag{16.63}$$

$$\Delta_a T_\gamma = T_{a\gamma} \Delta_a, \text{ where } T_\gamma \text{ is the translation by } \gamma \in \Gamma. \tag{16.64}$$

Combining (16.63) and (16.64), we get

$$(T_\gamma \Delta_a)(T_{\gamma'} \Delta_{a'}) = T_{\gamma + a\gamma'} \Delta_{aa'}, \tag{16.65}$$

which shows that $\pi(\gamma, a) = T_\gamma \Delta_a$ is a representation of the composition law $(\gamma, a)(\gamma', a') = (\gamma + a\gamma', aa')$, characteristic of a semidirect product. Indeed, we might say that the underlying structure is the semidirect product $G_\mathcal{A} = \Gamma \rtimes \mathcal{A}$ of the abelian group Γ by the multiplicative semigroup $\mathcal{A}$ (both discrete).

The natural dilation Δ_a is the analogue of the familiar dilation "à trous" used in the discrete WT [177], and it is also "full of holes", namely, zeros, a fact that prevents a nice sampling interpretation. Consider for instance Δ_2, the natural dilation by a factor 2 on $\ell^2(\mathbb{Z})$. The main difference between Δ_2 and the continuously defined D_2 lies in the fact that, by construction, half of the coefficients of a sequence dilated using Δ_2 vanish: $(\Delta_2 f)_{2k+1} = 0$, $\forall k \in \mathbb{Z}$. Therefore, a sequence dilated with Δ_2 can hardly be interpreted as a sampling of a continuously defined function, dilated with D_2.

More generally, the fundamental difference between these D_a and Δ_a is the following: Given a finite sequence $\{u_\gamma\}$, the measure of the support of a dilated sequence $\Delta_a u$, does not change, since the measure we use on Γ is the counting measure.

As in the case of $\mathbb{Z}_p$, the solution is to replace the natural dilations Δ_a by *pseudodilations*, defined in the same way.

Definition 16.5.5: *Let $\mathcal{A}$ be an abelian semigroup with unit, acting on a lattice $\Gamma \subset G$, in such a way that the action is compatible and one-to-one. A pseudodilation (or principal pseudodilation) on $\ell^2(\Gamma)$ is a bounded operator*

$$\mathcal{D}_a = K_a \Delta_a, \tag{16.66}$$

where K_a is a bounded linear operator acting on $\ell^2(\Gamma)$, in such a way that

$$\mathcal{D}_a \mathcal{D}_{a'} = \mathcal{D}_{aa'}, \quad a, a' \in \mathcal{A}, \tag{16.67}$$

16.5. A group-theoretical approach to discrete wavelet transforms

$$\mathcal{D}_a T_\gamma = T_{a\gamma} \mathcal{D}_a, \quad a \in \mathcal{A}, \gamma \in \Gamma. \tag{16.68}$$

Notice that (16.67) and (16.68) are identical to (16.63) and (16.64), and, thus, one may suspect the occurrence of a representation of $G_\mathcal{A}$. Indeed, exactly as for $\mathbb{Z}_p$ (and for the same reason), the operator K_a must be a convolution operator, as follows from the following lemma, the exact parallel of Lemma 16.5.1.

Lemma 16.5.6: *The operator $\mathcal{D}_a = K_a \Delta_a$ is a pseudodilation, in the sense of Definition 16.5.5, if and only if K_a is the convolution by some sequence $\{h_\gamma^{(a)}, \gamma \in \Gamma\}$, satisfying the compatibility condition*

$$h_\gamma^{(aa')} = \left(h^{(a)} * \Delta_a h^{(a')}\right)_\gamma = \sum_{\gamma'} h_{\gamma-a\gamma'}^{(a)} h_{\gamma'}^{(a')}, \quad \forall a, a' \in \mathcal{A}, \tag{16.69}$$

Thus, we have, for all $u \in \ell^2(\Gamma)$,

$$(\mathcal{D}_a u)_\gamma = \sum_{\gamma' \in \Gamma} h_{\gamma-a\gamma'}^{(a)} u_{\gamma'} = \left(h^{(a)} * (\Delta_a u)\right)_\gamma, \tag{16.70}$$

and we shall write:

$$\mathcal{D}_a = h^{(a)} * \Delta_a, \quad a \in \mathcal{A}. \tag{16.71}$$

Introduce now, as described in Section 4.4, the (group) Fourier transform $H_a = \widehat{h^{(a)}}$ of the sequences $h^{(a)}$, which is a function on the dual $\widehat{\Gamma}$ of Γ. Under the Fourier transform, the relations (16.70) and (16.71) read as

$$\widehat{\mathcal{D}_a u}(\chi) = H_a(\chi) \widehat{u}(a\chi), \quad \forall \chi \in \widehat{\Gamma}. \tag{16.72}$$

Therefore the compatibility condition (16.69), that simply translates (16.67), reads in Fourier space as

$$H_{aa'}(\chi) = H_a(\chi) H_{a'}(a\chi), \quad \forall \chi \in \widehat{\Gamma}. \tag{16.73}$$

In other words, we have recovered for the filters H_a exactly the cocycle equations (16.55) and (16.56).

In the usual multiresolution approach, one uses in fact two filters, a low-pass filter h and a high-pass filter g, satisfying some identities, for instance, the QMF relations discussed in Section 13.1.2. Thus, in addition to the pseudodilation $\mathcal{D}_a$, which leads to the the filters H_a, we need another ingredient. As mentioned above, we assume that, for all $a \neq 1$, Q_a is finite and nontrivial, so that $d(a) > 1$. We now introduce the following definition.

Definition 16.5.7: *Given a principal pseudodilation $\mathcal{D}$, an associated pseudodilation on $\ell^2(\Gamma)$ is a mapping $\widetilde{\mathcal{D}}$ that assigns to any pair $a, a' \in \mathcal{A}, a' \neq 1$, the family of linear operators $\widetilde{\mathcal{D}}_{a,a';\kappa}, \kappa = 1, \ldots, d(a') - 1$, of the form*

$$\widetilde{\mathcal{D}}_{a,a';\kappa} = \widetilde{K}_{a,a';\kappa} \Delta_{aa'}, \tag{16.74}$$

where $\tilde{K}_{a,a';\kappa}$ is a bounded linear operator acting on $\ell^2(\Gamma)$, such that

$$\tilde{D}_{a,a';\kappa}T_\gamma = T_{(aa')\gamma}\tilde{D}_{a,a';\kappa}, \quad a,a' \in \mathcal{A}, \gamma \in \Gamma, \quad (16.75)$$

$$\mathcal{D}_{a_0}\tilde{D}_{a_1,a';\kappa} = \tilde{D}_{a_0a_1,a';\kappa}, \quad a_0, a_1, a' \in \mathcal{A}. \quad (16.76)$$

An associated pseudodilation $\tilde{\mathcal{D}}$ is completely determined by the operators

$$\tilde{\mathcal{D}}_{a;\kappa} := \tilde{\mathcal{D}}_{1,a;\kappa}, \quad a \in \mathcal{A}, \, a \neq 1, \, \kappa = 1,\ldots d(a)-1,$$

and the relation (16.76).

Exactly as in the case of principal pseudodilations (Lemma 16.5.6), one then shows that each $\tilde{K}_{a,a';\kappa}$ must be the convolution operator by a sequence $g^{(a,a';\kappa)}$, so that the associated pseudodilation reads as

$$\tilde{\mathcal{D}}_{a,a';\kappa}u = g^{(a,a';\kappa)} * (\Delta_{aa'}u). \quad (16.77)$$

In addition, the sequences $\{g_\gamma^{(a,a';\kappa)}, \gamma \in \Gamma\}$ must satisfy (and may actually be generated by) the following compatibility equations:

$$g^{(a_0a_1,a';\kappa)} = h^{(a_0)} * \Delta_{a_0} g^{(a_1,a';\kappa)}. \quad (16.78)$$

Let $G_{(a,a';\kappa)}$ denote the Fourier transform of the sequence $g^{(a,a';\kappa)}$. In the Fourier domain, the compatibility relations (16.78) read as

$$G_{a_0a_1,a';\kappa}(\chi) = H_{a_0}(\chi)G_{a_1,a';\kappa}(a_0\chi). \quad (16.79)$$

These arguments motivate the following definition, generalizing the corresponding one for $\mathbb{Z}_p$ given in Section 16.5.2:

Definition 16.5.8: *A family $\{(H_a, G_{a,a';\kappa}), a, a' \in \mathcal{A}, a' \neq 1, \kappa = 1,\ldots, d(a')-1\}$ of bounded functions on $\hat{\Gamma}$ satisfying the conditions (16.73) and (16.79) is called a family of compatible filters.*

Therefore, we can summarize the results of this section as follows:

Theorem 16.5.9: *Let $\mathcal{D}$ be a principal pseudodilation and $\tilde{\mathcal{D}}$ be an associated pseudodilation. Then, they are necessarily filters of the form (16.71) and (16.77), and the corresponding functions on $\hat{\Gamma}$ $\{(H_a, G_{a,a';\kappa}), a, a' \in \mathcal{A}, \kappa = 1,\ldots, d(a')-1\}$ constitute a family of compatible filters.*

Since $h^{(a)}$ can be recovered from H_a by inverse Fourier transform, the complete description of the $\mathcal{D}_a$ is obtained if one can completely classify the solutions of (16.73).

To give a trivial example, consider the natural pseudodilations on Γ, defined by the sequences:

$$h_\gamma^{(a)} = \delta_{\gamma,0}, \quad g_\gamma^{(a;\kappa)} = \delta_{\gamma,\gamma_\kappa}, \quad \kappa = 1,\ldots d(a)-1, \quad (16.80)$$

and define the other sequences $g^{(a,a';\kappa)}$ using (16.78). It is readily verified that the corresponding family of functions $G_{a,a';\kappa}$, together with the functions $H_a = 1$ occuring in the case of natural dilations, yield a family of compatible filters. Other explicit examples may be found in [40].

16.5. A group-theoretical approach to discrete wavelet transforms

At this point, it is useful to introduce the adjoints of the principal and associated pseudodilation operators $\mathcal{D}_a$, $\widetilde{\mathcal{D}}_{a,a';\kappa}$, and define the *synthesis operators*,

$$\mathcal{S}_a = \mathcal{D}_a \mathcal{D}_a^* + \sum_{\kappa=1}^{d(a)-1} \widetilde{\mathcal{D}}_{a;\kappa} \widetilde{\mathcal{D}}_{a;\kappa}^* . \tag{16.81}$$

Then, one says that a family of compatible filters defines a *subband coding scheme* if $\mathcal{S}_a = 1$, for all $a \in \mathcal{A}$. Indeed, this condition guarantees that the signal may be reconstructed perfectly from its transform.

Hence, given such a subband coding scheme, one may decompose any $s \in \ell^2(\Gamma)$ as follows. Let $s^1 = s$, and define the sequences $s^a, a \in \mathcal{A}$ and $d^{a,a';\kappa}$ inductively by

$$s^a = \mathcal{D}_a^* s, \quad d^{a,a';\kappa} = \widetilde{\mathcal{D}}_{a';\kappa}^* s^a. \tag{16.82}$$

Then, we have the "wavelet-like" decomposition (compare Section 13.1.2):

$$\begin{aligned}
s &= \mathcal{D}_a s^a + \sum_{\kappa=1}^{d(a)-1} \widetilde{\mathcal{D}}_{a;\kappa} d^{a,1;\kappa} \\
&= \mathcal{D}_{a^2} s^{a^2} + \sum_{\kappa=1}^{d(a)-1} \widetilde{\mathcal{D}}_{a,a;\kappa} d^{a,a;\kappa} + \sum_{\kappa=1}^{d(a)-1} \widetilde{\mathcal{D}}_{a;\kappa} d^{1,a;\kappa} \\
&= \cdots .
\end{aligned}$$

As an example, let us take the natural pseudodilations. We easily verify that, with the sequences given in (16.80), we have

$$(\mathcal{D}_a \mathcal{D}_a^* s)_\gamma = \begin{cases} s_\gamma, & \text{if } a|\gamma, \\ 0, & \text{otherwise,} \end{cases}$$

and

$$(\widetilde{\mathcal{D}}_{a;\kappa} \widetilde{\mathcal{D}}_{a;\kappa}^* s)_\gamma = \begin{cases} s_\gamma, & \text{if } a|(\gamma + \gamma_\kappa), \\ 0, & \text{otherwise.} \end{cases}$$

In such a case, it is clear that $\mathcal{S}_a = 1$, so we have again a subband coding scheme. We notice that this example is the abstract version of the scheme used in [175] to generate families of perfect reconstruction quadrature mirror filters. Let us also note that, exactly as in the continuous case (Section 13.1.3), this analysis may be extended to a biorthogonal scheme, which consists in decoupling the decomposition and reconstruction steps by using different families of compatible filters in those two steps.

Compatible filters in the case $G = \mathbb{R}$, $\mathcal{A} \subset \mathbb{Z}_+$

We specialize now to the case of interest in classical 1-D wavelet theory, namely, $G = \mathbb{R}$ and $\Gamma = \mathbb{Z}$. Let $\mathcal{A}$ be any subsemigroup of the multiplicative semigroup $\mathbb{Z}_+^*$. In order to make more explicit the fact that we are now dealing with integers, we shall in this case denote the elements of $\mathcal{A}$ by n. The usual multiresolution case corresponds to $n = 2^j$. Let $\{(H_n, G_n), n \in \mathcal{A}\}$ be a system of compatible filters. After identifying the dual $\mathbb{T}$ of $\mathbb{Z}$ with the interval $[-\pi, \pi)$, the functions H_n and G_n on $\mathbb{T}$ are now considered as 2π-periodic functions.

We first notice that, given a 2π-periodic function Φ, and setting, for all $n \in \mathcal{A}$,

$$H_n(\theta) = \frac{\Phi(n\theta)}{\Phi(\theta)}, \qquad (16.83)$$

we immediately obtain a solution of (16.73), provided that the quotients are well defined and define bounded 2π-periodic functions for all $n \in \mathcal{A}$. We call such families *trivial*, for reasons that will become clear soon. We notice that (16.83) is reminiscent of the relationship (13.5), $\widehat{\phi}(2\xi) = m_0(\xi)\widehat{\phi}(\xi)$, between the Fourier transform of the scaling function and the low-pass filter in classical multiresolution theory. The identification is *not* perfect, however, for $\Phi(\theta)$ is 2π-periodic, unlike the Fourier transform of the scaling function. Nevertheless, any bounded function $\Phi \in L^\infty(\mathbb{R})$ such that the ratio (16.83) defines a bounded 2π-periodic function for all $n \in \mathcal{A}$ also yields a solution to the compatibility relation (16.73).

More generally, functions $H_n(\theta)$ of the form

$$H_n(\theta) = n^\alpha \frac{\Phi(n\theta)}{\Phi(\theta)}, \qquad \alpha \in \mathbb{R}, \qquad (16.84)$$

also satisfy the compatibility relations (16.73). This comes from the fact that $(nn')^\alpha = n^\alpha n'^\alpha$, in other words, that $n \mapsto n^\alpha$ is a character of the semigroup $\mathcal{A}$.

Another simple example of such a "trivial" family of compatible filters is provided by the theory of spline functions. Let us again consider any subsemigroup $\mathcal{A} \subset \mathbb{Z}_+^*$, and define, for all $n \in \mathcal{A}$,

$$h_k^{(n)} = \begin{cases} n^{-1}, & \text{if } 0 \leq k \leq n-1, \\ 0, & \text{otherwise.} \end{cases} \qquad (16.85)$$

A direct calculation yields

$$H_n(\theta) = \frac{1}{n} \frac{1 - e^{-in\theta}}{1 - e^{-i\theta}}. \qquad (16.86)$$

It is easily verified that such a function satisfies (16.73), since it is of the form (16.84), with $\Phi(\theta) = 1 - e^{-i\theta}$. Now, for each $\kappa = 1, \ldots, n-1$, let

$$G_{n;\kappa}(\theta) = H_n\left(\theta - \kappa \frac{2\pi}{n}\right). \qquad (16.87)$$

16.5. A group-theoretical approach to discrete wavelet transforms

Using the trigonometric identity $\sum_{\kappa=0}^{n-1} \sin(\theta + \kappa\pi/n)^{-2} = n^2 \sin(n\theta)^{-2}$, valid for all $n \in \mathbb{Z}_+^*$ and $\theta \in \mathbb{R}$, we easily see that, for all $n \in \mathcal{A}$,

$$|H_n(\theta)|^2 + \sum_{\kappa=1}^{d(n)-1} |G_{n;\kappa}(\theta)|^2 = 1. \tag{16.88}$$

This relation tells us that we have constructed a subband coding scheme, called the *Haar multiresolution analysis* in the classical wavelet literature [Dau, Mal, Vet].

Actually, the Haar MRA is the simplest instance of the so-called *spline multiresolution analyses* [Chu]. Spline filters of order N correspond to filters H_n of the form

$$H_n(\theta) = \frac{1}{n^N} \left(\frac{1 - e^{-in\theta}}{1 - e^{-i\theta}} \right)^N, \tag{16.89}$$

and $G_{a;\kappa}$ filters given in (16.87). Since the filters H_n again assume the form (16.84), the compatibility condition (16.73) is verified by construction. The perfect reconstruction condition (16.88), however, is not satisfied any more. This problem is usually bypassed by going to a biorthogonal scheme, as alluded to above.

Cohomological interpretation

Let us come back now to the general case. As for $\mathbb{Z}_p$, the relation (16.73) satisfied by the filters defining principal pseudodilations may be given an interesting cohomological interpretation. We refer to [40] for details and to [Asc] or [Hil] for the terminology and mathematical background.

As before, we assume that the semigroup $\mathcal{A}$ operates on $\widehat{\Gamma}$ in such a way that the Haar measure $\mu_{\widehat{\Gamma}}$ on $\widehat{\Gamma}$ is $\mathcal{A}$-quasi-invariant. Let $\mathcal{M}$ denote the commutative ring of all complex-valued functions Φ on $\widehat{\Gamma}$ such that $\Phi(\chi) \neq 0$ almost everywhere on $\widehat{\Gamma}$. Then, the mapping

$$(a, \Phi) \longmapsto a\Phi, \quad n \in \mathcal{A}, \, \Phi \in \mathcal{M},$$
$$\text{with} \quad (a\Phi)(\chi) = \Phi(a\chi), \text{ for all } \chi \in \widehat{\Gamma}, \tag{16.90}$$

defines a compatible action of $\mathcal{A}$ on $\mathcal{M}$.

For each integer $r \geq 1$, we define *r-cochains* as functions $H^r : \mathcal{A} \times \ldots \times \mathcal{A} \to \mathcal{M}$ that map ordered n-tuples of elements of $\mathcal{A}$ into $\mathcal{M}$. We denote by $\mathcal{C}^r(\mathcal{A}, \mathcal{M}) \equiv \mathcal{C}^r$ the set of all n-cochains. For $n = 0$, we define

$$\mathcal{C}^0(\mathcal{A}, \mathcal{M}) := \mathcal{M},$$

i.e., the 0-cochains are the elements of $\mathcal{M}$. $\mathcal{C}^r$ is an abelian group under the composition law $(U^r V^r)(a_1, \ldots, a_r) = U^r(a_1, \ldots, a_r) V^r(a_1, \ldots, a_r)$. The

380   16. The Discretization Problem: Frames, Sampling, and All That

semigroup $\mathcal{A}$ acts on $\mathcal{C}^r$ by
$$[a(H^r)](a_1,\ldots,a_r) = a\left(H^r(a_1,\ldots,a_r)\right),$$
and one can easily check that this action is compatible with the group law of $\mathcal{C}^r$.

Next, one introduces, in the standard way, the so-called differentials $\partial_r : \mathcal{C}^r(\mathcal{A},\mathcal{M}) \to \mathcal{C}^{r+1}(\mathcal{A},\mathcal{M})$, which are defined as follows:
$$\partial_r(H^r) = \prod_{i=0}^{r+1} [p^i_{r+1}(H^r)]^{(-1)^{r+i}}, \tag{16.91}$$

where $p^i_{r+1}(H^r)$ is given by
$$p^i_{r+1}(H^r)(a_1,\ldots,a_{r+1}) =$$
$$= \begin{cases} (a_1 H^r)(a_2,\ldots,a_{r+1}), & \text{if } i = 0, \\ H^r(a_1,\ldots,a_i a_{i+1},\ldots,a_{r+1}), & \text{if } 0 < i < r+1, \\ H^r(a_1,\ldots,a_r), & \text{if } i = r+1 \end{cases} \tag{16.92}$$

and the inverse of a cochain is simply the inverse of the function. It turns out that ∂_r is a group homomorphism for each integer $r \geq 0$, and
$$\text{Im}\,(\partial_r \circ \partial_{r-1}) = \{1\}.$$

We define
$$\mathcal{Z}^r(\mathcal{A},\mathcal{M}) = \text{Ker}\,\partial_r, \quad \mathcal{B}^r(\mathcal{A},\mathcal{M}) = \text{Im}\,\partial_{r-1}.$$

Following the usual terminology, the elements of $\mathcal{Z}^r(\mathcal{A},\mathcal{M})$ are called *r-cocycles* and those of $\mathcal{B}^r(\mathcal{A},\mathcal{M})$, *r-coboundaries*. For each $r \geq 1$, the quotient group
$$\mathcal{H}^r(\mathcal{A},\mathcal{M}) = \mathcal{Z}^r(\mathcal{A},\mathcal{M})/\mathcal{B}^r(\mathcal{A},\mathcal{M})$$
is the *rth cohomology group* of $\mathcal{A}$ in $\mathcal{M}$, and it measures the deviation from exactness of the sequence [compare Section 4.5.3(3)]
$$\mathcal{C}^0 \xrightarrow{\partial_0} \mathcal{C}^1 \xrightarrow{\partial_1} \mathcal{C}^2 \longrightarrow \ldots \longrightarrow \mathcal{C}^{r-1} \xrightarrow{\partial_{r-1}} \mathcal{C}^r \xrightarrow{\partial_r} \mathcal{C}^{r+1} \longrightarrow \ldots$$

In fact, for characterizing the compatible filters, we only need the first cohomology group. Indeed, an explicit calculation gives that, for all $\Phi \in \mathcal{M}$ and $a \in \mathcal{A}$,
$$[\partial_0 \Phi](a)(\chi) = \frac{\Phi(a\chi)}{\Phi(\chi)}, \quad \text{for all } \chi \in \widehat{\Gamma}, \tag{16.93}$$
and, for all $H \in \mathcal{C}^1(\mathcal{A},\mathcal{M})$ and $a, a' \in \mathcal{A}$,
$$[\partial_1 H](a,a')(\chi) = \frac{H_{aa'}(\chi)}{H_a(\chi) H_{a'}(a\chi)}, \quad \chi \in \widehat{\Gamma}. \tag{16.94}$$

Comparing (16.94) with (16.73), it is clear that each family $\{H_n : n \in \mathcal{A}\}$ of elements of $\mathcal{M}$ is uniquely identified with an element $H \in \mathcal{C}^1(\mathcal{A},\mathcal{M})$.

Then, according to (16.94), an element $H \in \mathcal{C}^1(\mathcal{A}, \mathcal{M})$ defines a compatible family of filters $\{H_n\}$, that is, a solution of the cocycle equation (16.73), iff $H \in \operatorname{Ker} \partial_1$, and a trivial solution iff $H \in \operatorname{Im} \partial_0$, by (16.93). In addition, we want the filters H_n to be bounded, which we write $H \in L^\infty(\mathcal{A}, \widehat{\Gamma})$. In conclusion, we obtain the following characterization.

Proposition 16.5.10: *The set of all families of compatible filters indexed by $\mathcal{A}$ is given by $\operatorname{Ker} \partial_1 \cap L^\infty(\mathcal{A}, \widehat{\Gamma})$.*

Given an element H of $\mathcal{C}^1$ that defines a family of compatible filters, it is interesting to know how to define explicitly the functions H_n for a given $a \in \mathcal{A}$. The answer is known for the particular class of solutions of the form (16.93), which we have called trivial. More generally, the complete classification of compatible filters would require us to compute $\mathcal{H}^1(\mathcal{A}, \mathcal{M})$, which indexes classes of solutions of the cocycle equation modulo trivial solutions. This problem is still open in abstract wavelet analysis, even in the simple case $G = \mathbb{R}$, $\mathcal{A} = \{2^j, j = 0, 1, 2, \ldots\}$. It is also interesting by itself from the point of view of cohomology theory. Conversely, one may also hope that the tools of cohomology theory will help to classify the wavelet filters.

Compatible filters and discretized wavelet transform

As a last point, we show how the pseudodilation operators and compatible filters occur naturally in the context of discretization of the continuous wavelet transform. We shall see that discretization appears as a sort of generalized intertwining operator between the canonical action of the affine group on $L^2(\mathbb{R})$ and the action of an affine semigroups on $\ell^2(\mathbb{Z})$. As a byproduct, this approach also provides a geometrical interpretation of the pyramidal algorithms that have been developed for the computation of the wavelet coefficients, and a generalization for arbitrary integral scale.

Up to now, we have only worked at the level of sequences defined on a lattice Γ. To proceed further, our first step will be to transport the action of the filters on $L^2(G)$. In order to do that, we need additional assumptions. From now on, we will assume that the semigroup $\mathcal{A}$ acts on G also, and that this latter action is compatible, one-to-one, and transitive; i.e., for all $g \in G$ and $a \in \mathcal{A}$, there is some $g' \in G$ such that $g = ag'$. Finally, we assume that the Haar measure $\mu = \mu_G$ on G is $\mathcal{A}$-quasi-invariant (see Section 4.1). Let us denote by $\rho_a(g) = d\mu(ag)/d\mu(g)$ the corresponding Radon–Nikodym derivative [formally, in the notation of (4.7), we could write $\rho_a(g) \equiv \lambda(a^{-1}, g)$, but this does not make sense, since $\mathcal{A}$ is only a semigroup; see the remark after Definition 16.5.3]. The function ρ_a satisfies the usual cocycle condition (4.8): For all $a, a' \in \mathcal{A}$ and $g \in G$,

$$\rho_{aa'}(g) = \rho_a(g) \rho_{a'}(ag) . \qquad (16.95)$$

The translation invariance of the Haar measure μ then implies that $\rho_{a'}(ag - \gamma) = \rho_{a'}(ag)$ for all $a, a' \in \mathcal{A}, g \in G, \gamma \in \Gamma$.

Let now $\mathcal{D}$ and $\widetilde{\mathcal{D}}$ be a principal and associated pseudodilations, and let H_a and $G_{a,a';\kappa}$ be the corresponding filters, assumed to be bounded. Let us introduce the following operators, acting on $L^2(G)$:

$$\mathcal{U}_a f(g) = \rho_a(g) \sum_\gamma h_\gamma^{(a)} f(ag - \gamma) \qquad (16.96)$$

$$\mathcal{V}_{a,a';\kappa} f(g) = \rho_{aa'}(g) \sum_\gamma g_\gamma^{(a,a';\kappa)} f(aa'g - \gamma). \qquad (16.97)$$

These operators are the $L^2(G)$ analogues of the discrete operators $\mathcal{D}_a$ and $\widetilde{\mathcal{D}}_{a;\kappa}$, and they enable us to transport the action of the pseudodilations onto $L^2(G)$. The following is a direct consequence of the compatibility relations and the cocycle condition (16.95).

Corollary 16.5.11: *The operators $\mathcal{U}_a$ and $\mathcal{V}_{a,a';\kappa}$ satisfy*

$$\mathcal{U}_{aa'} = \mathcal{U}_a \mathcal{U}_{a'}, \qquad \mathcal{V}_{a,a';\kappa} = \mathcal{U}_a \mathcal{V}_{1,a';\kappa}. \qquad (16.98)$$

In the Fourier domain, the operators $\mathcal{U}_a$ and $\mathcal{V}_{a;\kappa} \equiv \mathcal{V}_{1,a;\kappa}$ read as

$$\widehat{\mathcal{U}_a f}(a\chi) = H_a(\chi) \widehat{f}(\chi), \qquad \widehat{\mathcal{V}_{a;\kappa} f}(a\chi) = G_{a;\kappa}(\chi) \widehat{f}(\chi), \quad \chi \in \widehat{G}. \quad (16.99)$$

Finally, the counterpart of the perfect reconstruction property is given by the following corollary.

Corollary 16.5.12: *Assume that the filters H_a and $G_{a;\kappa}$ generate a subband coding scheme. Then, we have the equality*

$$\mathcal{U}_a^* \mathcal{U}_a + \sum_{\kappa=1}^{d(a)-1} \mathcal{V}_{a;\kappa}^* \mathcal{V}_{a;\kappa} = 1.$$

These properties motivate the following definition.

Definition 16.5.13: *Let $\{H_a, a \in \mathcal{A}\}$ be a family of filters satisfying the compatibility condition (16.73). A scaling function associated to the family $\{H_a, a \in \mathcal{A}\}$ is a square integrable function $\phi \in L^2(G)$ that is invariant under the operator $\mathcal{U}_a$, for all $a \in \mathcal{A}$.*

Given a scaling function $\phi(g)$ and an associated pseudodilation $\widetilde{\mathcal{D}}$, one can then construct the associated generalized wavelets, defined by

$$\psi^{a;\kappa}(g) = \mathcal{V}_{a;\kappa} \phi(g), \qquad (16.100)$$

and develop an abstract theory of wavelets. Instead of doing so, we rather come back to the usual situation $G = \mathbb{R}, \Gamma = \mathbb{Z}$.

Let $\mathcal{A} \subset \mathbb{Z}_+^*$ be a multiplicative semigroup, whose elements we denote again by n. The Lebesgue measure (which is also the Haar measure) on $\mathbb{R}$ is $\mathcal{A}$-quasi-invariant, and the corresponding Radon–Nikodym derivative is

$$\rho_n(t) = n, \quad n \in \mathcal{A}, t \in \mathbb{R}.$$

16.5. A group-theoretical approach to discrete wavelet transforms

Let $\{(H_n, G_{n,n';\kappa}) : n, n' \in \mathcal{A}, \kappa = 1,\ldots d(n') - 1\}$ be a system of compatible filters labeled by $\mathcal{A}$. Let us assume we have a scaling function $\phi \in L^1(\mathbb{R}) \cap L^2(\mathbb{R})$, with $\int \phi(t)dt = 1$, satisfying the following two requirements:

1. *Stability:* The functions $\phi(t-k), k \in \mathbb{Z}$ form a frame of their closed linear span V.
2. *Refinability:*

$$\phi(t) = n \sum_l h_l^{(a)} \phi(at-l), \quad \forall n \in \mathcal{A}. \tag{16.101}$$

Then, it follows from (16.101) that, for all θ [compare with (13.8)],

$$\widehat{\phi}(n\theta) = H_n(\theta)\widehat{\phi}(\theta), \quad \forall n \in \mathcal{A}, \theta \in \mathbb{R}; \tag{16.102}$$

that is, ϕ is a scaling function in the sense of Definition 16.5.13.

Formally, (16.102) and the normalization condition define the function ϕ: For all $n \in \mathcal{A}, n \neq 1$, we may write, like in (13.7),

$$\widehat{\phi}(\theta) = H_n\left(\frac{\theta}{n}\right) H_n\left(\frac{\theta}{n^2}\right) H_n\left(\frac{\theta}{n^3}\right) \cdots.$$

One should be careful with this formula, however, since the convergence of such infinite products to nice L^2 functions is often problematic, although it has been proved in the standard multiresolution case [see (13.7)].

Let us now consider an associated pseudodilation $\widetilde{D}$ with filters $\{g^{(n,n';\kappa)}\}$. According to (16.100), we introduce the family of (square integrable) functions $\{\psi^{n;\kappa}(t)\}$, defined by

$$\psi^{n;\kappa}(t) = (V_{n;\kappa}\phi)(t) = n \sum_l g_l^{(n;\kappa)} \phi(at-l). \tag{16.103}$$

In the Fourier domain, we have

$$\widehat{\psi^{n;\kappa}}(\omega) = G_{n;\kappa}\left(\frac{\omega}{n}\right) \widehat{\phi}\left(\frac{\omega}{n}\right), \tag{16.104}$$

where we have set

$$G_{n;\kappa}(\theta) = \sum_{k \in \mathbb{Z}} g_k^{(n;\kappa)} e^{-ik\theta}.$$

Using the scaling function ϕ and the wavelets $\psi^{n;\kappa}$, we shall now make the connection between compatible filters and the discretized version of the continuous wavelet transform.

As stated in (12.14), the continuous wavelet transform on $L^2(\mathbb{R})$ is given by

$$(T_\psi f)(b,a) = \langle T_b D_a l\psi \mid f \rangle = \frac{1}{\sqrt{|a|}} \int_\mathbb{R} \overline{\psi(a^{-1}(x-b))} f(x)\, dx. \tag{16.105}$$

Instead of the usual multiresolution discretization described in Section 12.5, we want to compute the following transforms, for any discrete semigroup $\mathcal{A} \subset \mathbb{Z}_+^*$:

$$(\mathcal{T}_{m;\kappa}f)(b,n) = \frac{1}{\sqrt{n}} \int_{\mathbb{R}} \overline{\psi^{m;\kappa}(n^{-1}(t-b))} f(t)\, dt, \qquad (16.106)$$

where $b \in \Gamma \equiv \mathbb{Z}$, $n \in \mathcal{A}$ is any multiple of $m \in \mathcal{A}$, and $\kappa = 1\ldots d(m)-1$. We are also interested in

$$(\mathcal{S}f)(b,n) = \frac{1}{\sqrt{n}} \int_{\mathbb{R}} \overline{\phi(n^{-1}(t-b))} f(t)\, dt, \quad n \in \mathcal{A},\ b \in \Gamma. \qquad (16.107)$$

The key point is as follows. Given any (finite energy) sequence $s \in \ell^2(\mathbb{Z})$, the stability (or frame) property tells us that one can always find $f \in L^2(\mathbb{R})$, such that $s_k = \int f(t)\phi(t-k)dt$. In other words, using the terminology of [175], the sequence $s = \{s_k, k \in \mathbb{Z}\}$ may be obtained by sampling imperfectly some f with respect to the lattice $\mathbb{Z}$ and the function ϕ; that is,

$$s = \Xi_{\mathbb{Z}}(\tilde{\phi} * f) = \Xi_{\mathbb{Z}}^{\phi} f, \qquad (16.108)$$

where $\Xi_{\mathbb{Z}} : L^2(\mathbb{R}) \to \ell^2(\mathbb{Z})$ is the (perfect) sampling operator, $(\Xi_{\mathbb{Z}} h)_k = h(k)$, $k \in \mathbb{Z}$, and $\tilde{\phi}(t) = \phi(-t)$. Let us consider the Fourier transform of s, given by $\hat{s}(\theta) = \sum_{k \in \mathbb{Z}} s_k e^{-ik\theta}$. Then, the Poisson summation formula (4.128) yields:

$$\hat{s}(\theta) = \sum_{\chi \in \mathbb{Z}^\perp} \widehat{\tilde{\phi} * f}(\theta + \chi) = \sum_{l \in \mathbb{Z}} \overline{\hat{\phi}(\theta + 2\pi l)} \hat{f}(\theta + 2\pi l). \qquad (16.109)$$

Let us now set

$$s_k^n = (\mathcal{D}_n^* s)_k, \qquad d_k^{n;\kappa} = \left(\tilde{\mathcal{D}}_{n;\kappa}^* s\right)_k, \qquad (16.110)$$

and, more generally, for $n, m \in \mathcal{A}$ and $\kappa = 1, \ldots, d(m)-1$,

$$d_k^{n,m;\kappa} = \left(\tilde{\mathcal{D}}_{n,m;\kappa}^* s\right)_k = \left(\tilde{\mathcal{D}}_{m;\kappa}^* \mathcal{D}_n^* s\right)_k. \qquad (16.111)$$

The action of the adjoint pseudodilation $\mathcal{D}_n^*$ in the Fourier domain reads as

$$\widehat{\mathcal{D}_n^* s}(\theta) = \frac{1}{n} \sum_{l=0}^{n-1} \overline{H_n(n^{-1}(\theta + 2\pi l))}\, \hat{s}(n^{-1}(\theta + 2\pi l)). \qquad (16.112)$$

Then, using (16.109) and the 2π-periodicity of H_n, we get

$$\begin{aligned} s_k^n &= \frac{1}{2\pi} \frac{1}{n} \int_{-\pi}^{\pi} e^{ik\theta} \sum_{l=0}^{n-1} \overline{H_n(n^{-1}(\theta + 2\pi l))}\, \hat{s}(n^{-1}(\theta + 2\pi l))\, d\theta \\ &= \frac{1}{2\pi} \int_{-\pi}^{\pi} e^{ikn\theta}\, \hat{f}(\theta) \overline{H_n(\theta)}\, \overline{\hat{\phi}(\theta)}\, d\theta \end{aligned}$$

16.5. A group-theoretical approach to discrete wavelet transforms

$$= \frac{1}{2\pi}\int_{-\pi}^{\pi} e^{ikn\theta} f(\theta)\overline{\widehat{\phi}(n\theta)}\,d\theta, \quad \text{by (16.102)}.$$

In other words

$$s_k^n = \frac{1}{n}\int_{\mathbb{R}} f(t)\overline{\phi(n^{-1}t - k)}\,dt = \frac{1}{\sqrt{n}}(\mathcal{S}f)(nk, n). \tag{16.113}$$

Similarly, we obtain, using (16.104),

$$d_k^{n;\kappa} = \frac{1}{2\pi}\int_{\mathbb{R}} e^{ink\theta}\widehat{f}(\theta)\,\overline{G_{n;\kappa}(\theta)}\,\overline{\widehat{\phi}(\theta)}\,d\theta = \frac{1}{n}\int_{\mathbb{R}} f(t)\,\overline{\psi^{n;\kappa}(n^{-1}t - k)}\,dt$$

$$= \frac{1}{\sqrt{n}}(\mathcal{T}_{n;\kappa}f)(nk, n). \tag{16.114}$$

More generally, we obtain from (16.103)

$$d_k^{n,m;\kappa} = \frac{1}{\sqrt{nm}}(\mathcal{T}_{m;\kappa}f)(nmk, nm) = \left(\widetilde{\mathcal{D}}^*_{n,m;\kappa}s\right)_k. \tag{16.115}$$

In other words, the wavelet coefficients $d_k^{n,m;\kappa}$ of a discrete signal $s \in \ell^2(\mathbb{Z})$, obtained in the discrete scheme based on $(\mathcal{A}, \mathbb{Z})$, may also be computed by sampling the continuous WT corresponding to the scaling function ϕ and the wavelets $\psi^{a;\kappa}$. Conversely, the sampled values of the continuous wavelet coefficients may be obtained by imperfect sampling by the function ϕ, followed by (the adjoint of) a pseudodilation. This makes the announced connection between compatible filters and continuously defined wavelets. We summarize this result in the following theorem.

Theorem 16.5.14: Let $\{(H_n, G_{n,m;\kappa}) : n, m \in \mathcal{A}, m \neq 1, \kappa = 1, \ldots d(m) - 1\}$ be a family of compatible filters, and assume there exists a (scaling) function $\phi \in L^1(\mathbb{R}) \cap L^2(\mathbb{R})$ such that

$$\widehat{\phi}(n\theta) = H_n(\theta)\widehat{\phi}(\theta), \quad \forall n \in \mathcal{A}, \forall \theta \in \mathbb{R}. \tag{16.116}$$

Let $\psi^{n;\kappa}$ be the wavelets defined by

$$\widehat{\psi}^{n;\kappa}(n\theta) = G_{n;\kappa}(\theta)\widehat{\phi}(\theta), \quad \forall \theta \in \mathbb{R}. \tag{16.117}$$

Then, the sampled values of the corresponding wavelet coefficients may be obtained by imperfect sampling by the function ϕ, followed by (the adjoint of) a pseudodilation,

$$\frac{1}{\sqrt{nm}}(\mathcal{T}_{m;\kappa}f)(nmk, nm) = \frac{1}{\sqrt{nm}}\langle T_{nmk}D_{nm}\psi^{m;\kappa}|f\rangle = \left(\widetilde{\mathcal{D}}^*_{n,m;\kappa}s\right)_k,$$

$$k \in \mathbb{Z}, n, m \in \mathcal{A}, \tag{16.118}$$

where $s = \Xi_{\mathbb{Z}}(\widetilde{\phi} * f) = \Xi_{\mathbb{Z}}^{\phi}f$. Similar statements hold for the projection on the scaling function,

$$\frac{1}{\sqrt{n}}(\mathcal{S}f)(nk, n) = \frac{1}{\sqrt{n}}\langle T_{nk}D_n\phi|f\rangle = (\mathcal{D}_n^*s)_k. \tag{16.119}$$

The theorem may also be reformulated by stating that the following diagram is commutative, where the top line lives in $L^2(\mathbb{R})$ and the bottom one in $\ell^2(\mathbb{Z})$:

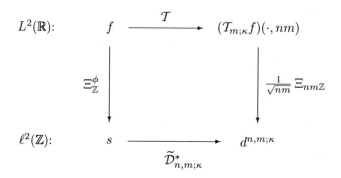

The commutative diagram above suggests that the (normalized) sampling operators $\frac{1}{\sqrt{a}}\Xi_{a\mathbb{Z}}$ act as generalized intertwining operators between the representation of the affine group on $L^2(\mathbb{R})$ and the representation of the affine semigroup $\mathbb{Z} \rtimes \mathcal{A}$ on $\ell^2(\mathbb{Z})$. This may offer a different point of view on the harmonic analysis of the affine group.

In conclusion, this approach yields a general setting for multiresolutions associated with the action of a semigroup $\mathcal{A}$ on a lattice Γ, especially when $\mathcal{A}$ is the semidirect product of the group of lattice translations by a discrete abelian semigroup of dilations. The tool is the notion of compatible filters, built on pseudodilations.

Transporting the action of the pseudodilations to the case of continuously defined functions yields structures that are extremely close to multiresolution analyses. In particular, the pseudodilations provide the algebraic setting for the pyramidal algorithms familiar in discrete wavelet analysis. Indeed, the pyramidal algorithm for computing recursively scaling function coefficients is nothing but a recursive use of a pseudodilation (more precisely, the adjoint of a pseudodilation). The corresponding wavelet coefficients are obtained via the (adjoint) associated pseudodilations.

The outcome is a reformulation, in a very general group-theoretical framework, of discrete wavelet transforms based on multiresolution. Such algebraic structures associated with multiresolution pave the way towards generalizations. The latter may, for example, include (separable or nonseparable) wavelet bases associated with nontrivial dilations in more than one dimension (see Section 14.3.4), or the wavelet bases associated with aperiodic tilings developed for studying quasicrystals in [72] and [141].

16.6 Conclusion

Discretization of the continuous wavelet transform, as for any CS transform, is unavoidable for numerical reasons. On the one hand, it leads to the theory of discrete frames, and many examples have been given in this chapter. On the other hand, purely discrete transforms are usually preferred in the digital world of signal processing. The results of the last section show that, here too, the (fast) algorithms have a group-theoretical backbone, which opens the door to future generalizations. Once more, group theory is a powerful unifying factor.

Conclusion and Outlook

As must have become clear from the last few chapters of this book, research on wavelets — both theoretical and applied — has currently been gathering increasing momentum, perhaps more so because of its numerous applications in today's cutting-edge technology. The use of coherent states, as an applied and theoretical tool in physical and mathematical research, is perhaps equally pervasive.

Present status of CS and wavelet research

Let us just list — of course, with no pretension of being exhaustive — a number of active directions in which work is presently moving.

A. In mathematics

- The use of coherent states associated to complex holomorphic structures, as illustrated in Chapter 6, and their applications to the study of the geometry of holomorphic vector bundles. Equally, there is also the question of nonholomorphic extensions of the theory.
- The search for general square integrability results on homogeneous spaces of groups and possible orthogonality relations in this setting.
- Wavelets on manifolds: Extension of the work described in Section 15.1 to hyperboloids or more general (pseudo-)Riemannian manifolds.

- The study of wavelets based on irrational numeration systems, such as $\tau, \beta, \ldots$, in 1-D and their extensions to higher dimensions.

B. In physics

- Developing systematic applications of time-frequency methods (e.g., Gabor transforms, wavelets, Wigner–Ville transforms) to laser–atom interactions (see Section 12.7) or in spectroscopy, in particular, NMR spectroscopy — this latter having potential applications in medical physics.
- Finding appropriate CS to describe specific packets of Rydberg states of atoms, with applications to semiclassical approximations or to explain quasi-classical behavior.
- Applications of wavelet analysis in structure calculations in solid-state physics (see [Be2, 49]).
- Applications of wavelet analysis in numerous areas of biophysics and psychophysics — e.g., in the analyses of EEG, ECG, as well as their use in medical diagnostic techniques [Ald, Tho].
- Further applications of wavelets in the analysis and understanding of fully developed turbulence in fluid dynamics [Be2].
- Applications of 2-D directional wavelets in quasicrystallography, both physical and mathematical (see Section 14.3.3).

C. In engineering and applied science

- Applications of time-frequency methods (Gabor, wavelets, or Wigner–Ville) in mechanical systems, e.g., in the analysis of the behavior of materials subjected to shocks [279].
- Novel applications of continuous wavelets, in particular, directional wavelets, in imaging problems, e.g., watermarking of images (that is, inserting in an image an invisible signature to guarantee its ownership), or detection and modeling of (oriented) textures, a special case of shape recognition, with applications in robotic vision.
- Military application of 3-D or space–time continuous wavelet transforms for tracking and identification of targets (missiles!).
- Application of wavelets or some generalizations of wavelets to speech processing and voice recognition by various devices, for instance, by a car!
- Further development of fast algorithms for 1-D and 2-D continuous wavelet transforms.

What is the future from our point of view?

In the spirit of the sort of mathematical analysis presented in this book, a number of open areas of work are discernible.

- Much work still remains to be done in the use of CS to study the quantization process, both along the lines of geometric quantization, including group representation theory and also using Feynman path integrals — subjects that we did not have the occasion to touch upon in this book.
- A study of possible deformations of the various structures introduced here, leading to q-deformed CS, CS for supersymmetric Lie algebras, p-adic CS or wavelets, and so on.
- An adequate mathematical tool to analyze material structures using contact microscopy still needs to be developed. In this kind of work, a material probe is brought and moved around in the immediate vicinity of the material surface being studied and it would appear that a CS or wavelet type of analysis would be far better adapted as a tool for such a study than the standard Fourier transform, which is the natural mathematical tool for analyzing material structures through diffraction experiments.
- Discretization problems, in connection with the availability of very fast computers, and parallelization of wavelet (pyramidal-like) algorithms.
- Systematic study of decoherence and quantum nondemolition experiments using coherent states.

Clearly, there is plenty of work waiting to be pursued, in all directions. The lesson we might draw at this point is that group theory remains a valuable guide and a source of mathematical inspiration, as well as a powerful unifying factor. Secondly, that it always pays to develop mathematical tools with sufficient precision and generality, not only for their intrinsic beauty, but also in view of the breadth of applications which they afford — a virtue that can be of immense practical value (and indeed, economically promising as well). We conclude, therefore, by characterizing the field that we have explored in this book as "applied mathematical physics."

References

A. Books and Theses

[Akh] N. I. Akhiezer, *The Classical Moment Problem and Some Related Questions in Analysis*, Oliver & Boyd, Edinburgh and London, 1965

[Ala] A. Alaux, *L'Image par Résonance Magnétique*, Sauramps Médical, Montpellier, 1994

[Ald] A. Aldroubi and M. Unser (eds.), *Wavelets in Medicine and Biology*, CRC Press, Boca Raton, FL, 1996

[Arn] A. Arnéodo, F. Argoul, E. Bacry, J. Elezgaray, and J. F. Muzy, *Ondelettes, multifractales et turbulences–De l'ADN aux croissances cristallines*, Diderot, Paris, 1995

[Asc] E. Ascher, *Extensions et cohomologie de groupes*, Lecture Notes, Enseignement du troisième cycle de la physique en Suisse Romande (CICP), 1972

[Bar] A. O. Barut and R. Rączka, *Theory of Group Representations and Applications*, PWN, Warszawa, 1977

[Be1] S. K. Berberian, *Notes on Spectral Theory*, Van Nostrand, Princeton, NJ, 1966

[Be2] J. C. van den Berg (ed.), *Wavelets in Physics*, Cambridge Univ. Press, Cambridge, 1999

[Bie] L. Biedenharn and J. D. Louck, *The Racah-Wigner Algebra in Quantum Theory*, Encyclopaedia of Mathematics, Vol. 9, Addison-Wesley, Reading, MA, 1981

[Bog] J. Bognar, *Indefinite Inner Product Spaces*, Springer-Verlag, Berlin, 1974

[Bor] A. Borel, *Représentations des groupes localement compacts*, Lect. Notes in Math., Vol. 276, Springer-Verlag, Berlin, 1972

[Bo1] K. Bouyoucef, *Sur des aspects multirésolution en reconstruction d'images: Application au Télescope Spatial de Hubble,* Thèse de Doctorat, Univ. P. Sabatier, Toulouse, 1993

[Bo2] A. Bouzouina, *[Groupe de Weyl-Heisenberg discret],* Thèse de Doctorat, Univ. Paris-Dauphine, 1997

[Bra] O. Bratteli and D. W. Robinson, *Operator Algebras and Quantum Statistical Mechanics I, II,* Springer-Verlag, Berlin, 1979

[Bus] P. Busch, P. J. Lahti, and P. Mittelstaedt, *The Quantum Theory of Measurement,* Springer-Verlag, Berlin and Heidelberg, 1991

[Chu] C. K. Chui, *An Introduction to Wavelets,* Academic Press, San Diego, CA, 1992

[Coh] C. Cohen-Tannoudji, B. Diu, and F. Laloë, *Mécanique Quantique, Tome I,* Hermann, Paris, 1977

[Com] J-M. Combes, A. Grossmann and Ph. Tchamitchian (eds.), *Wavelets, Time-Frequency Methods and Phase Space (Proc. Marseille 1987),* Springer-Verlag, Berlin, 1989; 2nd ed., 1990

[Dau] I. Daubechies, *Ten Lectures on Wavelets,* SIAM, Philadelphia, 1992

[DeV] R. De Valois and K. De Valois, *Spatial Vision,* Oxford Univ. Press, New York, 1988

[Dix1] J. Dixmier, *Les algèbres d'opérateurs dans l'espace hilbertien (Algèbres de von Neumann)* Gauthier-Villars, Paris, 1957

[Dix2] J. Dixmier, *Les C*-algèbres et leurs représentations,* Gauthier-Villars, Paris, 1964

[Duv] M. Duval-Destin, *Analyse spatiale et spatio-temporelle de la stimulation visuelle à l'aide de la transformée en ondelettes,* Thèse de Doctorat, Université d'Aix-Marseille II, 1991

[Fea] J-C. Feauveau, *Analyse multirésolution par ondelettes non orthogonales et bancs de filtres numériques,* Thèse de Doctorat, Université Paris-Sud, 1990

[Fei] H. G. Feichtinger and T. Strohmer (eds.), *Gabor Analysis and Algorithms– Theory and Applications,* Birkhäuser, Boston-Basel-Berlin, 1998

[Fen] D. H. Feng, J. R. Klauder, and M. Strayer (eds.), *Coherent States: Past, Present and Future (Proc. Oak Ridge 1993),* World Scientific, Singapore, 1994

[Fla] P. Flandrin, *Temps-Fréquence,* Hermès, Paris, 1993; Engl. transl. *Time Frequency/Time Scale Analysis,* Academic Press, San Diego, CA, 1998

[Fol] G. B. Folland, *A Course in Abstract Harmonic Analysis,* CRC Press, Boca Raton, FL, 1995

[Gaa] S. A. Gaal, *Linear Analysis and Representation Theory,* Springer-Verlag, Berlin, 1973

[Gel] I. M. Gelfand and N. Ya. Vilenkin, *Generalized Functions,* Vol.4, Academic Press, New York, 1964

[Got] K. Gottfried, *Quantum Mechanics. Vol. I: Fundamentals,* Benjamin, New York and Amsterdam, 1966

[Gro] S. de Groot, *La transformation de Weyl et la fonction de Wigner: une forme alternative de la mécanique quantique,* Les Presses de l'Université de Montréal, Montréal, 1974

[Gun] H. Günther, *NMR Spectroscopy,* 2nd ed., Wiley, Chichester, New York, 1994

References

[Gui] V. Guillemin and S. Sternberg, *Symplectic Techniques in Physics*, Cambridge Univ. Press, Cambridge, 1984

[Hel] S. Helgason, *Differential Geometry, Lie Groups, and Symmetric Spaces*, Academic Press, New York, 1978

[Hil] P. Hilton and U. Stammbach, *A Course in Homological Algebra*, Springer-Verlag, Berlin, 1971

[Hol] M. Holschneider, *Wavelets, An Analysis Tool*, Oxford Univ. Press, Oxford, 1995

[Hum] J. E. Humphreys, *Introduction to Lie Algebras and Representation Theory*, Springer-Verlag, Berlin, 1972

[Inö] E. Inönü, *A study of the unitary representations of the Galilei group in relation to Quantum Mechanics*, Ph.D. Thesis, Univ. Ankara, 1954

[Ino] A. Inomata, H. Kuratsuji, and C. C. Gerry, *Path Integrals and Coherent States of SU(2) and SU(1,1)*, World Scientific, Singapore, 1992

[Jac] N. Jacobson, *Lie Algebras*, Interscience Publ., New York and London, 1962

[Jaf] S. Jaffard and Y. Meyer, *Wavelet Methods for Pointwise Regularity and Local Oscillations of Functions*, Memoirs Amer. Math. Soc., Vol. 143, Amer. Math. Soc., Providence, RI, 1996

[Kah] J-P. Kahane and P. G. Lemarié-Rieusset, *Fourier Series and Wavelets*, Gordon and Breach, Luxembourg, 1995; French transl. *Séries de Fourier et ondelettes*, Cassini, Paris, 1998

[Kat] T. Kato, *Perturbation Theory for Linear Operators*, Springer-Verlag, Berlin, 1976

[Kem] E. C. Kemble, *Fundamental Principles of Quantum Mechanics*, McGraw-Hill, New York, 1937.

[Kim] Y. S. Kim and M. E. Noz, *Phase Space Picture of Quantum Mechanics–Group-Theoretical Approach*, World Scientific, Singapore, 1991

[Kir] A. A. Kirillov, *Elements of the Theory of Representations*, Springer-Verlag, Berlin, 1976

[Kl1] J. R. Klauder and E. C. G. Sudarshan, *Fundamentals of Quantum Optics*, Benjamin, New York, 1968

[Kl2] J. R. Klauder and B. S. Skagerstam, *Coherent States–Applications in Physics and Mathematical Physics*, World Scientific, Singapore, 1985

[Kna] A. W. Knapp, *Lie Groups Beyond an Introduction*, Birkhäuser-Verlag, Basel, 1996

[Lan] S. Lang, *Algebra*, 3rd ed., Addison-Wesley, Reading, MA, 1993

[Lip] R. L. Lipsman, *Group Representations*, Lect. Notes in Math., Vol. 388, Springer-Verlag, Berlin, 1974

[Lyn] P. A. Lynn, *An Introduction to the Analysis and Processing of Signals*, 2nd ed., MacMillan, London, 1982

[Ma1] G. W. Mackey, *Induced Representations of Groups and Quantum Mechanics*, Benjamin, New York, 1968

[Ma2] G. W. Mackey, *Theory of Unitary Group Representations*, Univ. Chicago Press, Chicago, 1976

[Mae] S. Maes, *The wavelet transform in signal processing, with application to the extraction of the speech modulation model features*, Thèse de Doctorat, Univ. Cath. Louvain, Louvain-la-Neuve, 1994

[Mag] W. Magnus, F. Oberhettinger, and R. P. Soni, *Formulas and Theorems for the Special Functions of Mathematical Physics*, Springer-Verlag, Berlin, 1966

[Mal] S. G. Mallat, *A Wavelet Tour of Signal Processing*, Academic Press, New York, 1998

[Mar] D. Marr, *Vision*, Freeman, San Francisco, CA, 1982

[Mes] H. Meschkowsky, *Hilbertsche Räume mit Kernfunktionen*, Springer-Verlag, Berlin, 1962

[Me1] Y. Meyer, *Algebraic Numbers and Harmonic Analysis*, North-Holland, Amsterdam, 1972

[Me2] Y. Meyer (ed.), *Wavelets and Applications (Proc. Marseille 1989)*, Masson, Paris, and Springer-Verlag, Berlin, 1991

[Me3] Y. Meyer, *Les Ondelettes, Algorithmes et Applications*, Armand Colin, Paris, 1992; Engl. transl. *Wavelets, Algorithms and Applications*, SIAM, Philadelphia, 1993

[Me4] Y. Meyer and S. Roques (eds.), *Progress in Wavelet Analysis and Applications (Proc. Toulouse 1992)*, Ed. Frontières, Gif-sur-Yvette 1993

[Mur] R. Murenzi, *Ondelettes multidimensionnelles et applications à l'analyse d'images*, Thèse de Doctorat, Univ. Cath. Louvain, Louvain-la-Neuve, 1990

[Pap] A. Papoulis, *Signal Analysis*, McGraw-Hill, New York, 1977

[Pau] Th. Paul, *Ondelettes et Mécanique Quantique*, Thèse de Doctorat, Univ. d'Aix-Marseille II, 1985

[Per] A. M. Perelomov, *Generalized Coherent States and Their Applications*, Springer-Verlag, Berlin, 1986

[Phe] R. R. Phelps, *Lectures on Choquet's Theorem*, Van Nostrand, Princeton, NJ, 1966

[Pru] E. Prugovečki, *Stochastic Quantum Mechanics and Quantum Spacetime*, Reidel, Dordrecht, 1986

[Ree] M. Reed and B. Simon, *Methods of Modern Mathematical Physics. I. Functional Analysis*, Academic Press, New York, 1980

[Ren] J. Renaud, *Contribution à l'étude mathématique de la quantification et des contractions. Courbure, vitesse de la lumière, masse et constante de Planck*, Thèse de Doctorat, Univ. Paris 7, 1994

[Rud] W. Rudin, *Fourier Analysis on Groups*, Interscience, New York, 1962

[Rus] M. B. Ruskai, G. Beylkin, R. Coifman, I. Daubechies, S. Mallat, Y. Meyer, and L. Raphael (eds.), *Wavelets and Their Applications*, Jones and Bartlett, Boston, 1992

[Sch] F. E. Schroeck, Jr, *Quantum Mechanics on Phase Space*, Kluwer, Dordrecht, 1996

[Scu] M. O. Scully and M. S. Zubairy, *Quantum Optics*, Cambridge Univ. Press, Cambridge, 1997

[Sho] J. A. Shohat and J. D. Tamarkin, *The Problem of Moments*, Amer. Math. Soc., Providence, RI, 1950

[StZ] S. Strătilă and L. Zsidó, *Lectures on von Neumann Algebras*, Editura Academiei, Bucharest and Abacus Press, Tunbridge Wells, Kent, 1979

[StW] R. F. Streater and A. S. Wightman, *PCT, Spin and Statistics and all That*, Benjamin, New York, 1964

[Sug] M. Sugiura, *Unitary Representations and Harmonic Analysis: An Introduction*, North-Holland/Kodansha Ltd., Tokyo, 1990
[Ta1] M. Takesaki, *Tomita's Theory of Modular Hilbert Algebras and its Applications*, Lect. Notes in Math., Vol. 128, Springer-Verlag, Berlin, 1970
[Ta2] M. Takesaki, *Theory of Operator Algebras.I*, Springer-Verlag, New York, 1979
[Tho] G. Thonet, *New aspects of time-frequency analysis for biomedical signal processing*, Thèse de Doctorat, EPFL, Lausanne, 1998
[Tor] B. Torrésani, *Analyse continue par ondelettes*, InterÉditions/CNRS Éditions, Paris, 1995; Engl. Transl., SIAM, Philadelphia (to appear)
[Unt] A. Unterberger, *Analyse harmonique et analyse pseudo-différentielle du cône de lumière*, Astérisque **156**, SMF, Paris, 1987
[Van] P. Vandergheynst, *Ondelettes directionnelles et ondelettes sur la sphère*, Thèse de Doctorat, Univ. Cath. Louvain, Louvain-la-Neuve, 1998
[Var] V.S. Varadarajan, *Geometry of Quantum Theory*, 2nd ed., Springer-Verlag, New York, 1985
[Vet] M. Vetterli and J. Kovačević, *Wavelets and Subband Coding*, Prentice-Hall, Englewood Cliffs, NJ, 1995
[vNe] J. von Neumann, *Mathematical Foundations of Quantum Mechanics*, English transl. by R.T. Byer, Princeton Univ. Press, Princeton, NJ, 1955
[vWe] C. von Westenholz, *Differential Forms in Mathematical Physics*, North-Holland, Amsterdam, 1986
[Wic] M. V. Wickerhauser, *Adapted Wavelet Analysis from Theory to Software*, A. K. Peters, Wellesley, MA, 1994
[Wis] W. Wisnoe, *Utilisation de la méthode de transformée en ondelettes 2D pour l'analyse de visualisation d'écoulements,* Thèse de Doctorat ENSAE, Toulouse, 1993
[Woj] P. Wojtaszczyk, *A Mathematical Introduction to Wavelets*, Cambridge Univ. Press, Cambridge, 1997
[Woo] N. J. M. Woodhouse, *Geometric Quantization*, 2nd ed., Clarendon Press, Oxford, 1992

B. Articles

[1] M. Alexandrescu, D. Gibert, G. Hulot, J-L. Le Mouel, and G. Saracco, Worldwide wavelet analysis of geomagnetic jerks, *J. Geophys. Res. B* **101** (1996) 21975–21994
[2] S. T. Ali, A geometrical property of POV-measures and systems of covariance, in *Differential Geometric Methods in Mathematical Physics*, 207–228; H. D. Doebner, S. I. Andersson, and H. R. Petry (eds.), Lect. Notes in Math., Vol. 905, Springer-Verlag, Berlin, 1982
[3] S. T. Ali, Commutative systems of covariance and a generalization of Mackey's imprimitivity theorem, *Canad. Math. Bull.* **27** (1984) 390–397
[4] S. T. Ali, Stochastic localisation, quantum mechanics on phase space and quantum space-time, *Riv. Nuovo Cim.* **8** (1985) 1–128
[5] S. T. Ali, A general theorem on square-integrability: Vector coherent states, *J. Math. Phys.* **39** (1998) 3954–3964
[6] S. T. Ali and G. G. Emch, Geometric quantization: Modular reduction theory and coherent states, *J. Math. Phys.* **27** (1986) 2936–2943

[7] S. T. Ali and E. Prugovečki, Mathematical problems of stochastic quantum mechanics: Harmonic analysis on phase space and quantum geometry, *Acta Appl. Math.* **6** (1986) 1–18; Extended harmonic analysis of phase space representation for the Galilei group, *ibid.* **6** (1986) 19–45; Harmonic analysis and systems of covariance for phase space representation of the Poincaré group, *ibid.* **6** (1986) 47–62

[8] S. T. Ali and S. De Bièvre, Coherent states and quantization on homogeneous spaces, in *Group Theoretical Methods in Physics*, 201–207; H-D. Doebner et al. (eds.), Lect. Notes in Phys., Vol. 313, Springer-Verlag, Berlin, 1988

[9] S. T. Ali and J-P. Antoine, Coherent states of 1+1 dimensional Poincaré group: Square integrability and a relativistic Weyl transform, *Ann. Inst. H. Poincaré* **51** (1989) 23–44

[10] S. T. Ali and H-D. Doebner, Ordering problem in quantum mechanics: Prime quantization and a physical interpretation, *Phys. Rev. A* **41** (1990) 1199–1210

[11] S. T. Ali, J-P. Antoine, and J-P. Gazeau, De Sitter to Poincaré contraction and relativistic coherent states, *Ann. Inst. H. Poincaré* **52** (1990) 83–111

[12] S. T. Ali, J-P. Antoine, and J-P. Gazeau, Square integrability of group representations on homogeneous spaces. I. Reproducing triples and frames, *Ann. Inst. H. Poincaré* **55** (1991) 829–855

[13] S. T. Ali, J-P. Antoine, and J-P. Gazeau, Square integrability of group representations on homogeneous spaces. II. Coherent and quasi-coherent states. The case of the Poincaré group, *Ann. Inst. H. Poincaré* **55** (1991) 857–890

[14] S. T. Ali and G. A. Goldin, Quantization, coherent states and diffeomorphism groups, in *Differential Geometry, Group Representations and Quantization*, , 147–178; J. D. Hennig et al. (eds.), Lect. Notes in Phys., Vol. 379, Springer-Verlag, Berlin, 1991

[15] S. T. Ali, J-P. Antoine, and J-P. Gazeau, Continuous frames in Hilbert space, *Ann. Phys.(NY)* **222** (1993) 1–37

[16] S. T. Ali, J-P. Antoine, and J-P. Gazeau, Relativistic quantum frames, *Ann. Phys.(NY)* **222** (1993) 38–88

[17] S. T. Ali, G. G. Emch, and A. El Gradechi, Modular algebras in geometric quantization, *J. Math. Phys.* **35** (1994) 6237–6243

[18] S. T. Ali and U. A. Mueller, Quantization of a classical system on a coadjoint orbit of the Poincaré group in 1+1 dimensions, *J. Math. Phys.* **35** (1994) 4405–4422

[19] S. T. Ali, J-P. Antoine, J-P. Gazeau, and U. A. Mueller, Coherent states and their generalizations: A mathematical overview, *Reviews Math. Phys.* **7** (1995) 1013–1104

[20] S. T. Ali, J-P. Gazeau, and M. R. Karim, Frames, the β-duality in Minkowski space and spin coherent states, *J. Phys. A: Math. Gen.* **29** (1996) 5529–5549

[21] S. T. Ali and E. Prugovečki, Systems of imprimitivity and representations of quantum mechanics on fuzzy phase spaces, *J. Math. Phys.* **18** (1977) 219–228

[22] S. T. Ali, N. M. Atakishiyev, S. M. Chumakov, and K. B. Wolf, The Wigner operator and Wigner function for general Lie groups, preprint, Cuernavaca, 1998

[23] P. Aniello, G. Cassinelli, E. De Vito, and A. Levrero, Square-integrability of induced representations of semidirect products, *Rev. Math. Phys.* **10** (1998) 301-313; Wavelet transforms and discrete frames associated to semidirect products, *J. Math. Phys.* **39** (1998) 3965–3973

[24] J-P. Antoine, Etude de la dégénérescence orbitale du potentiel coulombien en théorie des groupes. I, II, *Ann. Soc. Scient. Bruxelles* **80** (1966) 169–184; **81** (1967) 49–68

[25] J-P. Antoine, Quantum mechanics beyond Hilbert space, in *Irreversibility and Causality–Semigroups and Rigged Hilbert Spaces*, 3–33; A. Bohm, H-D. Doebner, and P. Kielanowski (eds.), Lect. Notes in Phys., Vol. 504, Springer-Verlag, Berlin, 1998

[26] J-P. Antoine and D. Speiser, Characters of irreducible representations of simple Lie groups, *J. Math. Phys.* **5** (1964) 1226–1234

[27] J-P. Antoine, M. Duval-Destin, R. Murenzi, and B. Piette, Image analysis with 2D wavelet transform: detection of position, orientation and visual contrast of simple objects, in [Me2], 144–159

[28] J-P. Antoine, P. Carrette, R. Murenzi and B. Piette, Image analysis with 2D continuous wavelet transform, *Signal Processing* **31** (1993) 241–272

[29] J-P. Antoine and U. Moschella, Poincaré coherent states: The two-dimensional massless case, *J. Phys. A: Math. Gen.* **26** (1993) 591–607

[30] J-P. Antoine and F. Bagarello, Wavelet-like orthonormal bases for the lowest Landau level, *J. Phys. A: Math. Gen.*, **27** (1994) 2471–2481

[31] J-P. Antoine, P. Vandergheynst, K. Bouyoucef, and R. Murenzi, Alternative representations of an image via the 2D wavelet transform. Application to character recognition, *Visual Information Processing IV, SPIE Proc.*, **2488** (1995) 486–497

[32] J-P. Antoine and R. Murenzi, Two-dimensional directional wavelets and the scale-angle representation, *Signal Processing* **52** (1996) 259–281

[33] J-P. Antoine, R. Murenzi, and P. Vandergheynst, Two-dimensional directional wavelets in image processing, *Int. J. Imaging Systems Technol.* **7** (1996) 152–165

[34] J-P. Antoine and R. Murenzi, Two-dimensional continuous wavelet transform as linear phase space representation of two-dimensional signals, in *Wavelet Applications IV, SPIE Proc.*, **3078** (1997) 206-217

[35] J-P. Antoine, D. Barache, R. M. Cesar, Jr, and L. da F. Costa, Shape characterization with the wavelet transform, *Signal Process.* **62** (1997) 265–290

[36] J-P. Antoine and P. Vandergheynst, Wavelets on the n-sphere and related manifolds, *J. Math. Phys.* **39** (1998) 3987–4008

[37] J-P. Antoine, Ph. Antoine, and B. Piraux, Wavelets in atomic physics, in *Spline Functions and the Theory of Wavelets*, 261–276; S. Dubuc and G. Deslauriers, (eds.), CRM Proceedings and Lecture Notes, Vol. 18, AMS, Providence, RI, 1999; Wavelets in atomic physics and in solid state physics, [Be2], Ch. 8

[38] J-P. Antoine, R. Murenzi, and P. Vandergheynst, Directional wavelets revisited: Cauchy wavelets and symmetry detection in patterns, *Appl. Comput. Harmon. Anal.* **6** (1999) 314–345

[39] J-P. Antoine and P. Vandergheynst, Wavelets on the 2-sphere: A group-theoretical approach, *Appl. Comput. Harmon. Anal.* **7** (1999) 1–30

[40] J-P. Antoine, Y.B. Kouagou, D. Lambert, and B. Torrésani, An algebraic approach to discrete dilations. Application to discrete wavelet transforms, *J. Fourier Anal. Appl.* (to appear)

[41] J-P. Antoine and I. Mahara, Galilean wavelets: Coherent states for the affine Galilei group, *J. Math. Phys.* **40** (1999) (to appear)

[42] J-P. Antoine, A. Coron, and J-M. Dereppe, Water peak suppression: Time-frequency vs. time-scale approach, preprint UCL-IPT-99-02; *J. Magn. Reson.* (submitted)

[43] J-P. Antoine, L. Jacques, and R. Twarock, Wavelet analysis of a quasiperiodic tiling with fivefold symmetry, *Phys. Lett. A* **262** (1999) (to appear)

[44] J-P. Antoine, L. Jacques, and P. Vandergheynst, Penrose tilings, quasicrystals, and wavelets, in *Wavelet Applications in Signal and Image Processing VII, Proc. SPIE* **3813** (1999) (in press)

[45] Ph. Antoine, B. Piraux, and A. Maquet, Time profile of harmonics generated by a single atom in a strong electromagnetic field, *Phys. Rev. A* **51** (1995) R1750–R1753

[46] Ph. Antoine, B. Piraux, D. B. Milošević, and M. Gajda, Generation of ultrashort pulses of harmonics, *Phys. Rev. A* **54** (1996) R1761–R1764; Temporal profile and time control of harmonic generation, *Laser Phys.* **7** (1997) 594–601

[47] F. T. Arecchi, E. Courtens, R. Gilmore, and H. Thomas, Atomic coherent states in quantum optics, *Phys. Rev. A* **6** (1972) 2211–2237

[48] F. Argoul, A. Arnéodo, J. Elezgaray, G. Grasseau, and R. Murenzi, Wavelet analysis of the self-similarity of diffusion-limited aggregates and electrodeposition clusters, *Phys. Rev. A* **41** (1990) 5537–5560

[49] T. A. Arias, Multiresolution analysis of electronic structure: semicardinal and wavelet bases, *Rev. Mod. Phys.* **71** (1999) 267–312

[50] A. Arnéodo, F. Argoul, E. Bacry, J. Elezgaray, E. Freysz, G. Grasseau, J. F. Muzy, and B. Pouligny, Wavelet transform of fractals, in [Me2], 286–352

[51] A. Arnéodo, E. Bacry, P. V. Graves, and J. F. Muzy, Characterizing long-range correlations in DNA sequences from wavelet analysis, *Phys. Rev. Lett.* **74** (1996) 3293–3296; A. Arnéodo, Y. d'Aubenton, E. Bacry, P. V. Graves, J. F. Muzy and C. Thermes, Wavelet based fractal analysis of DNA sequences, *Physica D* **96** (1996) 291–320

[52] A. Arnéodo, E. Bacry, and J. F. Muzy, The thermodynamics of fractals revisited with wavelets, *Physica A* **213** (1995) 232–275

[53] A. Arnéodo, E. Bacry, and J. F. Muzy, Oscillating singularities in locally self-similar functions, *Phys. Rev. Lett.* **74** (1995) 4823–4826

[54] A. Arnéodo, E. Bacry, S. Jaffard, and J. F. Muzy, Oscillating singularities on Cantor sets. A grand canonical multifractal formalism, *J. Stat. Phys.* **87** (1997) 179–209; Singularity spectrum of multifractal functions involving oscillating singularities, *J. Fourier Anal. Appl.* **4** (1998) 159–174

[55] N. Aronszajn, Theory of reproducing kernels, *Trans. Amer. Math. Soc.* **66** (1950) 337–404

[56] E. W. Aslaksen and J. R. Klauder, Unitary representations of the affine group, *J. Math. Phys.* **9** (1968) 206–211; Continuous representation theory using the affine group, *ibid.* **10** (1969) 2267–2275

[57] D. Astruc, L. Plantié, R. Murenzi, Y. Lebret, and D. Vandromme, On the use of the 3D wavelet transform for the analysis of computational fluid dynamics results, in [Me4], 463–470

[58] H. Bacry and J-M. Lévy-Leblond, Possible kinematics, *J. Math. Phys.* **9** (1968) 1605–1614

[59] H. Bacry, A. Grossmann, and J. Zak, Proof of the completeness of lattice states in kq representation, *Phys. Rev. B* **12** (1975) 1118–1120

[60] L. Baggett and K. F. Taylor, Groups with completely reducible regular representation, *Proc. Amer. Math. Soc.* **72** (1978) 593–600

[61] M. Bander and C. Itzykson, Group theory and the hydrogen atom. I, II, *Rev. Mod. Phys.* **38** (1966) 330–345, 346–358

[62] D. Barache, J-P. Antoine, and J-M. Dereppe, The continuous wavelet transform, a tool for NMR spectroscopy, *J. Magn. Reson.* **128** (1997) 1–11

[63] V. Bargmann, P. Butera, L. Girardello, and J. R. Klauder, On the completeness of coherent states, *Reports Math. Phys.* **2** (1971) 221–228

[64] A. O. Barut and H. Kleinert, Transition probabilities of the hydrogen atom from noncompact dynamical groups, *Phys. Rev.* **156** (1967) 1541–1545

[65] A. O. Barut and L. Girardello, New "coherent" states associated with non compact groups, *Commun. Math. Phys.* **21** (1971) 41–55

[66] A. O. Barut and B. W. Xu, Non-spreading coherent states riding on Kepler orbits, *Helv. Phys. Acta* **66** (1993) 711–720

[67] G. Battle, Wavelets: A renormalization group point of view, in [Rus], 323–349

[68] P. Bellomo and C. R. Stroud, Jr, Dispersion of Klauder's temporally stable coherent states for the hydrogen atom, *J. Phys. A: Math. Gen.* **31** (1998) L445–L450

[69] F. A. Berezin, Quantization, *Math. USSR Izvestija* **8** (1974) 1109–1165

[70] S. Bergman, Über die Kernfunktion eines Bereiches und ihr Verhalten am Rande. I, *Reine Angew. Math.* **169** (1933) 1–42

[71] D. Bernier and K. F. Taylor, Wavelets from square-integrable representations, *SIAM J. Math. Anal.* **27** (1996) 594–608

[72] C. Bernuau, Wavelet bases associated to a self similar quasicrystal, *J. Math. Phys.* **39** (1998) 4213–4225

[73] A. Bertrand, Développements en base de Pisot et répartition modulo 1, *C. R. Acad. Sci. Paris* **285** (1977) 419–421

[74] J. Bertrand and P. Bertrand, Classification of affine Wigner functions via an extended covariance principle, in *Group Theoretical Methods in Physics (Proc. Sainte-Adèle 1988)*, 1380–383; Y. Saint-Aubin and L. Vinet (eds.), World Scientific, Singapore, 1989

[75] Z. Bialynicka-Birula and I. Bialynicki-Birula, Space-time description of squeezing, *J. Opt. Soc. Am.* **B4** (1987) 1621–1626

[76] R. Bluhm, V. A. Kostelecký, and J. A. Porter, The evolution and revival structure of localized quantum wave packets, *Am. J. Phys.* **64** (1996) 944–953

[77] G. Bohnké, Treillis d'ondelettes associés aux groupes de Lorentz, *Ann. Inst. H. Poincaré* **54** (1991) 245–259

[78] K. Bouyoucef, D. Fraix-Burnaix, and S. Roques, Interactive Deconvolution with Error Analysis (IDEA) in astronomical imaging: Application to aberrated HST images on SN1987A, M87 and 3C66B, *Astron. Astroph. Suppl. Series* **121** (1997) 1–6

[79] A. Bouzouina and S. De Bièvre, Equipartition of the eigenfunctions of quantized ergodic maps on the torus, *Commun. Math. Phys.* **178** (1996) 83–105

[80] A. Briguet, S. Cavassila, and D. Graveron-Demilly, Suppression of huge signals using the Cadzow enhancement procedure, *NMR Newsl.* **440** (1995) 26

[81] C. M. Brislawn, Fingerprints go digital, *Notices Amer. Math. Soc.* **42** (1995) 1278–1283

[82] F. Bruhat, Sur les représentations induites des groupes de Lie, *Bull. Soc. Math. France* **84** (1956) 97–205

[83] Č. Burdik, Ch. Frougny, J-P. Gazeau, and R. Krejcar, Beta-integers as natural counting systems for quasicrystals, *J. Phys. A: Math. Gen.* **31** (1998) 6449–6472

[84] A. R. Calderbank, I. Daubechies, W. Sweldens, and B. L. Yeo, Wavelets that map integers to integers, *Appl. Comput. Harmon. Anal.* **5** (1998) 332–369

[85] A. L. Carey, Square integrable representations of non-unimodular groups, *Bull. Austr. Math. Soc.* **15** (1976) 1–12

[86] A. L. Carey, Group representations in reproducing kernel Hilbert spaces, *Reports Math. Phys.* **14** (1978) 247–259

[87] D. P. L. Castrigiano and R. W. Henrichs, Systems of covariance and subrepresentations of induced representations, *Lett. Math. Phys.* **4** (1980) 169–175

[88] U. Cattaneo, Densities of covariant observables, *J. Math. Phys.* **23** (1982) 659–664

[89] C. Cishahayo and S. De Bièvre, On the contraction of the discrete series of SU(1,1), *Ann. Inst. Fourier (Grenoble)* **43** (1993) 551–567

[90] A. Cohen, I. Daubechies, and J-C. Feauveau, Biorthogonal bases of compactly supported wavelets, *Commun. Pure Appl. Math.* **45** (1992) 485–560

[91] L. Cohen, General phase-space distribution functions, *J. Math. Phys.* **7** (1966) 781–786

[92] R. Coifman, Y. Meyer, and M. V. Wickerhauser, Wavelet analysis and signal processing, in [Rus], 153–178; Entropy-based algorithms for best basis selection, *IEEE Trans. Inform. Theory* **38** (1992) 713–718

[93] R. Coquereaux and A. Jadczyk, Conformal theories, curved spaces, relativistic wavelets and the geometry of complex domains, *Rev. Math. Phys.* **2** (1990) 1–44

[94] S. Dahlke and P. Maass, The affine uncertainty principle in one and two dimensions, *Comp. Math. Appl.* **30** (1995) 293–305

[95] T. Dallard and G. R. Spedding, 2-D wavelet transforms: generalisation of the Hardy space and application to experimental studies, *Eur. J. Mech., B/Fluids* **12** (1993) 107–134

[96] C. Daskaloyannis, Generalized deformed oscillator and nonlinear algebras, *J. Phys. A: Math. Gen.* **24** (1991) L789–L794; C. Daskaloyannis and K. Ypsilantis, A deformed oscillator with Coulomb energy spectrum, *J. Phys. A: Math. Gen.* **25** (1992) 4157–4166

[97] I. Daubechies, A. Grossmann, and Y. Meyer, Painless nonorthogonal expansions, *J. Math. Phys.* **27** (1986) 1271–1283

[98] I. Daubechies, The wavelet transform, time-frequency localisation and signal analysis, *IEEE Trans. Inform. Theory* **36** (1990) 961–1005

[99] I. Daubechies, Orthonormal bases of compactly supported wavelets, *Commun. Pure Appl. Math.* **41** (1988) 909–996

[100] I. Daubechies and S. Maes, A nonlinear squeezing of the continuous wavelet transform based on auditory nerve models, in [Ald], 527–546

[101] R. De Beer, D. Van Ormondt, F. T. A. W. Wajer, S. Cavassila, D. Graveron-Demilly, and S. Van Huffel, SVD-based modelling of medical NMR signals, in *SVD and Signal Processing, III: Algorithms, Architectures and Applications*, 467–474; M. Moonen and B. De Moor (eds.), Elsevier (North-Holland), Amsterdam, 1995

[102] S. De Bièvre, Coherent states over symplectic homogeneous spaces, *J. Math. Phys.* **30** (1989) 1401–1407

[103] S. De Bièvre and J. A. Gonzalez, Semi-classical behaviour of the Weyl correspondence on the circle, in *Group Theoretical Methods in Physics (Proc. Salamanca 1992)*, 343–346; M. del Olmo, M. Santander, and J. Mateos Guilarte (eds.), CIEMAT, Madrid, 1993

[104] S. De Bièvre and A. El Gradechi, Quantum mechanics and coherent states on the anti-de Sitter space–time and their Poincaré contraction, *Ann. Inst. H. Poincaré* **57** (1992) 403–428

[105] J. Deenen and C. Quesne, Dynamical group of collective states, I, II, III, *J. Math. Phys.* **23** (1982) 878–889, 2004–2015; **25** (1984) 1638–1650; Partially coherent states of the real symplectic group, *ibid.* **25** (1984) 2354–2366; Boson representations of the real symplectic group and their application to the nuclear collective model, *ibid.* **26** (1985) 2705–2716

[106] R. Delbourgo, Minimal uncertainty states for the rotation and allied groups, *J. Phys. A: Math. Gen.* **10** (1977) 1837–1846

[107] R. Delbourgo and J. R. Fox, Maximum weight vectors possess minimal uncertainty, *J. Phys. A: Math. Gen.* **10** (1977) L233–L235

[108] N. Delprat, B. Escudié, Ph. Guillemain, R. Kronland-Martinet, Ph. Tchamitchian, and B. Torrésani, Asymptotic wavelet and Gabor analysis: Extraction of instantaneous frequencies, *IEEE Trans. Inform. Theory* **38** (1992) 644–664

[109] R. H. Dicke, Coherence in spontaneous radiation processes, *Phys. Rev.* **93** (1954) 99–110

[110] D. L. Donoho, Nonlinear wavelet methods for recovery of signals, densities, and spectra from indirect and noisy data, in *Different Perspectives on Wavelets*, 173–205; Proc. Symp. Appl. Math. Vol. 38, I. Daubechies (ed.), Amer. Math. Soc., Providence, RI, 1993

[111] A. H. Dooley, Contractions of Lie groups and applications to analysis, in *Topics in Modern Harmonic Analysis*, Vol. I, 483–515; Istituto Nazionale di Alta Matematica Francesco Severi, Roma, 1983
[112] A. H. Dooley and J. W. Rice, Contractions of rotation groups and their representations, *Math. Proc. Camb. Phil. Soc.* **94** (1983) 509–517; On contractions of semisimple Lie groups, *Trans. Amer. Math. Soc.* **289** (1985) 185–202
[113] R. J. Duffin and A. C. Schaeffer, A class of nonharmonic Fourier series, *Trans. Amer. Math. Soc.* **72** (1952) 341–366
[114] M. Duflo and C. C. Moore, On the regular representation of a nonunimodular locally compact group, *J. Funct. Anal.* **21** (1976) 209–243
[115] M. Duval-Destin and R. Murenzi, Spatio-temporal wavelets: Application to the analysis of moving patterns, in [Me4], 399–408
[116] M. Duval-Destin, M-A. Muschietti, and B. Torrésani, Continuous wavelet decompositions, multiresolution, and contrast analysis, *SIAM J. Math. Anal.*, **24** (1993) 739–755
[117] G. G. Emch, Prequantization and KMS structures, *Intern. J. Theoret. Phys.* **20** (1981) 891–904
[118] G. G. Emch, Geometric quantization, regular representations and modular algebras, in *Group-Theoretical Methods in Physics (Proc. XVIII Intern. Coll., Moscow 1990)*, 356; V. V. Dodonov and V. I. Man'ko (eds.), Springer-Verlag, Berlin, 1991
[119] Q. Fan, Phase space analysis of the identity decompositions, *J. Math. Phys.* **34** (1993) 3471–3477
[120] M. Farge and Th. Philipovitch, Coherent structure analysis and extraction using wavelets, in [Me4], 477–481
[121] M. Farge, Wavelet transforms and their applications to turbulence, *Annu. Rev. Fluid Mech.*, **24** (1992) 395–457
[122] H. G. Feichtinger, Coherent frames and irregular sampling, in *Recent Advances in Fourier Analysis and Its Applications*, 427–440; J. S. Byrnes and J. L. Byrnes (eds.), Kluwer, Dordrecht, 1990
[123] H. G. Feichtinger and K. H. Gröchenig, Banach spaces related to integrable group representations and their atomic decompositions. I, *J. Funct. Anal.* **86** (1989) 307–340; id. II., *Mh. Math.* **108** (1989) 129–148
[124] M. Flensted-Jensen, Discrete series for semisimple symmetric spaces, *Ann. Math.* **111** (1980) 253–311
[125] K. Flornes, A. Grossmann, M. Holschneider, and B. Torrésani, Wavelets on discrete fields, *Appl. Comput. Harmon. Anal.* **1** (1994) 137–146
[126] J. Froment and S. Mallat, Arbitrary low bit rate image compression using wavelets, in [Me4], 413–418, and references therein
[127] H. Führ, Wavelet frames and admissibility in higher dimensions, *J. Math. Phys.* **37** (1996) 6353–6366
[128] H. Führ and M. Mayer, Continuous wavelet transforms from cyclic representations: A general approach using Plancherel measure, *J. Math. Phys.* (submitted)
[129] D. Gabor, Theory of communication, *J. Inst. Electr. Eng.(London)* **93** (1946) 429–457
[130] J-P. Gazeau, Four Euclidean conformal group in atomic calculations: Exact analytical expressions for the bound-bound two-photon transition

matrix elements in the H atom, *J. Math. Phys.* **19** (1978) 1041–1048; On the four euclidean conformal group structure of the sturmian operator, *Lett. Math. Phys.* **3** (1979) 285–292; Technique Sturmienne pour le spectre discret de l'équation de Schrödinger, *J. Phys. A: Math. Gen.* **13** (1980) 3605–3617; Four Euclidean conformal group approach to the multiphoton processes in the H atom, *J. Math. Phys.* **23** (1982) 156–164.

[131] J-P. Gazeau, A remarkable duality in one particle quantum mechanics between some confining potentials and $(R + L_\epsilon^\infty)$ potentials, *Phys. Lett.* **75A** (1980) 159–163.

[132] J-P. Gazeau, Coherent states for De Sitterian and Einsteinian relativities, in *Selected Topics in Quantum Field Theory and Mathematical Physics (Proc. Liblice 1989)*, 180–187; J. Niederle and J. Fischer (eds.), World Scientific, Singapore, 1990.

[133] J-P. Gazeau, On two analytic elementary systems in quantum mechanics, in *Proc. Colloq. Géométrie Analytique (Paris 1992)*; F. Norguet and S. Ofman (eds.), Hermann, Paris, 1995

[134] J-P. Gazeau, $SL(2,\mathbb{R})$-coherent states and integrable systems in classical and quantum physics, in *Quantization, Coherent States, and Complex Structures*, 147–158; J-P. Antoine, S. T. Ali, W. Lisiecki, I. M. Mladenov and A. Odzijewicz (eds.), Plenum Press, New York and London, 1995

[135] J-P. Gazeau and V. Hussin, Poincaré contraction of $SU(1,1)$ Fock-Bargmann structure, *J. Phys. A: Math. Gen.* **25** (1992) 1549–1573

[136] J-P. Gazeau and J. Renaud, Lie algorithm for an interacting $SU(1,1)$ elementary system and its contraction, *Ann. Phys. (NY)* **222** (1993) 89–121

[137] J-P. Gazeau and J. Renaud, Relativistic harmonic oscillator and space curvature, *Phys. Lett. A* **179** (1993) 67–71

[138] J-P. Gazeau and J. Patera, Tau-wavelets of Haar, *J. Phys. A: Math. Gen.* **29** (1996) 4549–4559

[139] J-P. Gazeau and V. Spiridonov, Toward discrete wavelets with irrational scaling factor, *J. Math. Phys.* **37** (1996) 3001–3013

[140] J-P. Gazeau and S. Graffi, Quantum harmonic oscillator: a relativistic and statistical point of view, *Boll. Unione Mat. Ital.* **11**-A (1997) 815–839

[141] J-P. Gazeau, J. Patera, and E. Pelantová, Tau-wavelets in the plane, *J. Math. Phys.* **39** (1998) 4201–4212

[142] J-P. Gazeau and J. R. Klauder, Coherent states for systems with discrete and continuous spectrum, preprint, U. Paris 7, *J. Phys. A: Math. Gen.* **32** (1999) 123–132

[143] J-P. Gazeau and B. Champagne, The Fibonacci-deformed harmonic oscillator, in *Algebraic methods in physics–A symposium for the 60th birthday of Jiří Patera and Pavel Winternitz*; Y. Saint-Aubin and L. Vinet (eds.), CRM Series Theor. Math. Phys., Vol. 3, Springer-Verlag, Berlin (to appear)

[144] R. Gilmore, Geometry of symmetrized states, *Ann. Phys. (NY)* **74** (1972) 391–463; On properties of coherent states, *Rev. Mex. Fis.* **23** (1974) 143–187

[145] R. J. Glauber, The quantum theory of optical coherence, *Phys. Rev.* **130** (1963) 2529–2539; Coherent and incoherent states of radiation field, *ibid.* **131** (1963) 2766–2788

[146] J. Glimm, Locally compact transformation groups, *Trans. Amer. Math. Soc.* **101** (1961) 124–138

[147] J-F. Gobbers and P. Vandergheynst, A fast continuous wavelet transform, preprint UCL-IPT-99-07, Louvain-la-Neuve (in preparation).

[148] R. Godement, Sur les relations d'orthogonalité de V. Bargmann, *C.R. Acad. Sci. Paris* **255** (1947) 521–523; 657–659

[149] A. Goldberg, I. Doghri, E. Van Vyve, J-P. Antoine, and P. Vandergheynst, Instrumented falling weight impact testing: Continuous wavelet analysis and modeling, preprint UCL-IPT-99-05, Louvain-la-Neuve

[150] C. Gonnet and B. Torrésani, Local frequency analysis with two-dimensional wavelet transform, *Signal Process.* **37** (1994) 389–404

[151] J. A. Gonzalez and M. A. del Olmo, Coherent states on the circle, *J. Phys. A: Math. Gen.* **31** (1998) 8841–8857

[152] P. Goupillaud, A. Grossmann, and J. Morlet, Cycle-octave and related transforms in seismic signal analysis, *Geoexploration* **23** (1984) 85–102

[153] K. H. Gröchenig, A new approach to irregular sampling of band-limited functions, in *Recent Advances in Fourier Analysis and Its Applications*, 251–260; J. S. Byrnes and J. L. Byrnes (eds.), Kluwer, Dordrecht, 1990

[154] K. H. Gröchenig, Gabor analysis over LCA groups, in [Fei], 211–231

[155] A. Grossmann, Parity operator and quantization of δ-functions, *Commun. Math. Phys.* **48** (1976) 191–194

[156] A. Grossmann and J. Morlet, Decomposition of Hardy functions into square integrable wavelets of constant shape, *SIAM J. Math. Anal.* **15** (1984) 723–736

[157] A. Grossmann and J. Morlet, Decomposition of functions into wavelets of constant shape, and related transforms, in *Mathematics + Physics, Lectures on Recent Results. I*, 135–166; L. Streit (ed.), World Scientific, Singapore, 1985

[158] A. Grossmann, R. Kronland-Martinet, and J. Morlet, Reading and understanding the continuous wavelet transform, in [Com], 2–20

[159] A. Grossmann, J. Morlet, and T. Paul, Integral transforms associated to square integrable representations. I. General results, *J. Math. Phys.* **26** (1985) 2473–2479

[160] A. Grossmann, J. Morlet, and T. Paul, Integral transforms associated to square integrable representations. II. Examples, *Ann. Inst. H. Poincaré* **45** (1986) 293–309

[161] Ph. Guillemain, R. Kronland-Martinet and B. Martens, Estimation of spectral lines with help of the wavelet transform. Application in NMR spectroscopy, in [Me2], 38–60

[162] E. A. Gutkin, Overcomplete subspace systems and operator symbols, *Funct. Anal. Appl.* **9** (1975) 260–261

[163] R. Haag, N. Hugenholtz, and M. Winnink, On the equilibrium states in quantum statistical mechanics, *Commun. Math. Phys.* **5** (1967) 215–236

[164] J. He and H. Liu, Admissible wavelets associated with the affine automorphism group of the Siegel upper half-plane, *J. Math. Anal. Appl.* **208** (1997) 58–70

[165] J. He and L. Peng, Wavelet transform on the symmetric matrix space, preprint, Beijing, 1997

[166] D. M. Healy, Jr, and F E. Schroeck, Jr, On informational completeness of covariant localization observables and Wigner coefficients, *J. Math. Phys.* **36** (1995) 453–507

[167] C. Heil and D. Walnut, Continuous and discrete wavelet transforms, *SIAM Rev.* **31** (1989) 628–666

[168] K. Hepp and E. H. Lieb, On the superradiant phase transition for molecules in a quantized radiation field: the Dicke maser model, *Ann. Phys. (NY)* **76** (1973) 360–404

[169] K. Hepp and E. H. Lieb, Equilibrium statistical mechanics of matter interacting with the quantized radiation field, *Phys. Rev. A* **8** (1973) 2517–2525

[170] J. A. Hogan and J. D. Lakey, Extensions of the Heisenberg group by dilations and frames, *Appl. Comput. Harmon. Anal.* **2** (1995) 174–199

[171] M. Holschneider, On the wavelet transformation of fractal objects, *J. Stat. Phys.* **50** (1988) 963–993

[172] M. Holschneider, Inverse Radon transforms through inverse wavelet transforms, *Inverse Probl.* **7** (1991) 853–861

[173] M. Holschneider, Localization properties of wavelet transforms, *J. Math. Phys.* **34** (1993) 3227–3244

[174] M. Holschneider, General inversion formulas for wavelet transforms, *J. Math. Phys.* **34** (1993) 4190–4198

[175] M. Holschneider, Wavelet analysis over abelian groups. *Applied Comput. Harmon. Anal.* **2** (1995) 52–60

[176] M. Holschneider, Continuous wavelet transforms on the sphere, *J. Math. Phys.* **37** (1996) 4156–4165

[177] M. Holschneider, R. Kronland-Martinet, J. Morlet, and Ph. Tchamitchian, A real-time algorithm for signal analysis with the help of wavelet transform, in [Com], 286–297

[178] M. Holschneider and Ph. Tchamitchian, Pointwise analysis of Riemann's 'nondifferentiable' function, *Invent. Math.* **105** (1991) 157–175

[179] W-L. Hwang and S. Mallat, Characterization of self-similar multifractals with wavelet maxima, *Appl. Comput. Harmon. Anal.* **1** (1994) 316–328

[180] W-L. Hwang, C-S. Lu, and P-C. Chung, Shape from texture: Estimation of planar surface orientation through the ridge surfaces of continuous wavelet transform, *IEEE Trans. Image Process.* **7** (1998) 773–780

[181] E. Inönü and E. P. Wigner, On the contraction of groups and their representations, *Proc. Nat. Acad. Sci. USA* **39** (1953) 510–524

[182] C. J. Isham and J. R. Klauder, Coherent states for n-dimensional Euclidean groups $E(n)$ and their application, *J. Math. Phys.* **32** (1991) 607–620

[183] C. Johnston, On the pseudo-dilation representations of Flornes, Grossmann, Holschneider, and Torrésani, *Appl. Comput. Harmon. Anal.* **3** (1997) 377–385

[184] G. Kaiser, Phase-space approach to relativistic quantum mechanics. I. Coherent state representation for massive scalar particles, *J. Math.Phys.* **18** (1977) 952–959; id. II. Geometrical aspects, *ibid.* **19** (1978) 502–507

[185] C. Kalisa and B. Torrésani, N-dimensional affine Weyl-Heisenberg wavelets, *Ann. Inst. H. Poincaré* **59** (1993) 201–236

[186] M. R. Karim and S. T. Ali, A relativistic windowed Fourier transform, preprint, Concordia Univ., Montreal, 1997
[187] T. Kawazoe, Wavelet transforms associated to a principal series representation of semisimple Lie groups. I. II, *Proc. Japan Acad., Ser. A-Math. Sci.* **71** (1995) 154–157, 158–160
[188] T. Kawazoe, Wavelet transform associated to an induced representation of $SL(n+2,\mathbb{R})$, *Ann. Inst. H. Poincaré* **65** (1996) 1–13
[189] J. R. Klauder, Continuous-representation theory. I. Postulates of continuous-representation theory, *J. Math. Phys.* **4** (1963) 1055–1058; II. Generalized relation between quantum and classical dynamics, *ibid.* 1058–1073
[190] J. R. Klauder, Path integrals for affine variables, in *Functional Integration, Theory and Applications*, 101–119; J-P. Antoine and E. Tirapegui (eds.), Plenum Press, New York and London, 1980
[191] J. R. Klauder, Are coherent states the natural language of quantum mechanics? in *Fundamental Aspects of Quantum Theory*, 1-12; V. Gorini and A. Frigerio (eds.), NATO ASI Series, Vol. B 144, Plenum Press, New York, 1986
[192] J. R. Klauder, Coherent states without groups: quantization on nonhomogeneous manifolds, *Mod. Phys. Lett. A* **8** (1993) 1735–1738
[193] J. R. Klauder, Quantization without quantization, *Ann. Phys. (N.Y.)* **237** (1995) 147–160
[194] J. R. Klauder, Coherent states for the hydrogen atom, *J. Phys. A: Math. Gen.* **29** (1996) L293–L296
[195] J. R. Klauder and R. F. Streater, A wavelet transform for the Poincaré group, *J. Math. Phys.* **32** (1991) 1609–1611
[196] J. R. Klauder and R. F. Streater, Wavelets and the Poincaré half-plane, *J. Math. Phys.* **35** (1994) 471–478
[197] A. Kleppner and R. L. Lipsman, The Plancherel formula for group extensions, *Ann. Ec. Norm. Sup.* **5** (1972) 459–516
[198] S. Kobayashi, Irreducibility of certain unitary representations, *J. Math. Soc. Japan* **20** (1968) 638–642
[199] K. Kowalski, J. Rembieliński, and L. C. Papoulas, Coherent states for a quantum particle on a circle *J. Phys. A: Math. Gen.* **29** (1996) 4149–4167
[200] R. Kunze, On the Frobenius reciprocity theorem for square integrable representations, *Pacific J. Math.* **53** (1974) 465–471
[201] Y. Kuroda, A. Wada, T. Yamazaki, and K. Nagayama, Postacquisition data processing method for suppression of the solvent signal. I, *J. Magn. Reson.* **84** (1989) 604–610; id. II. The weighted first derivative, *ibid.* **88** (1990) 141–145
[202] J-P. Leduc, F. Mujica, R. Murenzi, and M. J. T. Smith, Missile-tracking algorithm using target-adapted spatio-temporal wavelets, *Automatic Object Recognition VII, SPIE Proc.* **3069** (1997) 400–411; Spatio-temporal wavelet transforms for motion tracking, *IEEE ICASSP '97*, IEEE Computer Soc. Press, Los Alamitos, CA, 1997, Vol. 4, 3013–3016; Spatio-temporal continuous wavelets applied to missile warhead detection and tracking, *Visual Communications and Image Processing '97, SPIE Proc.* **3024** (1997) 787–798; F. Mujica, R. Murenzi, M. J. T. Smith, and J-P. Leduc, Robust tracking in compressed image sequences, *J. Elec. Imaging* **7** (1998) 746–754

[203] J-M. Lévy-Leblond, Galilei group and non-relativistic quantum mechanics, *J. Math. Phys.* **4** (1963) 453–507

[204] J-M. Lévy-Leblond, Galilei group and Galilean invariance, in *Group Theory and Its Applications,* Vol.II, 221–299; E. M. Locbl (cd.), Academic Press, New York, 1971

[205] J-M. Lévy-Leblond, On the conceptual nature of the physical constants, *Riv. Nuovo Cim* **7** (1977) 187–214

[206] E. H. Lieb, The classical limit of quantum spin systems, *Commun. Math. Phys.* **31** (1973) 327–340

[207] W. Lisiecki, Kähler coherent states orbits for representations of semisimple Lie groups, *Ann. Inst. H. Poincaré* **53** (1990) 857–890

[208] H. Liu and L. Peng, Admissible wavelets associated with the Heisenberg group, *Pacific J. Math.* **180** (1997) 101–123

[209] G. Mack, All unitary ray representations of the conformal group $SU(2,2)$ with positive energy, *Commun. Math. Phys.* **55** (1977) 1–28

[210] G. W. Mackey, Imprimitivity for representations of locally compact groups I, *Proc. Nat. Acad. Sci.* **35** (1949) 537–545

[211] S. G. Mallat, Multifrequency channel decompositions of images and wavelet models, *IEEE Trans. Acoust., Speech, Signal Process.* **37** (1989) 2091–2110

[212] S. G. Mallat, A theory for multiresolution signal decomposition: the wavelet representation, *IEEE Trans. Pattern Anal. Machine Intell.* **11** (1989) 674–693

[213] S. Mallat and W. L. Hwang, Singularity detection and processing with wavelets, *IEEE Trans. Inform. Theory* **38** (1992) 617–643

[214] S. Mallat and Z. Zhang, Matching pursuits with time frequency dictionaries, *IEEE Trans. Signal Process.* **41** (1993) 3397–3415

[215] S. Mallat and S. Zhong, Wavelet maxima representation, in [Me2], 207–284

[216] D. Marion, M. Ikura, and A. Bax, Improved solvent suppression in one- and two-dimensional NMR spectra by convolution of time-domain data, *J. Magn. Reson.* **84** (1989) 425–430

[217] Y. Meyer, Principe d'incertitude, bases hilbertiennes et algèbres d'opérateurs, *Séminaire Bourbaki,* 38ème année, 1985–1986, n° 62, 209–223; *Astérisque* **145–146** (1987); P. G. Lemarié and Y. Meyer, Ondelettes et bases hilbertiennes, *Rev. Math. Iberoamer.* **2** (1986) 1–18

[218] Y. Meyer, Quasicrystals, diophantine approximation and algebraic numbers, in *Beyond Quasicrystals, Les Houches 1994,* 3–16; F. Axel and D. Gratias (eds.), Les Editions de Physique, Paris, and Springer-Verlag, Berlin, 1995

[219] Y. Meyer and H. Xu, Wavelet analysis and chirps, *Appl. Comput. Harmon. Anal.* **4** (1997) 366–379

[220] L. Michel, Invariance in Quantum mechanics and group extensions, in *Group Theoretical Concepts and Methods in Elementary Particle Physics,* 135–200; F. Gürsey (ed.), Gordon and Breach, New York and London, 1964

[221] J. Mickelsson and J. Niederle, Contractions of representations of the de Sitter groups, *Commun. Math. Phys.* **27** (1972) 167–180

[222] V. F. Molchanov, Harmonic analysis on homogeneous spaces, in A. A. Kirillov (ed.), *Representation Theory and Noncommutative Harmonic Analysis II*, Springer-Verlag, Berlin, 1995

[223] H. Moscovici, Coherent states representations of nilpotent Lie groups, *Commun. Math. Phys.* **54** (1977) 63–68

[224] H. Moscovici and A. Verona, Coherent states and square integrable representations, *Ann. Inst. H. Poincaré* **29** (1978) 139–156

[225] R. Murenzi, Wavelet transforms associated to the n-dimensional Euclidean group with dilations: signals in more than one dimension, in [Com], 239–246

[226] B. Nagel, Generalized eigenvectors in group representations, in *Studies in Mathematical Physics (Proc. Istanbul 1970)*, 135–154; A. O. Barut (ed.), Reidel, Dordrecht and Boston, 1970; G. Lindblad and B. Nagel, Continuous bases for unitary irreducible representations of SU(1,1), *Ann. Inst. H. Poincaré* **13** (1970) 27–56

[227] M. A. Naĭmark, *Dokl. Akad. Nauk. SSSR* **41** (1943) 359–361; see also B. Sz-Nagy, *Extensions of Linear Transformations in Hilbert Space Which Extend Beyond this Space*, Appendix to F. Riesz and B. Sz-Nagy, *Functional Analysis*, Frederick Ungar, New York, 1960

[228] M. Nauenberg, Quantum wave packets on Kepler elliptic orbits, *Phys. Rev. A* **40** (1989) 1133–1136

[229] M. Nauenberg, C. Stroud, and J. Yeazell, The classical limit of an atom, *Scient. Amer.* (1994) 24–29

[230] H. Neumann, Transformation properties of observables, *Helv. Phys. Acta* **45** (1972) 811–819

[231] U. Niederer, The maximal kinematical invariance group of the free Schrödinger equation, *Helv. Phys. Acta* **45** (1972) 802–810

[232] M. M. Nieto and L. M. Simmons, Jr, Coherent states for general potentials. I. Formalism; II. Confining one-dimensional examples; III. Nonconfining one-dimensional examples, *Phys. Rev. D* **20** (1979) 1321–1331; 1332–1341; 1342–1350

[233] M. M. Nieto and L. M. Simmons, Jr, Coherent states for general potentials, *Phys. Rev. Lett.* **41** (1987) 207–210

[234] A. Odzijewicz, On reproducing kernels and quantization of states, *Commun. Math. Phys.* **114** (1988) 577–597

[235] A. Odzijewicz, Coherent states and geometric quantization, *Commun. Math. Phys.* **150** (1992) 385–413

[236] A. Odzijewicz, Quantum algebras and q-special functions related to coherent states maps of the disc, *Commun. Math. Phys.* **192** (1998) 183–215

[237] G. Olafsson and B. Ørsted, The holomorphic discrete series for affine symmetric spaces, *J. Funct. Anal.* **81** (1988) 126–159

[238] E. Onofri, A note on coherent state representations of Lie groups, *J. Math. Phys.* **16** (1975) 1087–1089

[239] E. Onofri, Dynamical quantization of the Kepler manifold, *J. Math. Phys.* **17** (1976) 401–408

[240] L. C. Papaloucas, J. Rembieliński, and W. Tybor, Vectorlike coherent states with noncompact stability group, *J. Math. Phys.* **30** (1989) 2406–2410

[241] W. Parry, On the β-expansions of real numbers, *Acta Math. Acad. Sci. Hungary* **11** (1960) 401–416

[242] Z. Pasternak-Winiarski, On the dependence of the reproducing kernel on the weight of integration, *J. Funct. Anal.* **94** (1990) 110–134

[243] Z. Pasternak-Winiarski, On reproducing kernels for holomorphic vector bundles, in *Quantization and Infinite Dimensional Systems (Proc. Białowieza, Poland, 1993)*, 109–112; J-P. Antoine, S. T. Ali, W. Lisiecki, I. M. Mladenov, and A. Odzijewicz (eds.), Plenum Press, New York and London, 1994

[244] Th. Paul and K. Seip, Wavelets in quantum mechanics, in [Rus], 303–322

[245] A. M. Perelomov, On the completeness of a system of coherent states, *Theor. Math. Phys.* **6** (1971) 156–164

[246] A. M. Perelomov, Coherent states for arbitrary Lie group, *Commun. Math. Phys.* **26** (1972) 222–236

[247] M. Perroud, Projective representations of the Schrödinger group, *Helv. Phys. Acta* **50** (1977) 233–252

[248] J. Phillips, A note on square-integrable representations, *J. Funct. Anal.* **20** (1975) 83–92

[249] W. W. F. Pijnappel, A. van den Boogaart, R. de Beer, and D. van Ormondt, SVD-Based quantification of Magnetic Resonance signals, *J. Magn. Reson.* **97** (1992) 122–134

[250] E. Prugovečki, Consistent formulation of relativistic dynamics for massive spin-zero particles in external fields, *Phys. Rev. D* **18** (1978) 3655–3673, Appendix C

[251] E. Prugovečki, Relativistic quantum kinematics on stochastic phase space for massive particles, *J. Math. Phys.* **19** (1978) 2261–2270

[252] C. Quesne, Coherent states of the real symplectic group in a complex analytic parametrization, I, II, *J. Math. Phys.* **27** (1986) 428–441, 869–878; Generalized vector coherent states of $sp(2N,\mathbb{R})$ vector operators and of $sp(2N,\mathbb{R}) \supset u(N)$ reduced Wigner coefficients, *J. Phys. A: Math. Gen.* **24** (1991) 2697–2714

[253] J. M. Radcliffe, Some properties of spin coherent states, *J. Phys. A: Math. Gen.* **4** (1971) 313–323

[254] J. H. Rawnsley, Coherent states and Kähler manifolds, *Quart. J. Math. Oxford* (2) **28** (1977) 403–415

[255] J. Renaud, The contraction of the $SU(1,1)$ discrete series of representations by means of coherent states, *J. Math. Phys.* **37** (1996) 3168–3179

[256] A. Rényi, Representations for real numbers and their ergodic properties, *Acta Math. Acad. Sci. Hungary* **8** (3–4) (1957) 477–493

[257] S. Roques, F. Bourzeix, and K. Bouyoucef, Soft-thresholding technique and restoration of 3C273 jet, *Astrophys. Space Sci.* (1996) 297-304

[258] D. J. Rotenberg, Application of Sturmian functions to the Schroedinger three-body problem: Elastic e^+–H scattering, *Ann. Phys. (NY)* **19** (1962) 262–278

[259] H. Rossi and M. Vergne, Analytic continuation of the holomorphic discrete series for a semi-simple Lie group, *Acta Math.* **136** (1976) 1–59

[260] D. J. Rowe, Coherent state theory of the noncompact symplectic group, *J. Math. Phys.* **25** (1984) 2662–2271

[261] D. J. Rowe, Microscopic theory of the nuclear collective model, *Rep. Prog. Phys.* **48** (1985) 1419–1480

[262] D. J. Rowe, G. Rosensteel, and R. Gilmore, Vector coherent state representation theory, *J. Math. Phys.* **26** (1985) 2787–2791

[263] D. J. Rowe and J. Repka, Vector-coherent-state theory as a theory of induced representations, *J. Math. Phys.* **32** (1991) 2614–2634

[264] J. Saletan, Contraction of Lie groups, *J. Math. Phys.* **2** (1961) 1–21

[265] G. Saracco, A. Grossmann, and Ph. Tchamitchian, Use of wavelet transforms in the study of propagation of transient acoustic signals across a plane interface between two homogeneous media, in [Com], 139–146

[266] E. Schrödinger, Der stetige Übergang von der Mikro- zur Makromechanik, *Naturwiss.* **14** (1926) 664–666

[267] H. Scutaru, Coherent states and induced representations, *Lett. Math. Phys.* **2** (1977) 101–107

[268] R. Simon, E. C. G. Sudarshan, and N. Mukunda, Gaussian pure states in quantum mechanics and the symplectic group, *Phys. Rev.* **A37** (1988) 3028–3038

[269] P. Schröder and W. Sweldens, Spherical wavelets: Efficiently representing functions on the sphere, *Computer Graphics Proc. (SIGGRAPH95)*, ACM Siggraph, 1995, 161–175

[270] E. Slezak, A. Bijaoui, and G. Mars, Identification of structures from galaxy counts. Use of the wavelet transform, *Astron. Astroph.* **227** (1990) 301–316

[271] M. Spera, On a generalized Uncertainty Principle, coherent states, and the moment map, *J. Geom. Phys.* **12** (1993) 165–182

[272] E. C. G. Sudarshan, Equivalence of semiclassical and quantum mechanical descriptions of statistical light beams, *Phys. Rev. Lett.* **10** (1963) 277–279

[273] W. Sweldens, The lifting scheme: a custom-design construction of biorthogonal wavelets, *Appl. Comput. Harmon. Anal.* **3** (1996) 1186–1200; The lifting scheme: a construction of second generation wavelets, *SIAM J. Math. Anal.* **29** (1998) 511–546

[274] G. Thonet, O. Blanc, P. Vandergheynst, E. Pruvot, J-M. Vesin, and J-P. Antoine, Wavelet-based detection of ventricular ectopic beats in heart rate signals, *Applied Sig. Process.* **5** (1998) 170–181

[275] B. Torrésani, Wavelets associated with representations of the affine Weyl-Heisenberg group, *J. Math. Phys.* **32** (1991) 1273–1279

[276] B. Torrésani, Time-frequency representation: wavelet packets and optimal decomposition, *Ann. Inst. H. Poincaré* **56** (1992) 215–234

[277] D. A. Trifonov, Generalized intelligent states and squeezing, *J. Math. Phys.* **35** (1994) 2297–2308

[278] L. Vanhamme, R. D. Fierro, S. Van Huffel, and R. de Beer, Fast removal of residual water in proton spectra, *J. Magn. Reson.* **132** (1998) 197–203

[279] E. Van Vyve, P. Vandergheynst, A. Goldberg, J-P. Antoine, and I. Doghri, Modelling and simulation of an impact test using wavelets, analytical solutions and finite elements, preprint UCL-IPT-99-05, Louvain-la-Neuve (submitted).

[280] J. Ville, Théorie et applications de la notion de signal analytique, *Câbles et Transm.* **2ème A.** (1948) 61–74

[281] J. Voisin, On some unitary representations of the Galilei group. I. Irreducible representations, *J. Math. Phys.* **6** (1965) 1519–1529

[282] D. F. Walls, Squeezed states of light, *Nature* **306** (1983) 141–146
[283] Y. K. Wang and F. T. Hioe, Phase transition in the Dicke maser model, *Phys. Rev. A* **7** (1973) 831–836
[284] E. P. Wigner, On the quantum correction for thermodynamic equilibrium, *Phys. Rev.* **40** (1932) 749–759
[285] W. Wisnoe, P. Gajan, A. Strzelecki, C. Lempereur, and J-M. Mathé, The use of the two-dimensional wavelet transform in flow visualization processing, in [Me4], 455–458
[286] K.B. Wolf, Wigner distribution function for paraxial polychromatic optics, *Opt. Commun.* **132** (1996) 343–352
[287] J. A. Yeazell, M. Mallalieu, and C. R. Stroud, Jr, Observation of the collapse and revival of a Rydberg electronic wave packet, *Phys. Rev. Lett.* **64** (1990) 2007–2010; J. A. Yeazell and C. R. Stroud, Jr, Observation of fractional revivals in the evolution of a Rydberg atomic wave packet, *Phys. Rev. A* **43** (1991) 5153–5156; M. Mallalieu and C. R. Stroud, Jr, Rydberg wave packets: fractional revivals and classical orbits, in [Fen], 301–314
[288] J. Zak, Balian-Low theorem for Landau levels, *Phys. Rev. Lett.* **79** (1997) 533–536; Orthonormal sets of localized functions for a Landau level, *J. Math. Phys.* **39** (1998) 4195–4200, and references quoted there
[289] W-M. Zhang, D. H. Feng, and R. Gilmore, Coherent states: Theory and some applications, *Rev. Mod. Phys.* **26** (1990) 867–927
[290] I. Zlatev, W-M. Zhang, and D. H. Feng, Possibility that Schrödinger's conjecture for the hydrogen atom coherent states is not attainable, *Phys. Rev. A* **50** (1994) R1973–R1975

Index

ν-selection, 96

action of semigroup on a lattice, 373
action-angle variables, 123
algorithm
  fast wavelet, 289, 330
  matching pursuit, 359
  pyramidal, 286, 371
application of CS
  in atomic physics, 29, 137, 196
  in nuclear physics, 140

bundle
  (co)tangent, 205
  normal, 205
  parallel, 222

canonical commutation relations (CCR), 14
cocycle, 49, 55, 57, 213, 371, 380
cocycle equation, 375, 381
coherent sectors, 162
coherent states (CS), 100
  action-angle, 123
  atomic, 196
  Barut–Girardello, 3
    of $SU(1,1)$, 251
  canonical, 2, 13, 16, 131
  classical theory, 136
  covariant, 128, 142
  for semidirect products, 214
  generalized, 11
  Gilmore–Perelomov, 9, 130
    of $SU(1,1)$, 66, 249, 250
  holomorphic, 116
  nonholomorphic, 122
  of affine Galilei group, 346
  of affine Poincaré group, 349
  of affine Weyl–Heisenberg G_{aWH}, 339
  of compact semisimple Lie groups, 136
  of Galilei $\widetilde{\mathcal{G}}(1,1)$, 242
  of isochronous Galilei group, 191
  of non-semisimple Lie groups, 141
  of noncompact semisimple Lie groups, 139
  of Poincaré $\mathcal{P}_+^\uparrow(1,1)$, 236
    massless case, 241
  of Poincaré $\mathcal{P}_+^\uparrow(1,3)$, 232
  of Schrödinger group, 348
  quasi-coherent states, 145
  spin or $SU(2)$, 137

square integrable covariant, 128, 142, 217
vector (VCS), 59, 133
weighted, 144, 361
    of affine Weyl–Heisenberg G_{aWH}, 339
    of Poincaré $\mathcal{P}_+^\uparrow(1,1)$, 237
cohomology group, 371, 380
continuous WT
    in n-D, 310
    in 1-D, 262
        applications, 278
        continuous wavelet packets, 289
        discretization, 274
        localization properties, 273
    in 2-D, 309
        applications, 322
        as a symmetry scanner, 324
        continuous wavelet packets, 330
        interpretation, 319
        representations, 319
    on manifolds, 336
    on the sphere S^2, 331
        Euclidean limit, 334

decomposition
    Cartan, 61, 80
    Gauss, 82, 332
    Iwasawa, 81, 332
dilation
    natural, 374
    pseudodilation
        associated, 375
        on $\mathbb{Z}_p$, 370
        principal, 374
discrete WT
    in 1-D, 283
        applications, 288
        generalizations, 287
    in 2-D, 327
        applications, 329
Duflo–Moore operator, 158, 174, 176, 178, 180, 193, 239

evaluation map, 22, 98, 101

factor, 166
Fibonacci numbers, 120, 293
filters, 286
    compatible, 376
    on $\mathbb{Z}_p$, 370
    pseudo-QMF, 290
    QMF, 286
Fourier transform, 260
    relativistic, 177
fractals, 274
frame, 36
    affine Weyl–Heisenberg CS, 339
    discrete, 42, 275
        for affine Poincaré group, 360
        for affine Weyl–Heisenberg group, 359
        for Poincaré group $\mathcal{P}_+^\uparrow(1,1)$, 364
        for semidirect products, 362
        Gabor or canonical CS, 354
        wavelet, 357
    discretization, 353
    for semidirect product, 222
    Galilei CS, 244
    holomorphic CS, 119
    Poincaré CS, 232
    tight, 40, 265
    unbounded, 144
free orbit, 172

gaussons, 14, 200
golden mean τ, 120, 292, 324
    τ-Haar basis, 296
    τ-integers, 293
group (abstract)
    direct product, 82
    extension, 83
    locally compact abelian (LCA), 70
        lattice in, 72
        sampling in, 74
    semidirect product, 83, 171
    unimodular, 48
group (explicit)
    $GL(n, \mathbb{R})$, 172
    G_{ut} (upper-triangular matrices), 49
    $SL(2, \mathbb{R}) \simeq SO_o(1,2) \simeq SU(1,1)$, 3
    affine, 11, 261
    affine Galilei, 345
    affine Poincaré, 349
    affine Poincaré $SIM(1,1)$, 177
    affine Weyl–Heisenberg G_{aWH}, 338
    anti-de Sitter $SO_o(1,2)$, 245

connected affine or $ax + b$, 11, 53, 153, 264
discrete Weyl–Heisenberg $G_{dWH}^{(N)}$, 356
Euclidean $E(2)$, $E(n)$, 219
Galilei, 186, 244
isochronous Galilei, 187
metaplectic $Mp(2n, \mathbb{R})$, 200
Poincaré $\mathcal{P}_+^\uparrow(1,1)$, 235
Poincaré $\mathcal{P}_+^\uparrow(1,3)$, 225
Schrödinger, 347
similitude group $SIM(2)$, 176
similitude group $SIM(n)$, 179, 308
symplectic $Sp(2n, \mathbb{R})$, 200
Weyl–Heisenberg G_{WH}, 18, 19, 160, 200, 337

Hilbert space
    direct integral of, 93
    measurable field of, 92, 98
    reproducing kernel, 9, 20, 90, 94, 98, 103, 106, 142, 149, 182, 184
    rigged, 28, 131, 144, 266
holomorphic map, 105

KMS states, 167

Lie algebra, 75
    contraction, 85, 246
    root, 75
    weight, 76
Lie group, 77
    (co)adjoint action, 78, 204, 207
        of $Sp(2n, \mathbb{R})$, 202
    coadjoint orbit, 79, 161, 206, 249
        of $SIM(n)$, 310
    contraction, 86
        $SO_o(3,1)$ to $SIM(2)$, 334
        $SU(1,1)$ to $\mathcal{P}_+^\uparrow(1,1)$, 247
    exponential map, 78
localization, 267, 273
localization operators, 22

manifold
    Kähler, 26
    symplectic, 26, 205
measure
    (quasi-)invariant, 48, 62, 142
        on cotangent bundle, 210

Borel, 34
Haar, 147
positive operator-valued (POV), 23, 34
    commutative, 36
    examples of, 37, 41
    Naĭmark extension theorem, 39
    projection-valued (PV), 36
    scale-angle, 324
minimal uncertainty states, 14
model
    Dicke, 29
    Hepp-Lieb, 137
    nuclear collective, 140
modular function, 48
modular structure, 167
moment problem, 116
multiresolution analysis, 284, 371, 372, 377

NMR spectroscopy, 279

orthogonality relations, 155, 174, 183, 192, 238, 239
overcomplete set, 21

phase space, 22, 227, 235, 237, 242, 244, 249, 272
POV function, 99, 103

quantization, 27

reconstruction
    formula, 152, 159, 265, 311
    operator, 265, 269, 270
representation
    discrete series, 148
        of $SU(1,1)$, 61
    Fock–Bargmann, 25
    induced, 56, 182, 184, 214, 217
        of semidirect products, 213
        of $SU(1,1)$, 63
    regular, 67, 149, 151
    square integrable, 148, 149, 174, 214, 261, 309
    square integrable mod (H, σ), 129, 142, 182, 217, 222, 242, 333

reproducing kernel, 9, 65, 100, 101, 142, 149, 182, 184, 265, 270, 311
  holomorphic, 114
  square integrable, 110
resolution of the identity, 3, 17, 65, 147, 149, 182, 184, 195, 217, 265, 311
resolution operator, 35, 142
  for Poincaré $\mathcal{P}_+^\uparrow(1,3)$ CS, 229
ridges in WT, 276

sampling, 292, 369, 384
  in LCA group, 74
scaling function, 284, 382
Schrödinger, 1, 347
section, 4, 49, 128, 142
  affine, 228
  affine admissible, 220
  Galilean, 228, 233, 238, 243
  Lorentz, 234, 238
  principal, 209, 234
  quasi-section, 145
  symmetric, 234, 238
skeleton of WT, 276
space
  coset or homogeneous, 4, 19, 48, 128, 142
  Fock–Bargmann, 25, 248
  Hardy, 264
  Krein, 240
square integrability, 9
squeezed states, 15, 200
subband coding scheme, 286, 377
subspace
  V-admissible, 184
  cyclic, 133
symbols, 5
system of covariance, 57
system of imprimitivity, 57

tempered distributions, 266
theorem
  Balian–Low, 355
  Mackey imprimitivity, 57
  Naĭmark extension, 39
  Pontrjagin duality, 72
  Schur lemmas, 68
time-frequency representation, 258

uncertainty relations, 14, 316, 354

vector
  α-admissible, 181
  admissible, 148, 261, 309
  admissible mod (H, σ), 129, 243, 333
von Neumann algebra, 166

wavelet transform (WT), 259
  n-D continuous (CWT), 310
  1-D continuous (CWT), 262
  1-D discrete (DWT), 283
  2-D discrete (DWT), 327
  of distributions, 270
  of Schwartz functions, 269
  on $\mathbb{Z}_p$, 291, 370
wavelet(s), 5, 154, 262
  n-D Mexican hat, 312
  n-D Morlet, 313
  τ-wavelets, 292
  1-D Mexican hat, Marr, 263
  1-D Morlet, 263
  Cauchy, 314
  conical, 313
  difference, 312
  directional, 312
  kinematical, 342
  minimal uncertainty, 317
  on the sphere S^2, 331
  Pisot, 303
Wigner distribution, 160
Wigner map, 161
Wigner transform, 161, 239
Wigner–Ville transform, 259
windowed Fourier or Gabor transform, 258

Zak transform, 356